Archives of Toxicology, Supplement 17

Toxicology in Transition

Proceedings of the
1994 EUROTOX Congress
Meeting Held in Basel, Switzerland, August 21–24, 1994

Edited by the Publication Committee
G.H. Degen, J.P. Seiler and P. Bentley

With 104 Figures and 59 Tables

Springer

Gisela H. Degen
Institut für Arbeitsphysiologie an der Universität Dortmund
Ardeystr. 67
D-44139 Dortmund

Jürg P. Seiler
Interkantonale Kontrollstelle für Heilmittel
Erlachstr. 8
CH-3000 Bern 9

Philip Bentley
Ciba-Geigy Ltd.
K-136.3.23
CH-4002 Basel

ISBN-13: 978-3-642-79453-7 e-ISBN-13:978-3-642-79451-3
DOI:10.1007/978-3-642-79451-3

CIP data applied for

Softcover reprint of the hardcover 1st edition 1995

Typesetting: Camera ready by authors
SPIN 10487327 27/3130 – 5 4 3 2 1 0 – Printed on acid-free paper

Contents

In Vitro tests in regulatory toxicology

Neurobehavioral abnormalities induced by substance abuse

Late Manuscripts

Young Scientists Poster Award

Genetically Modified Cells and Animals in Toxicological Research

(Chair: Dr. K. Bürki, Switzerland)

Concepts, Potentials and Methods for Constructing Transgenic Cells and Animals

K. Buerki
Preclinical Research, Sandoz Pharma Ltd., CH-4002 Basel, Switzerland

The spectre of methods to stably alter the genetic constitution of cells and animals is currently being extended by methods of unprecedented precision. Whereas classical transgene constructs are used to randomly insert additional sequences into the genome, todays targeting vectors allow the modification of a selected endogenous gene in a precisely predetermined way. This has opened new ways to experimentally investigate the molecular components of cells and of complex *in vivo* systems (Hanahan 1989). With regard to toxicology, genetically modified cells and animals will contribute to a better understanding of the primary events and of the mechanisms leading to the manifestation of damage by toxic agents (Liggitt and Reddington 1992). This in turn will eventually lead to refined *in vitro* and *in vivo* systems to monitor for toxic effects.

Methods

Transgenetics. Cloned transgenes or targeting vectors are stably transfected into cells by various different methods: use of viral vectors, calcium phosphate coprecipitation, lipofection, electroporation, or by direct microinjection into the cell nucleus (Keown et al. 1990). This latter method is by far the most efficient. It is widely applied to transfect DNA into adherent cells and into animal zygotes (Hogan et al. 1986, Wassarman and DePamphilis 1993).

Whereas the expression of transgenes *in vitro* is readily achieved using simple viral or eukaryotic promoters, the correct temporal and tissue specific expression of transgenes *in vivo* remains in many instances a considerable problem. This is mainly due to the lack of essential regulatory sequences in transgene constructs. The availability of yeast artificial chromosomes (YACs) harboring 200-300 kilobases of genomic sequences, and therefore likely to comprise all regulatory elements of a particular gene, may solve some difficulties in reliably expressing transgenes *in vivo* (Schedllet al. 1993). To overcome difficulties in inducing or repressing transgene activity, a tetracycline dependent binary transactivator system has recently been developed (Gossen and Bujard 1992, Gossen et al. 1994). *In vitro*, this system allows the temporal modulation of transgene activities by varying the tetracycline concentration in the medium.

Gene Targeting. The development of techniques for the selection and screening of rare homologous recombination events in mammalian cells after transfection with so-called targeting vectors has made it possible to alter endogenous genes in a predetermined way. Depending on the design of the targeting vector, gene targeting methods allow the introduction of insertions, deletions or of subtle mutations into a gene of interest (Bradley 1991, Joyner 1993). When applied to murine embryonic stem cells, it is possible to introduce the induced genetic modification into the germ line of chimaeric animals, and eventually breed mice heterozygous or homozygous for the modified gene (knock-out or knock-in mice). Such mice carry the experimentally modified gene in all their cells. However, clever new approaches allow the induction of gene-specific recombination events in restricted cell types (Kilby et al. 1993, Gu et al. 1994, Barinaga 1994). The approach involves the generation of mice carrying targeted recombinase-specific sequences in their genome, and the generation of a transgenic mouse line with tissue specific expression of the corresponding recombinase. When combined, the expressed recombinase will lead to the induction of a tissue specific recombination event, for example a liver specific knock-out of an appropriately marked gene.

Concepts

Technically, the experimental modification of characterized genes of cells or animals has become routine. The limits are set by the size restrictions for transgenes or targeting vectors due to the limits of current cloning strategies, and by the availability of suitable target cells. For instance, germ line transmission of genes modified by gene targeting is limited to the laboratory mouse, as germ-line competent embryonic stem cells are not available for any other species. In the pronuclei of zygotes, on the other hand, homologous recombination leading to gene targeting has not been observed. Genetic modifications can be induced to simply integrate marker sequences or reporter genes into a genome, or to probe for the function of a particular gene product. To this extent, a selected gene may be amplified in the form of transgenes, be silenced by a targeted disruption, or be modified by a targeted replacement of sequences. A number of strategies have been developed to induce so-called gain-of-function or loss-of-function modifications, both by transgenetics as well as by gene targeting (for review see Buerki and Ledermann 1994).

In toxicology, genetically modified cells and animals may be used for the following reasons: (1) To better understand at the molecular level the primary events and the pathogenesis induced by drugs or toxic agents. Examples are transgenic mice developing neoplasia, the so-called oncomice (Hanahan 1988). (2) To introduce into cells or animals human metabolic enzymes or drug receptors in order to increase the predictability of toxic effects to humans (Doehmer 1993). (3) To modify metabolic pathways in order to increase the sensitivity when monitoring for toxic effects. Examples are tumor prone mice, like the *pim* oncogene transgenic mice developed by Breuer et al. (1989), or the tumor prone knock-out mice deficient in the tumor suppressor gene *p53* (for review see Jacks et al. 1994). (4) To introduce specific reporter genes to facilitate the detection of genotoxic agents. Prominent examples are transgenic mice carrying easy recoverable reporter genes, the Muta Mouse (*lacZ*) and the Big Blue (*lacI*) mouse (for reviews see Gossen and Vijg 1993, Morrison and Ashby 1994).

Generating a genetically modified cell line or animal is just a first step in a long process to characterize and validate the expression and significance of a novel trait which has been introduced into a complex living system. Very defined genetic alterations may have highly pleiotropic effects which render the interpretation of the model difficult. Moreover, although they may carry human genes, transgenic animals remain animals, and therefore some of the factors limiting the use of traditional models will persist.

Potentials

A wealth of genetically modified cells and animals, in particular mice, are currently available or are being generated and studied in order to gain insight into the molecular mechanisms governing cell proliferation , differentiation and cell-cell interactions. This ulimately will lead to a better understanding of complex cellular functions and *in vivo* systems, in the normal and in the pathological state (Smithies 1993). Despite the difficulty in applying linear genetic thinking to complex interactions (Strohman 1994), genetically modified cells and animals will help to elucidate many of the not yet understood biological processes, including the induction and manifestation of damage induced by toxic agents. The insight into mechanistic relationships will eventally lead to the design of refined, specific test systems *in vitro, ex vivo* and *in vivo*, by combining metabolic mutants or repair deficient mutants with genetic traits that facilitate the readout of toxic effects at low threshold levels. As a result, toxicology could well undergo a change from a "high-dose discipline" to a "low-dose discipline" (Marshall 1994). However, while such a development is foreseeable, it will progress most likely at little steps and at time intervals determined by the careful evaluation and validation of any new system. Nevertheless, the exploitation of the potential of genetically modified cells and animals in experimental toxicology and safety assessment is a challenge to creative experts in the field.

References

Barinaga M (1994) Knockout mice: round two. Science 265:26-28

Breuer M, Slebos R, Verbeek S, van Lohuizen M, Wientjens E, Berns A (1989) Very high frequency of lymphoma induction by a chemical carcinogen in *pim*-1 transgenic mice. Nature 340:61-63

Bradley A (1991) Modifying the mammalian genome by gene targeting. Current Op Biotech 2:823-829

Buerki K, Ledermann B (1994) Transgenic animals as pharmacological tools. Adv Drug Res 26 (in press)

Doehmer J (1993) V79 Chinese hamster cells genetically engineered for cytochrome P450 and their use in mutagenicity and metabolism studies. Toxicology 82:105-118

Gossen J, Vijg J (1993) Transgenic mice as model systems for studying gene mutations *in vivo*. Trends Genet 9:27-31

Gossen M, Bujard H (1992) Tight control of gene expression in mammalian cells by tetracycline-responsive promoters. Proc Natl Acad Sci USA 89:5547-5551

Gossen M, Bonin AL, Bujard H (1994) Control of gene activity in higher eukaryotic cells by prokaryotic regulatory elements. Trends Biotech 12:58-62

Gu H, Marth JD, Orban PC, Mossmann H, Rajewsky K (1994) Deletion of a DNA polymerase β gene segment in T cells using cell type-specific gene targeting. Science 265:103-106

Hanahan D (1988) Dissecting multistep tumorigenesis in transgenic mice. Ann Rev Genet 22:479-519

Hanahan D (1989) Transgenic mice as probes into complex systems. Science 246: 1265-1275

Hogan B, Constantini F, Lacy E (1986) Manipulating the mouse embryo. A laboratory manual. Cold Spring Harbor Laboratory

Jacks T, Remington L, Williams BO, Schmitt EM, Halachmi S, Bronson RT, Weinberg RA (1994) Tumor spectrum analysis in *p53*-mutant mice. Curr Biol 4:1-7

Joyner AL (1993) Gene targeting. A practical approach. IRL Press, Oxford

Keown WA, Campbell CR, Kucherlapati RS (1990) Methods for introducing DNA into mammalian cells. Methods Enzymol 185:527-537

Kilby NJ, Snaith MR, Murray JAH (1993) Site-specific recombinases: tools for genome engineering. Trends Genet 9:413-421

Liggitt HD, Reddington GM (1992) Transgenic animals in the evaluation of compound efficacy and toxicity: will they be as useful as they are novel? Xenobiotica 22:1043-1054

Marshall E (1993) Toxicology goes molecular. Science 259:1394-1398

Morrison V, Ashby J (1994) A preliminary evaluation of the performance of the Muta Mouse (*lacZ*) and Big Blue (*lacI*) transgenic mouse mutation assays. Mutagenesis 9:367-376

Schedl A, Larin Z, Montoliu L, Thies E, Kelsey G, Lehrach H, Schuetz G (1993) A method for the generation of YAC transgenic mice by pronuclear microinjection. Nucl Acids Res 21:4783-4787

Smithies O (1993) Animal models of human genetic diseases. Trends Genet 9:112-116

Strohman R (1994) Epigenesis: The missing beat in biotechnology? Bio/Technology 12:154-164

Wassarman PM, DePamphilis ML (1993) Guide to techniques in mouse development. Methods Enzymol 225, Academic Press, San Diego

Use of Recombinant Cells in Metabolic and Mechanistic Studies

W.A. Schmalix and J. Doehmer

Institut für Toxikologie und Umwelthygiene, Technische Universität München, Lazarettstr. 62, 80636 München, Germany

Gene technology had an extensive impact on the understanding of xenobiotics metabolism over the last decade. This includes detection of novel cytochromes P450 and conjugating enzymes, elucidation of the chromosomal organisation of these genes, linking evolutionary relationships and classification of these enzymes by sequence alignment, identification of regulatory elements for controlled gene expression, and last but not least heterologous expression of these genes. In the mid-1980s the first successful heterologous expression of a cloned mammalian cytochrome P450 gene in yeast cells was reported (Oeda et al. 1985). Since then, this area has seen a tremendous expansion in terms of numbers of cytochromes P450 and conjugating enzymes expressed, and expression systems employed (Fig. 1).

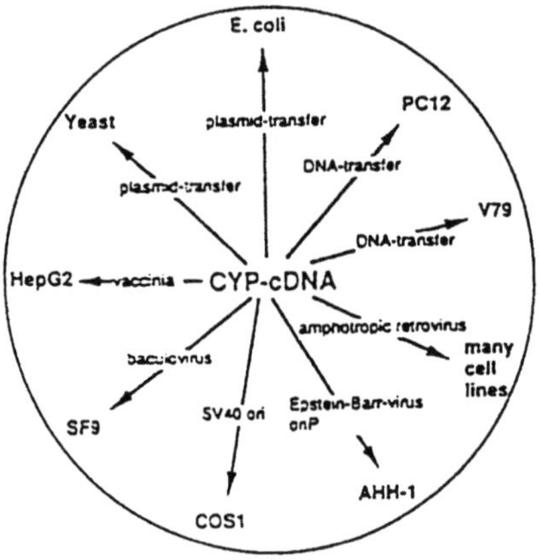

Fig. 1. cDNA directed expression of cytochromes P450 in bacterial, yeast, insect and mammalian cells, and the vectors used for gene transfer and maintenance.

Those expression systems are based on bacterial, yeast, insect, and mammalian cells. These systems have been extensively described in several publications (Waterman and Johnson, 1991; Schenkman and Greim, 1993). The technological procedures are basically the same for all expression systems, and have not changed over the years in principle. There are three major steps (Fig. 2): (i) obtaining a full length cDNA; (ii) constructing a system compatible recombinant expression vector; (iii) transfer and maintenance of the recombinant expression vector into the recipient cell.

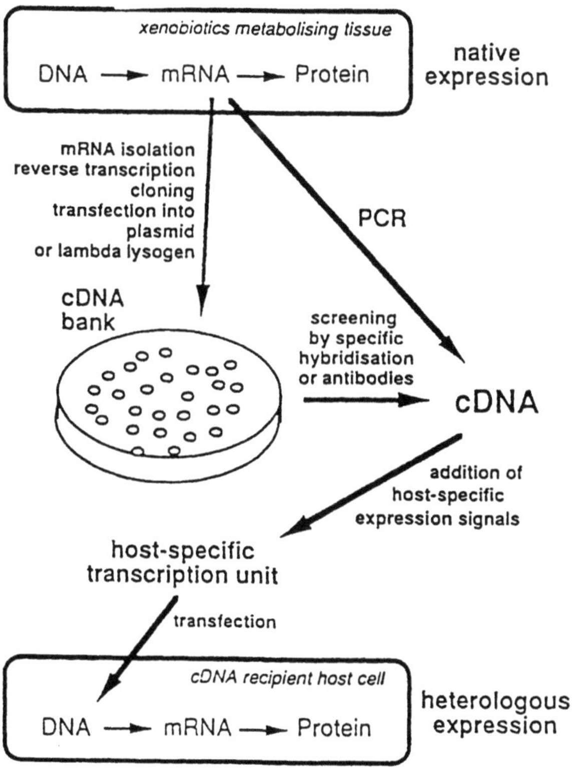

Fig. 2. Experimental steps leading from native to heterologous expression, including generation of a full length cDNA, construction of a compatible recombinant expression vector, gene transfer into the host cell for expression of the cDNA encoded protein. Protein should be as authentic as in the native expression situation, unless protein has been designed for experimental or industrial reasons.

Once these expression systems are validated for their cDNA encoded enzymatic activity, they may be applied in order to solve toxicological and pharmacological problems related to metabolism. It is the biology of a given expression system, which determines their experimental range. There is no such thing like an ultimate expression system. Usually, advantages and disadvantages of these systems overlap. It takes understanding of the gene technology applied and the biological features and endpoints of a given expression system, in order to make the most appropiate choice.

The bacterial expression systems

Heterologous expression in E. coli is an inexpensive and fast means for producing large quantities of the cDNA encoded protein. This is a particular feasible way, if there are no requirements for posttranslational protein modifications such as glycosylation, which is the case for most if not all cytochromes P450. However, several precautions and adjustments on the cDNA have to be considered, before efficient cDNA directed expression is achieved. This includes recombination with powerful bacterial promoters and changing the cDNA sequence to the coding demands of a bacterial cell. First bacterial expression of a mammalian CYP was reported for CYP2E1 (Larson et al. 1991). Efficient expression has been achieved after adjustment of the cDNA sequence (Barnes et al. 1991). More than 1% of the total bacterial protein content may result from an efficient cDNA expression (Fisher et al. 1992). But there is no guarantee, that this approach works for all cDNAs equally well. For unknown reasons, efficient expression of CYP1A1 cDNA in E.coli has not yet been achieved. Proteolysis is certainly one of the major obstacles for a successful heterologous expression in E.coli. Bacterial expressed mammalian CYP has to be purified, which is then to be reconstituted in the presence of lipids, the reductase, and other cofactors. Reconstitution may pose a problem for yielding the authentic molecular structure and substrate specificity of a CYP. This is already known from the classical situation, where biochemically isolated and purified CYP also has to be mixed in phospholipids with cofactors in order to obtain an enzymatically active preparation. After all, biological endpoints, which are relevant in mammalian cells, are naturally missing in E.coli.

Nevertheless, to obtain large quantities of CYP in the mg range, as they are needed e.g. for attempts on crystallization, bacterial expression is the choice of approach.

The yeast expression systems

Yeast cells may compromise advantages and disadvantages of heterologous expression in bacterial and mammalian cells (Sakaki et al. 1987; Pompon 1988; Eugster et al. 1990). Yeast cells, like *Saccharomyces cerevisiae*, are genetically well defined, and readily transfected and selected like bacterial cells. Yeast cells grow in cheap medium at a doubling time of 4 hours. Expression vectors contain yeast compatible promoters, e.g. the alcohol dehydrogenase promoter (Fig. 3). Yields of heterologously expressed CYP are usually higher than in any mammalian cell expression system, but lower than in any bacterial expression system. For maximum activity, yeast cells were engineered for mammalian CYP reductase, which appears to provide better coupling than the endogenous yeast CYP reductase. At the same time, yeast cells have some morphological characteristics similiar to mammalian cells, such as cell nucleus, and other cellular organelles. Yeast cells feature some unique biological characteristics, which may serve as valuable toxicological endpoints, e.g. mating. Disadvantages may result from the application of yeast cells, because these cells have a very rigid cell envelope, which may pose a barrier for the uptake of chemicals.

The mammalian cell expression systems

Mammalian cells came second for heterologous expression of CYPs. The first report was on the expression of bovine CYP17A in COS monkey cells using the classical SV40-T-Antigen expression vector system (Zuber et al.1986). Since then, the expression of CYPs in cultivated cells became a highly favored approach to CYPs and metabolic activation, although total activities are much lower than in the bacterial and yeast expression system. However, this is compensated by the fact, that mammalian cells provide relevant toxicological endpoints not present in the bacterial and yeast cells. Those endpoints are characteristic biological features

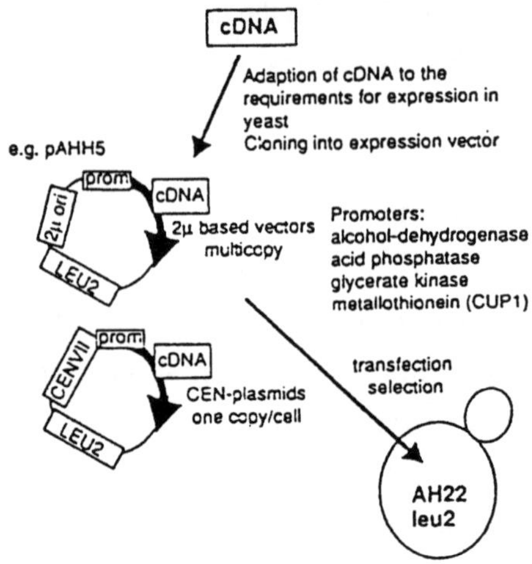

Fig. 3. Experimental procedures for achieving cDNA directed expression in yeast cells.

of the various cell lines. Among those endpoints are cytotoxicity, mutagenicity, chromosomal abberation, micronucleus formation, morphological transformation, differentiation, and many more. Expressing CYPs in these cells allows a direct allocation of metabolites, and their effects on a selected endpoint to a particular CYP under highly defined conditions. Second, if the background CYP activity is low or zero in the parental cell, the engineered cell will be defined for the cDNA encoded CYP. In this way, engineered cell lines are ideal analytical tools for sorting out complex situation, where a multitude of CYPs may play key roles in activating chemicals. If analytical procedures allow sensitive detection of metabolites, these cell lines may be applied for establishing metabolite profiles, and sorting out pathways. None of the genetically engineered cell lines are perfect for every application. Each cell line and expression system has its unique characteristics, and thus, have their advantages and disadvantages depending on the experimental question to be solved.

Since 1986 various approaches for cDNA directed expression in mammalian cells (Figs. 4,5,6,7,8) were developed. They differ in two ways, by the expression vector and the recipient cell. This determines mostly the level and stability of activity, and the range of applications.

Expression in COS cells is one of the most classical mammalian expression systems making use of a green monkey kidney cell (COS) expressing constantly the Large-T-antigen of the DNA-tumorvirus SV40, which maintaines incoming episomal plasmids carrying the SV40 origin of replication. However, the system yields only transient expression within 4 days after transfection. But, expression level is massive for a few days after transfection, and therefore, does not need further selection procedures. The system is best suited for structure/function studies, where no prolonged expression period is mandatory.

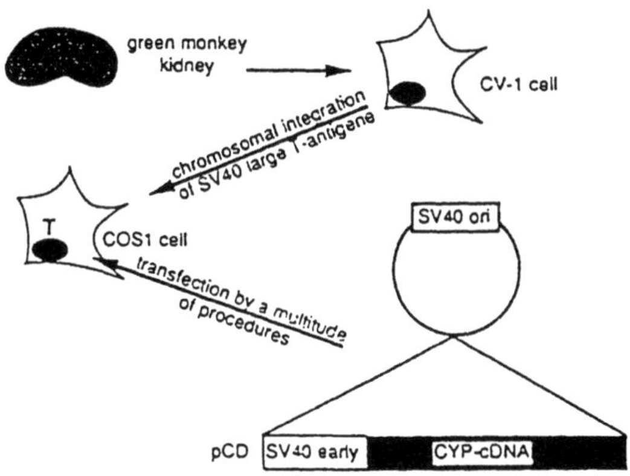

Fig. 4. Maintenance and Expression under the control of the SV40 origin of replication and SV40 Early promoter in COS cells.

Other expression systems rely on infectious transmission of live recombinant virus. Prominent examples are either based on Vaccinia (Fig. 5) (Battula et al.

1987), and RNA-tumor virus (Fig. 6) (Battula 1989). In both instances, these viruses have some peculiarities, which make them useful vectors for one or the other reason.

The Vaccinia virus containes a large genome consisting of about 180,000 base pair DNA, which would be difficult to insert CYP cDNA. Therefore, a cellular situation is created in CV1 cells, where a cDNA containing plasmid and wild type Vaccinia DNA join. Both DNAs share some homology by the thymidine kinase (TK) gene. Thus, homologous recombination between plasm id and viral DNA may occur. CV1 cells producing viral particles containing recombinant viral DNA by homologous recombination are selected by resistance towards bromodeoxyuridine (BUdR). These viral particles can infect any other cultivated cell and express the cDNA encoded CYP. Preferred host cells in this approach were human hepatoma derived cells, the socalled HepG2 cells. HepG2 cells have ample CYP reductase activity in order to facilitate maximum CYP activity. However, Vaccinia virus is

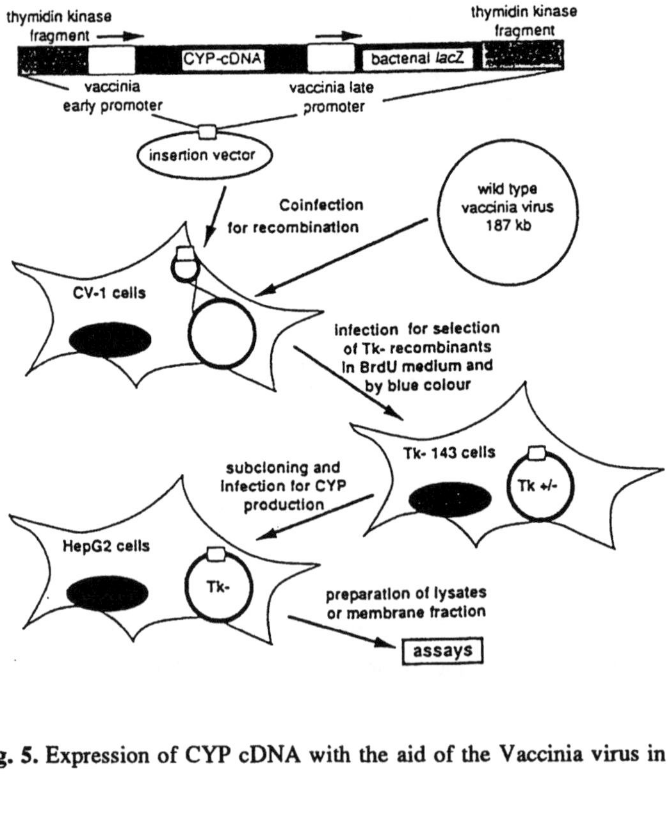

Fig. 5. Expression of CYP cDNA with the aid of the Vaccinia virus in HepG2 cells.

lytic, and so is the recombinant Vaccinia virus. Lysis occurs 3 and 4 days after infection. This is a sufficient expression period for producing large quantities of CYP with good activity. Therefore, this system provides essentially a defined and highly active cell homogenate, which is best for studying and analyzing CYP dependent metabolites (Aoyama et al. 1990). The homogenate may also serve as a substitute for the S9 liver mix used in the Ames test.

Infectious transfer and stable expression at the same time is achieved using a RNA-tumorvirus (Fig. 6). As is the case in Vaccinia mediated transfer, infectious recombinant viral particles have to be generated. It starts with inserting the cDNA into the genomic DNA of a retrovirus. The recombinant retroviral DNA is transfected into a socalled packaging cell line. This cell line contains already a retroviral genomic copy, which provides all proteins and functions for retroviral replication and production of viral particles. For safety reasons, this retroviral copy has been deleted for its packaging signal, and therefore, will not yield infectious particles. The cDNA recombinant genomic DNA is defective for other functions but not for the packaging signal. Both retroviral genomes complement each other by ending up with infectious particles, which contain the cDNA recombinant retroviral genome only, but have the surface structure of a wild type retrovirus. This combination is called a pseudo type virus. This allows one round of infection only in compatible host cells, because the cDNA recombinant retroviral genome itself cannot yield replication competent retroviral genomes. However, the cDNA recombinant retrovirus will function for stable chromosomal integration as any other retrovirus does. This integrated copy may from then on serve as a template for CYP expression. Limitations of these systems are given by the range of host cells for the various RNA-tumorvirus, and by the time needed for obtaining an effective packaging cell line.

A widely used and commercially available CYP expression system is based on a human lymphoblastoid cell line (Crespi 1994) (AHH1), which has been constructed with a vector system similiar to the SV40 viral based system in COS cells (Crespi 1989). AHH1 cells like COS cells had been infected with the Epstein Barr Virus (EBV) and continue to produce the viral antigen (EBNA) after an abortive infection. The EBNA antigen corresponds with an origin of replication

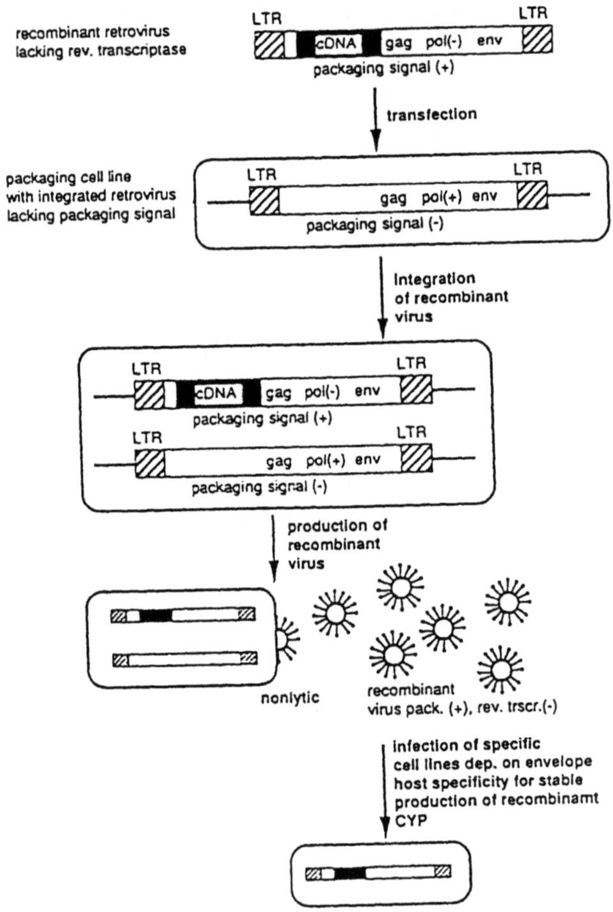

Fig. 6. The retroviral system for stable expression of CYPs.

(oriP). Any plasmid containing this sequence will be maintained episomally. Episomal replication of plasmid may yield high copy numbers per cell. For the AHH1 system two different EBV compatible plasmid vectors are available yielding a high a low copy number of recombinant plasmid per cell. Application of both vectors was used for constructing a AHH1 derived cell line expressing five different human CYPs at almost identical molarities (Crespi et al. 1991). Although extrachromosomal replication in a mammalian cells is known to cause stability problems, the AHH1 system has a stable expression level. Many different promoter constructs and transfer procedures were tried, resulting into AHH derived cell lines with 10 to 100fold higher CYP content and CYP activity than any other mammalian expression system.

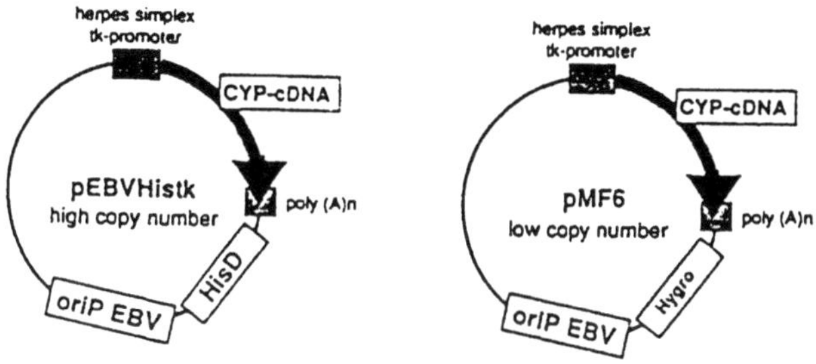

Fig. 7. Recombinant CYP plasmid vectors as used for the EBV derived episomal expression system based on the human lymphoblastoid cell line AHH1.

A simple system for stable expression has been achieved for the V79 Chinese hamster cells (Doehmer et al. 1988). The CYP full length cDNA was placed under control of the SV40 Early promoter in order to ensure high and constitutive expression upon DNA mediated transfection into the V79 cells. Cells were selected by cointroduction of the selectable marker gene for neomycin phosphotransferase (neo). Under these circumstances, the CYP recombinant plasmid stably integrated into the chromosomal DNA of the V79 cells, where it serves as a template for a constant production of the cDNA encoded CYP. V79 cells have characteristics unusual for established cell lines, as a stable karyotype, fast growth with a doubling time of 12 hours, a suitable marker gene HPRT on their single X-chromosome, sturdy morphology, no special requirements for cultivation, extremely low cell number limit for growth, good cloning efficiency. These and other characteristics made V79 cells a preferred indicator cells for a variety of endpoints including cytotoxicity, mutagenicity, chromosomal stability, and several other endpoints. Another advantage, which was formerly a drawback, is the lack of any endogenous CYP activity. Therefore, these cells are defined for the cDNA encoded CYP activity, and makes them a perfect analytical tool for dissecting complex CYP situations. Genetically engineered V79 cells have been shown to be capable of producing quantities of metabolites sufficient for analytical procedures. Under the very same growing conditions biological endpoints can be measured. This allows comparative studies tying together CYPs, metabolites, and effects (Doehmer 1993).

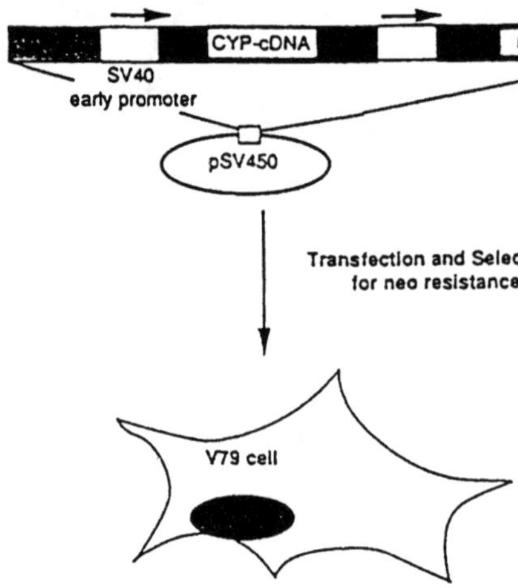

Fig. 8. Recombinant CYP cDNA directed expression in V79 Chinese hamster cells.

Typical CYP activities as observed in genetically engineered V79 cells are listed in Table 1. These activities were compared to those seen in HepG2 cells. There are many different variant HepG2 cell lines, with no or constitutive expression of CYPs. These variant HepG2 cell lines also differ extensively in terms of doubling time, some with 20 hours, others with 40 hours. There is no such thing like an universal HepG2 cell line. Thus, results obtained by applying a HepG2 cell lines have to be accepted and interpreted with a grain of salt. The HepG2 cell line available to us expresses CYP1A1 after overnight exposure to inducers. Interestingly, the EROD/AHH activity ratio comes close to 2 in the case of the genetically engineered V79 cells, and close to 1 with the HepG2 cells, indicating a different metabolic situation in HepG2 cells, which is far from being completely understood and defined.

Table 1. EROD and AHH activities in V79MZ parental cells, V79MZh1A1, V79MZr1A1, and HepG2 cells. Values give the mean ± standard deviation in pmol/min/mg total protein. Number of independent preparations and measurements in parentheses. HepG2 cells were pretreated with 20 μmol benzo[a]anthracene for 22 hours prior homogenization. The ration of EROD/AHH activity for each cell line is shown.

Cell line	EROD	AHH	EROD/AHH
V79MZ	0.0 ± 0 (3)	0.0 ±0 (3)	-
V79MZh1A1	48.8 ± 5.9 (5)	27.4 ±6.3 (5)	1.8
V79MZr1A1	52.1 ± 6.9 (4)	26.5 ±8.6 (4)	2.0
HepG2	5.9 ± 3.9 (3)	7.4 ±2.9 (3)	0.8

Expressing homologous CYP isoforms from different species in the same *in vitro* system allows an immediate comparison of metabolite profiles. A striking difference was seen on the metabolism of benzo[a]pyrene in the CYP1A1 expressing V79 cells listed in Table 1. Rat CYP1A1 oxidized benzo[a]pyrenein the pre-bay and K-region to almost the same extent, whereby the human CYP1A1 oxidized preferentially the pre-bay region, known to form the ultimate mutagenic and carcinogenic metabolites (Jacob et al. to be published).

Conclusions

Recombinant systems have consistently proven over the years to be valuable analytical tools for analyzing, and understanding the complex *in vivo* metabolism of xenobiotics (Gonzalez et al. 1990). Some of these systems have interesting biological endpoints, which can be easily measured. Thus, CYPs, metabolites, and effects can be linked directly under very defined and experimentally controlled conditions. In more complex *in vitro* systems, e.g. primary hepatocytes, there are only indirect means by antibody inhibition studies, or CYP induction studies. Thus, the gene technological approach not only creates new experimental opportunities, but also allows the study of CYPs, which would otherwise be difficult to isolate in the classical way, because of low expression levels in native tissues, or low abundance of good quality tissue, as it is usually the case with human tissue.

The potential of the gene technological approach in metabolism studies has not yet been fully exploited. As the already established systems continue to be applied, new systems are being developed, in some cases with new exciting experimental possibilities. But, no matter the system is well established or brand new, it needs critical evaluation in terms of activity present, and of feasibility in the routine testing. Some of these developments will be leading nowhere, although they look fancy. An example is the approach to immortalize primary hepatocytes by oncogenes and maintaining full differentiation at the same time, which, unfortunately, frequently fails. Nevertheless, this approach gets always a revival every few years, and a lot of supportive attention from uninformed referees.

Whenever a decision is to be taken, a comparative listing of the system specific features should help to find the one system, which matches the experimental problem best.

References

Aoyama T, Yamano S, Guzelian PS, Gelboin HV, Gonzalez FJ (1990a) Five of 12 forms of vaccinia virus-expressed human hepatic cytochrome P450 metabolically activate aflatoxin B1. Proc Natl Acad Sci USA 87: 4790-4793

Barnes HJ, Arlotto MP, Waterman MR (1991) Expression and enzymatic activity of recombinant cytochrome P450 17a-hydroxylase in Escherichia coli. Proc Natl Acad Sci USA 88: 5597-5601

Battula N, Sagara J, Gelboin HV (1987) Expression of P1-450 and P3-450 DNA coding sequences as enzymatically active cytochromes P-450 in mammalian cells. Proc Natl Acad Sci USA 84: 4073-4077

Battula N (1989) Transduction of cytochrome P3-450 by retroviruses: Constitutive expression of enzymatically active microsomal hemoprotein in animal cells. J Biol Chem 264: 2991-2996

Crespi CL (1994) Human cytochrome P450 products. Catalogue of GenTest Corp., 6 Henshaw St. Woburn, MA 01801, USA

Crespi CL, Langenbach R, Rudo K, Chen YT, Davies RL (1989) Transfection of a human cytochrome P-450 gene into the human lymphoblastoid cell line AHH-1, and use of the recombinant cell line in gene mutation assays. Carcinogenesis 10:295-301

Crespi CL, Gonzalez FJ, Steimel DT, Turner TR, Gelboin HV, Penman BW, Langenbach R (1991) A metabolically competent human cell line expressing five cDNAs encoding procarcinogen-activating enzymes: Application to mutagenicity testing. Chem Res Toxicol 4: 566-572

Doehmer J, Dogra S, Friedberg T, Monier S, Adesnik M, Glatt HR, Oesch F (1988) Stable expression of rat cytochrome P-450IIB1 cDNA in Chinese hamster cells (V79) and metabolic activation of aflatoxin B1. Proc Natl Acad Sci USA 85: 5769-5773

Doehmer J (1993) V79 Chinese hamster cells genetically engineered for cytochrome P450 and their use in mutagenicity and metabolism studies. Toxicology 82: 105-118

Eugster HP, Sengstag C, Meyer UA, Hinnen A, Würgler FE (1990) Constitutive and inducible expression of human cytochrome P450IA1 in yeast saccharomyces cerevisiae: An alternative enzyme source for in vitro studies. BBRC 172: 737-744

Fisher CW, Caudle DL, Martin-Wixtrom CA, Quattrochi LC, Tukey RH, Waterman MR, Estabrook RW (1992) High-level expression of functional human cytochrome P450 1A2 in Escherichia coli. FASEB J 6: 759-764

Gonzalez FJ, Crespi CL, Gelboin HV (1990) cDNA-expressed human cytochrome P450s: a new age of molecular toxicology and human risk assessment. Mutat. Res. 247: 113-127

Larson JR, Coon MJ, Porter TD (1991) Alcohol-inducible cytochrome P-450IIE1 lacking the hydrophobic NH2-terminal segment retains catalytic activity and is membrane-bound when expressed in Escherichia coli. J Biol Chem 266: 7321-7324

Oeda K, Sakaki T, Ohkawa H (1985) Expression of rat liver cytochrome P-450 MC cDNA in Saccharomyces cerevisiae. DNA 4: 203-210

Pompon D (1988) cDNA cloning and expression in yeast S. cerevisiae of beta-naphtoflavone induced rabbit liver P-450 LM4 and LM6. Eur J Biochem 177: 285-293

Sakaki T, Shibata M, Yabusaki Y, Ohkawa H (1987) Expression in Saccharomyces cerevisiae of chimeric cytochrome P450 cDNAs constructed from cDNAs for rat cytochrome P450c and P450d. DNA 6: 31-39

Schenkman JB, Greim H (1993) Cytochrome P450. Handbook of Experimental Pharmacology. Vol. 105, Springer Verlag, Berlin, Heidelberg, New York, London, Paris, Tokyo, Hong Kong, Barcelona Budapest.

Waterman MR, Johnson EF (1991) Cytochrome P450. Methods in Enzymologie. Vol. 206, Academic Press, San Diego, New York, Boston, London, Sydney, Tokyo, Toronto.

Zuber MX, Simpson ER, Waterman MR (1986) Expression of bovine 17 a-hydroxylase cytochrome P-450 cDNA in nonsteroidogenic (COS 1) cells. Science 234: 1258-1261

Applications for Transgenic Animals in Toxicology

D. Brusick
Hazleton Washington Inc., 9200 Leesburg Pike, Vienna, Virginia, U.S.A.

The development of transgenic animals has already had a significant impact in biomedical research. Non-human models of human diseases (i.e., sickle cell disease, cystic fibrosis and AIDS) have been produced in mice and are being used to screen potential therapies (Breslow, 1994). Domesticated animals genetically engineered to: (a) resist disease, (b) produce meat or eggs with less cholesterol, or (c) accelerate growth may be introduced into agriculture in the near future (Sedlak, 1989). Transgenic animals have also been proposed as living bioreactors for the commercial production of pharmaceuticals and other biologicals not easily synthesized by other methods (Van Brunt, 1988).

Transgenic models which fit into chemical safety assessment schemes consist of those used to screen for toxicity (i.e., detection of carcinogens and mutagens) and those used to elucidate mechanisms of toxicity. Table 1 summarizes these two roles.

Table 1. Current Roles for Transgenic Animal Models in Toxicological Assessment

Role	Process	Examples
Screening	Development of susceptible animals which can quickly detect toxic agents	OncoMouse MutaMouse™ PIM® Mouse
Elucidation of Mechanisms	Use of gene targeting to study toxicity response mechanisms	*Ah* complex *Hox* complex

To date, all commercial models are limited to screens for mutation (Gossen and Vijg, 1993) or tumors (Sullivan et al., 1993) and have been constructed in mice. The use of the mouse as the host animal for the transgenes in toxicology screening models is the consequence of several critical factors:

1. The mouse pronucleus is more accessible than the rat pronucleus.
2. Embryonic stem cell cultures can only be produced in mice.
3. Large gene libraries are available for mice, but not for rats.
4. Short gestation time and large litter size facilitates the use of mice.

From a toxicological perspective, rat models would be substantially more useful, as a larger body of toxicity data is available in the rat. Comparative risk assessments and mechanistic extrapolations could be more informative if results from rat transgenic models of carcinogenesis and mutagenesis were also available. A rat model of Stratagene's Big Blue™ transgenic model for mutation detection is reportedly close to commercialization (Sullivan et al., 1993).

Methods for the Production of Transgenic Animals

Transgenic animals are produced by the introduction of foreign DNA (genes) into their genomes using three different approaches. Depending upon the method used, the transgene will become stably integrated into the host genome at selective sites or at random. Three methods used in the production of transgenic models are summarized in Table 2.

Engineered animals can be made homozygous for the integrated DNA transgene through selected breeding, with the result that the transgene is transmitted to future generations. Integration of the foreign DNA can be confirmed through the use of probes specific for the transgene DNA.

The first method uses viral shuttle vectors for the production of transgenes, and is limited to the introduction of relatively small segments of DNA (8-10 kilobases). This method has been virtually replaced by the introduction of direct pronuclear injection and recombinational insertion into embryonic stem cell (ES) cultures, which can introduce large segments of DNA.

In the second method, the DNA construct is injected into either the male or female pronucleus in fertilized ova (Figure 1). Single or multiple tandem concatemers of the transgene are integrated at sites located randomly along the mouse chromosome. (Figure 2). Expression of the transgene along the chromosome (active or inactive) is dependent upon either the insertion site or the presence of regulatory elements (tissue specific or general). Selection of a tissue specific controlling element for combination with the transgene (fusion genes) will determine the specific tissue or organ in which the transgene is expressed.

The third method involves the use of embryonic stem (ES) cells. ES cells are pluripotent embryonic cultures that can be reintroduced into a normal mouse blastocyst resulting in chimeric mice with ES cells forming with somatic and germinal tissues (Figure 3). This property of ES cells makes them ideal for producing transgenic mice. ES cell cultures can be genetically modified by transfection of the cells with transgene constructs. The transfected DNA will undergo recombination with the host cell genomic DNA resulting in its insertion into the host cell genome (Figure 4). Some advantages in using ES cells for constructing transgenic mice are that the transgene, including regulatory elements, can be constructed to: (a) undergo homologous recombination with endogenous genes resulting in their replacement, and (b) limit the host cell to a single integrated transgene. Taking advantage of these two characteristics of ES cells, it is now possible to carry out gene-targeting experiments.

Table 2. Common Methods for Production of Transgenic Animals

Method	Description	Characteristics	Citation
Viral Transfer	Transgene inserted into viral DNA. The virus serves as the vector to carry the transgene into the host embryonic cell. The transgene becomes inserted into the host genome as a result of viral integration.	Viruses are effective vectors for DNA, but transgene size is limited to 8-10 kilobases.	Jaenish, 1976
Microinjection	Transgene is physically injected into mouse zygote pronucleus. Injected eggs are returned to female for development. The transgene becomes inserted into the host genome as a result of "illegitimate" recombination.	Effective method for production of transgenic mice. Insertion sites are typically random and often in multiple copies of the transgene.	Gordon, *et al.*, 1980
Embryonic Stem Cell Culture	Transgene constructs are transfected into pluripotent embryonic stem cells. The engineered cells are placed in a mouse blastocyst and returned to the female for development. The transgene becomes inserted into the host genome by recombination.	This method allows single copies of transgenes to be inserted into host cells. Greater manipulation of the transgene and its expression are possible with this method.	Kuehn, *et al.*, 1987

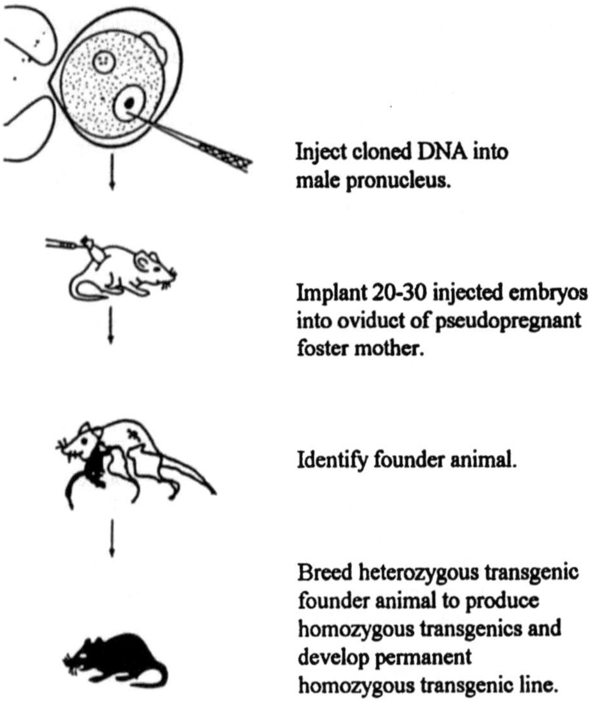

Fig. 1. Basic procedures involved in the construction of transgenic animals using a microinjection method for introducing the transgene.

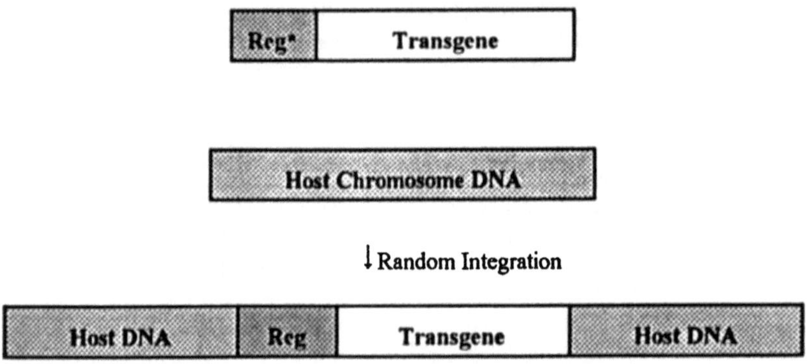

* Regulatory element might be a nonspecific (metallothionein) tissue specific (insulin-T-antigen) promoter responsible for expression of the transgene.

Fig. 2. Insertion of the transgene into the host DNA by illegitimate recombination processes following microinjection. Segments containing multiple head to tail copies of the transgene are frequently insertions.

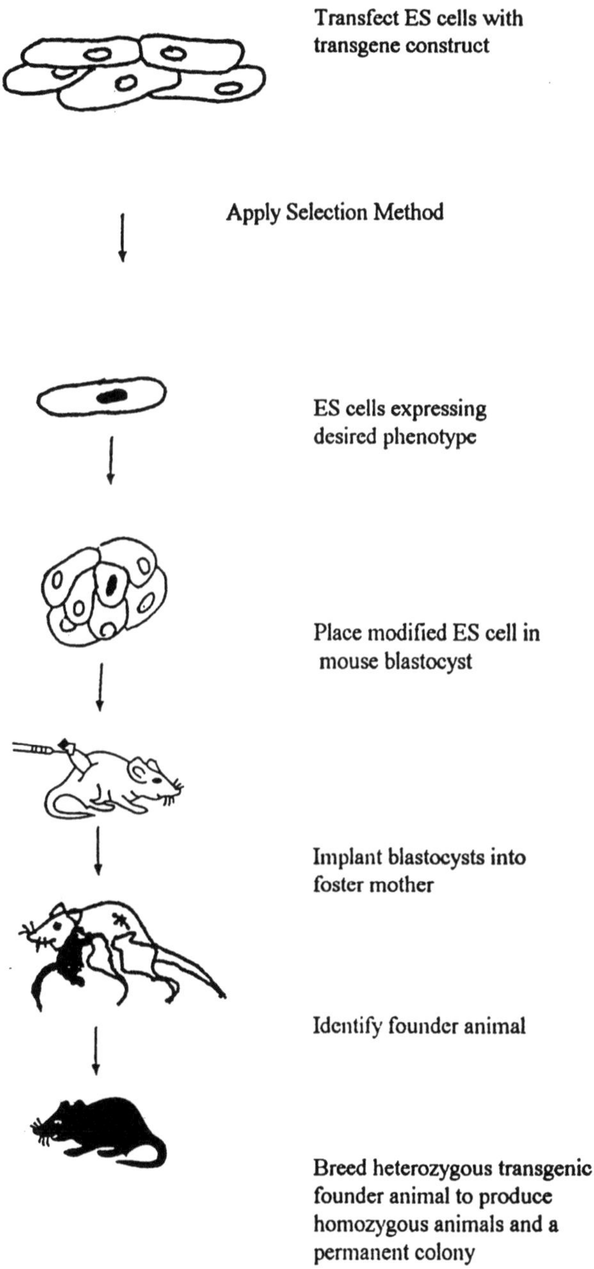

Fig. 3. Basic pronucleus involved in the construction of transgenic mice using embryonic stem cells to introduce the transgene.

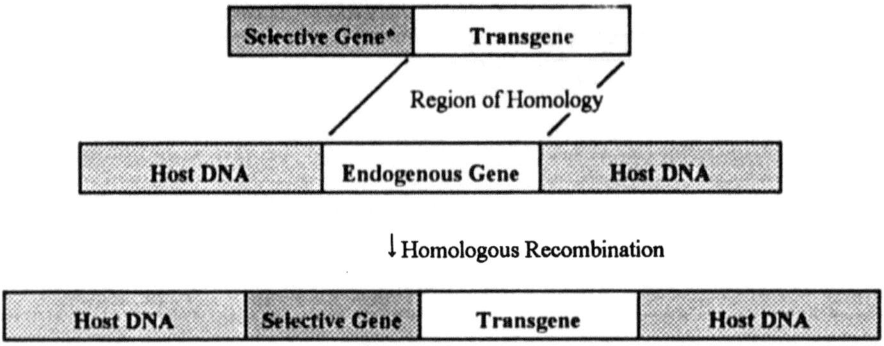

* A gene which enables host cells that have acquired the transgene cassette to survive an exogenously applied selective chemical (i.e., gene for resistance to a toxic chemotherapeutic agent neomycin)

Fig. 4. Insertion of transgene into the host chromosome through homologous recombination.

Gene-targeting is the process through which the function of a gene is determined through selective mutagenesis of the targeted gene and its replacement of the endogenous in a transgenic animal. To carry out gene-targeting, the gene of interest must be cloned. The cloned gene is inactivated by a mutation prior to introduction into the ES cell. A selective gene and, if desired, a reporter gene are added to the transgene selection cassette. The transgene cassette is introduced into ES cell cultures by transfection or other appropriate techniques. Following an expression time, selective conditions are introduced to eliminate cells that have not incorporated the transgene complex into the genome replacing the endogenous gene. Expression of the reporter gene, if included, establishes that the newly introduced transgene complex is responding to intracellular regulatory processes. ES cells with the appropriate genotype are then used to construct transgenic animals. By replacing the endogenous gene with one which is irreversibly inactivated, the role of the endogenous gene in maintaining homeostatic properties can be accurately identified and studied (Mark, 1992; Saffer, 1992).

Transgenic Models for Use in Toxicological Screening

The development and commercialization of Big Blue™, MutaMouse™, Oncomouse, PIM®, and the TSG-p53® transgenic mice raised expectations that transgenic mice engineered to screen for immunotoxic, developmental and neurotoxic agents might soon be available. While many transgenic animals involving immunologic, neurologic and developmental diseases have been constructed, none have been proposed for use in toxicological screening.

While nothing similar to the transgenic animal engineered for mutagen and carcinogen detection has appeared, some transgenic mouse models adaptable to testing have been constructed. A transgenic mouse developed through targeted inactivation of the RAG-2 gene of mice results in a model with reduced ability to produce mature T or B lymphocytes (Shinkai et al., 1992). These mice are susceptible to infection and have small or absent thymus glands. The mice (TIM™) are currently available through GenPharm and could be used in screening biological pharmaceuticals and pesticides which might otherwise stimulate severe immunological reactions. The model may also be used as a hypersensitive model for infectivity protocols that have been suggested as part of batteries for immunotoxicity testing.

A transgenic mouse model has been developed at the University of Southern California which expresses a gene causing neurons to over express monoamine oxidase-B (Gen, 1994). The consequence is excessive production of free radicals in the strains of the mice, resulting in toxicity. This mouse model might be used as a sensitive screen for agents which enhance the production of toxicity in CNS tissues. Screening transgenic models, such as those used for carcinogen and mutagen detection, are being modified to have human genes used as the target sites for chemical interaction. This modification can reduce some of the variability associated with extrapolation across species boundaries.

Transgenic mice with altered skeletal development have been produced. However, the plethora of mechanisms associated with the production of developmental anomalies precludes construction of a model acceptable for general screening. Perhaps the use of one of the models for mutation induction can be applied to teratology screening since chemicals can be administered during gestation. Target tissues identified by production of anomalies can also be screened for concomitant increases in *Lac I* or *Lac Z* mutations as one means to elucidate the mechanisms responsible for the terata. Behavioral toxicity screening may also be addressed through the development of hypersensitive transgenic models.

The usefulness of transgenic models for mutagen and carcinogen assessment is presently being assessed by industrial and government toxicology laboratories. Neither model has been recommended for use in regulatory testing programs at this time. Critics have raised a number of issues regarding the reliability of the current transgenic models for carcinogenesis and mutagenesis, unrelated to the toxicological relevance of the host species. The two commercially available models for mutation detection rely upon target genes which are unexpressed *in situ*. Data are accumulating which suggests that DNA repair, and hence mutant fixation and expression, differ between expressed (transcribed) and unexpressed (untranscribed) genes. If this is the situation, risk extrapolations based upon these models may be inaccurate. The mouse models for carcinogenesis consist of insertion of single oncogenes (*myc*, *ras*) or tumor suppressor genes (p53). Data from several sources suggests that oncogenes do not act in isolation to induce tumor cell induction or proliferation, and caution must be exercised to make the models as relevant as possible. Some of the tumor models involve over or under production of specific gene products to make the animals tumor susceptible (i.e., *c-myc* and *pim*-1). The degree of production in the transgenic model must mimic that associated with levels found in human tumors or cells of predisposed humans. Vast over production of oncogene products in transgene models may produce a distorted picture of the carcinogenesis responsiveness of the host animal (Sullivan et al., 1993).

Most other transgenic models associated with drug and pesticide developments have been generated for the purpose of acting as disease models with which therapeutic treatments may be produced. Overall, the development of transgenic models for toxicological screening appears to be very limited.

Transgenic Models for Use in Elucidation of Toxic Mechanisms

A more productive use of transgenic models in toxicological evaluation has been the application of targeted mutagenesis to better understand molecular mechanisms associated with toxicity (Figure 5). Targeted mutagenesis can be accomplished by selectively inserting DNA into the target gene, knocking out the gene function (insertion mutation). It can also be produced by sequence replacement with homologous DNA containing mutations (Figure 6). Once mechanisms are understood, screening transgenics might follow.

Gene targeting transgenic models have already played an important role in the elucidation of how dioxins and other PCBs exert their affects on drug metabolism enzymes (Nebert, personal communication). Discovery and elucidation of the binding, transport and regulatory processes associated with the dioxin TCDD has been elucidated using targeted mutagenesis in mice which are both susceptible and resistant. This research has resulted in the identification of comparable genes in humans, leading to an understanding of the comparative toxicity of hydrocarbons in humans, and the development of transgenic mice containing constructs composed of human Ah receptor genes from susceptible and resistant phenotypes (Nebert, personal communication). These mice may ultimately be used as screening models for environmental toxicants.

A second example demonstrating the potential for using target mutagenesis as a toxicological tool has been the work of Whiting, et al. (1991). Their work focuses on the role of a family of genes called the homeobox (*Hox*) genes. This family of genes influences embryonic development by sequencing the expression of various regulators of embryonic development. Transgenic mice containing specific inactivated *Hox* locus prove to be valuable tools for identifying the developmental domain controlled and the spacial expression for each locus. Sequence timing was determined by including a reporter gene (*Lac Z*) that would produce a histologically visible signal when the gene is activated. From this research, it may be possible to: (a) identify susceptible periods during gestation for chemical exposures, or (b) determine the types of developmental damage produced by chemicals which interfere with the expression of the *Hox* gene family. Thus, transgenic mice may serve to define developmental organization and its responsivity to chemical disruption.

Another gene family which might be a candidate for the production of transgenic screens in developmental toxicology is the transforming growth factor-B (TGF-B). This "superfamily" of genes contains several subfamilies including growth/differentiation factors (GDF gene family), and the bone morphogenetic proteins (BMP gene family).

Mutations which exhibit changes in mesenchyme aggregation, chondrogenesis and differentiation of bone map to members of these gene families. Some of the genes in the TGF-B superfamily have been cloned and sequenced (Storm et al., 1994).

A candidate gene family that should be amenable to the targeted mutagenesis approach is the loose collection of genes collectively referred to as stress genes. Members of this family produce "stress proteins" in response to toxic exposures (Sanders, 1994). This gene family comprises the epigenetic repair system of an organism and protects organisms from damage as a result of exposure to a wide variety of exogenous exposures. Elucidation of the genes in this family, and determination of their role in toxicity manifestation will not only provide better insight into hazard assessment, but could also lead to a new set of transgenic models containing cassettes of stress genes that could be used to screen for toxic agents. For example, animals which produce easily detected signals confirming induction of specific stress proteins might enable target tissues to be quickly identified (i.e., rats exposed to mercuric chloride exhibit synthesis of stress proteins before overt renal injury can be detected).

"Super transgenic" models consisting of two or more transgenes are possible. Transgenic mice containing both *ras* and *myc* oncogenes show a high degree of synergism (Van Brunt, 1990). Other possible "super transgenic" models might be constructed using combinations of MutaMouse™ and transgenics with specific DNA repair deficiencies (Matsukuma et al., 1989). The combination could increase the sensitivity and provide useful information regarding the role of different repair pathways on mutation production. One could also envision adding transgenes which regulate chemical metabolism to the repair deficient MutaMouse™ transgenic.

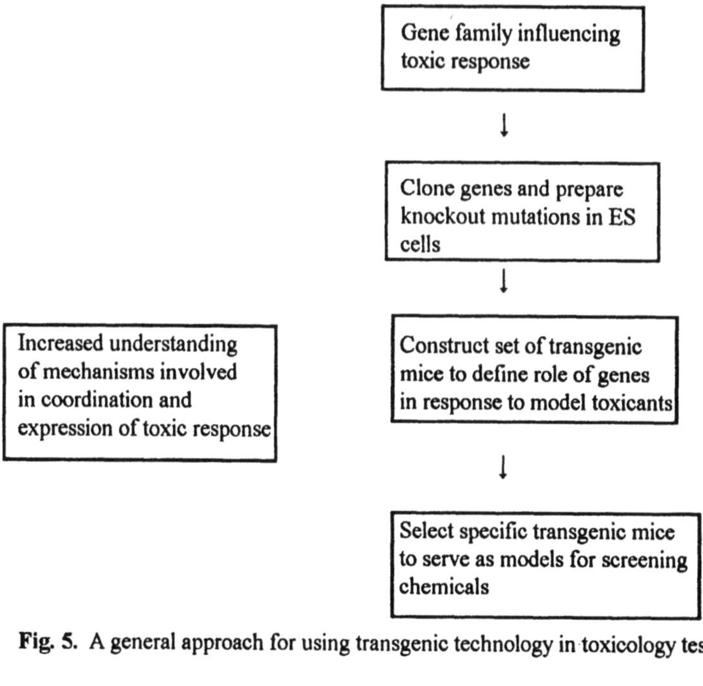

Fig. 5. A general approach for using transgenic technology in toxicology testing.

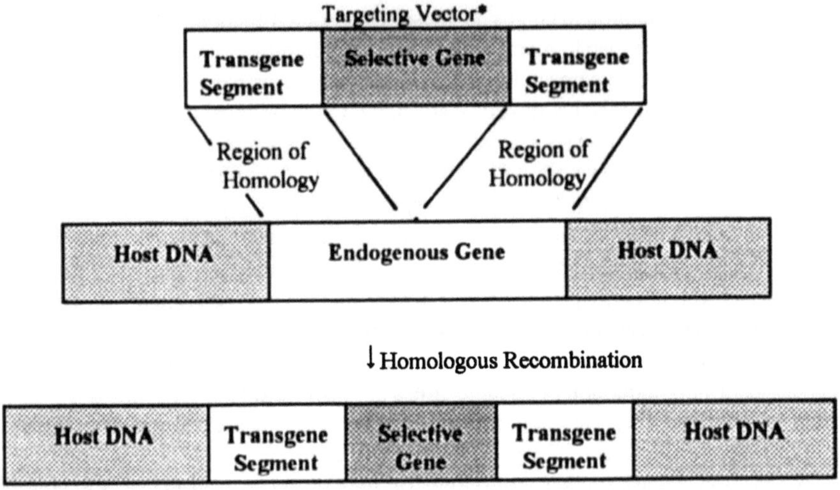

* The transgene is inactivated by insertion of the selective gene and following homologous recombination the resultant transgenic animal carries a "null" mutation in place of the functional endogenous gene.

Fig. 6. Gene targeting using a selection cassette.

Other Considerations Influencing the Use of Transgenic Models

At least two other issues are likely to affect the development of transgenic models for toxicological testing.

Animal Welfare Concerns

Critics of transgenic development argue that animal patents, such as the one issued to Harvard University in 1988 for OncoMouse, would devalue animal life by reducing the animal being to a commercial commodity (Raines, 1988). It is also argued that development of animals with phenotypes that interfere with normal functioning or life-expectancy could be viewed as leading unnecessarily to greater animal suffering.

Patent Concerns

Transgenic animal models are being patented to protect the scientific and financial interests of the developer. Proponents of this process argue that the ability to patent transgenic animals stimulates innovation and research in this technology.

On the other hand, it can be argued that the use of patented animal models could increase the cost of safety evaluation through higher animal prices and licensing fees. Regulatory agencies which review toxicological data may be accused of financial bias if data from one patented transgenic model is preferred over another commercially available model.

Conclusions

Transgenic animal models have been, and will continue to be, introduced into animal toxicological assessments. This technology, along with *in vitro* alternative models, will significantly influence the evolution of toxicology during the next decade resulting in a better understanding of mechanisms of toxicity, and a general reduction in the numbers of animals used in this scientific discipline.

References

Breslow J (1994) Lipoproteins and heart disease. Bio/Technology 12:365-370

Camper SA (1987) Research applications of transgenic mice. Bio Techniques 5: 638-650

Genetic Engineering News, University Gen Beat, June 1, 1994

Gordon J, Scangos GA, Plotkin DJ, Barbosa JA, Ruddle F (1980) Genetic transformation of mouse embryos by microinjection of purified DNA. Proc. Natl. Acad. Sci. 77:7380-7384

Gossen J, Vijg J (1993) Transgenic mice as model systems for studying gene mutations *in vivo*. TIG 9:27-31

Jaenisch R (1976) Germ line integration and Mendelian transmission of the exogenous Moloney leukemia virus. Proc. Natl. Acad. Sci. 73:1260-1264

Kuehn MR, Bradley A, Robertson EJ, Evans MJ (1987) A potential animal model for Lesch-Nyhan syndrome through introduction of HPRT mutations into mice. Nature 326:295-298

Mark WH (1992) Manipulation of the mouse genome using embryonic stem cells. Lab Animal 21:41-53

Matsukuma S, Nakatsuru Y, Nakagawa K, Utakoji T, Sugano H, Kataoka H, Sekiguchi M, Ishikawa T (1989) Enhanced O^6-methylguanine-DNA methyltransferase activity in transgenic mice containing an integrated *E. coli ada* repair gene. Mutation Research 218:197-206

Raines LH (1988) The mouse that roared. Issues in Science and Technology 4:64-70

Saffer JD (1992) Transgenic mice in biomedical research. Lab Animal 21:30-38

Sanders B (1994) Stress proteins as molecular chaperons: implications for toxicology. Environmental Health Perspective 102:538-540

Sedlak BJ (1989) Transgenics enters commercial arena with high hopes for the 1990s. The Medical Business Journal 5:61, 67-70

Shinkai Y, Rathbun G, Lam KP, Oltz EM, Stewart V, Mendesohn M, Charron J, Datta M, Young F, Stall AM, Alt FW (1992) RAG-2-deficient mice lack mature lymphocytes owing to inability to initiate V(D)J rearrangement. Cell 6:68

Storm EE, Huynh TV, Copeland NG, Jenkins NA, Kingsley DM, Lee S (1994) Limb alterations in *brachypodism* mice due to mutations in a new member of the TGF-B-superfamily. Nature 368:639-642

Sullivan N, Gatehouse D, Tweats D (1993) Mutation, cancer and transgenic models: relevance to the toxicology industry. Mutagenesis 8:167-174

Van Brunt J (1990) Transgenics primed for research. Bio/Technology 8:725-728

Van Brunt J (1988) Molecular farming: transgenic animals as bioreactors. Bio/Technology 6:1149-1154

Whiting S, Marshall H, Coole M, Krumlanf R, Rigby PWJ, Stott D, Allenmann RK (1991) Multiple spatially specific enhancers are required to reconstruct the pattern of *Hox*-2.6 gene expression. Genes Dev. 5:2048-2059

The Benchmark Dose Concept

(Chair: Dr. M.A. Martens, Belgium)

Principles, Benefits and Limitations of the NOEL Approach

David J. ESDAILE
RHONE-POULENC - Secteur Agro, B.P. 153, 06903 Sophia Antipolis - France

1 - Overview

The concept of the NOEL approach is, in itself, very simple. Many types of toxicology studies are performed utilizing a graded range of dose levels of test compound. The NOEL (No Observed Effect Level) is the highest dose level at which and below which no effects of the test compound are observed amongst the evaluated parameters. A refinement of this concept is the use of the NOAEL (No Observed Adverse Effect Level) which only takes into account effects which are regarded as adverse.

2 - Need for the NO(A)EL Approach

The NOEL (No Observed Effect Level) or NOAEL (No Observed Adverse Effect Level) is often the single statistic taken from a toxicology study, or series of studies, for use in risk assessment calculations. A toxicology package (normally including chronic rodent, non-rodent and teratology studies, etc...) is evaluated and the lowest NO(A)EL identified. The objective being to select the most sensitive species and test system to provide a conservative estimate of an exposure level which does not cause adverse effects, on which to base the risk assessment.

The NOEL value provides an estimate of the "safe" dose-level for the test species in the test system. In order to extrapolate to an estimate of the "safe" dose level in man (e.g. Acceptable Daily Intake, ADI, or Threshold Limit Value, TLV) the NOEL is divided by a Safety Factor (SF) or Uncertainty Factor (UF). Hence, the current need for the NOEL (or an equivalent) is based on its routine use in risk assessment.

3 - Definitions of Terms used in the NO(A)EL Approach and in Risk Assessment

The following terms are those most frequently encountered in publications encompassing the concept of the NO(A)EL.

• NOEL* (no observed effect level) - An experimentally determined dose at which no statistically or biologically significant indication of toxic effects are observed.

• NOAEL* (no observed adverse effect level) - An experimentally determined dose at which no statistically or biologically significant indication of the toxic effect of concern is observed.

• Adverse (effect) - Functional impairment or pathological lesion which may affect the performances of the organism or which reduces its ability to respond to additional challenge.

* See notes 1, 2, 3 and 4 which refer to principles and concepts relating to the NO(A)EL.

> Note 1
> The concept of a "dose-response" is generally accepted.
> "Empirical observations have generally revealed that as the dosage of a toxicant is increased, the toxic response (in terms of severity and/or incidence of effects also increases. This dose-response relationship is well-founded in the theory and practice of toxicology and pharmacology. Such behavior is observed in the following instances : in quantal responses, in which the proportion of responding individuals in a population increases with dose ; in graded responses, in which the severity of the toxic response within an individual increases with dose ; and in continuous responses, in which changes in a biological parameter (e.g., body or organ weight) vary with dose."
> Barns and Dourson (1988) [US EPA]

> Note 2
> Further, the concept of a "threshold" is also accepted.
> "In the case of [non-oncogenic] toxicity, organic homeostatic, compensating and adaptive mechanisms exist that must be overcome before a toxic endpoint is manifested. For example, there could be a large number of cells performing the same or similar function whose population must be significantly depleted before the effect is seen.
> The threshold concept is important in the regulatory context. The individual threshold hypothesis holds that a range of exposures from zero to some finite value can be tolerated by the organism with essentially no chance of expression of the toxic effect."
> Barns and Dourson (1988) [US EPA]

Note 3
A number of concepts are common to the NOEL and NOAEL definitions :
• The NO(A)EL can only be based on the range of parameters *observed* in a study.
• The *effect* must be of true biological significance and related to administration of compound. Chance statistical differences between control and treated groups should be disregarded.
• The value corresponds with one of the doses used in the study.

Note 4
The concept of the NOAEL is not accepted worldwide by all regulatory authorities, hence many organisations prefer to use the term NOEL when possible. Some effects of treatment can be clearly identified as not being adverse, examples include colouring agents which cause a temporary colouration of the intestinal mucosa, and reduced food intake in chronic rodent toxicity studies leading to a reduced incidence of obesity and a prolongation of life-span.

• ADI† (acceptable daily intake) - The amount of a chemical to which a person can be exposed on a daily basis over an extended period of time (usually a lifetime) without suffering a deleterious effect.

• RfD† (reference dose) - An estimate (with uncertainty spanning perhaps an order of magnitude) of a daily exposure to the human population (including sensitive sub-groups) that is likely to be without appreciable risk of deleterious effects during a lifetime.

• SF† (safety factor) - The divisor applied to the NOAEL to calculate the ADI; i.e., ADI = NOAEL/SF.

• UF† (uncertainty factor) - The divisor which, along with the modifying factor (MF), is applied to the NOAEL to calculate the RfD; i.e., Rfd= NOAEL/(UFxMF). The magnitude of the UF is generally determined by the considerations detailed in Table 1.

• MF† (modifying factor) - An additional factor sometimes used in the derivation of the RfD to reflect the professional judgment of the assessor in evaluating the peculiarities of the data base for a particular chemical.

• EED (estimated exposure dose) - The chemical dose anticipated to result from human exposure under a prescribed set of conditions.

• MOE or MOS (margin of exposure or margin of safety respectively) - The ratio between the NOAEL and the EED; i.e., MOE = NOAEL/EED.

• Critical endpoint (adverse effect) - Specific adverse effect on which the risk assessment is based.

• Critical study - The study yielding the NOAEL which is used as the basis of the RfD.

† The ADI and RfD describe the same concept and differ only in their terminology, as do the SF and the UF (combined with the MF) .

Table 1. Guidelines for the use of Uncertainty Factors and Modifying Factors in Deriving Reference Doses

Standard uncertainty factors (UFs)

Use a 10-fold factor when extrapolating from valid experimental results in studies using prolonged exposure to average healthy humans. This factor is intended to account for the variation in sensitivity among the members of the human population and is referenced as "10H".

Use an additional 10-fold factor when extrapolating from valid results of long-term studies on experimental animals when results of studies of human exposure are not available or are inadequate. This factor is intended to account for the uncertainty involved in extrapolating from animal data to humans and is referenced as "10A".

Use an additional 10-fold factor when extrapolating from less than chronic results on experimental animals when there are no useful long-term human data. This factor is intended to account for the uncertainty involved in extrapolating from less than chronic NOAELs to chronic NOAELs and is referenced as "10S".

Use an additional 10-fold factor when deriving a RfD from a LOAEL, instead of a NOAEL. This factor is intended to account for the uncertainty involved in extrapolating from LOAELs to NOAELs and is referenced as "10L".

Modifying factor (MF)

Use professional judgment to determine the MF, which is an additional uncertainty factor that is greater than zero and less than or equal to 10. The magnitude of the MF depends upon the professional assessment of scientific uncertainties of the study not explicitly treated above, e.g., the completeness of the overall data base and the number of species test. The default value for the MF is 1.

If all the above factors were used to the maximum, it would give a total factor of 10^5, however, in practice factors of greater than 10^3 are rarely employed.

• LO[A]EL - (lowest observed [adverse] effect level) - The lowest dose which produces either statistically or biologically significant increases in either frequency or severity of [adverse] effects between the exposed population and its appropriate control.

• FEL - (frank effect level) - Less rigorously defined ; distinguished from the LOAEL by a greater frequency and/or severity of the adverse effect.

• $NOAEL_{subchr}$ - NOAEL from a subchronic study (usually 90 days).

• $NOAEL_{chr}$ - NOAEL from a chronic study (usually 2 years).

Definitions are rephrased slightly from Dourson and Stara (1983) and Barns and Dourson (1988).

4 - Factors to Consider when Evaluating a NOAEL

a) Statistical differences between control and treated groups are only relevant *where the differences are of true biological significance.* Take historic control data into account where appropriate.

b) For quantal responses of low incidence, historic control data and the "normal expected range" of results should be considered.

c) For quantal responses, ensure terminology and classification of observation allows a full and logical interpretation of results.

Critical issues in the identification of the NOAEL are the determination of when a difference between control and treated groups reflect an "effect" (as opposed to a chance difference), and when an effect can be regarded as an "adverse effect". A combination of probability (determined by statistical means) and expert judgment are used in the evaluation of the significance of differences between control and treated groups for both continuous data and incidence data. The principle is simple, but the determination of the NOAEL can be difficult, particularly when results involve a low incidence of relatively rare observations, where effects do not follow a standard dose-response or where effects seen are of marginal significance.

5 - Importance of Dose Selection and Study Design

Regulatory guidelines generally indicate the use of a control group (untreated and/or vehicle control) and at least three treated groups so that information on the dose-response relationship can be obtained.
The desirable goal is that the highest dose selected meets the objectives of maximally enhancing the sensitivity of the test without introducing distortions induced by excessive toxicity. Thus, the highest dose level "should result in toxic effects", or for non-toxic materials be at the relevant "limit" dose. The criteria for selection of this dose vary with the guideline consulted. Confronted with this situation the current practice for selecting the highest dose for agrochemicals or industrial compounds is generally to meet the requirement of the most stringent guideline ; in practice this means the US-EPA "Maximum Tolerated Dose" (MTD) requirements.
The lowest dose should not cause any significant toxicity and should not interfere with the normal growth, development or longevity of the animals. Most regulatory guidelines also state that the low dose should not be less than 10% of the high dose. This statement is potentially contradictory in that most biologically active test compounds cause some toxicity at doses of 10% of a dose causing frank toxic effects. Hence, in practice this latter recommendation is largely ignored.
The intermediate dose (or doses) are fixed mid-range between the other levels. Most regulatory guidelines suggest that the intermediate dose levels should ideally exhibit minimally toxic effects. In practice this is very difficult to predict and often the expected NOEL is selected with the lower dose used as a "fall back". In selecting the mid - and low dose levels, there is the practical need that the NOEL must cover the human exposure level with an adequate safety margin.

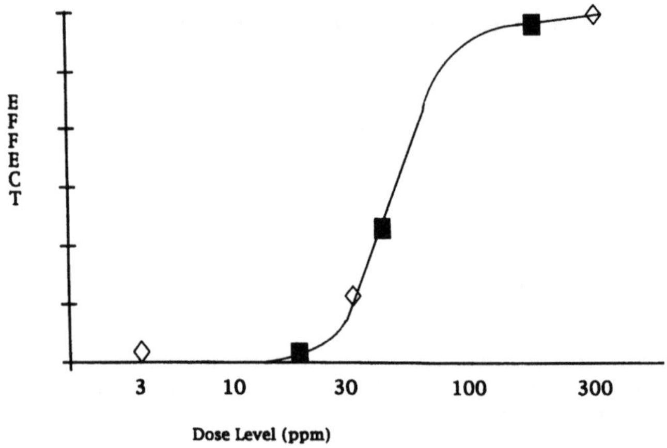

Fig.1. Hypothetical dose response curve showing 'ideal' dose selection (■) where lowest dose level would be indistinguishable from control, and a dose selection (◇) designed to be certain of obtaining toxic effects at the highest dose and no effect at the lowest dose.

If the dose response curve for the critical toxic effect were known, "ideal" dose levels could be selected (see filled squares, fig.1). However, this is very rarely the case, consequently, to be certain of achieving both an MTD, and a dose without effects it is likely that more widely spread dose levels would be selected (see open diamonds, fig.1).

In order to obtain the highest possible NOAEL, either a larger number of dose groups would need to be used or extensive pre-study research would be required to determine the dose-response relationship. These approaches are not generally economical, and would involve a much increased use of animals.

The NO(A)EL "... represents a statistical no effect level that is dependent on the power of the study design" (Kimmel and Gaylor, 1988). Improvements to study design, such as increasing group sizes, reducing inter-animal variability and the use of sophisticated statistical procedures all tend to reduce the NOAEL. Hence, it can be argued that adherence to the NOAEL approach inhibits the improvement or refinement of toxicological test models in regulatory toxicology. It is an inevitable consequence that many regulatory studies are conducted according to the minimum requirements laid down in guidelines.

6 - NOAEL in Risk Assessment as Practiced by Industry and Regulatory Authorities

Since the NOAEL is the principal single statistic used in risk assessment and the calculations of "safe" exposure levels for man, it is of critical importance. Risk assessments based on the NOAEL can be the basis for regulatory acceptance or rejection of a new pharmaceutical or plant protection product, or for the continued use of existing products. Additionally, it is used for risk assessment of environmental pollutants (such as hydrocarbons in ground water). However, it is recognized that the NOAEL is an imprecise statistic, highly dependent on dose level selection and study design.

7 - Benefits and Limitations

Benefits of the NOAEL approach :

- Standard approach for many years,
- Simple concept,
- Widely understood,
- Pragmatically, the approach appears to have protected public health for many years,
- Rigid study design (on which the NOAEL concept is based) is easy to implement.

Limitations of the NOAEL approach :

- Shape of the dose-response curve ignored,
- NOAEL is an imprecise measure,
- Highly dependent on selection of dose levels,
- Dependent on statistical procedures used and expert interpretation of results,
- Requires at least one dose without adverse effects,
- Poor study design (i.e. inter-animal variability not minimised) and smaller group sizes lead to a higher NOAEL,
- No estimate of risk above the RfD.

REFERENCES

Barnes, D.G. and Dourson M.L., (1988). Reference Dose (RfD): Description and use in health risk assessments. Reg. Toxicol. Pharmacol. 8:471-486.

Dourson, M.L. and Stara, J.F. (1983). Regulatory History and Experimental Support of Uncertainty (Safety) Factors. Ref. Tox. Pharm. 3:224-238.

Kimmel, C.A. and Gaylor, D.W. (1988). Issues in qualitative and quantitative risk analysis for developmental toxicology. Risk Analysis 8:15.20.

Comparison of BMD with NOAEL and LOAEL Values Derived from Subchronic Toxicity Studies

Rashmi S. Nair[1], *Michel W. Stevens*[1], *Mark A. Martens*[2] *and Jethro Ekuta*[3]

[1]Monsanto Company, 800 N. Lindbergh Blvd., St. Louis, MO 63167. [2]Monsanto Europe SA, Technical Center, rue Laid Burniat, B1348 Louvain-La-Neuve, Belgium.
[3]University of Mississippi, Department of Pharmacology, University, MS 38677.

1. INTRODUCTION

Health-based exposure standards for non-cancer endpoints are currently derived based on a premise that an acceptable level can be determined at which there will be no significant risk of an adverse human health effect. This assumes the existence of a threshold exposure level for the adverse effect in question. The approach most widely used for non-cancer endpoints in the United States is to determine the no-observed-adverse-effect-level (NOAEL) for the critical effect and then to apply uncertainty factors (UFs) to account for scientific uncertainties in the total data base (Barnes and Dourson, 1988). The resulting value is referred to as a reference dose (RfD) for oral exposure, or reference concentration (RfC) for inhalation exposure. This approach is similar to other methods used worldwide, e.g., derivation of Acceptable Daily Intake or the ADI.

There are a number of deficiencies inherent in the present NOAEL approach (Crump, 1984). Recognition of these deficiencies has stimulated development of alternative methodologies (Crump, 1984, Hertzberg, 1989, Jarabek and Hasselblad, 1992). One proposed methodology is the Benchmark Dose (BMD); this method fits a dose-response model to experimental data and allows calculation of a dose corresponding to a given incidence of effect, e.g., dose associated with a 10% incidence. A BMD is defined as a lower statistical confidence limit for the dose corresponding to a specified increase in effect over background (Crump, 1984). This approach takes into account sample size and the shape of the dose-response curve. The BMD is not strongly dependent on the mathematical model used because the method does not involve extrapolation far below the experimental range. Thus, the method avoids much of the model dependency often associated with extrapolation of carcinogenicity data to low doses. Commercially available statistical programs for BMD can be applied to either "quantal" data, in which only the incidence of an effect is recorded, or "continuous" data, in which the graded severity of the effect is noted.

Since there is interest in replacing the NOAEL with a BMD for developing reference doses, several attempts have been made to compare a benchmark dose with the NOAEL from the same study. Shoaf (1993) examined quantitative exposure concentration-response data for the critical studies considered in the development of reference concentrations (RfCs) for 49 air toxics. Concentration-response data were available for only 12 of the 49 compounds. In these studies, the 10% effective concentration (EC_{10}) compared well with the lowest-observed-adverse-effect-level (LOAEL) of the critical study and the lower bound on dose of the 10% effective concentration (LEC_{10}) compared favorably with the NOAEL. Thus, Shoaf concluded that the calculated EC_{10} and LEC_{10} (i.e., BMD_{10}) values may be acceptable in development of RfCs. Similarly, Faustman (ILSI, 1994) compared results of the BMD to the NOAEL for developmental toxicity data. Faustman observed that continuous BMD_{05}s were similar to corresponding continuous NOAELs, whereas quantal BMD_{05}s were almost always smaller than corresponding quantal NOAELs. Faustman also concluded that among the three models used to compute the BMD for developmental toxicity studies, the log-logistic model fits the data sets best.

The present study was undertaken to expand the database comparing NOAEL and benchmark dose for subchronic toxicity studies. Phase I of the study was a comparison of overall study NOAELs or LOAELs to the BMD and the maximum likelihood estimate (MLE) of the benchmark dose. Data for calculation of BMD were obtained from various subchronic studies conducted by Monsanto Company. The proposed analyses utilized several available mathematical models appropriate to each set of data and evaluated the influence of these models on the BMD value.

Phase II of the study examined the influence of the number of dose levels used in a study on the development of a BMD. An overall BMD_{10} was obtained from all datasets with more than 3 dose levels for a given endpoint. This overall BMD_{10} was compared with BMD_{10}s on the same endpoint obtained by excluding datasets from one or more dose levels.

2. MATERIALS AND METHODS

2.1 Selection of Studies

Fifty-one subchronic toxicity studies conducted by Monsanto Chemical Group between 1979 and 1988 were considered for the Phase I study. Thirty studies which showed endpoints exhibiting a dose-response relationship, i.e., one or more dose levels showed a statistically significant, dose-related change, were selected for BMD calculation. The study duration for 29 of the studies was 90 days or more; one study was a 4-week inhalation study with complete histopathological evaluation. The routes of exposure represented among these studies were: gavage (5), dietary (14), inhalation (9) and drinking water (2).

In Phase II of the project, 8 subchronic toxicity studies were obtained from projects conducted by Monsanto Agricultural Group from 1988 to 1990. These were all the studies conducted during that time period that had at least 5 dose levels; these 8 studies were used to evaluate the effect of number of dose levels on BMD. The route of exposure for all of these studies was dietary, and all studies were 90 days in duration.

2.2 Endpoints Used For BMD Calculations

The endpoints used to calculate BMDs varied with the type of study. Only endpoints concluded to have dose-related changes were used to develop BMDs and were derived from data based on mortality, hematology, clinical chemistry, gross/histopathology, body weight and absolute/relative organ weight.

2.3 Models Used For BMD Calculations

Four different computer programs were purchased from ICF Kaiser International, KS Crump Group, 1201 Gaines Street, Ruston, Louisiana 71270. These programs were used to compute the BMDs and are described in the subsections that follow.

2.3.1. THRESHW

THRESHW fits the Weibull or extreme model to quantal animal toxicity data using the maximum likelihood method. THRESHW calculates the maximum likelihood estimates and statistical confidence limits for either the risk or the dose corresponding to a given risk level. THRESHW calculates lower confidence limits on dose corresponding to a given value of risk by computing directly the smallest dose corresponding to a fixed risk which is consistent with the data (as a lower limit on dose). The goodness of fit of the model is determined with a Chi-square statistic.

2.3.2. THRESH

THRESH fits a quantal polynomial regression model to quantal animal toxicity data using the maximum likelihood method. The THRESH model can be fitted either by using a linear or a multi-stage option. With the multi-stage option, a user can either set the degree of the polynomial or use a default, which is dose levels minus one. Like THRESHW, THRESH calculates upper confidence limits on risk and lower confidence limits on dose corresponding to a given value of risk.

2.3.3 THC

THC fits a polynomial mean response regression model to continuous animal toxicity data using the maximum likelihood method. Actual values for each animal, or the mean and corrected sum of squares for each group (assuming a normal distribution), can be used for input data. The user can set the degree of the polynomial or use the multistage option (degree equals the number of dose levels minus one).

2.3.4 THWC

THWC fits a power mean response regression model to continuous animal toxicity data by the maximum likelihood method. As with THC model, actual values for each animal, or the mean and corrected sum of squares for each group, can be used for input data. The upper and lower bounds for the power term (q2), were set to 1 and 18, respectively, these being the limits specified in the program.

2.4 Comparisons Conducted

2.4.1 The NOAEL and LOAEL were compared with benchmark doses (BMD_{01}, BMD_{05}, BMD_{10}) from quantal endpoints. In 15 of 16 studies evaluated, these quantal endpoints are the critical endpoint in the study. Thus, this is also a comparison of overall study NOAEL to the overall benchmark doses. Additionally, the BMDs were compared to the MLE of the benchmark dose at 1, 5, and 10% response rates to determine the extent of variability between the BMD and MLE determined for the BMD.

2.4.2 For two studies, continuous data were converted to quantal data (responders considered to be greater than 2 standard deviations from the mean), and corresponding quantal BMDs were determined for comparison of quantal BMD_{10} to continuous BMD_{10}.

2.4.3 Endpoint NOAEL were compared to the corresponding endpoint benchmark doses, (BMD_{01}, BMD_{05} and BMD_{10}) for continuous datasets.

2.4.4 The overall study BMD_{10} obtained from studies with more than 5 dose levels was compared to BMD_{10} calculated from data from 3 or 4 dose levels in order to study the influence of number of dose levels on the variability of BMD.

3.0 Results

3.1 Evaluation of Quantal Data

The quantal BMDs for all studies presented in this section were calculated using the THRESHW model. For one study, where BMDs were obtained using THRESH and THRESHW models, THRESH (linear option) gave the highest BMDs, while THRESH (multistage option) gave the lowest. Overall, using the THRESHW model resulted in BMD values somewhere between the values obtained with the two THRESH options.

Tables 1a and 1b show 16 studies with quantal data (14 studies from Phase I and two studies from Phase II). Two studies from Phase II are included in this section because they contained quantal data and had actual NOAELs. An actual NOAEL was identified for the critical endpoint in only 7 of these studies. For the studies for which a NOAEL was available, the arithmetic mean ratio of the NOAEL to BMD (for critical endpoint) were: NOAEL/BMD_{10}--0.83, and NOAEL/BMD -- 3.11 (Table 1a). These ratios were similar when all quantal datasets (N = 41 datasets) from

Table 1a - Comparison of BMD10, BMD05 and BMD01 for critical endpoint to NOAEL for quantal data - studies with NOAEL

Study ID	Quantal Endpoint	Units	NOAEL	LOAEL	BMD10	BMD05	BMD01	NOAEL/ BMD10	NOAEL/B MD05	NOAEL/ BMD01
C	Periportal Vacu-olization	ppm diet	280	3000	1380	671	131	0.20	0.42	2.14
F	Hepatocellular Coudy Swelling	mg/kg	10	250	36	18	3	0.28	0.57	2.94
J	Inflammation of Gastric Mucosa	mg/kg	30	200	48	29	9	0.63	1.03	3.26
P	Tubulointerstitial Nephropathy	ppm diet	743	1910	1097	820	304	0.68	0.91	2.44
S	Centrilobular Vacuolization	ppm diet	100	500	208	123	33	0.48	0.81	3.02
T	Enlarged hepato-cyte	ppm diet	100	300	178	135	95	0.56	0.74	1.05
U	Hepatocelluar degeneration	ppm diet	100	300	152	74	14	0.66	1.36	6.94
							Mean[a]	0.50	0.83	3.11

[a] = Arithmetic mean

each of these studies were included. In all cases, the NOAEL was less than the BMD_{10}. The NOAEL was between the BMD_{05} and BMD_{01} for 5 of 7 studies and between the BMD_{05} and BMD_{10} for 2 of 7 studies.

Table 1b - BMD10, BMD05 and BMD01 for critical endpoint (quantal data) and LOAEL for studies with no NOAEL

Study	Quantal Endpoint	Units	LOAEL	BMD10	BMD05	BMD01
B	Necrosis of Olefactory epithelium.	mg/cu m	105	68	33	6.5
D	Hepatocellular Cloudy Swelling	mg/cu m	15	9	4	0.8
I	Tubular degen/tubular dilatation	ppm diet	15	5	2	0.5
K	Urinary Urobilinogen	mg/kg	3	1	1	0.2
L	Splenic congestion	mg/kg	3	2	1	0.2
O	Hypertrophy /Hyper-plasia	ppm	140	560	273	53.0
Q	Squamous Metaplasia	mg/cu m	10	1	1	0.1
R	Hepatocellular Hyper-trophy	mg/kg	250	15	8	1.5
V	Mortality	mg/kg	25	34	17	3.2

For all 16 studies, the mean ratio of the LOAEL to the BMD for the 10, 5 and 1% response were 3.01, 5.47 and 23.36, respectively. As shown in Tables 1a and 1b, the BMD_{10} was lower than the LOAEL in all but 2 cases, BMD_{05} was lower than the LOAEL in all but 1 case and BMD_{01} was lower than LOAEL in all cases.

Table 2 presents the ratio of the maximum likelihood estimate (MLE) to the BMD_{10}, BMD_{05} or BMD_{01}. As expected, the deviation of the BMD values from the MLE was the highest at the BMD_{01} level. The mean ratios of the MLE to the BMD at the 10, 5 an 1 % response were 2.21, 2.79 and 6.90, respectively. Since one study (Study Q) appeared to be an outlier, means for the above ratios were also calculated excluding this dataset. The mean ratio of the MLE to the BMD at the 10, 5 and 1% response level after excluding data from Study Q were 1.98, 2.32 and 3.99, respectively.

Table 2 - Comparison of maximum likelihood estimate to the lower confidence limit of benchmark doses

Study ID	Quantal Endpoint	MLE10	MLE10/ BMD10[a]	MLE5	MLE5/ BMD5[a]	MLE1	MLE1/ BMD1[a]
B	Necrosis of Olefactory epithelium.	105.7	1.55	52.0	1.56	10.4	1.60
C	Periportal Vacuolization	2589.3	1.88	1257.3	1.87	245.9	1.88
D	Hepatocellular Cloudy Swelling	14.6	1.72	7.1	1.73	1.4	1.75
F	Hepatocellular Coudy Swelling	65.5	1.81	40.9	2.32	22.1	6.50
I	Tubular degen/ tubular dilatation	7.9	1.68	4.1	1.78	0.9	1.80
J	Inflammation of Gastric Mucosa	140.4	2.93	123.3	4.25	94.8	10.30
K	Urinary Urobilinogen	3.0	2.50	2.3	2.88	1.2	6.00
L	Splenic congestion	2.3	1.44	1.1	1.38	0.2	1.00
O	Hypertrophy /Hyperplasia	1195.7	2.14	687.0	2.52	202.8	3.83
P	Tubulointerstitial Nephropathy	1567.0	1.43	1302.6	1.59	1067.5	3.51
Q	Squamous Metaplasia	6.8	5.63	5.9	9.83	5.1	50.50
R	Hepatocellular Hypertrophy	39.3	2.55	21.5	2.87	5.5	3.67
S	Centrilobular Vacuolization	500.0	2.40	405.7	3.30	266.8	8.06
T	Enlarged hepatocyte	319.5	1.80	284.4	2.11	257.6	2.71
U	Hepatocelluar degeneration	290.6	1.92	160.1	2.17	46.5	3.23
V	Mortality	66.1	1.95	40.2	2.44	13.0	4.06
		Mean[b]	2.21		2.79		6.90
			1.98[c]		2.32[c]		3.99[c]

a = Actual values for BMD10, BMD05 and BMD01 and the units can be found in Table 1a and 1b.
b = arithmetic mean; c = arithmetic mean excluding value for study Q

3.2 Evaluation of Continuous Data

For continuous endpoints, benchmark dose values obtained with the THC model were almost always the same as those obtained with the THWC model for corresponding endpoints. In nearly all cases of continuous data studied, the BMD_{10} was double the BMD_{05} and 10 times the BMD_{01}. This relationship, however, is not true for the MLEs.

Table 3 presents 12 studies which had continuous data for calculation of benchmark doses. A total of 117 datasets were evaluated from these studies. Endpoint NOAELs were unavailable for 20% of these datasets. The NOAEL was equal to or less than the BMD_{01} for 9% of the datasets; for 44% of the datasets, the NOAEL was between the BMD_{01} and BMD_{05}; for 12% of the datasets, the NOAEL was between the BMD_{05} and BMD_{10}; and for 15% of the datasets the NOAEL was greater than BMD_{10}. Thus, for continuous data, no consistent pattern was observed between the endpoint NOAEL and the BMD when mean values of various endpoints were used to calculate the BMD.

Table 3 - Comparison of NOAEL to BMD01, BMD05 and BMD10 for continuous endpoints

Study ID	# Cont. Endpt.	NOAEL< BMD01	NOAEL= BMD01	NOAEL> BMD01 <BMD05	NOAEL< BMD05 <BMD10	NOAEL =>BMD 10	NO NOAEL
A	26	-	-	11	2	8	5
E	2	-	-	-	0	2	-
G	1	-	-	1	-	-	-
H	1	-	-	1	-	-	-
I	3	-	-	3	-	-	-
K	36	-	1	22	6	-	7
L	14	1	-	3	-	2	8
M	7	-	-	2	4	-	1
N	1	-	-	1	0	-	-
O	16	5	-	5	2	3	1
P	7	-	-	3	-	3	1
W	3	3	-	-	-	-	-
Total Endpt.	117	9	1	52	14	18	23
% of Total		8	1	44	12	15	20

3.3 Comparison of Quantal and Continuous Benchmark Doses for the Same Endpoint.

Table 4 presents the comparison of quantal benchmark doses to the continuous benchmark dose for the same endpoint. For the 5 data sets that were evaluated, the BMD_{10} values from quantal data were lower than BMD_{10} values from continuous data. For the MLE comparison, there appears to be a difference in the two types of endpoints studied. For the methemoglobin data, the quantal MLE values were 3- to 4-fold higher than continuous MLE values, while, for the cholinesterase activity, the quantal MLE values were equivalent to continuous MLE values. Examination of the data for methemoglobin levels from Study K showed that a steep dose response was obtained when continuous data were converted to quantal data, e.g., percent response rates for males at 90 days in the low, mid and high-dose were 30%, 100% and 100%, respectively, when the mean methemoglobin levels in the same three groups were 1.80, 3.31 and 5.76%, respectively. Conversion of the cholinesterase data from continuous to quantal data produced a more graded dose-response curve. These analyses illustrate a need for better methods to evaluate continuous data.

Table 4 - Comparison of continuous benchmark doses to quantal benchmark doses

Study ID	Endpoint	Cont. BMD[a]	Quan. BMD	QBMD/ CBMD	Cont. MLE	Quan. MLE	QMLE/ CMLE
K	Methemoglobin-day 45, males	0.7	0.07	0.10	0.7	2.5	3.57
K	Methemoglobin-day 90, males	0.7	0.63	0.90	0.9	3.6	4.00
F	Plasma choline-esterase - males	222.1	117.8	0.53	286.5	298.6	1.04
F	RBC choline-esterase-males	416.1	146.3	0.35	748.7	361.3	0.48
F	RBC choiline-esterase-females	265.1	123.1	0.46	340.2	337.7	0.99
			Mean[b]	0.47		Mean	2.02

a = 10% response level
b = arithmetic mean

3.4 Effect of Number of Dose Levels Tested on BMD (Phase II)

Eight studies were evaluated in Phase II of this study. Each of the studies had at least 5 dose levels. In six of these studies, only continuous data were evaluated, and in two of these studies both continuous and quantal endpoints were evaluated. As discussed below, the BMDs were lower when 3 or 4 levels are used if, and only if, at least one of the levels excluded from computation actually has a response.

In the first study, reducing the number of dose levels to 3 from the original 7 levels produced a remarkable decrease in BMD_{10} (approximately 30% of that obtained from all dose levels) by removing data from the 4 lower dose levels. Note removal of data from 3 lower dose levels did not strongly influence the BMD_{10}. Removal of data from the 4 lower levels had an even more remarkable effect on the BMD_{05} and BMD_{01} which were reduced to 20 and 5% of the value obtained from all dose levels.

The second study had 3 continuous endpoints and 5 dose levels. Removal of the highest dose from computation reduced the BMD_{10}s to 61 - 69% of the corresponding values with all 5 dose levels. This highest dose had the most significant effects. However, including the highest dose and removing the two lowest doses (which showed no effect) did not alter the BMD_{10}s. Therefore, as is true of quantal data, removal of a dose level that has an effect appears to lower the BMD_{10}s for continuous data.

For the remainder of the studies, reducing the number of dose levels did not have any influence on the BMD_{10}, either due to lack of effect at the lower dose levels or because of large variation in the datasets.

4. DISCUSSION

Overall, this study shows that the BMD approach rewards datasets which have good dose-response. For studies with data showing good dose-response relationship, there was less variability between the maximum likelihood estimate and the benchmark dose. Also for studies with good quantal dose-response data, the BMD_{05} was in the same range as the NOAEL for a majority of the studies. For two studies, BMD_{10} also approximated the NOAEL. A recent report recommending the use of BM05 or BM10 as a starting point for developing an RfD value was in concert with our finding (ILSI, 1994).

For continuous data, there does not appear to be any consistent relationship between the NOAEL and the BMD values when mean values are compared. For example, in terms of equivalency, the NOAEL was found to range from below the BMD_{01} to greater than BMD_{10}; the majority of the values were found to be between BMD_{01} and BMD_{05}. The results of the continuous data evaluation in this report are consistent with findings of other investigators (ILSI report, 1994).

For both continuous and quantal models, reducing the number of dose levels does not affect the BMDs, unless a dose level dropped contributed greatly to the shape of the dose-response curve. In this case, the BMDs were lower. This effect is larger with quantal than continuous data. Obtaining a lower BMD when fewer datasets are used has two ramifications. First, the lower BMD observed fits with the current conservative default approach in evaluating health effects when there is less knowledge, i.e., increased conservativeness. Second, the dose level selection for toxicity studies which will produce a gradation of response rather than focus on obtaining a NOAEL can improve the accuracy of the BMD. This is critical if regulatory authorities adopt use of benchmark dose approach for developing health-based exposure standards.

While the comparisons of BMDs to NOAEL are necessary to determine utility in the regulatory process for developing health-based criteria, consideration should be given to other data that are obtained from the BMD approach. For example, modeling the dose-response data also provides information on the slope of the dose-response curve. Use of the slope data for extrapolating to an RfD value should be considered. Recently, Alexeeff, et al. (1993) have proposed use of the slope of the dose-response curve for developing short-term exposure limits. Their proposal has possible application to developing other health-based criteria.

In summary, the benchmark dose method allows selection of a dose from which to derive an ADI or RfD, especially from studies that do not exhibit a NOAEL. The BMD methodology also offers an acceptable alternative to the use of a NOAEL in developing health-based exposure criteria from studies which have good dose-response data. The BMD approach should be used cautiously for studies with poor dose-response data. The BMD approach is more technically demanding than the NOAEL/LOAEL approach because it requires mathematical computation. It is scientifically more challenging because it requires selection of a critical data endpoint and support for computing the BMD. Still, the appropriate consideration of dose response inherent in the BMD method for evaluation of toxicity data is fully consistent with the basic principles of toxicology.

REFERENCES

Alexeeff, G.V., Lewis, D.C. and Ragle, N.L. (1993). Estimation of Potential of Health Effects from Acute Exposure to Hydrogen Fluoride Using a "Benchmark Dose" Approach. Risk Analysis **13**: 63-69.

Barnes, D.G. and Dourson, M.L. (1988). Reference dose (RfD): Description and use in Health risk assessments. Regul. Toxicol. Pharmacol., **8**: 471-486.

Crump, K.S. (1984) A new method for determining allowable daily intakes. Fundamental and Applied Toxicology **4**: 854-871.

Hertzberg, R.C. (1989) Fitting a model to categorical response data with application to special extrapolation to toxicity. Health Physics, **57** Supplement 1): 405 - 409.

Jarabek, A.M. and Hasselblad, V. (1992) Application of a Bayseian Statistical Approach to Response Analysis of Noncancer Toxic Effects. TheToxicologist, **12**: 98.

ILSI Report (1994). Report of the Benchmark Dose Workshop. International Life Sciences Institute. Risk Science Institute. Washington, D.C.

Shoaf, C.R. (1993) Comparison of benchmark and NOAEL approaches in risk assessment of air toxics. The Toxicologist, **13**: 142.

Bioindicators in Ecotoxicology

(Chair: Prof. Dr. J. Lopez-Barea, Spain)

Biomarkers in Ecotoxicology: an Overview

J. López-Barea
Departamento de Bioquímica y Biología Molecular (Facultad de Veterinaria) e Instituto de Biología Básica y Aplicada, Universidad de Córdoba, Avenida. de Medina Azahara s/n, 14071 Córdoba, Spain

1. The Need for Biomarkers in Ecotoxicology

Human activities induce stress on ecological systems by importing pollutants, modifying habitats, introducing exotic species or removing the native ones and by changing climate. These activities not only affect the survival or the performance of individual organisms, but also the structure and function of natural ecosystems and the diversity of life at several levels of organisation, including the number of species, the genetic composition, and the variety of ecosystems and landscapes (Steinberg et al. 1994).

The OECD estimates that near 1,500 new chemicals are introduced annually to the more than 100,000 already in use (Sheehan 1994; Steinberg et al. 1994). It has been estimated that some 11,000 compounds are manufactured in sufficient amounts to pose an environmental threat, thus requiring some form of risk assessment before release into the global ecosystem. Environmentally relevant foreign chemicals (xenobiotics) include polyaromatic hydrocarbons (PAHs), polychlorinated biphenyls (PCBs), dioxins, dibenzofurans, tributil tins, nitroaromatics, organophophates, phatalate esters, metals (e.g. Cd, Hg, Cr, Ni, Pb, Zn, Ti, Cu, Fe) and organochlorines (e.g. dieldrin, DDT, hexachlorohexanes, hexachlorobenzenes and chlorophenols) (Livingstone 1993; Sheehan 1994). Some of these xenobiotics are carcinogens, genotoxic, or in some other way pose an environmental threat to human health.

The need to detect and assess the impact of pollution, particularly low concentrations of increasingly complex mixtures of contaminants, on environmental quality has led to the study and development of indicators of the biological effect(s) of contaminants on organisms. Several indicators respond to a wide variety of xenobiotics and, therefore, detect the presence of both known and unknown contaminants, at difference with chemical analysis which measure only a fraction of the xenobiotics present in a particular environment and reveal nothing about their adverse effect.

Prior to death or sickness, organisms and populations respond to natural or anthropogenically induced stressors by changing different parameters (so-called biomarkers) at biochemical, histological, immunological, physiological, or organismic levels (Moore et al. 1987; Haux and Förlin 1988; McCarthy and Shugart 1990; Moore and Simpson 1992; Stegeman et al. 1992; Walker 1992; Livingstone 1993; Peakall 1994; Timbrell et al. 1994). A comprehensive definition of biomarker would be: *"Measurements of body fluids, cells, tissues or whole animals that indicate in biochemical, cellular, physiological, behavioural or energetic terms the presence of contaminants or the extent of the host response"* (McCarthy and Shugart 1990; Widdows and Donkin 1991).

Ecotoxicology is the science which identifies and assess the effects of anthropogenically released xenobiotics upon ecosystems, and predicts the chemical load that a specific community can carry without being altered. Until very recently, many ecotoxicological studies have been using endpoints derived from "classical" Toxicology, such as death or different types of injury of organisms, which do not take into account the complex structure and functions of ecosystems. Nevertheless, at the ecosystem level the risks can be only assessed by simultaneously using structural (diversity, connectivity and distribution of species) and functional ecosystem parameters (production, consumption, nutrient and energy flows, evolution) (Steinberg et al. 1994). A framework based in a wide variety of indicators has been proposed for evaluation of the ecological status of different US resource categories (Hunsaker 1993).

Ecological assessment of environmental toxicity often use resident biota as indices of habitat quality (so-called bioindicators). Xenobiotic exposure alters populations in characteristic ways, thus evaluation of such changes allows inferences of the nature of the stressor. Ecotoxicology is currently using the toxic responses of sentinel organisms for the detection of environmental contamination. A sentinel species should have wide geographical distribution, be easily collected, display a sessile lifestyle or a restricted territory, and finally being well understood in biochemical, physiological and biological terms (Lowe and Kendall 1990). The use of biomarkers allows to assess the effects of sublethal stress to the bioindicators, and to obtain insight into cause-effect relationships at the community and ecosystem level (Steinberg et al, 1994).

Biomarkers that reflect the health status of organisms at lower organisation levels (molecular or cellular) respond rapidly to stress, have high toxicological relevance, and can be used as early-warning indicators of environmental alterations before irreversible damage to the ecosystems occur. Those reflecting health at higher organisational levels (populations, communities, ecosystems) respond much more slowly and have lower toxicological relevance but are more ecologically relevant (Adams et al. 1989), although they often detect damages at an irreversible stage. However, there is a need for more work demonstrating that biological responses to pollutants in the laboratory are comparable with field data. In addition, the range in which organisms respond to natural stressors (heat, drought, oxygen tension, parasites, infections, reproductive stress, etc.) has to be evaluated prior to application of biomarkers for the evaluation of multiple stresses.

2. Types and Uses of Biomarkers

It has been proposed a three tiered approach to the use of biomarkers for the detection of environmental pollutants. At the hazard identification level, the biomarkers would indicate exposure to xenobiotics. The so-called *biomarkers of exposure* determine in body fluids or tissues either the xenobiotic in question or a product of its reaction with a biological molecule. They have been subdivided into markers of internal dose, indicating the occurrence and extent of exposure of the bioindicator, and markers of effective dose, signalling the extent of exposure of the target molecule, structure or cell. Thus, exposure biomarkers have a *diagnostic value* to assess whether a bioindicator has been in contact with a potentially damaging xenobiotic.

Table 1 summarises some of the most popular exposure biomarkers, including the detection of the parent xenobiotic or derived metabolite, and the determination of several types of damages to DNA or proteins, which can be assayed by a wide variety of methodologies that will be dealed with at a later section. The detection of DNA strand breaks by alkaline elution and either specific or unspecific mutations by new molecular methodologies, such as PCR, are also included. Oxidative damage can be also detected by alteration in the intracellular glutathione redox status or the formation of malondialdehyde or thiobarbituric acid reactive substances produced by lipid peroxidation, as well as by the development of new superoxide dismutase isoenzymes (Rodriguez-Ariza et al. 1993a, 1994; Pedrajas et al. 1993).

Table 1. Biomarkers of exposure to environmental xenobiotics

Exposure to	Biomarker
Xenobiotics	• Parent compound or derived metabolites detected by ^{1}H-NMR • Mercapturic acids detected in urine
Alkylating compounds	• DNA adducts: by ^{32}P postlabeling or HPLC with EC detection • Protein adducts: alkyl-Val derivatives of hemoglobin • Strand breaks: alkaline elution • Unspecific mutations: DNA fingerprint with random primers • Restriction site mutagenesis by PCR
Oxidative stress	• DNA damage: ^{8}OHdG in urine by HPLC with EC detection • Protein damage: oxidised-His derivatives • Lipid damage: MDA or thiobarbiturate reactive substances • Intracellular redox status: GSH, GSSG, ProtSSG, NAD(P)H • Formation of new superoxide dismutase isoenzymes

A second step would require hazard assessment, that is, biomarkers indicating the extent of the problem and identifying the xenobiotics. The so-called *biomarkers of effect* indicate the response of the exposed organism to the particular xenobiotic or complex mixture. Logically, there are a large number of potential biomarkers for the biological effects of chemicals (McCarthy and Shugart 1990; Hugett et al. 1992; Moore and Simpson 1992; Walker 1992; Livingstone 1993; Peakall 1994; Timbrell et al. 1994). Table 2 summarises some of the better studied effect biomarkers, including several methodologies indicating exposure to organophosphates, lead, polyaromatic hydrocarbons, polychlorinated biphenyls, dioxins, organochlorines, transition metals and conditions inducing oxidative stress.

With increasing understanding of the molecular responses, including a detailed knowledge of the dose-response relationships, and by linking them to deleterious biological events at higher organisation levels, molecular biomarkers becomes *prognostic*, e.g. induction of cytochrome P450 can be linked via formation of DNA-adducts to the initiation of contaminant-caused cancers. The ultimate long-term goal, *risk prediction*, would involve the use of biomarkers which could be linked to human epidemiology and to effects at the population, community and ecosystem level (Stegeman and Lech 1991; Peakall and Shugart 1993; Livingstone 1993). While there have been quite extensive studies on biomarkers at the molecular, cellular, and organismic level, and several of them

Table 2. Biomarkers of effect of environmental xenobiotics

Exposure to	Biomarker
Organophosphates	• Acetylcholinesterase inhibition
Lead	• δ-Aminolevulinate dehydratase inhibition • Zn-protoporphyrin levels in erithrocytes • δ-Aminolevulinate levels in urine
PAHs, PCBs, dioxins organochlorines	• Induction of cytochrome P450 system • Induction of Ah-gene battery enzymes: GSH-transferase, UDP glucuronyl transferase, DT-diaphorase • Induction of specific GSH-transferase isoenzymes
Xenobiotics	• Induction of multidrug resistance system • Heat-schock proteins: hsp90, hsp72, hsp70, hsp60, ubiquitin
Transition metals	• Metallothioneins and phytochelatins induction
Oxidative stress	• Induction of antioxidative enzymes: superoxide dismutase, catalase, GSH peroxidase, glucose-6-P dehydrogenase, 6-P-gluconate dehydrogenase, etc.

have been widely validated, biomarkers at the population, community and ecosystem level are still at the research stage (Peakall 1994). However, a generally held viewpoint is that a contaminant-related change in a molecular or cellular biomarker indicates "cause of concern" and is sufficient in itself to merit attention and action. Thus in the following sections we will concentrate on biomarkers used at molecular and cellular levels.

Figure 1 summarizes the pathway leading from exposure to a particular xenobiotic (X) to the adverse effects that it can produce at the levels of cells, organisms, populations, communities and ecosystems. Some of these changes can be used as exposure biomarkers to identify the threatening hazards, while other changes can be exploited as effect biomarkers. Once the dose-response relationships have been well established, these biomarkers can be combined for their use for risk assessment and risk management. All interactions with xenobiotics start at the molecular level, at which most organisms are quite similar, although with some exceptions. Most molecular biomarkers can be applied to a wide range of organisms (Stegeman and Lech 1991; Peakall and Shugart 1993; Livingstone 1993). An early response to contaminant exposure is one of the most useful properties of a molecular biomarker, as early-warning indicator of con-

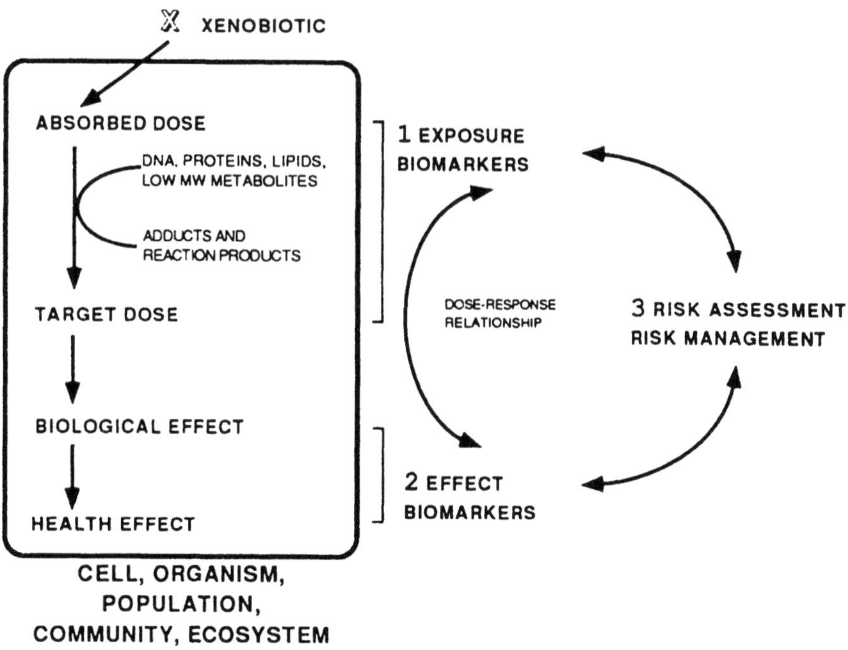

Fig. 1. Biological effects of xenobiotics and their use as biomarkers

taminant exposure and effect, but indicating also that pollutants may have consequences at the ecosystem level. Unlike most toxicity bioassays which use a relatively restricted range of test organisms, biomarkers are applicable to both laboratory and field studies (Murphy and Kaputska 1990).

In addition, according to their degree of interference there are two types of biomarker, *invasive* and *non invasive*, depending on whether the organism studied has to be severely damaged (or even killed) or it is only slightly disturbed for analysis. Logically, there is a general trend towards the use non invasive biomarkers (Fossi and Leonzio 1994). Table 3 summarises the classification of biomarkers according to their different categories previously discussed.

Up to now we have described the different types of biomarkers according to their signification and the levels of organisation to which they respond. Biomarkers can be also classified in terms of their relevance for the assessment of ecological stress. A biomarker indicating genotoxicity implies a high ecotoxic potential, because threshold values cannot be defined for these substances: any concentration is considered hazardous. Next in hierarchy would be these biomarkers indicating interference with the immunological system, since immunosuppressive effects may be synergic or potentiating with other xenobiotics. The next category include the non-specific detoxifying enzymes (such as MFOs, glutathione transferases, UDP-glucuronyl transferases), whose levels are induced in response to a great variety of organic xenobiotics, thus indicating a history of ecosystem contamination. Finally, stress proteins (such as heat shock proteins, metallothioneins, phytochelatins) may also be induced by naturally occurring stressors (e.g. heat, corticoids, etc.) and in consequence have the lowest ecological relevance.

According to their design, the tests used in ecotoxicological studies can be classified in two categories. The first imply *experimental exposure under controlled conditions* of different bioindicators to complex mixtures (e.g. water, sediments) from the studied ecosystem. They are often carried out by acute exposure, and use a set of experimental organisms, selected by their ecological rele-

Table 3. Classification of biomarkers

Signification	Organisation level	Interference
• Exposure	• Molecular	• Invasive
-Internal dose	• Cellular	• Non-invasive
-Effective dose	• Physiological	
• Effect	• Organism	
• Risk assessment	• Population	
	• Community	
	• Ecosystem	

vance or their simplicity for laboratory management. The most popular bioindicators are the immobilisation of *Daphnia magna* or mosquito larvae, and the alteration of the swimming behaviour of several small fish. The second category imply the assessment of the *biomarkers status in sentinel bioindicators* from the studied ecosystem. In this case, biochemical and physiological biomarkers are studied (Peakall 1994; Steinberg et al 1994). The sessile molluscs and bentonic fish living within the sediments are usually used.

Chemical analyses can be useful in determining body-burdens, unless of course the xenobiotic is biotransformed, but they do not provide information about effects. Ecological analyses are useful in decribing differences or changes if historical earlier data are available. These approaches do not tell us what caused the change although chemical body burden might provide some pointers. This is where biomarkers may help in the resolution of these problems. Biomarkers are used for routine long-term surveillance programmes, hazard assessment at specific discharge sites, enforcement of compliance with regulatory environmental standards and monitoring of the effectiveness of remediation actions (McCarthy and Shugart 1990). They should be applied as part of a monitoring programme including chemical analysis of contaminant body-burdens, general biomarkers of animal condition, and specific biomarkers of contaminant effect, including physiological changes (Livingstone 1993).

3. Molecular Biomarkers

Some biomarkers are so well defined and validated, and are so specific that can be used without a back up by chemical analysis. They include inhibition of acetyl cholinesterase by organophosphates, inhibition of δ-aminolevulinate dehydrase by lead, and eggshell thinning caused by DDT and related compounds (Peakall 1994). Other validated biomarkers include induction of mixed function oxidases, such as cytochrome P450s, the covalent binding of PAHs to form DNA-adducts, the formation of oxidatively-damaged nucleosides in response to oxidative stress and the detection of other DNA alterations.

3.1 Cytochrome P4501A1 Induction

Cytochrome P4501A1 is a phase I oxidative enzyme catalysing the first step in biotransformation of many organic xenobiotics, such as PAHs, PCBs and dioxins (Goksøyr and Förlin 1992; Nebert et al. 1990). In mammals, cytochrome P4501A1 is induced by organic xenobiotics via its binding to the Ah receptor, the basis of its use as a biomarker (Okey 1990; Nebert et al. 1990). This enzyme can activate xenobiotics, such as particular PHAs, to mutagenic metabolites, thus its induction enhances the carcinogenicity of such compounds (Stegeman and Lech 1991; Livingstone 1993). It is encoded by the *CYP1A1* gene, of the P450 multigene family (Nebert et al. 1991). Cytochrome P4501A has also been extensively studied in fish, its properties and inducibility being similar to its mammalian counterpart (Stegeman and Lech 1991; Goksøyr and Förlin 1992; Livingstone 1993). This biomarker has been also successfully applied to birds (Fossi et al. 1986) and feral rodents (Lubet et al. 1992).

Induction can be followed by enzymatic activity, 7-ethoxy-resorufin *O*-deethylase (EROD) and aryl hydrocarbon hydroxylase (AHH), although inhibition by pollutants (Boon et al. 1992) and susceptibility to denaturation or proteolysis (Goksøyr and Förlin 1992) is a serious drawback. Immunodetection of CYP1A protein has been recently proposed as the more suitable alternative overcoming most of the disadvantages inherent to EROD activity (Goksøyr and Förlin 1992), although immunodetection requires specific antibodies, not available to every laboratory. Levels of P4501A mRNA have been also measured by dot blot and Northern blot techniques (Goksøyr and Förlin 1992; Courtenay et al. 1993) using CYP1A cDNA probes from rainbow trout (Heilman et al. 1988).

The use of CYP1A induction as biomarker requires characterisation of several factors which can increase variability. Variations with age, sex, and season have been seen in fish (Goksøyr and Förlin 1992) and other vertebrates (Rattner et al. 1989). Other factors are species differences (Goksøyr and Förlin 1992), diet (Lemaire et al. 1992), temperature (Andersson and Förlin 1992), and most importantly reproductive condition (Jiménez et al. 1990).

Biotransformation of PAHs and other organic xenobiotics to carcinogenic metabolites (Stegeman 1989; Stegeman and Lech 1991; Goksøyr and Förlin 1992) implies that P4501A induction may have important toxicological consequences. Positive correlations have been found between contaminants, hepatic cytochrome P4501A, bile metabolites, DNA-adducts and hepatic neoplasms and related lesions (Varanasi et al. 1987; Myers et al. 1991).

3.2 Metallothionein Induction

Metallothioneins (MT) are low MW proteins (6-7 kDa) rich in Cys but containing few hydrophobic or aromatic aminoacids present in most animals which bind metals from groups Ib and IIb, via cysteinyl residues (Kagi and Kojima 1987; Kille et al. 1992; Roesijadi and Robinson 1994). They regulate endogenous metals (Cu and Zn), detoxicate excess of these and pollutant metals (Cd, Hg and Ag), act as free radical scavengers, and are involved in some general stress responses (Kagi and Kojima 1987; Kille et al. 1992).

Several metals, including Cd, Cu, Zn, Co, Ni, Bi, and Ag, activate transcription of MT genes, increasing synthesis of these proteins, the basis of its use as a biomarker for metal exposure (Haux and Förlin 1988; Stegeman et al. 1992;

Kille et al. 1992). Quantitation of MT relies on assessment of gene expression at different levels. The protein can be determined after separation by size-exclussion or ion-exchange chromatography and specific metal determination by atomic absorption spectroscopy. Metal substitution with radiolabeled Cd or analysis of protein thiols can be used instead (Engel et al. 1987; Stegeman et al. 1992; Pavicic et al. 1993). ELISA and RIA immunodetection has been also used for MT quantification. Antibodies against rainbow trout MT react with MT from many fish species, although they do not crossreact with mammalian MTs (Norey et al. 1990), due probably to differences of their N-terminal sequences where the primary epitope is believed to be (Kille et al. 1992).

Elucidation of the nucleotide sequences of piscine MTs (Chan et al. 1989) allowed the development of probes to detect and quantitate metallothionein mRNA levels by northern blot or in slot blot (Zafarullah et al. 1989; Kille et al. 1992). Since MT synthesis is also induced by glucocorticoids, progesterone, and other circulatory factors, such as interleukin-1 and interferon (Kille et al. 1992; Livingstone 1993), MT status can change with species, reproductive condition and diet, and these factors must be considered when using MT as a biomarker (Livingstone 1993).

3.3 DNA and Protein Adducts

Many organic xenobiotics (e.g. PAHs, arylamines or aflatoxins) are metabolically activated by phase I or phase II enzymes to electrofilic metabolites which bind to nucleic acids or proteins, forming covalent adducts. DNA-adduct formation integrates xenobiotic uptake, metabolism and repair, and is one of the initial event in mutations and carcinogenesis (Maccubbin 1994). It is being currently used in humans as a biomarker for exposure to environmental and occupational carcinogens (Santella 1991; Perera et al. 1992), and a similar role as biomarker of organic contaminant exposure and damage has been proposed for several other organisms (Chipman and Marsh 1991; Dunn 1991; Jones and Parry 1992; Maccubbin 1994).

The most widely used technique for measuring DNA-adduct formation is ^{32}P post-labeling. It can detect bulky DNA adducts formed by PAHs and aflatoxins metabolites (Gupta and Randerath 1988). DNA is extracted, hydrolysed to dNps and radiolabeled to dpNps using [γ-^{32}P]ATP and T_4 polynucleotide kinase. The adducted dpN*ps are separated from normal dpNps by multidimensional ion exchange thin layer chromatography, and visualised by autoradiography. When used to the limit of its sensitivity it can detect one adduct in 10^{10} normal nucleotides (Gupta and Randerath 1988; Jones and Parry 1992).

Most of the field studies carried out with ^{32}P post-labeling show correlation between DNA-adducts and contaminant exposure. The radioactive spots are usually located in diagonal patterns (Kurelec et al. 1989; Jones and Parry 1992). Some concern has been raised by the observation of adducts of apparently natural or endogenous origin, present in both fish (Kurelec et al. 1989) and marine invertebrates (Marsh et al. 1993), which are species specific, endogenously regulated and independent of pollution exposure. The main problem of ^{32}P post-labeling is its difficulty to identify the different adducts.

ELISA techniques using monoclonal or polyclonal antibodies have been used to identify specific adducts (Poirier 1984; Maccubbin 1994). HPLC separation coupled to electrochemical detection (HPLC-EC) has been developed to detect DNA-adducts derived from oxidative attack (Park et al. 1989). Immunoaffinity columns with antibodies showing high affinity against 8-OHdGuanosine and 8-OHGuanine coupled to HPLC-EC allows the detection of oxidative DNA products in urine of rats and humans (Degan et al. 1991). Hydroxyl radical-induced DNA lesions of Gua and Ade have been found in neoplastic and microscopically normal livers of fish exposed to environmental carcinogens, and shown to be a useful biomarker with a putative link to cancer development (Malins 1993).

Several alkylating compounds and metabolites from arylamines and PAHs bind to high levels to human hemoglobin following environmental exposures, being particularly well characterised for the potent bladder carcinogen, 4-aminobiphenyl (Tannenbaum 1990). Serum albumin adducts have also been investigated, in particular for aflatoxin B_1 exposures yielding excellent dose-response relationships (Wild et al. 1992). The measurement and quantification of AFB_1-Alb adducts has been extensively validated in experimental and human sample analyses, as a rapid and facile approach for screening very large number of people which reflect longer exposure periods than urinary adduct markers (Wild et al. 1990). Oxidative modification of proteins by free radicals can be monitored by the formation of oxidized histidine, detectable by HPLC-EC, which has been proposed as an useful biomarker for assessing protein modification under oxidative stress (Uchida and Kawakishi 1993).

3.4 Other Molecular Biomarkers

Human and animal *cholinesterases* have been used as biomarkers of exposure to organophosphates and carbamates, since inhibition (>60%) of brain acetylcholinesterase is directly related to their toxic effects (Walker 1992; Peakall 1994). Before lethal effects are seen, organophosphorous compounds cause several sublethal effects, including physiological changes when the inhibition of brain cholinesterase activity exceeds 40% (Grue et al. 1992). In fact, the degree of inhibition of brain cholinesterase activity provides evidence for poisoning of mammals and birds by organophosphorous insecticides. The disadvantage of brain cholinesterase as biomarker for field studies is that it requires destructive sampling (Fossi and Leonzio 1994). Thus, blood esterases (cholinesterases and carboxylesterases) are increasingly being used as biomarkers, although not without problems, which can be circumvented by the use of specific antibodies for particular esterases (Walker 1992).

The enzyme *δ-aminolevulinate dehydratase* catalyses the formation of porfobilinogen, a Hb precursor. The enzyme is relatively easy to measure and highly sensitive to Pb, thus inhibition of this enzymatic activity has long been used as a biomarker for lead exposure in humans (Granick et al. 1972) and waterfowl (Dieter 1979). It has been recently shown that this biomarker can be also successfully applied to marine organisms, such as fish (Schmitt et al. 1993) and oysters (Brock 1993).

The activity of several detoxifying and antioxidative enzymes, such as glutathione-S-transferases, glutathione peroxidase, catalase, superoxide dismutase, glucose-6-phosphate dehydrogenase, and glutathione reductase, increases in fish and molluscs from Spanish littoral areas highly polluted with metals and organic xenobiotics (Rodriguez-Ariza et al. 1992, 1993a; Martínez-Lara et al. 1992; Pedrajas et al. 1993). It has been also shown that fish from these polluted areas undergo oxidative stress, as shown by their oxidized glutathione redox status (Rodriguez-Ariza et al. 1993b, 1994b).

New procedures for the rapid separation and quantification of the isoenzymes of several detoxifying and antioxidative enzymes have been developed. Glutathione-S-transferase isoenzymes have been separated by ion-exchange HPLC with on-line detection of activity. It has been shown that specific GST isoenzymes are induced in fish living in polluted environments (Martínez-Lara et al. 1992). Superoxide dismutase isoenzymes have been separated by polyacrylamide gel electrophoresis or isoelectrofocusing, and stained *in situ* for enzymatic activity. New Cu,Zn-SOD isoenzymes of intermediate pI appear in extracts of fish from polluted sites (Pedrajas et al. 1993).

4. Cellular and Physiological Biomarkers

From a physiological point of view the response of an organism to a chemical may be seen in terms of a scale of increasing severity by exposure degree. Thus, the effects of low exposure levels may be within the limits of homeostasis, but as exposure increases the animal progress from a stressed state, in which the repair mechanisms still function due to compensatory mechanisms, to a reversible diseased state with disturbed functions, and subsequently to a irreversible disease state leading to death. Different cellular and physiological biomarkers are required to measure these different stages (Depledge 1989; Walker 1992).

At the physiological level, there may be biomarkers not related to any particular mode of action, but indicating compensatory responses, for which there will be an energy cost. In fact, variations in the adenylate energy charge have been proposed to monitor the energy costs produced by environmental pollution (Moal et al. 1991). In such cases individual organisms may fail to survive in a competitive environment or fail to reproduce. Thus, a biomarker may indicate a change which is harmful in an ecological sense, although no specific toxic mechanism is involved (Walker 1992).

Molecular cell biology indicates many possible biomarkers relating to perturbation of cellular processes or defensive reactions involving those systems that protect the cell against hostile environmental agents. They include cell surface receptors, heat-shock proteins, enzymes for xenobiotic detoxication and activation, cellular oncogenes and suppresor genes, internal membranous components, such as endoplasmic reticulum, Golgi, transport vesicles and lysosomes, and cytoskeleton building blocks of the (Darnell et al. 1990).

4.1 Pathobiology of Cell Membranes

The cell surface or plasma membrane is the main interface with the environment. Xenobiotics enter the cell by diffusion across plasma membrane or endocytosis associated with LDL lipoproteins (Darnell et al. 1990). The plasma membrane plays also many crucial physiological functions, such as endocytosis and exocytosis. Thus, perturbation of these functions by xenobiotics will profoundly affect cells and, consequently, organs and whole animals. In addition to molecular receptors, the cell membrane has many proteins associated, including ion channels, gap junction and desmosomal complexes, adhesion molecules, molecular pumps and signal transduction systems (Darnell et al. 1990).

Pathological changes associated with pollution has been described in various types of cell membranes. Disturbance of endocytic processes has been described in fish liver cells (Moore and Simpson 1992). The lysosomal system is also affected by xenobiotics, undergoing volume increases, lipid and lipofuscin accumulation, destabilisation of lysosomal membrane which impairs function, and inhibition of the lysosomal propton pump (Moore 1990).

In addition to be the major protein synthesis site, the endoplasmic reticulum (ER) is central to many reactions involved in detoxication or activation of xenobiotics. All these enzymes are integral components of the ER and some of them can be induced by particular types of xenobiotic. Following exposure to many xenobiotics, the ER undergoes marked proliferation altering the internal organisation of the cell (Hinton and Lauren 1990). Activated derivatives of certain xenobiotics can also be retained within ER membranes entering a redox cycle. This produces reactive oxygen species wich react with many biological molecules, including DNA (Di Giulio et al. 1989).

4.2 Multixenobiotic Resistance

Many aquatic organisms resistant to pollution are simultaneously resistant to mutiple toxic xenobiotics. The molecular mechanism of multixenobiotic resistance (MXR) is similar to that of multidrug resistance previously found in tumor cell lines (Endicott and Ling 1989). It involves a transmembrane 170 kDa glycoprotein (p170) which binds a xenobiotic and facilitates its efflux by an energy-dependent process leading to a decreased accumulation and the resistance to xenobiotics (Kurelec 1992).

In aquatic organisms exposed to pollutants the MXR mechanism pumps out also man-made xenobiotics (Kurelec and Pivcevic 1991). There are compounds, such as verapamyl a competitive inhibitor of p170, that impairs the function of MXR and reverse the multixenobiotic resistance Kurelec 1992). Most important, nontoxic lipophilic agents (natural or man made) may also be recognized and expelled by this pumping mechanism, and at high concentrations might saturate the system and thereby reverse MXR (Kurelec 1992). Since MXR is a general biological defense mechanism for protection of the organism from natural endogenous and environmental toxins, its inhibition by such chemosensitisers may cause a serious disease (Kurelec 1992; Moore et al. 1993). Compounds with these properties obviously deserve the top position among toxic xenobiotics and therefore emphasise the importance of recognising pollutants' potential to reverse the MXR mechanism.

4.3 Stress (Heat-Shock) Proteins

The cellular stress response protects organisms from damage by different stressors, including elevated temperatures, ultraviolet light, heavy metals and xenobiotics. The stress promotes the rapid synthesis of various proteins which play a pivotal role in regulating protein homeostasis. Exposure to adverse physical and chemical conditions denaturates proteins, resulting in missfolding and aggregation. Thus, under adverse environmental conditions the synthesis of stress proteins increase to repair denatured proteins and protect cellular proteins from environmentally induced damage (Sanders 1993).

Most of the stress proteins are also synthesised under normal conditions, where they play a central role in catalysing protein folding. Stress-90, stres-70 and chaperonin are a subset of ubiquitous proteins called molecular chaperons which direct the folding and assembly of other cellular proteins and regulate the kinetic partitioning between protein folding , translocation reactions, and protein aggregation (Rothman 1989; Ellis 1990; Hartl et al. 1992; Sanders 1993). The sequences of these three stress proteins are highly conserved , while other group with molecular weight 16-24 kDa called low MW stress proteins are highly species specific. In addition, ubiquitin is a 7 kDa protein involved in non-lysosomal degradation of intracellular proteins (Sanders 1993).

Stress proteins play important roles in physiological processes involving the rapid breakdown and reorganisation of tissues, such as larval settling, molting, resorption of gametes, etc. They also may be involved in triggering changes in life stages in response to changes in environmental conditions (Sanders 1993). Because many aquatic organisms live in environments characterised by rapid cell lines (Endicott and Ling 1989). It involves a transmembrane 170 kDa glycoprotein (p170) which binds a xenobiotic and facilitates its efflux by an energy-dependent process leading to a decreased accumulation and the resistance to xenobiotics (Kurelec 1992).

In aquatic organisms exposed to pollutants the MXR mechanism pumps out also man-made xenobiotics (Kurelec and Pivcevic 1991). There are compounds, such as verapamyl a competitive inhibitor of p170, that impairs the function of MXR and reverse the multixenobiotic resistance Kurelec 1992). Most important, nontoxic lipophilic agents (natural or man made) may also be recognized and expelled by this pumping mechanism, and at high concentrations might saturate the system and thereby reverse MXR (Kurelec 1992). Since MXR is a general biological defense mechanism for protection of the organism from natural endogenous and environmental toxins, its inhibition by such chemosensitisers may cause a serious disease (Kurelec 1992; Moore et al. 1993). Compounds with these properties obviously deserve the top position among toxic xenobiotics and therefore emphasise the importance of recognising pollutants´ potential to reverse the MXR mechanism.

4.3 Stress (Heat-Shock) Proteins

The cellular stress response protects organisms from damage by different stressors, including elevated temperatures, ultraviolet light, heavy metals and xenobiotics. The stress promotes the rapid synthesis of various proteins which play a pivotal role in regulating protein homeostasis. Exposure to adverse physical and chemical conditions denaturates proteins, resulting in missfolding and aggregation. Thus, under adverse environmental conditions the synthesis of stress proteins increase to repair denatured proteins and protect cellular proteins from environmentally induced damage (Sanders 1993).

Most of the stress proteins are also synthesised under normal conditions, where they play a central role in catalysing protein folding. Stress-90, stres-70 and chaperonin are a subset of ubiquitous proteins called molecular chaperons which direct the folding and assembly of other cellular proteins and regulate the kinetic partitioning between protein folding , translocation reactions, and protein aggregation (Rothman 1989; Ellis 1990; Hartl et al. 1992; Sanders 1993). The sequences of these three stress proteins are highly conserved , while other group with molecular weight 16-24 kDa called low MW stress proteins are highly species specific. In addition, ubiquitin is a 7 kDa protein involved in non-lysosomal degradation of intracellular proteins (Sanders 1993).

Stress proteins play important roles in physiological processes involving the rapid breakdown and reorganisation of tissues, such as larval settling, molting, resorption of gametes, etc. They also may be involved in triggering changes in life stages in response to changes in environmental conditions (Sanders 1993). Because many aquatic organisms live in environments characterised by rapid environmental fluctuations, the role of stress proteins in conferring tolerance to such harsh conditions is particularly relevant.

Synthesis of heat-induced stress proteins is tightly coupled with the presence of damaged or denatured proteins. Induction of the stress response by chemical contaminants is slower than induction by heat-shock, reflecting the fact that heat denatures proteins more rapidly than chemicals, which are dependent on biological availability, uptake, mechanisms of toxicity, sequestration, or even detoxification (Sanders 1993). Stress proteins have been shown as useful biomarkers of exposure to different contaminants, such as tributyl tin, arsenite, chromate, lindane, and diazinon, each of which has a different mode of action (Sanders 1993). They have also been useful as biomarkers of contaminant exposure in archived environmental samples, since some of them remain elevated in tissues for long periods of time (Sanders 1993; Sanders and Martin 1993).

Some precautions must, however be taken into account when using the stress response as an indicator of adverse biological effects in organisms exposed to environmental stressors in the field. To minimise false positives, the response should not be induced by the handling of organisms during sample collection. It should be also noticed that some stress proteins are induced by steroid hormones in some target tissues. Natural variations in environmental temperature and, even, dietary influences also appear to modify the stress response (Sanders 1993).

4.4 Genotoxic Damages

Some of the environmental pollutants are genotoxins inducing inheritable DNA damages, thus, their detection should be a main objective of Ecotoxicology. Tests for genotoxins are essential components of the strategies employed for assessing possible human risks arising from exposure to such compounds. The mutagenesis process involve distinct cellular events which can be studied with different organisms or cells (Mohn and de Raat 1993).

Ecotoxicological evaluation of mutagenic risks differs from human risk assessment. For ecosystems only the maintenance of their integrity at population level is important. However, the genetic stability of some species with low reproductive rates and small populations, may be endangered by environmental genotoxins (Mohn and de Raat 1993). Induction of DNA lesions may initiate structural and point mutations, leading to abnormality in both somatic and germ cells (Jones and Parry 1992). Somatic mutations, particularly those affecting protooncogenes and/or tumour suppressor genes, are closely associated with tumorigenesis (Mohn and de Raat 1993). Identification of environmental genotoxins becomes prioritary when significant exposure occurs. The measurement of DNA strand breaks (alkaline elution procedure), cell proliferation and cell death by apoptosis can also be combined for assessing exposure to genotoxic and non-genotoxic carcinogenic stimuli (Moore et al. 1993).

Genotoxins present in the environment can be detected by various strategies. The most simple is to assess them in air, water or sediments samples, after chemical or physical concentration. Different biological endpoints are used to determine the genotoxic activity, although bacterial mutagenicity tests are the most popular (Maron and Ames 1983).

A better alternative is the detection of genotoxins in bioindicators (often the same used in other ecotoxicological studies) since they concentrate many pollutants and biotransform them to highly potent genotoxins. Thus, ethanolic extracts of three mollusc species from areas with different pollution levels contained genotoxins detectable by strains of *Salmonella typhimurium* and *Escherichia coli* (Rodriguez-Ariza et al. 1992, 1993). The genotoxic compounds detected were direct-acting (mammalian metabolic activation not required), of polar nature (directly eluting from XAD-2 columns), and of oxidative type (active only on *S. typhimurium* TA102 and *E. coli* catalase-deficient strains).

Enhanced metabolic activation of promutagens by extracts of bioindicator organisms, assessed with bacterial genotoxic assays, is another valuable biomarker. Bioactivation requires the simultaneous action of different pathways, and reproduces more closely the *in vivo* situation, thus being a more comprehensive biomarker than specific assays of individual activating enzymes (Rodriguez-Ariza et al. 1994a). Increased metabolic activation of known promutagens in polluted animals has been reported in fish (Kurelec et al. 1979) and molluscs (Britvic and Kurelec 1986). Hepatic S9 fractions from polluted fish show 5-, 19- and 43-fold higher capability to activate benzo(a)pyrene, 2-acetylaminofluorene and 2-amino-anthracene, respectively, than those from reference animals (Rodriguez-Ariza et al. 1994a).

A further step in genotoxin detection is the direct assessment of mutations in bioindicator organisms. DNA fingerprint with random primers shows new PCR products in B(a)P exposed rats, a procedure proposed as a new biomarker (Castellani et al. 1993). Recently, the Restriction Site Mutagenesis assay has been developed for detecting mutations at the nucleotide level *in situ*. It involves DNA isolation, extensive restriction with the selected endonuclease, and amplification of preselected DNA segments containing the target site. Molecules mutated at the restriction site are resistant and will be amplified (Parry et al. 1990). Probes for detecting RSMs in several *Mytilus edulis* genes, and on the globin gene of *Xenopus laevis* are available (Jones and Parry 1992).

Oncogene proteins are promising biomarkers in human cancer epidemiology. The most successful marker is the increased expression of the p21 *ras* oncoprotein in tumor tissue, cytological specimens and serum (Brandt-Rauf 1991). Deletion or mutation of tumor suppressor genes is essential in tumorigenesis. The p53 gene encodes a nuclear protein vital in the regulatory control of cell proliferation: loss of wild-type p53 function is a key event for malignant transformation in many cancers (Levine et al. 1991), and some p53 mutations confer oncogenic properties to this protein, becoming more stable than the wild type, and their accumulation can be detected with antibodies (Oren 1992). Increased levels of p53 protein have been detected in the nuclei of tissues with precancerous and cancerous lesions in human esophagus (Wang et al. 1993), solar keratosis (Taguchi et al. 1993), bronchial neoplasia (Bennet et al. 1993) and head and neck cancers (Shin et al. 1994). Research in fish molecular oncology is one decade behind that in mammalian models, but studies in fish oncogenes and suppressor genes have recently provided new biomarkers sensitive to environmentally induced carcinogenesis (Van Beneden and Ostrander 1994).

An extensive literature has described the suitability of the amphibian micronucleus assay for the detection of genotoxins in freshwater samples leading to chromosomal mutations. This includes different amphibian species such as *Xenopus laevis* (Jaylet et al. 1986). The assay has been successfully adapted to mussels (Majone et al. 1987) and fish (De Flora et al. 1993).

References

Adams SM, Shepard KL, Greely MSJr, Jimenez BD, Ryon MG, Shugart LR, McCarthy JF (1989) Mar Environ Res 28:459-464

Andersson T, Förlin L (1992) Regulation of the cytochrome P450 enzyme system in fish. Aquat Toxicol 24:1-19

Bennet WP, Colby TV, Travis WD, Borkowski A, Jones RT, Lane DP, Metcalf RA, Samet JM, Takeshima Y, Gu JR, Vähäkangas KH, Soini Y, Pääkkö P, Welsh JA, Trump BF, Harris CC (1993) p53 protein accumulates frequently in early bronchial neoplasia. Cancer Res 53:4817-4822

Boon JP, Everaarts JM, Hillebrand MTJ, Eggens MJ, Pijnenburg J, Goksøyr A (1992) Changes in levels of hepatic biotransformation enzymes and hemoglobin levels in female plaice (*Pleuronectes platessa*) after oral administration of a technical polychlorinated biphenyl mixture (Clophen A40). Sci Total Environ 114:113-133

Britvic S, Kurelec B (1986) Selective activation of carcinogenic aromatic amines to bacterial mutagens in the marine mussel *Mytilus galloprovincialis*. Comp Biochem Physiol 85C:111-114

Brandt-Rauf PW (1991) Oncogene proteins as biomarkers in the molecular epidemiology of occupational carcinogenesis. The example of the *ras* oncogene-encoded p21 protein. Int Arch Occup Environ Health 63:1-8

Brock V (1993) Effects of mercury on physiological condition and content of the biomarker ALA in the oyster *Ostrea edulis*. Mar Ecol Progr Ser 96:169-175

Castellani S, Mattei N, Renzoni A, Walker CH, Savva D (1993) Preliminary studies on the use of PCR and DNA fingerprints to detect the genotoxic effects of environmental chemicals. Proceedings of the First SETAC World Congress, Lisbon, pp. 247-247

Chan KM, Davidson WS, Hew CL, Fletcher GL (1989) Molecular cloning of metallothionein cDNA and analysis of metallothionein gene expression in winter flounder tissues. Can J Zool 67:2520-2529

Chipman JK, Marsh JW (1991) Bio-techniques for the detection of genetic toxicity in the aquatic environment. J Biotechnol 17:199-208

Courtenay S, Grunwald C, Kreamer G-L, Alexander R, Wirgin I (1993) Induction and clearance of cytochrome P4501A mRNA in Atlantic tomcod caged in bleached kraft mill effluent in the Miramichi River. Aquat Toxicol 27:225-244

Darnell J, Lodish H, Baltimore D (1990) *Molecular Cell Biology*, Scientific American Books, Inc

De Flora S, Viganò L, D'Agostini F, Camoirano A, Bagnasco M, Bennicelli C, Melodia F, Arillo A (1993) Multiple genotoxicity biomarkers in fish exposed in situ to polluted river water. Mutation Res 319:167-177

Degan P, Shigenaka MK, Park E-M, Alperin P, Ames BN (1991) Immunoaffinity isolation of urinary 8-hydroxy-2'-deoxyguanosine and 8-hydroxyguanine and quantification of 8-hydroxy-2'-deoxyguanosine in DNA by polyclonal antibodies. Carcinogenesis 12:865-871

Depledge MH (1989) The rational basis for detection of early effects of marine pollutants using physiological indicators. Ambio 18:301-302

Dieter MP (1979) Blood δ-aminolevulinic acid dehydratase (ALA-D) to monitor lead contamination in canvasback ducks (*Aythya valisineria*). In: *Animals as Monitors of Environmental Pollutants*. National Academy of Sciences, Washington DC, p.177

Di Giulio RT, Washburn PC, Wenning RJ, Winston GW, Jewell CS (1989) Biochemical responses in aquatic animals: a review of determinants of oxidative stress. Environ Toxicol Chem 8:1103-1123

Dunn BP (1991) Carcinogen adducts as an indicator for the public health risks of consuming carcinogen-exposed fish and shellfish. Environ Health Perspect 90:111-116

Ellis RJ (1990) The molecular cheperon concept. Semin Cell Biol 1:1-13

Endicott JA, Ling V (1989) The biochemistry of P-glycoprotein-mediated multidrug resistance. Annu Rev Biochem 58:137-165

Engel DW, Roesijadi G (1987) Metallothioneins: a monitoring tool. In: *Pollution and Physiology of Estuarine Organisms*, Vernberg WB, Thurberg FP, Calabrese A, and Vernberg WB (eds), University of South Carolina Press, Columbia, pp. 421-438

Fossi C, Leonzio C, Focardi S (1986) Increase in organochlorines and MFO activity in water birds wintering in an Italian lagoon. Bull Environ Contam Toxicol 37:538-543

Fossi C, Leonzio C (eds) (1994) *Nondestructive Biomarkers in Vertebrates*, Lewis Publishers, Boca Ratón

Goksøyr A, Förlin L (1992) The cytochrome P-450 system in fish aquatic toxicology and environmental monitoring. Aquatic Toxicol. 22:287-312

Granick S, Sassa S, Granick JL, Levere RD, Kappas A (1972) Assays for porphyrins, δ-aminolevulinic acid dehydratase, and porphyrinogen synthase in microliter samples of whole blood: Application to metabolic defects involving the heme pathway. Proc Natl Acad Sci USA 69:2381-2385

Grant SG, Bigbee WL (1993) In vivo somatic mutation and segregation at the human glycophorin A (*GPA*) locus: phenotypic variation encompassing both gene-specific and chromosomal mechanisms. Mutation Res 288:163-172

Grue CE, Hart ADM, Mineau P (1992) Biological consequences of depressed brain cholinesterase activity in wild life. In: *Cholinesterase-inhibiting Insecticides: Impact on Wildlife and the Environment*, Mineau P (ed), Elsevier, Amsterdam, pp. 151-212

Gupta RC, Randerath K (1988) Analysis of DNA adducts by ^{32}P labelling and thin-layer chromatography. In: *DNA Repair Vol 3*, Friedberg E, Hanawalt PH (eds), Marcel Dekker, New York, pp. 399-418

Hartl FU, Martin J, Neupert W (1992) Protein folding in the cell: the role of molecular chaperones Hsp70 and Hsp60. Annu Rev Biophys Biomol Struct 21:293-325

Haux C, Förlin L (1988) Biochemical methods for detecting effects of contaminants in fish. Ambio 17:376-380

Heilman LJ, Sheen Y-Y, Bigelow SW, Nebert DW (1988) Trout P450IA1: cDNA and deduced protein sequence, expression in liver, and evolutionary significance. DNA 7:379-387

Hinton DE, Laurén DJ (1990) Liver structural alterations accompanying chronic toxicity in fishes: potential biomarkers of exposure. In: *Biomarkers of Environmental Contamination*, McCarthy JF and Shugart LR (eds), Lewis Publishers, Boca Ratón, pp. 17-57

Huggett RA, Kimerle RA, Mehrle PM Jr, Bergman HL (eds) (1992) *Biomarkers: Biochemical, Physiological, and Histological Markers of Anthropogenic Stress*. Lewis Publishers, Boca Ratón

Hunsaker CT (1993) New concepts in environmental monitoring: the question of indicators. Sci Total Env suppl 1993: 77-95

Jaylet A, Depario P, Ferrier V, Grinfeld S, Siboulet R (1986) A new micronucleus test using peripheral blood erythrocytes of the newt *Pleurodeles walte* to detect mutagens in fresh water, Mutation Res 164:245-257

Jiménez BD, Oikari A, Adams SM, Hinton DE, McCarthy JF (1990) Hepatic enzymes as biomarkers: interpreting the effects of environmental, physiological and toxicological variables. In: *Biomarkers of Environmental Contamination*. McCarthy JF, Shugart LR (eds), Lewis Publishers, Boca Ratón, pp. 123-142

Jones NJ, Parry JM (1992) The detection of DNA adducts, DNA base changes and chromosome damage for the assessment of exposure to genotoxic pollutants. Aquat Toxicol 22:323-344

Kagi JHR, Kojima Y (1987) Chemistry and biochemistry of metallothioneins. In: *Metallothionein II*. Kagi JHR, Kojima Y (eds), Birkhauser-Verlag, Basel, pp. 25-52

Kille P, Kay J, Leaver M, George S (1992) Induction of piscine metallothionein as a primary response to heavy metal pollutants: applicability of new sensitive molecular probes. Aquat Toxicol 22:279-286

Kurelec B, Matijasevic Z, Rijavec M, Alacevic M, Britvic S, Müller WG, Zahn RK (1979) Induction of benzo(a)pyrene monooxygenase in fish and the Salmonelle test as a tool for detecting mutagenic/carcinogenic xenobiotics in the aquatic environment. Environ Contam Toxicol 21:799-807

Kurelec B, Garg A, Krca S, Chacko M, Gupta RC (1989) Natural environment surpasses polluted environments in inducing DNA damage in fish. Carcinogenesis 10:1337-1339

Kurelec B, Pivcevic B (1991) Evidence for a multixenobiotic resistance mechanism in the mussel *Mytilus galloprovincialis*. Aquat Toxicol 19:291-302

Kurelec B (1992) The multixenobiotic resistance mechanism in aquatic organisms. Crit Rev Toxicol 22:23-43

Lemaire P, Mathieu A, Giudicelli J, LaFaurie M (1992) Effect of diet on the responses of hepatic biotransformation enzymes to benzo(a)pyrene in the European sea bass (*Dicentrarchus labrax*). Comp Biochem Physiol 102:413-420

Levine AJ, Momand J, Finlay CA (1991) The p53 tumour suppressor gene. Nature 351:453-456

Livingstone DR (1993) Biotechnology and pollution monitoring: use of molecular biomarkers in the aquatic environment. J Chem Tech Biotechnol 57:195-211

Lowe WR, Kendall RJ (1990) Sentinel species and sentinel bioassay. In: *Biomarkers of Environmental Contamination.* McCarthy JF, Shugart LR (eds), Lewis Publishers, Boca Ratón, pp. 309-331

Lubet RA, Nims RW, Beebe LE, Fox SD, Issaq HJ, McBee K (1992) Induction of hepatic CIP1A activity as a biomarker for environmental exposure to Aroclor 1254 in feral rodents. Arch Environ Contam Toxicol 22:339-344

Maccubbin AE (1994) DNA adduct analysis in fish: laboratory and field studies. In: *Aquatic Toxicology. Molecular, Biochemical and Cellular Perspectives.* Malins DC, Ostrander GK (eds), Lewis Publishers, Boca Ratón, pp. 267-294

Majone F, Brunetti R, Gol I, Levis AA (1987) Persistence of micronuclei in the marine mussel, *Mytilus galloprovincialis* after treatment with mitomycin C. Mutation Res 191:157-161

Malins DC (1993) Identification of hydroxyl radical-induced lesions in DNA base structure: biomarkers with a putative link to cancer development. J Toxicol Environ Health 40:247-261

Maron DM, Ames BN (1983) Revised methods for the Salmonella mutagenicity test. Mutat Res 113:173-215

Marsh JW, Chipman JK, Livingstone DR (1993) Formation of DNA adducts following laboratory exposure of the mussel, *Mytilus edulis*, to xenobiotics. Sci Total Environ suppl 1993:567-572

Martínez-Lara E, Pascual P, Toribio F, López-Barea J, Bárcena JA (1992) A rapid method for the quantification of glutathione transferase isoenzymes in crude extracts. J Chromatogr 609:141-146

McCarthy JF, Shugart LR (eds) (1990) *Biomarkers of Environmental Contamination.* Lewis Publishers, Boca Ratón

Moal J, le Coz JR, Samain JF, Daniel JY (1991) Adenylate energy charge: a possible trophic index for management of oyster intensive aquaculture. Comp Biochem Physiol 100C:201-205

Mohn GR, de Raat WK (1993) Ecological significance of mutagens. Sci Total Environ suppl 1993:1771-1778

Moore MN, Livingstone DR, Widdows J, Lowe DM, Pipe RK (1987) Molecular, cellular and physiological effects of oil-derived hydrocarbons on molluscs and their use in impact assessment. Phil Trans R Soc London B 316:603-623

Moore MN (1990) Lysosomal cytochemistry in marine environmental monitoring. Histochem J 22:187-191

Moore MN, Simpson MG (1992) Molecular and cellular pathology in environmental impact assessment. Aquat Toxicol 22:313-322

Moore MN (1993) Biomarkers of contaminant exposure and effect: a way forward in marine environmental toxicology. Sci Total Environ suppl 1993:1335-1343

Moore MN, Chipman JK, den Besten PJ, Kurelec B, Bergman A (1993) Sci Total Env suppl 1993:1767-1770

Murphy TA, Kaputska L (1990) Capabilities and limitations of approaches to in situ ecological evaluation. In: *In situ Evaluation of Biological Hazards of Environmental Pollutants*. Sandhu SS, Lower WR, deSerres FJ, Suk WA, Tice RR (eds). Plenum Press, New York

Myers MS, Landahl JT, Krahn M, McCain BB (1991) Relationships between hepatic lesions and exposure to toxic chemicals in marine fish from the US west coast. Environ Health Perspect 90:7-15

Nebert DW, Petersen DD, Fornace AJ (1990) Cellular responses to oxidative stress: the [Ah] gene battery as a paradigm. Environ Health Perspect 88:13-25

Nebert DW, Nelson DR, Coon MJ, Estabrook RW, Feyereisen R, Fujii-Kuriyama Y, González FJ, Guengerich FP, Gunsalus IC, Johnson EF, Loper JC, Sato R, Waterman MR, Waxman DJ (1991) The P450- superfamily: update on new sequences, gene mapping, and recommended nomenclature. DNA Cell Biol 10:1-14

Norey CG, Lees WE, Darke BM, Stark JM, Baker TS, Cryer A, Kay J (1990) Immunological distinction between piscine and mammalian metallothioneins. Comp Biochem Physiol 95B:597-601

Okey AB (1990) Enzyme induction in the cytochrome P-450 system. Pharmacol Therapeut 45:241-298

Oren O (1992) *p53*: the ultimate tumor suppressor gene? FASEB J 6:3169-3176

Park J-W, Cundy KC, Ames BN (1989) Detection of DNA adducts by high performance liquid chromatography with electrochemical detection. Carcinogenesis 10:827-832

Parry JM, Shamsher M, Skibinski DOF (1990) Restriction site mutation analysis, a proposed methodology for the detection and study of DNA base changes following mutagen exposure. Mutagenesis 5:209-212

Pavicic B, Raspor D, Martincic D (1993) Quantitative determination of metallothionein-like proteins in mussels. Methodological approach and field evaluation. Marine Biol 115:435-444

Peakall DB (1994) Biomarkers. The way forward in environmental assessment. Toxicol and Ecotoxicol News 1:55-60

Peakall DB, Shugart LR (1993) *Biomarkers. Research and Application in the Assessment of Environmental Health*. Springer Verlag, Berlin

Pedrajas JR, Peinado J, López-Barea J (1993) Purification of isoforms of Cu,Zn-superoxide dismutase from fish liver: appearance of new isoforms as consequence of pollution. Free Rad Res Commun 19:29-41

Perera F, Brenner D, Jeffrey A, Mayer J, Tang D, Warburton D, Young T-I, Wazneh L, Latriano L, Motykiewicz G, Grzybowska E, Chorazy M, Hemminki K, Santella R (1992) DNA adducts and related biomarkers in populations exposed to environmental carcinogens. Environ Health Perspect 98:133-137

Poirier MC (1984) The use of carcinogen -DNA adduct antisera for quantitation and localisation of genomic damage in animal models and human population. Environ Mutagen 6:879-887

Rattner BA, Hoffman DJ, Marn CM (1989) Use of mixed-function oxygenases to monitor contaminant exposure in wildlife. Environ Toxicol Chem 8:1093-1102

Rodriguez-Ariza A, Abril N, Navas JI, Dorado G, López-Barea, Pueyo C (1992) Metal, mutagenicity and Biochemical studies on bivalve molluscs from Spanish coasts. Environ Molec Mutagen 19:112-124

Rodriguez-Ariza A, Martínez-Lara E, Pascual P, Pedrajas JR, Abril N, Dorado G, Toribio F, Bárcena JA, Peinado J, Pueyo C, López-Barea J (1993a) Biochemical and genetic indices of marine pollution in Spanish littoral. Sci Total Environ suppl 1993:109-116

Rodriguez-Ariza A, Peinado J, Pueyo c, López-Barea J (1993b) Biochemical indicators of oxidative stress in fish from polluted littoral areas. Can J. Fish Aquat Sci 50:2568-2573

Rodriguez-Ariza A, Dorado G, Navas JI, Pueyo C, López-Barea J (1994a) Promutagen activation by fish liver as biomarker of littoral pollution. Environ Molec Mutagen, in press

Rodriguez-Ariza A, Toribio F, López-Barea J (1994b) Rapid determination of glutathione status in fish using HPLC and electrochemical detection. J Chromatogr Biomed Appl 656:311-318

Roesijadi G, Robinson WE (1994) Metal regulation in aquatic animals: mechanisms of uptake, accumulation, and release. In: *Aquatic Toxicology. Molecular, Biochemical and Cellular Perspectives.* Malins DC, Ostrander GK (eds), Lewis Publishers, Boca Ratón, pp. 387-420

Rothman JE (1989) Polypeptide chain binding proteins: catalysts of protein folding and related processes in cell. Cell 59:591-602

Sanders BM (1993) Stress proteins in aquatic organisms: an environmental perspective. Crit Rev Toxicol 23:49-75

Sanders BM, Martin LS (1993) Stress proteins as biomarkers of contaminant exposure in archived environmental samples. Sci Total Environ 139/140:459-470

Santella RM (1991) DNA adducts in humans as biomarkers to environmental and occupational carcinogens. Environ Carcino Ecotox Revs C9:57-81

Schmitt CJ, Wildhaber ML, Hunn JB, Nash T, Tieger MN, Steadman BL (1993) Biomonitoring of lead-contaminated Missouri streams with an assay for erythrocyte δ-aminolevulinic acid dehydratase activity in fish blood. Arch Environ Contam Toxicol 25:464-475

Sheehan D (1994) Environmental Biochemistry. The Biochemist June/July:34-37

Shin DM, Kim J, Ro J, Hittelman J, Roth JA, Hong WK, Hittelman WN (1994) Activation of *p53* gene expression in premalignant lesions during head and neck tumorigenesis. Cancer Res 54:321-326

Stegeman JJ (1989) Cytochrome P-450 forms in fish: catalytic, immunological and sequence similarities. Xenobiotica 19:1093-1110

Stegeman JJ, Lech JJ (1991) Cytochrome P-450 monooxygenase systems in aquatic species: carcinogen metabolism and biomarkers for carcinogen and pollutant exposure. Environ Health Perspect 90:101-109

Stegeman JJ, Brouwer M, DiGiulio RT, Förlin L, Fowler B, Sanders BM, Van Veld PA (1992) Molecular responses to environmental contamination: enzyme and protein synthesis as indicators of chemical exposure and effect. In: *Biomarkers: Biochemical, Physiological, and Histological Markers of Anthropogenic Stress.* Huggett RA, Kimerle RA, Mehrle PM Jr, Bergman HL (eds), Lewis Publishers, Boca Ratón, pp 235-335

Steinberg CEW, Geyer HJ, Kettrup AAF (1994) Evaluation of xenobiotic effects by ecological techniques. Chemosphere 28:357-374

Taguchi M, Watanabe S, Sato Y, Kameya T, Munakata N, Ishihara K, Nakane PK, Hisatome H, Ikeda S (1993) Immunohistochemical examination of tumor-suppressor gene p53 product and pyrimidine dimer in solar keratosis. J Cancer Res Clin Oncol 119:260-262

Tannenbaum SR (1990) Hemoglobin-carcinogen adducts as molecular biomarkers in epidemiology. In: *Princess Takamatsu Symposia Vol 21*, Japan Scientific Society Press, Tokyo, pp. 351-360

Timbrell JA, Draper R, Waterfield CJ (1994) Biomarkers in Ecotoxicology. New uses for some old molecules? Toxicol and Ecotoxicol News 1:4-14

Uchida K, Kawakishi S (1993) 2-Oxo-histidine as a novel biological marker for oxidatively modified proteins. FEBS Lett 332:208-210

Van Beneden RJ, Ostrander GK (1994) Expression of oncogenes and tumor suppressor genes in teleost fishes. In: *Aquatic Toxicology. Molecular, Biochemical and Cellular Perspectives.* Malins DC, Ostrander GK (eds), Lewis Publishers, Boca Ratón, pp. 295-325

Varanasi U, Stein JE, Nishimoto M, Reichert WL, Collier TK (1987) Chemical carcinogenesis in feral fish: uptake, activation, and detoxication of organic xenobiotics. Environ Health Perspect 71:155-170

Walker CH (1992) Biochemical responses as indicators of toxic effects of chemicals in ecosystems. Toxicol Lett 64/65:527-533

Wang L-D, Hong J-Y, Qiu S-L, Gao H, Yang CS (1993) Accumulation of p53 protein in human esophageal precancerous lesions: a possible early biomarker for carcinogen-esis. Cancer Res 53:1783-1787

Widdows J, Donkin P (1991) Role of physiological energetics in ecotoxicology. Comp Biochem Physiol 100C:69-75

Wild CP, Jiang YZ, Allen SJ, Jansen LA, Hall AJ, Montesano R (1990) Aflatoxin-albumin adducts in human sera from different regions of the world. Carcinogenesis 11:2271-2274

Wild CP, Hudson G, Sabbioni G, Wogan GN, Whittle H, Montesano R, Groopman JD (1992) Correlation of dietary intake of aflatoxins with the level of albumin bound aflatoxin in peripheral blood in The Gambia, West Africa. Cancer Epidemiol, Biomarkers Prev 1:229-234

Zafarullah M, Olsson P-E, Gedamu L (1989) Endogenous and heavy metal ion induced metallothionein gene expression in salmonid tissues and cell lines. Gene 83:85-93

Use of cytochrome P450 1A (CYP1A) in fish as a biomarker of aquatic pollution

Anders Goksøyr
Laboratory of Marine Molecular Biology, University of Bergen, HIB, N-5020 Bergen, Norway

Development of diagnostic and predictive biomarkers for use in pollution monitoring depends on a fundamental characterization and understanding of the mechanisms and regulations of the molecular response. Since first proposed as a pollution indicator in the mid-1970s (Payne & Penrose, 1975; Ahokas et al., 1975; Payne, 1976), the insight in the cytochrome P450 (CYP) system in fish, its molecular basis, regulation, and the mechanisms underlying the induction response to organic xenobiotics, has increased tremendously. Although the picture is still far from complete, we now have the knowledge and experience to put its potential as a pollution biomarker into use. However, understanding the influences of factors such as water temperature, season, sexual maturation, developmental status and diet, is an important and critical contribution to the full implementation of CYP1A responses in fish as a biomarker of aquatic pollution.

The cytochrome P450 system

The cytochrome P450 (CYP or P450) system comprise a large superfamily of heme proteins involved in the oxidative metabolism of numerous lipophilic exogenous and endogenous substrates, including drugs, aromatic hydrocarbons, pesticides, fatty acids, prostaglandins and steroids. It is found in all phyla studied to date. The known membership of this superfamily is steadily increasing, with the latest score numbering 221 genes divided into 36 subfamilies based on sequence similarities (Nelson et al., 1993). It is estimated that as many as 60 and perhaps up to 200 different P450 genes are present within a given (mammalian) species (Puga and Nebert, 1989). Due to its importance in pharmacology and carcinogenesis, the major part of the research efforts into this enzyme system is being performed on laboratory rodents and humans. A number of review chapters on this system have been compiled in the recent monograph by Schenkman & Greim (1993).

Although the fundamental features of the monooxygenase reactions are the same in all P450-mediated biotransformations, large differences in the chemistry of both substrate and product exist, from large, complex and bulky molecules to small and planar ones. Different isozymes, belonging to different subfamilies, appear to have varying degrees of substrate specificities, but with some subfamilies dominating in the biotransformation of xenobiotics (Table 1), whereas others are more specifically metabolizing endogenous substrates. Large differences also exist in the regulatory mechanisms of the different isoenzyme forms, ranging from specific cytosolic receptors regulating gene transcription, to posttranscriptional and, in some cases, posttranslational control (Nebert et al., 1989). Certain subfamilies are particularly responsive to xenobiotic compounds, prominent examples being the induction of the CYP1A subfamily by planar aromatic and chlorinated hydrocarbons, and the response of the CYP2B subfamily to the bulky phenobarbital drugs and non-planar PCBs (Table 1). The former response has been observed in all vertebrate species studied, whereas the latter is not detected in fish, amphibia or reptiles (Nebert et al., 1989; Stegeman and Hahn, 1994).

Table 1. Prominent inducible cytochrome P450 subfamilies and their most common substrates[a]

Subfamily	Prominent inducers	Common substrates
CYP1A	PAH, BNF, planar PCBs, dioxins	PAH, ethoxyresorufin
CYP2B	Barbiturates, non-planar PCBs, DDT	Barbiturates, steroids
CYP2E	Ethanol, ketones	Ethanol, alkylnitrosamines
CYP3A	PCN, glucocorticoids	Steroids
CYP4A	Clofibrate, phthalates, 2,4,5-T	Fatty acids

a) Based on Stegeman and Hahn (1994). Abbreviations used: PAH, polyaromatic hydrocarbons; BNF, ß-naphthoflavone; PCB, polychlorinated biphenyl; DDT, dichlorodiphenyltrichloroethane; 2,4,5-T, 2,4,5-trichloro??

Regulation of CYP1A via the Ah receptor

It is becoming increasingly clear that the CYP1A subfamily is the most important subfamily in the metabolic activation of chemical carcinogens, in that it always directs metabolism towards the formation of reactive, genotoxic intermediates (Ioannides, 1990). Because of this role, the CYP1A subfamily is among the best characterized and well studied in the whole P450 superfamily.

Extensive research has established that planar aromatic hydrocarbons like 2,3,7,8-tetrachlorodibenzo-p-dioxin (TCDD), benzo(a)pyrene (BaP), and certain non- and mono-ortho substituted polychlorinated biphenyls (PCBs), are able to induce the transcription of the *CYP1A* genes through the aryl hydrocarbon (Ah) receptor. This receptor has recently been shown to belong to a new class of nuclear receptors, distinct from the steroid and thyroid hormone receptor super-family.

Studies have identified a class of response elements, xenobiotic response elements (XREs), upstream of the *CYP1A* gene through which aryl hydrocarbons activate transcription of the gene. Subsequent to binding a hydrocarbon ligand in the cytosol, the Ah receptor forms a heterodimer with the Ah receptor nuclear translocator (ARNT) protein. This ligand·receptor·translocator complex enters the nucleus where it interacts specifically with the XREs, initiating the transcription of, among others, the *CYP1A* genes, as depicted in Fig. 1.

In all mammals studied to date, two distinct *CYP1A* genes, resulting in two different protein products, CYP1A1 and 1A2, are known to exist. The two genes, although both under control of the Ah receptor, are apparently independently regulated by diferent xenobiotic inducers (*CYP1A2* lack XREs), but the mechanisms behind this phenomenon have not been clarified (see chapters in Schenkman & Greim, 1993).

CYP1A in fish

The duplication event leading to the divergence of mammalian *CYP1A1* and *CYP1A2* from a single ancestral *CYP1A* gene is suggested to have occurred less than 250 million years ago (Nebert and Gonzalez, 1987), subsequent to divergence of the fish and mammalian lines of evolution. Whether multiple *CYP1A* genes occur in fish has been a matter of discussion since the first P450 proteins were purified from fish.

Williams and Buhler (1982; 1984) purified two similar proteins from BNF-treated rainbow trout (*Oncorhynchus mykiss*) with high activity towards benzo(a)pyrene (P450LM4a and P450LM4b). P450LM4b has later been confirmed to belong to the CYP1A subfamily, based on catalytic properties, inducibility with aromatic hydrocarbons, and immunological relationships (see Stegeman and Hahn, 1994). A single CYP1A-like protein was obtained from scup (*Stenotomus chrysops*) in John Stegeman's laboratory (Klotz et al., 1983), and also from cod (*Gadus morhua*)

(Goksøyr, 1985), and perch (*Perca fluviatilis*) (Zhang et al., 1991) (see reviews by Goksøyr and Förlin, 1992; Stegeman, 1993; Stegeman and Hahn, 1994). In studies using polyclonal and monoclonal antibodies towards these forms, only a single CYP1A cross-reacting protein has been detected in most fish species (ibid.; Goksøyr et al., 1991).

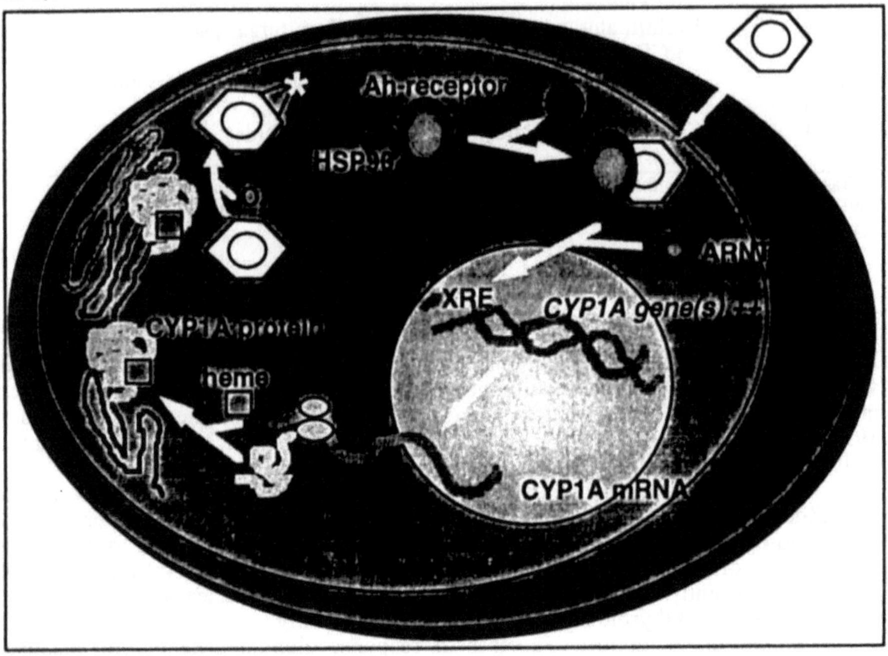

Figure 1. The CYP1A induction response involves binding of a planar aromatic ligand (e.g. benzo(a)pyrene or TCDD) to the cytosolic Ah-receptor, translocation of the complex (including the Ah-receptor nuclear translocator, ARNT, protein) into the nucleus, and activation of the *CYP1A* gene(s) by binding to upstream xenobiotic response elements (XREs). The CYP1A mRNA is translated to protein in the ribosomal machinery, binds heme, and is inserted into the membrane of the endoplasmic reticulum, where it performs its monooxygenase activity.

Observations that typical CYP1A activities such as aryl hydrocarbon hydroxylase (AHH), 7-ethoxyresorufin O-deethylase (EROD), and 7-ethoxycoumarin O-deethylase (ECOD) may be differentially induced by CYP1A1 and CYP1A2-type inducers (diesel oil, isosafrole, piperonyl butoxide), have stimulated the discussion that there might be multiple CYP1A proteins expressed in fish (see Buhler and Williams, 1989; Stegeman, 1989; Stegeman, 1993). Another possible explanation for these observations could be different susceptibilities of a single CYP1A protein's catalytic activities towards modulating effects of various exogenous or endogenous agents, as discussed previously (Goksøyr and Förlin, 1992).

When the first *CYP1A* gene was isolated from fish, Heilmann et al. (1988) reported that only a single gene belonging to this subfamily could be detected in rainbow trout. This cDNA sequence showed 57 to 59% similarity with mammalian *CYP1A1* forms, and 51 to 53% similarity with *1A2* forms, suggesting that this form represents the fish *CYP1A1* gene. A single *CYP1A* gene was also reported from plaice (*Pleuronectes platessa*) (Leaver, Pirrit & George, EMBL AC X73631), butterflyfish (Vrolijk & Chen, 1994), toadfish and scup (Morrison et al., 1994). However, in a recent study, Berndtson and Chen (1994) report the isolation and complete genomic sequence of two distinct, but closely related *CYP1A* genes from rainbow trout. Transcripts of both genes were expressed in liver in response to treatment with 3-methylcholanthrene (3-MC). The two genes were similar in size and had 96% sequence identity, however with significant differences in the 5' upstream region as well as in the introns, supporting the conclusion that these genes are not allelic variants but true distinct genes. These findings also make it unlikely that the two genes could be due to polymorphism at the *CYP1A1* locus, as salmonid species like rainbow trout are tetraploid. It should be made clear that although the second gene is called *CYP1A2*, it is not orthologous to mammalian *CYP1A2*, suggesting a different number should be found. In fact, careful analysis of the sequences indicates that the names of the two genes should be switched, due to a suggested insertion that may have been missed (Morrison et al., 1994). The high similarity between the two rainbow trout genes make it plaucible that they result from a quite recent duplication event, probably not more than 5-10 million years ago, and certainly later than the divergence of *Salmoniformes* from the rest of the fish orders (ibid.). Whether other fish species possess one or more *CYP1A* genes will not be known until more cloning and sequencing work has been performed with fish.

Induction of CYP1A in fish by environmental pollutants

Laboratory studies

The fundamental property of the fish CYP1A system that makes it interesting as a biomarker for aquatic pollution, is its inducibility with numerous foreign chemicals acting as pollutants in the environment. Although not clarified in detail, the induction response in fish proceeds presumably by largely the same mechanisms as in mammals, i.e. via the Ah receptor, resulting in increased levels of CYP1A mRNA transcripts, increased levels of *de novo* synthesized protein, and increased levels of CYP1A-associated catalytic activities. The latter is the most commonly measured level, and is normally assessed by measuring EROD or AHH activity in fish liver preparations. With the advent of specific molecular tools, both immunochemical and nucleotide-based detection systems have been developed, increasing our understanding of the induction response, as well as giving additional quality control to induction studies under specific situations. We will return to this later. With regard to the Ah receptor in fish, recent studies have demonstrated its presence in all bony fish and sharks, whereas it could not be discovered in jawless fish and invertebrates (see Stegeman and Hahn, 1994; Hahn et al., 1994).

Organic pollutants of major concern in the aquatic environment are polyaromatic hydrocarbons (PAHs), oil compounds, PCBs, polychlorinated dibenzo-p-dioxins and dibenzofurans (PCDD/PCDF), and other halogenated compounds such as pesticides and herbicides. A number of these compounds have been evaluated as inducers of CYP1A in controlled laboratory studies with different fish species. In a few cases, such studies have also been performed with fish hepatocytes. The response has generally been assessed by measuring catalytic activity, but more recently protein and mRNA levels have also been studied. The structural features associated with CYP1A induction in fish are similar to those seen in mammals, with large PAH compounds as strong inducers, whereas small aromatic and aliphatic hydrocarbons are non-inducing. A planar configuration is also important for the chlorinated aromatics, with non-ortho substituted PCBs being the most potent inducers in this class, along with 2,3,7,8-substituted dioxins and dibenzofurans. Mono-ortho substituted PCBs are much weaker, and sometimes inactive, as inducers in fish (Table 2). Whereas PCB mixtures generally give induction, a notable exception is the lack of clear response in cod (Beyer et al., in prep.; Egaas et al., 1994).

A confounding factor has been detected in the ability of some of these compounds also to inhibit or inactivate CYP1A activity. Therefore, complementary analyses of protein and/or mRNA levels sometimes give additional information on the induction response, a response that could have gone undetected with catalytic measurements alone (see Goksøyr and Förlin, 1992; Stegeman and Hahn, 1994).

Although there is an increasing body of information on the induction effect of many of these compounds, there is still a need for a systematic effort to define more precise structure-activity relationships, to investigate the mechanistic basis for induction, and to assess variations between different species of fish.

Table 2. Examples of xenobiotics effective as inducers of CYP1A in fish[a]

Compound	Method[b]
ß-naphthoflavone	A, P, R
3-Methylcholanthrene	A, P, R
Benzo(a)pyrene	A, P
Benzanthracene	A
Dibenzanthracene	A
Chrysene	A, P
Fluoranthene	A, P
Pyrene	A, P
Petroleum	A, P
PCB mixtures	A, P
3,3',4,4'-TCB (#77)	A, P, R
3,3',4,4',5-PCB (#126)	A, P
3,3',4,4',5,5'-HCB (#169)	A, P
2,3,3',4,4'-PCB (#105)	A, P
2,3',4,4',5-PCB (#118)	A, P
2,2',4,4',5,5'-HCB (#153)	A, P
2,3,3',4,4',5-HCB (#156)	A, P
Dioxin mixtures	A, P
2,3,7,8-TCDD	A, P
1,2,3,7,8-PCDD	A, P
OCDD	A, P
2,3,7,8-TCDF	A, P, R
2,3,4,7,8-PCDF	A, P
OCDF	A, P
Pulp mill effluents	A
Isosafrole	A, P
Piperonyl butoxide	A, P, R
Endosulfan	A, P
Cyclopropenoid fatty acids	A

a) Note: the induction potential may vary between different species of fish. Also, in some cases the different analytical techniques have given different results (see text). This table was compiled from the following sources: Goksøyr and Förlin, 1992; Stegeman and Hahn, 1994; van der Weiden, 1993; Egaas et al., 1994; Grøsvik et al., 1993; Husøy & Goksøyr, unpubl. results.

b) A: catalytic activity assay (usually EROD or AHH), P: protein determination (ELISA or Western blots), R: mRNA determination (Northern blots or slot blots).

Field studies

Evaluation through field studies is a central part in developing the CYP1A response in fish as a biomarker for aquatic pollution. Experiences from such field studies, although often of varying quality in study design, give important information on the validity and applicability of a biomarker. Indeed, the interest in studying the CYP1A response in fish initially came from reports that fish living in contaminated environments had elevated levels of CYP1A enzymatic activity (Payne & Penrose, 1975). In that work, increased AHH activity was observed for the first time in brown trout from a small urban lake in Newfoundland with a history of petroleum contamination. Since then, a number of field studies have supported the concept of CYP1A in fish as a biomarker for contamination in coastal waters, rivers, and lakes around the world, and a number of workshops and reviews have discussed these (see Payne et al., 1987; Haux and Förlin, 1988; Stegeman and Lech, 1991; Goksøyr and Förlin, 1992; Stebbing et al., 1992; Stegeman et al., 1992; Förlin et al., 1994; Stegeman and Hahn, 1994).

The general conclusion from these studies is that fish from coastal regions near urban and industrial centers have induced levels of CYP1A in their livers. This response is normally negatively correlated with the distance from such areas. Other local sources as well as acute pollution incidences are also associated with CYP1A induction. However, the persistence and bioaccumulating nature of these chemicals result in a global distribution, e.g. of organochlorines (Walker & Livingstone, 1992). Regional dispersion patterns of these compounds may also be reflected by CYP1A levels, as was shown in deep sea fish by Stegeman et al. (1986), although the "noise" from other factors in studies over large areas may sometimes obscur this observation, as discussed below.

Studies where chemical analyses have been performed and correlated with CYP1A analyses, have shown that levels of contaminants such as PCBs and PAHs (measured in the fish themselves or in other environmental matrices) generally can be linked to the induction response (e.g. Van Veld et al., 1990; Addison and Edwards, 1988; Goksøyr et al., 1994; Förlin et al., 1991; Collier et al., 1994). Recent studies that include contaminant measurements in individual fish from coastal areas in the US, indicate that PAHs may be more important than PCBs as inducers in feral fish from the Puget Sound, whereas winter flounder (*Pseudopleuronectes americanus*) from the Northeast US show a better relationship with PCBs than PAHs, and white croaker (*Pseudosciaena manchurica*) from California show a better relationship with total pesticides (Collier et al., 1993a and personal communication). In all cases it should be taken into account that the analyzed contaminants do not appear alone, but normally in a mixture with several other compounds with potentially inducing or modulating properties. In such cases delineating the inducing agent may be a difficult task. A strategy of isolation and fractionation of the pollutant source, combined with controlled laboratory studies, has been employed with great success e.g. in the case of Swedish pulp mill effluents and sediment-associated contaminants in Puget Sound, USA (Andersson et al., 1988; Collier and Varanasi, 1991; Pesonen and Andersson, 1992).

Large acute pollution incidences sometimes occur, with a need to define the spatial and temporal range of biological effects in the environment. Experiences from several oil spill incidences in the last five years demonstrate the potential of the CYP1A response in fish to serve as a biomarker in this task. In the case of the "Exxon Valdez" tanker accident in Alaska, March 1989, CYP1A measurements, along with other biomarker studies, have revealed far-reaching and long-lasting effects of the oil spill on several fish species (Collier et al., 1993b). After the 1991 Gulf War oil spill, an intensive survey was carried out to determine coastal and environmental consequences of the spill (Price & Robinson, 1993). These included CYP1A measurements in local fish species. Although the results indicated pollution effects, the sampling design used was less than adequate, giving data with little statistical hold (Beyer and Goksøyr, 1993). However, after the "Braer" tanker accident in Shetland, January 1993, biological monitoring studies were performed that included measurements of CYP1A induction. The results show that induction was evident in farmed salmon for about 4 months after the oil spill (Ritchie & O'Sullivan, 1994), and also in inshore species like wild rockling until 5 months after the spill (George et al., 1994).

Analytical techniques for detecting CYP1A induction in fish

The induction response, involving all stages from gene transcription to mature enzyme, can be studied at all intermediate levels (see Table 3). Most studies have focussed on the final product of this process, i.e. the CYP1A mediated monooxygenase (previously 'mixed function oxidase') activity, usually measured as EROD or AHH activity. Both of these activities are more or less specifically catalyzed by CYP1A (Stegeman and Kloepper-Sams, 1987; Goksøyr et al., 1991), and their technical details have been described elsewhere (e.g. Förlin et al., 1994). EROD can be measured both spectrophotometrically and fluorometrically, and simple plate reader assays have been developed for the latter (Eggens & Galgani, 1992). AHH can be assayed by fluorometric or radiometric methods, the two giving somewhat different results.

Although being more complicated, AHH seem to be less prone to inherent variability than the EROD assay (Collier et al., 1994). In addition to the technical aspects of the methods, research has shown that they are both susceptible to modulation by factors present in the sample, like contaminants (PCBs or PAHs) or endogenous compounds (steroids) (see Goksøyr & Förlin, 1992).

After the CYP1A proteins had been purified from different sources, antibody probes became available and have been employed in different immunochemical techniques, including Western blotting (e.g. Stegeman et al., 1986; Goksøyr et al., 1991), ELISA (Goksøyr, 1991), and immunohistochemistry (Goksøyr et al., 1987; Husøy et al., 1994a; Smolowitz et al., 1989; Stegeman et al., 1991). Although it is now more than 10 years since the first CYP1A proteins were purified, there is still only a limited set of antibody probes available. Polyclonal antibodies have been prepared against the four CYP1A species purified, and a set of monoclonal antibodies were prepared against the scup CYP1A protein by Stegeman and coworkers (Park et al., 1986). Their MAb 1-12-3 has shown an impressively broad species cross-reactivity which includes humans, rodents, marine mammals, reptiles, amphibians and fish (see Stegeman & Hahn, 1994). Recently, we have prepared a monoclonal antibody (NP-7) against cod CYP1A, which was shown to be efficient in detecting CYP1A in diverse fish samples including hagfish, but with somewhat lower reactivity towards perciform fishes and flatfish (Husøy et al., 1994b; 1994c). This antibody is now made commercially available.

Probes for detecting CYP1A mRNA levels in fish became available when the first fish *CYP1A* gene was cloned (Heilmann et al., 1988). This and similar probes have since been used in different types of assays to monitor the transcript levels of CYP1A in induced fish (Haasch et al., 1989; Renton & Addison, 1992; Grøsvik et al., 1993). Recently, the highly sensitive reverse transcriptase-polymerase chain reaction (RT-PCR) assay has been applied to CYP1A messenger detection in fish (Wirgin et al., 1993; Campbell & Devlin, 1994). Although sample handling is more tedious and critical in mRNA analyses, these techniques may be useful supplements to catalytic and immunochemical CYP1A detection methods mentioned above.

Table 3. Detection methods for CYP1A in fish[a]

Detection level	Probe	Assay
CYP1A enzyme	Catalytic assay	EROD, AHH
CYP1A protein	Antibody	Western blotting, ELISA, immunohistochemistry
CYP1A mRNA	cDNA, oligonucleotide	Northern blotting, RT-PCR, in situ hybridization

a) Abbreviations used: EROD, 7-ethoxyresorufin O-deethylase; AHH, aryl hydrocarbon hydroxylase; ELISA, enzyme-linked immunosorbent assay; RT-PCR, reverse transcriptase-polymerase chain reaction.

Factors affecting the CYP1A response in fish

Several endogenous processes (e.g. sexual maturation and developmental stage) as well as physical or abiotic factors (e.g. temperature and season) are known to cause background variations in CYP1A levels and activities in natural fish populations (for a recent review, see Andersson and Förlin, 1992). These have to be well characterized and taken into account when employing the CYP1A response in biomarker studies. Below I will briefly summarize the most important of these factors. In addition to those mentioned here, a range of other factors exists that may also influence this system. These include light (photoperiod), pressure, salinity (osmotic pressure), and stress factors like handling, crowding, noise, ionizing radiation etc. Studies on the effects of these factors are virtually non-existent, and their impact on background variabilities similarly difficult to evaluate.

Temperature

Fish inhabit waters covering a wide range of temperatures, from the freezing point of sea water in Arctic regions (- 1.9°C) to hot tropical ponds (35-40°C). Although many species minimize their exposure to different temperatures by migration, most fish will experience large fluctuations in water temperature over a period of time. To compensate for these changes, cells have various options for regulating the levels of enzymes and the microenvironment they work in. Temperature compensatory responses of CYP1A, generally including elevated enzyme activities and/or CYP1A levels in fish acclimated at lower temperatures, have been reported in several laboratory studies (see Andersson and Förlin, 1992; Sleiderink et al., 1994a), although the response is less pronounced in some species, e.g. turbot, *Scopthalmus maximus* (Skjegstad et al., 1994).

In field studies the situation is more complex, involving other factors discussed below as well. However, in large-scale surveys of CYP1A in dab from the North Sea, Sleiderink et al.,(1994a) observed a clear temperature-dependence of CYP1A levels in summer months, when stratification of the water column at some stations resulted in large differences in temperature. In spite of this effect, a correlation with tissue levels of PCBs could be extracted from the data (ibid., Fig. 2), using careful statistical analyses.

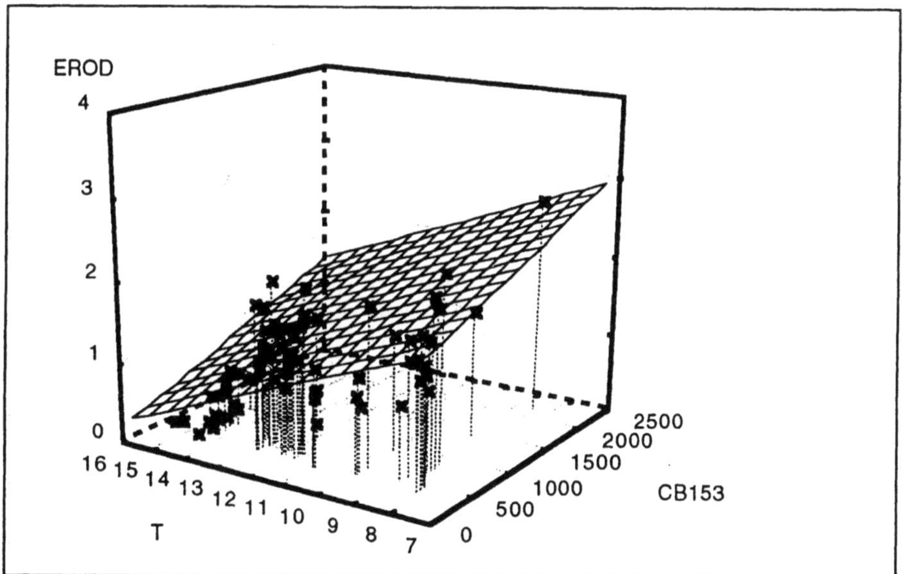

Figure 2: Relationship between EROD activity (nmol/min/mg protein) in liver fractions of mature male dab from the southern North Sea, PCB-153 levels in muscle (ng/g lipid), and bottom water temperature (data from Sleiderink et al., 1994a) The data show an inverse relationship between water temperature and EROD activity, whereas PCB-153 levels are positively related to EROD activity. Multiple regression analysis gives a relationship between the data represented by the grid.

Season and Sex

Seasonal changes have been observed in the CYP1A system of several species under natural conditions (Lindström-Seppä, 1985; Tarlebø et al., 1985; Edwards et al., 1988; George et al., 1990; Larsen et al., 1992). In other studies, fish have been kept at laboratory conditions with constant temperature, but with a natural photoperiod (Förlin and Haux, 1990, Arukwe & Goksøyr, 1994). Many of these studies have included the period of sexual maturation, when sex differences also become apparent. It is obvious from these studies that seasonal variations and sexual maturation affect background levels of CYP1A, with a general depression of CYP1A levels that is most pronounced in females during spawning time. Although several laboratory studies have shown that the CYP1A system of fish responds to inducers even at the peak of reproduction (Collier et al., 1986; Goksøyr and Larsen, 1991, Arukwe & Goksøyr, 1994), it is not clear how the induction response is affected by these factors in field situations with more realistic levels of inducers (e.g. Elskus et al., 1989; Collier et al., 1992a).

Based on data from several sampling surveys in the North Sea, Sleiderink et al. (1994b) have evaluated the sensitivity of the CYP1A responses in dab at different periods of the year. The conclusion with this species, is that highest sensitivity is obtained during autumn, when stratification of the water column in the North Sea is lacking, and the influence of reproduction is still low. The use of male fish is recommended as a precautionary rule.

Diet

Diet and nutrition are known to influence the P450 system in mammals (e.g. Yang et al., 1992). In fish, only a few reports have focused on conncetions between dietary status and xenobiotic metabolism and effects in fish. Effects of dietary factors on the fish CYP1A system observed so far include lipids (Ankley et al., 1989; George & Henderson, 1992), vitamins and antixoidants (Ankley et al., 1989), trace minerals (iron, Goksøyr et al., 1994), terpenoids (Vrolijk & Chen, 1994) as well as carbohydrates (Goksøyr & Drægni, unpublished results) and pigments such as astaxanthin (Husøy, Kristiansen, Torrisen & Goksøyr, unpublished results). In field studies and surveys, these effects may signficantly influence observations of site differences, as fish in different areas may feed on different prey organisms with different levels of such dietary constituents.

Species differences

Although the number of fish species investigated to date with respect to their P450 system is small compared with the total number of species thought to exist (approx. 25,000), notable diferences between species in both basal CYP1A levels and activities, as well as induction responses to certain pollutants, have been observed (Goksøyr et al., 1987). Examples include the low response of cod to PCBs and TCDD, as opposed to other species (Beyer et al., manuscript in prep.; Egaas et al., 1994; Hektoen et al., 1994), the refractoriness of toadfish CYP1A to PAH induction in the field (Collier et al., 1993c), as well as differences in biotransformation activities between two flatfish species from Puget Sound (Collier et al., 1992b). Genetic differences between populations within a given species may also become apparent when these issues are addressed in future investigations.

Quality control in field studies

Taken together, the influence of the factors mentioned above may significantly affect results obtained from field studies. If not carefully planned, observations and conclusions resulting from such studies may be totally non-informative or, worse, even misleading. It is therefore important that planning and implementation of field studies include an evaluation of possible influencing factors in the area. If the study area involves a gradient from a local pollution source, sites (including relatively pristine reference sites) should be selected so as to represent similar physical environments in terms of water temperature, depth, and salinity. Bottom conditions, and thereby feeding possibilities, should be similar, and enough time and resources should be set aside for the collection of fish, so that a sufficient number of individuals of similar size, sex and age of the target species are caught. Experienced field workers know the difficulties in catching the right number of the right species in a short time, and sometimes a less than optimal catch has to be accepted.

In addition to the problems of the catch, field studies involving sampling of feral fish include another inherent uncertainty. The fish caught at a given station may not necessarily represent fish living at the site more permanently. Several fish species, although temporarily stationary, may migrate over certain distances e.g. in conncetion with season and reproduction (e.g. Rijnsdorp et al., 1992). This problem is more difficult to resolve, but can be attacked by a different strategy. Caging of fish has been used in a few field studies over the past years, both in freshwater and marine areas (e.g. Lindström-Seppä and Oikari, 1989; Haasch et al., 1993; Goksøyr et al., 1994; Beyer et al., in prep.). By placing subpopulations of a genetically homogeneous stock into cages moored at sites with similar physical characteristics, this approach ensures that fish has been stationary over the period of the study, and that the fish caught are of the right size and species. Since the fish used will start out with the same clean prehistory, the resulting response will reflect the bioavailability of contaminants at the site and their inducing effects. In prolonged exposure situations, cages should be designed so as to allow feeding, e.g. by allowing contact with benthic organisms, or by attracting prey organisms to the cage by light.

Extrapolation from molecular biomarker to higher level reponses

A challenge of biomarker development is to define the significance of environmental induction. What can be interpreted from the CYP1A induction response in terms of the health of the individual, of the population, and of the ecosystem? It is as yet uncertain how far we will be able to go in answering these questions. Clearly, there is still a lack of information as to how responses at molecular levels are integrated and transferred upwards in the biological hierarchy. At the individual level, the effect is dual: Increased CYP1A activity means increased capability of disposing with contminants that are substrates of the enzyme system. However, at the same time, this means an increased risk of producing the activated intermediates involved in adduct formation and carcinogenesis. Higher prevalences of cancers may affect population structure, and thereby have effects at the ecosystem level. Significant correlations have been observed between CYP1A activities and early/preneoplastic hepatic lesions in English sole from the Puget Sound area (Myers et al., 1994).

Other links may be through the endocrine system, signal transduction or immune competence. Depending on the target cells of CYP1A induction, intricate balances of hormone or signal production or immune system functioning may be disabled. These may include the metabolism of steroids and fatty acids (arachidonic acid), as well as the control of nitric oxide (NO) levels. For example, using immunohistochemistry, CYP1A induction has been shown to occur in fish pituitary, a central organ for hormonal control of growth and reproduction (Andersson et al., 1993). Responses have also been observed in endocrine cells of the kidney and in the spleen of cod (Husøy et al., 1994a), and in dermal chemoreceptors and the olfactory epithelium in the nose of topminnows (Smolowitz et al., 1992). A general target cell type has been found in the endothelium of many different organs (Stegeman & Hahn, 1994; Husøy et al., 1994a), serving important functions in the communication of the interior cells with body fluids. Other studies have shown that hormonal metabolism may indeed be affected by induction, and Nebert (Nebert, 1991) has proposed a role for the CYP1A response in regulation of growth, differentiation, homeostasis, and neuroendocrine functions. If correct, the significance of CYP1A induction is only a question of dose and strength. However, a lot more work needs to be done before we can correctly describe all the mechanisms involved.

Conclusions

Almost 20 years of research has provided a fundament of knowledge and experience in the use of CYP1A induction in fish as an indicator for aquatic pollution. Laboratory experiments and field studies have shown its promises as a biomarker as well as its limitations. Bearing these in mind, carefully controlled investigations looking for a CYP1A response in fish is becoming a useful component both of environmental monitoring programmes as well as ecotoxicological testing with sublethal endpoints. Although we still have limited knowledge about the ecological implications of this response, with support from the precautionary principle the CYP1A induction response itself should be sufficient for raising alarm and concern about the health of the environment, allowing for more detailed (and expensive) chemical and biological analyses to paint the full picture of the situation.

Acknowledgements

I want to thank the Norwegian Research Council (NFR), Statoil, VISTA, the State Pollution Authority (SFT) and NUFU for research grants to projects and studies referred to in the text. The collaboration of my coworkers at the Laboratory of Marine Molecular Biology and a number of colleagues elsewhere is greatly appreciated.

References

Addison, R.F. & A.J. Edwards, 1988. Hepatic microsomal mono-oxygenase activity in flounder *Platichthys flesus* from polluted sites in Langesundfjord and from mesocosms experimentally dosed with diesel oil and copper. Mar. Ecol. Prog. Ser. 46: 51-54.

Ahokas, J.T., O. Pelkonen & N.T. Kärki, 1975. Metabolism of polycyclic hydrocarbons by a highly active aryl hydrocarbon hydroxylase system in the liver of a trout species. Biochem. Biophys. Res. Comm. 63: 635-641.

Andersson, T. & L. Förlin, 1992. Regulation of the cytochrome P450 enzyme system in fish. Aquat. Toxicol. 24: 1-20.

Andersson, T., L. Förlin, J. Härdig & Å. Larsson, 1988. Physiological disturbances in fish living in costal water polluted with bleached kraft pulp mill effluents. Can. J. Fish. Aquat. Sci. 45: 1525-1536.

Andersson, T., L. Förlin, S. Olsen, A. Fostier & B. Breton, 1993. Pituitary as a target organ for toxic effects of P4501A1 inducing chemicals. Mol. Cell. Endoc. 91: 99-105.

Ankley, G.T., V.S. Blazer, S.M. Plakas & R.E. Reinert, 1989. Dietary lipid as a factor modulating xenobiotic metabolism in channel catfish (*Ictalurus punctatus*). Can. J. Fish. Aquat. Sci. 46: 1141-1146.

Arukwe, A. & A. Goksøyr, 1994. Changes in turbot (*Scophthalmus maximus*) hepatic cytochrome P450 system during a reproductive cycle. Abstract, 15th Annual Conf. of ESCPB, Genova, Italy, Sep. 20-23, 1994.

Berndtson, A.K. & T.T. Chen, 1994. Two unique *CYP1* genes are expressed in response to 3-methylcholantrene treatment in rainbow trout. Arch. Biochem. Biophys. 310: 187-195.

Beyer, J. & A. Goksøyr, 1993. Cytochrome P450 Observations in Gulf Fish. Marine Pollution Bulletin 27: 293-296.

Buhler, D.R. & D.E. Williams, 1989. Enzymes involved in metabolism of PAH by fishes and other aquatic animals: Oxidative enzymes (or phase I enzymes). Metabolism of polycyclic aromatic hydrocarbons in the aquatic environment. Boca Raton, Florida, CRC Press. pp. 151-184.

Campbell, P.M. & R.H. Devlin, 1994. Induction of P450 1A1 in the livers and gonads of juvenile chinook. Abstract, 3rd Int. Marine Biotechnology Conference, Tromsø, Norway, Aug. 7-12, 1994.

Collier T.K., B.F. Anulacion, J.E. Stein, A. Goksøyr & U. Varanasi, 1994. A field evaluation of cytochrome P450 1A as a biomarker of contaminant exposure in three species of flatfish. Environ. Toxicol. Chem., in press.

Collier T.K., B.F. Anulacion, A. Goksøyr, C. Grunwald, I. Wirgin & U. Varanasi, 1993a. Relationships between induction of cytochrome P450 1A and contaminant residues in feral teleosts. Abstract, 7th Int. Symp. on Responses of Marine Org. to Pollutants. Gøteborg, Sweden, April 20-22, 1993.

Collier, T.K., M.M. Krahn, C.A. Krone, L.L. Johnson, M.S. Myers, S.L. Chan & U. Varanasi, 1993b. Oil exposure and effects in subtidal fish following the *Exxon Valdez* oil spill. 1993 International Oil Spill Conference, Tampa, Florida, US,

Collier, T.K., J.E. Stein, A. Goksøyr, M.S. Myers, J.W. Gooch, R.J. Huggett & U. Varanasi, 1993c. Biomarkers of PAH exposure in oyster toadfish (*Opsanis tau*) from the Elizabeth River, Virginia. Environ. Sci. 2: 161-177.

Collier, T.K., J.E. Stein, H.R. Sanborn, T. Hom, M.S. Myers, and U. Varanasi, 1992a. Field studies of reproductive success in English sole (*Parophrys vetulus*): Correlations with bioindicators of maternal contaminant exposure. Sci. Total Environ. 116:169-185.

Collier, T.K., S.V. Singh, Y.C. Awasthi & U. Varanasi, 1992b. Hepatic xenobiotic metabolizing enzymes in two species of benthic fish showing different prevalences of contaminant-associated liver lesions. Toxicol. Appl. Pharmacol. 113: 319-324.

Collier, T.K., J.E. Stein, R.J. Wallace & U. Varanasi, 1986. Xenobiotic metabolizing enzymes in spawning English sole (*Parophrys vetulus*) exposed to organic-solvent extracts of marine sediments from contaminated and reference areas. Comp. Biochem. Physiol. 84C: 291-298.

Collier, T.K. & U. Varanasi, 1991. Hepatic activities of xenobiotic metabolizing enzymes and biliary levels of xenobiotics in English sole (*Parophrys vetulus*) exposed to environmental contaminants. Arch. Environ. Contam. Toxicol. 20: 462-473.

Edwards, A.J., R.F. Addison, D.E. Willis & K.W. Renton, 1988. Seasonal Variation of hepatic mixed function oxidases in winter flounder (*Pseudopleuronectes americanus*). Marine Environ. Res. 26: 299.309.

Egaas E., N.O. Svendsen, M. Sandvik, J.U. Skåre, A. Bernhoft, H. Hektoen, A. Goksøyr, J. Beyer & K. Hylland, 1994. Biotransformation enzymes as biomarkers for pesticides in fish, a comparative study. Abstract, 3rd Int. Marine Biotechnology Conference, Tromsø, Norway, Aug. 7-12, 1994.

Eggens M. & F. Galgani, 1992. Ethoxyresorufin O-deethylase (EROD) activity in flatfish: Fast determination with a fluorescence plate reader. Marine Environ. Res. 33: 213-221.

Elskus, A.A., J.J. Stegeman, L.C. Susani, D. Black, R.J. Pruell & S.J. Fluck, 1989. Polychlorinated biphenyls concentration and cytochrome P-450E expression in winter flounder from contaminated environments. Marine Environ. Res. 28: 25-30.

Förlin, L., L. Balk, M. Celander, S. Bergek, M. Hjelt, C. Rappe, C. de Witt & B. Jansson, 1991. Biotransformation enzyme activities, and PCDD and PCDF levels in pike caught in a Swedish lake. Marine Environ. Res. 34: 169-174.

Förlin, L., A. Goksøyr & A.-M. Husøy, 1994. Cytochrome P450 monooxygenase as indicator of PCB/dioxin like compounds in fish. Biological Monitoring of Estuarine and Coastal Waters. Boca Raton, Florida, CRC Press. (In press).

Förlin, L. & C. Haux, 1990. Sex differences in hepatic cytochrome P-450 monooxygenase activities in rainbow trout during an annual reproductive cycle. J. Endocrinol. 124: 207-213.

George, S., P. Young, M. Leaver & D. Clarke, 1990. Activities of pollutant metabolising and detoxication systems in the liver of the plaice, Pleuronectes platessa: Sex and seasonal variations in non-induced fish. Comp. Biochem. Physiol. 96C: 185-192.

George S. & J. Henderson, 1992. Influence of dietary PUFA content on cytochrome P450 and tyransferase activities in Atlantic salmon (*Salmo salar*). Marine Environ. Res. 34: 127-131.

George S., J. Wright, J. Conroy & J. Christiansen, 1994. Validation of piscine cytochrome P450 1A (CYP1A) responses in low temperature environments to North Sea crude oil pollution. Abstract, 3rd Int. Marine Biotechnology Conference, Tromsø, Norway, Aug. 7-12, 1994.

Goksøyr, A., 1985. Purification of hepatic microsomal cytochromes P-450 from ß-naphthoflavone-treated Atlantic cod *(Gadus morhua)*, a marine teleost fish. Biochim. Biophys. Acta 840: 409-417.
Goksøyr, A., 1991. A semi-quantitative cytochrome P450IA1 ELISA: A simple method for studying the monooxygenase induction response in environmental monitoring and ecotoxicological testing of fish. Sci. Tot. Envir., 101: 255-262.
Goksøyr, A., T. Andersson, D.R. Buhler, J.J. Stegeman, D.E. Williams & L. Förlin, 1991. Immunochemical cross-reactivity of ß-naphtoflavone-inducible cytochrome P-450 (P4501A) in liver microsomes from different fish species and rat. Fish. Physiol. Biochem. 9: 1-13.
Goksøyr, A., T. Andersson, T. Hansson, J. Klungsøyr, Y. Zhang & L. Förlin, 1987. Species characteristics of the hepatic xenobiotic and steroid biotransformation systems of two teleost fish, Atlantic cod *(Gadus morhua)* and rainbow trout *(Salmo gairdneri)*. Toxicol. Appl. Pharmacol. 89: 347-360.
Goksøyr, A., J. Beyer, A.M. Husøy, H.E. Larsen, K. Westrheim, S. Wilhelmsen & J. Klungsøyr, 1994. Accumulation and effects of aromatic and clorinated hydrocarbons in juvenile Atlantic cod (*Gadus morhua*) caged in a polluted fjord (Sørfjorden, Norway). Aquat. Toxicol. 29: 21-35.
Goksøyr, A., M. Bjørnevik & A. Maage, 1994. Effects of dietary iron concentrations on the cytochrome P450 system of Atlantic salmon (*Salmo salar*). Can. J. Fish. Aquat. Sci. 51: 315-320.
Goksøyr, A. & L. Förlin, 1992. The cytochrome P450 system in fish, aquatic toxicology, and environmental monitoring. Aquat. Toxicol. 22: 287-312.
Goksøyr, A. & H.E. Larsen, 1991. The cytochrome P450 system of Atlantic salmon *(Salmo salar)*: I. Basal properties and induction of P450 1A1 in liver of immature and mature fish. Fish Physiol. Biochem. 9: 339-349.
Grøsvik B.E., H.E. Larsen & A. Goksøyr, 1993. Effect of piperonyl butoxide and ß-naphthoflavone on cytochrome P450 1A expression and activity in Atlantic salmon (*Salmo salar*). Abstract, 7th Int. Symp. on Responses of Marine Org. to Pollutants. Gøteborg, Sweden, April 20-22, 1993.
Haasch, M. L.,Wejksnora, P. J.,Stegeman, J. J., & Lech, J. J. (1989). Cloned rainbow trout liver $P_1$450 complementary DNA as a potential environmental monitor. Toxicol. Appl. Pharmacol., 98, 362-368.
Hahn M.E., A. Poland, E. Glover & J.J. Stegeman, 1994. Photoaffinity labelling of the Ah-receptor: phylogenetic survey of diverse vertebrate and invertebrate species. In press.
Haux, C. & L. Förlin, 1988. Biochemical methods for detecting effects of contaminants on fish. Ambio 17: 376-380.
Heilmann, L.J., Y.-Y. Sheen, S.W. Bigelow & D.W. Nebert, 1988. Trout P450IA1: cDNA and deduced protein sequence, expression in liver, and evolutionary significance. DNA 7: 379-387.
Husøy, A.-M., M.S. Myers, M.L. Willis, T.K. Collier, M. Celander & A. Goksøyr, 1994a. Immunohistochemical localization of CYP1A and CYP3A isozymes in hepatic and extrahepatic tissues of Atlantic cod. Toxicol. Appl. Pharmacol., in press.
Husøy, A.-M., N.P. Maaseide & A. Goksøyr, 1994b. Monoclonal antibodies to cod CYP1A: a tool in aquatic toxicology. Abstract, 3rd Int. Marine Biotechnology Conference, Tromsø, Norway, Aug. 7-12, 1994.
Husøy A.-M., K.O. Strætkvern, N.P. Maaseide, S.O. Olsen & A. Goksøyr, 1994c. Polyclonal and monoclonal antibodies against immunoaffinity-purified cod cytochrome P450 1A. Marine Mol. Biol. Biotech., submitted.
Haasch, M.L., R. Prince, P.L. Wejksnora, K.L. Cooper & J.L. Lech, 1993. Caged and wild fish: Induction of hepatic cytochrome P-450 (CYP1A1) as an environmental biomonitor. Environmental Toxicol. Chem. 12: 885-95.
Ioannides, C., 1990. Induction of cytochrome P-450 I and its influences in chemical carcinogenesis. Biochem. Soc. Trans. 18: 32-34.

Klotz, A.V., J.J. Stegeman & C. Walsh, 1983. An aryl hydrocarbon hydroxylating hepatic cytochrome P-450 from the marine fish Stenotomus chrysops. Arch. Biochem. Biophys. 226: 578-592.
Larsen, H.E., M. Celander & A. Goksøyr, 1992. The cytochrome P450 system of Atlantic salmon (*Salmo salar*): II. Variations in hepatic catalytic activities and isozyme patterns during an annual reproductive cycle. Fish Physiol. Biochem. 10: 291-301.
Lindström-Seppä, P., 1985. Seasonal variation of the xenobiotic metabolizing enzyme activities in the liver of male and female vendace (*Coregonus albula* L.). Aquat. Toxicol. 6: 323-331.
Lindström-Seppä, P. & A. Oikari, 1989. Biotransformation and other physiological responses in whitefish caged in a lake receiving pulp and paper mill effluents. Ecotoxicol. Environ. Safety 18: 191-203.
Morrison, H., M. Olekziac, M. Sogan & J.J. Stegeman, 1994. Identification of two new fish CYP1A sequences and phylogenetic analysis of the CYP1A subfamily. J. Mol. Evol., submitted.
Myers, M.S., L.L. Johnson, T. Hom, T.K. Collier, J.E. Stein & U. Varanasi, 1994. Toxicopathic hepatic lesions in subadult English sole (Pleuronectes vetulus) from Puget Sound, Washington, USA: relationships with other biomarkers of contaminant exposure. Submitted.
Nebert, D.W., 1991. Proposed role of drug-metabolizing enzymes: Regulation of steady state levels of the ligands that effect growth, homeostasis, differentiation and neuroendocrine functions. Mol. Endocrin. 5: 1203-1214.
Nebert, D.W. & F.J. Gonzalez, 1987. P450 genes: Structure, evolution, and regulation. Ann. Rev. Biochem. 56: 945-993.
Nebert, D.W., D.R. Nelson & R. Feyereisen, 1989. Evolution of the cytochrome P-450 genes. Xenobiotica 19: 1149-1160.
Nelson, D.R., T. Kamataki, D.J. Waxman, F.P. Guengerich, R.W. Estabrook, R. Feyereisen, F.J. Gonzalez, M.J. Coon, I.C. Gunsalus, O. Gotoh, K. Okuda & D.W. Nebert, 1993. The P450 superfamily - update on new sequences, gene mapping, accession numbers, early trivial names of enzymes, and nomenclature. DNA Cell Biol. 12: 1-51.
Park, S. S., Miller, H., Klotz, A. V., Kloepper-Sams, P. J., Stegeman, J. J., & Gelboin, H. V., 1986. Monoclonal antibodies to liver microsomal cytochrome P-450E of the marine fish *Stenotomus chrysops* (scup): Cross reactivity with 3-methylcholanthrene induced rat cytochrome P-450. Arch. Biochem. Biophys. 249: 339-350.
Payne, J.F., 1976. Field evaluation of benzopyrene hydroxylase induction as a monitor for marine pollution. Science 191: 945-946.
Payne, J.F. & W.R. Penrose, 1975. Induction of aryl hydrocarbon (benzo(a)pyrene) hydroxylase in fish by petroleum. Bull. Env. Cont. Toxicol. 14: 112-116.
Payne, J.F., L.L. Fancey, A.D. Rahimtula & E.L. Porter, 1987. Review and perspective on the use of mixed-function oxygenase enzymes in biological monitoring. Comp. Biochem. Physiol. 86C: 233-245.
Pesonen, M. & T. Andersson, 1992. Toxic effects of bleached and unbleached paper mill effluents in primary cultures of rainbow trout hepatocytes. Ecotox. Environ. Safety 24: 63-71.
Price, A.R.G. & J.H. Robinson (eds.), 1993. The 1991 Gulf War: Coastal and Marine Environmental Consequences. Marine Poll. Bull. vol. 27.
Puga, A. & D.W. Nebert, 1989. Evolution of the P-450 gene superfamily and regulation of the murine Cyp1a1 gene. Biochem. Soc. Trans. 18: 7-10.
Renton, K.W. & R.F. Addison, 1992. Hepatic microsomal mono-oxygenase activity and P450IA mRNA in North Sea dab *Limanda limanda* from contaminated sites. Mar. Ecol. Prog. Ser. 91: 65-69.
Rijnsdorp, A.D., A.D. Vethaak & P.I. van Leeuwen, 1992. Population biology of dab Limanda limanda in the southeastern North Sea. Mar. Ecol. Prog. Ser. 91: 19-35.
Ritchie, W. & M. O'Sullivan (eds.), 1994. The Environmental Impact of the Wreck of the Braer. Report from the Scottish Office, Edinburgh.

Schenkman, J.B. & H. Greim (eds.), 1993. Cytochrome P450. Handb. Exp. Pharm. vol. 105. Springer-Verlag, Berlin.

Skjegstad, N.L., O. Garatun-Tjeldstø, O. Grahl-Nielsen & A. Goksøyr, 1994. Temperature adaptation of turbot (*Scophthalmus maximus* L.): biochemical studies of cytochrome P450 and fatty acid profiles in liver membranes. Abstract, 15th Annual Conf. of ESCPB, Genova, Italy, Sep. 20-23, 1994.

Sleiderink, H.M., J. Beyer, E. Scholtens, A. Goksøyr, J. Nieuwenhuize, J.M. Van Liere, J.M. Everaarts & J.P. Boon, 1994a. Influence of temperature and environmental contaminants on CYP1A levels in North Sea dab (*Limanda limanda*). Aquatic Toxicology, submitted.

Sleiderink, H.M., I. Oostingh, A. Goksøyr & J.P. Boon, 1994b. Sensitivity of cytochrome P450 1A induction in dab (*Limanda limanda*) of different age and sex as a biomarker for environmental contaminants in the southern North Sea. Arch. Environ. Cont. Toxicol., submitted.

Smolowitz, R.M., Moore, M.J., & Stegeman, J.J., 1989. Cellular distribution of cytochrome P-450E in Winter flounder liver with degenerative and neoplastic disease. Marine Environ. Res. 28: 441-446.

Smolowitz, R.M., Schultz, M.E. & Stegeman, J.J., 1992. Cytochrome P4501A induction in tissues, including olfactory epithelium, of topminnows (*Poeciliopsis spp.*) by waterborne benzo(a)pyrene. Carcinogenesis 13: 2395-2402.

Stebbing, A.R.D., V. Dethlefsen & M. Carr (eds.), 1992. Biological effects of contaminants in the North Sea. Results of the ICES/IOC Bremerhaven Workshop. Mar. Ecol. Prog. Ser. vol. 91, no. 1-3.

Stegeman, J.J., 1989. Cytochrome P450 forms in fish: catalytic, immunological and sequence similarities. Xenobiotica 19: 1093-1110.

Stegeman, J.J., 1993. Cytochrome P450 forms in fish. In: Cytochrome P450 (Schenkman J.B. & H. Greim, eds.). Handb. Exp. Pharm. vol. 105: 279-291. Springer-Verlag.

Stegeman, J.J., Smolowitz, R.M., & Hahn, M.E., 1991. Immunohistochemical localization of environmentally induced cytochrome P450IA1 in multiple organs of the marine teleost *Stenotomus chrysops* (Scup). Toxicol. Appl. Pharmacol. 110: 1-17.

Stegeman, J.J., M. Brouwer, T.D.G. Richard, L. Förlin, B.A. Fowler, B.M. Sanders & P.A. Van Veld, 1992. Molecular responses to environmental contamination: Enzyme and protein systems as indicators of chemical exposure and effect. In: Biomarkers: Biochemical, physiological and histological markers of anthropogenic stress (Huggett, R.J. et al., eds.), pp. 235-335. Lewis Publishers.

Stegeman, J.J. & M.E. Hahn, 1994. Biochemistry and molecular biology of moonoxygenases: Current perspectives on forms, functions and regulation of cytochrome P450 in aquatic species. In: Aquatic toxicology, molecular, biochemical, and cellular perspectives (Malins D.M. & G.K. Ostrander, eds.), pp. 87- 204. Lewis Publishers.

Stegeman, J.J. & P.J. Kloepper-Sams, 1987. Cytochrome P-450 isoenzymes and monooxygenase activity in aquatic animals. Environ. Health Persp. 71: 87-95.

Stegeman, J.J., P.J. Kloepper-Sams & J.W. Farrington, 1986. Monooxygenase induction and chlorobiphenyls in the deep-sea fish *Coryphaenoides armatus*. Science 231: 1287-1289.

Stegeman, J.J. & J.J. Lech, 1991. Cytochrome P-450 monooxygenase systems in aquatic species: Carcinogen metabolism and biomarkers for carcinogen and pollutant exposure. Environ. Health. Persp. 90: 101-109.

Tarlebø, J., J.E. Solbakken & K.H. Palmork, 1985. Variation in hepatic aryl hydrocarbon hydroxylase activity in flounder, *Platichtys flesus*: A baseline study. Helgol. Meer. Unters. 39: 187-199.

van der Weiden, M.E.J., 1993. Cytochrome P450 1A induction in carp as a biological indicator for the aquatic contamination of chlorinated polyaromatics. Ph. D. Thesis, Universiteit Utrecht, the Netherlands.

Van Veld, P.A., D.J. Westbrook, B.R. Woodin, R.C. Hale, C.L. Smith, R.J. Huggett & J.J. Stegeman, 1990. Induced cytochrome P-450 in intestine and liver of spot

(*Leiostomus xanthurus*) from a polycyclic aromatic hydrocarbon contaminated environment. Aquatic Toxicol. 17: 119-132.
Vrolijk, N.H. & T.T. Chen, 1994. Characterization and expression of CYP1A in the butterflyfish, Chaetodon capistratus, and its response to ß-naphthoflavone and terpenoid treatment. Abstract, 3rd Int. Marine Biotechnology Conference, Tromsø, Norway, Aug. 7-12, 1994.
Walker, C.H. & D.R. Livingstone (eds.), 1992. Persistent pollutants in marine ecosystems. Pergamon Press, Oxford.
Williams, D.E. & D.R. Buhler, 1982. Purification of cytochromes P-448 from ß-naphthoflavone-treated rainbow trout. Biochim. Biophys. Acta 717: 398-404.
Williams, D.E. & D.R. Buhler, 1984. Benzo(a)pyrene-hydroxylase catalyzed by purified isozymes of cytochrome P-450 from ß-naphthoflavone-fed rainbow trout. Biochem. Pharmacol. 33: 3743-3753.
Wirgin, I.I., C. Grunwald & S. Courtenay, 1993. Time and dose response of CYP1A mRNA induction in Atlantic tomcod exposed to 2,3,7,8-TCDD, 3,3',4,4'-TCB and BNF. Abstract, 7th Int. Symp. on Responses of Marine Org. to Pollutants. Gøteborg, Sweden, April 20-22, 1993.
Yang, C., J.F. Brady & J.-Y. Hong, 1992. Dietary effects on cytochromes P450, xenobiotic metabolism, and toxicity. FASEB J. &: 737-744.

Receptor-mediated toxicity

(Chair: Prof. Dr. K. Savolainen, Finland)

Cellular and Molecular Biology of Aryl Hydrocarbon (Ah) Receptor - Mediated Gene Expression

S. Safe and V. Krishnan
Veterinary Physiology and Pharmacology, Texas A&M University, College Station, TX 77843-4466

Abstract

2,3,7,8-Tetrachlorodibenzo-*p*-dioxin (TCDD) and related compounds elicit diverse toxic and biochemical responses in laboratory animals and mammalian cells in culture. TCDD induces CYP1A1 gene expression and results of extensive research have delineated the molecular mechanism of this response. In target cells, TCDD initially binds to the aryl hydrocarbon (Ah) receptor which accumulates in the nucleus as an Ah-receptor:aryl hydrocarbon nuclear translocator (Arnt) protein heterodimeric complex. The nuclear Ah receptor complex acts as a ligand-induced transcription factor which binds to *trans*-acting genomic dioxin/xenobiotic responsive elements (DREs/XREs) located in the 5'-regulatory region upstream from the initiation start site and this interaction results in transactivation of gene transcription. DREs have been identified in several other genes which are induced by TCDD, including CYP1A2, aldehyde-3-dehydrogenase, NAD(P)H quinone oxidoreductase, and glutathione S transferase Ya and similar induction response pathways have been observed or proposed. However, TCDD and other Ah receptor agonists also inhibit expression of several genes and research in this laboratory has investigated inhibition of estrogen (E2)-induced genes including uterine epidermal growth factor, c-*fos* protooncogene, and the progesterone receptor, estrogen receptor (ER) and cathepsin D genes in human breast cancer cell lines. In MCF-7 human breast cancer cells, E2 induces cathepsin D gene expression and this is associated with formation of an ER/Sp1 complex at the sequence in the promoter region (-199/-165) of this gene. Within 30 min TCDD causes a rapid inhibition of E2-induced cathepsin D gene expression in MCF-7 cells. Moreover, using a series of synthetic oligonucleotides which include the wild-type ER/Sp1 and various mutants, it was shown by gel electromobility shift and transient transfection assays that the nuclear Ah receptor complex binds to an imperfect DRE located between the ER and Sp1 binding sequences. This interaction results in disruption of the ER/Sp1 complex and inhibition of E2-

induced gene expression. These results illustrate that the nuclear Ah receptor complex also exhibits activity as a negative transcription factor via a mechanism which is similar to that reported for Ah receptor-mediated induction of gene expression.

Introduction

2,3,7,8-Tetrachlorodibenzo-*p*-dioxin (TCDD) and related halogenated aromatic compounds are formed as by-products in numerous industrial and combustion processes (Fiedler, 1993) and have been identified as contaminants in almost every component of the global ecosystem (Rappe, 1992, 1993). Human exposure to halogenated aromatics is characterized by development of chloracne and induction of some hepatic enzyme activities (Kimbrough, 1985, 1987). Halogenated aromatics elicit a diverse spectrum of toxic and biochemical responses in laboratory animals and mammalian cells in culture including lethality, dermal toxicity (including chloracne), hepatotoxicity and porphyria, immunotoxicity, developmental and reproductive toxicity, carcinogenesis, tissue-specific hypo- and hyperplastic responses, body weight loss, modulation of endocrine response pathways, and induction and inhibition of diverse enzyme activities (Poland and Knutson, 1982; Whitlock, 1986, 1987, 1990; Safe, 1986, 1990; Goldstein and Safe, 1989). The induction of these biochemical and toxic responses by halogenated aromatics are sex-, strain-, species-, age- and tissue/organ-specific. For example, in long term dietary feeding studies with TCDD in male and female Sprague-Dawley rats, hepatocellular carcinomas were observed only in female rats (Kociba et al., 1978). In contrast, TCDD is carcinogenic to both male and female Osborne-Mendel rats (NIH, 1982). The response-specificity of halogenated aromatics is not fully understood and is being investigated in several laboratories.

Characterization of the Ah Receptor

Poland, Nebert and coworkers investigated the effects of halogenated and polynuclear aromatic hydrocarbons in genetically inbred mice (Poland et al., 1975; Poland and Glover, 1975; Nebert et al., 1976) and their results demonstrated that responsiveness to aryl hydrocarbons (Ahs) such as 3-methylcholanthrene (MC) and TCDD were strain-dependent. For example, in C57BL/6 mice, both TCDD and MC induced hepatic microsomal aryl hydrocarbon hydroxylase (AHH) activity whereas in DBA/2 mice only TCDD was active as an inducer at a dose approximately ten times higher than required for C57BL/6 mice. In genetic crosses and backcrosses, Ah-responsiveness is inherited as a simple autosomal dominant trait suggesting that regulation of Ah-responsiveness is controlled by a single locus (or closely linked loci). It was hypothesized that responsiveness to TCDD, MC and related compounds was due to their interaction with an intracellular receptor and this suggestion was also supported by the structure-activity relationships observed for TCDD and related halogenated aromatics for induction of AHH and δ-aminolevulinic acid synthetase activities in chick embryo liver (Poland and Glover, 1973). Poland and coworkers (1976) first identified the Ah receptor by demonstrating that radiolabeled TCDD specifically-bound with high affinity to a hepatic protein from C57BL/6 mice and this report triggered a multitude of studies on the structure, function and molecular biology of this protein which was designated as the Ah receptor.

The Ah receptor has been identified in multiple species, organs/tissues and cells and has been extensively characterized using [^{3}H]TCDD and other radioligands (reviewed in Safe, 1988; Landers and Bunce, 1991; Okey et al., 1994). The properties of the Ah receptor are variable and dependent, in part, on the assay conditions; however, the data summarized in Table 1 highlight some of the important characteristics of this protein. Photoaffinity labeling studies indicate that the molecular weight of the Ah receptor is highly variable between species (95- to 124-kDa) and there are also significant intraspecies differences in apparent molecular weights (Poland and Glover, 1987, 1989;

Table 1. Properties of the Ah receptor.

Receptor	Property
Cytosolic Ah receptor complex	270- to 310-kDa; $K_D < 1$ nM; low capacity; 9-10.5S; low DNA binding activity; species-dependent transformation to a DNA-binding form; binds different structural classes of polyaromatic compounds
Transformed cytosolic or nuclear Ah receptor complex	170- to 220-kDa; 5-6.5S; high DNA binding activity
Ah receptor protein	species- and strain-dependent variation in apparent molecular weight (Mr 95 to 124-kDa)

Poland et al., 1987). Velocity sedimentation analysis of the liganded cytosolic Ah receptor on sucrose gradients gives a 9-10.5S specifically-bound peak with apparent molecular masses which vary from 270- to 310-kDa. The unbound cytosolic Ah receptor is a low capacity, high affinity binding protein which exhibits saturable binding to [^{3}H]TCDD and related radioligands. The K_D values for ligand binding to the Ah receptor are generally < 1 nM and results of kinetic and protein dilution assays suggest that the K_D values may be as low as 1 pM (Bradfield and Poland, 1988; Bunce et al., 1988). The 9-10.5S cytosolic Ah receptor binds weakly to DNA; however, both temperature and salt convert or transform the cytosolic Ah receptor into a 5-6.5S form (Mr 170- to 220-kDa) which binds with high affinity to DNA. The transformed cytosolic and nuclear Ah receptor complexes are indistinguishable and have been characterized as a heterodimer containing the Ah receptor and aryl hydrocarbon nuclear translocator (Arnt) proteins (Elferink et al., 1990; Gasiewicz et al., 1991).

The Ah receptor gene from several species has been cloned and this protein has been identified as a member of a family of proteins which include the Sim and Per proteins from *Drosophila* and the Arnt protein (Burbach et al., 1992; Ema et al., 1992; Dolwick et al., 1993; Schmidt et al., 1993). Deletion analysis studies of the murine Ah receptor showed that there are several domains within the Ah receptor which are required for activity and these include: an N-terminal basic helix-loop-helix region which is required for DNA binding; a PAS region common to the Per, Arnt and Sim proteins which also contains A/B repeats; a Q-rich C-terminal region which is important for DNA binding and formation of the transformed Ah receptor complex; and a ligand-binding domain and other regions which enhance receptor function. Hankinson and coworkers (1991) have cloned the gene for the Arnt protein which is required for formation of the nuclear Ah receptor heterodimeric complex and this protein is also a basic helix-loop-helix transcription factor which contains several structural domains which are similar to those reported for the Ah receptor and other members of this group of nuclear proteins. The results of several studies demonstrate that the nuclear Ah receptor complex is a ligand-activated transcription factor which is required for Ah receptor-mediated induction of gene expression (reviewed in Whitlock, 1986, 1987, 1990, 1993; Dolwick and Bradfield, 1993).

Modulation of Gene Expression by TCDD

Induction of CYP1A1 Gene Expression. Induction of CYP1A1 gene expression and related enzyme activities by TCDD and related Ah receptor agonists is one of the most common and sensitive indicators of exposure to these compounds. Early studies with genetically-inbred mice and mutant cell lines indicated that formation of the nuclear Ah receptor was required for subsequent induction of CYP1A1 mRNA levels and protein by Ah receptor agonists (Tukey et al., 1982; Miller et al., 1983). Deletion analysis studies of 5'-upstream sequences in the *CYP1A1* gene promoter and their derived chimeric gene constructs indicated that specific genomic sequences designated as dioxin or xenobiotic responsive elements (DREs/XREs) are required for ligand-induced gene transcription (Gonzalez and Nebert, 1985; Jones et al.,1985, 1986a; Neuhold et al., 1989; Sogawa et al., 1986, Fujisawa-Sehara et al., 1986, 1988). These *cis*-acting enhancer sequences bind the Ah receptor-Arnt heterodimer and this interaction is required for transactivation of the *CYP1A1* gene. A core nucleotide sequence of 7 bases has been identified in most XREs/DREs and mutational analysis studies indicate that a 4-base sequence is required for binding of the nuclear heterodimer (Whitlock, 1993).

5'-T-GCGTG-3'	5'-CGTG-3'
3'-A-CGCAC-5'	3'-GCAC-5'

Molecular biology studies have demonstrated that Ah receptor-mediated induction of CYP1A1 gene expression is accompanied by several other events including binding of proteins to other 5'-upstream sequences, increased accessibility of the promoter sequences to nuclear proteins, changes in nucleosome and chromatin structure and DNA bending (reviewed in Whitlock, 1993). The molecular biology of CYP1A1-induced gene expression (Fig. 1) has been used as a model for describing Ah receptor-mediated biochemical and toxic responses; however, it is likely that mechanisms associated with modulation of target gene expression by TCDD is dependent on the gene, organization of its corresponding 5'-upstream flanking DNA, and the tissue/cell specific expression of other *trans*-acting factors.

Induction of Other Genes and Enzyme Activities by Ah Receptor Agonists. TCDD, MC and related compounds induce aldehyde dehydrogenase, glutathione S-transferase, glucuronosyl transferase, NAD(P)H quinone oxidoreductase, and CYP1A2-dependent activities. These induction responses are accompanied by enhanced gene expression and the results are consistent with an Ah receptor-mediated response pathway. Moreover, genetic analyses of the 5'-upstream region of the aldehyde dehydrogenase, CYP1A2, glutathione S-transferase, and NAD(P)H quinone oxidoreductase genes have identified XRE sequences that play a role in Ah receptor-mediated induction of gene expression (Rushmore et al., 1990; Quattrochi et al., 1994; Asman et al., 1993;

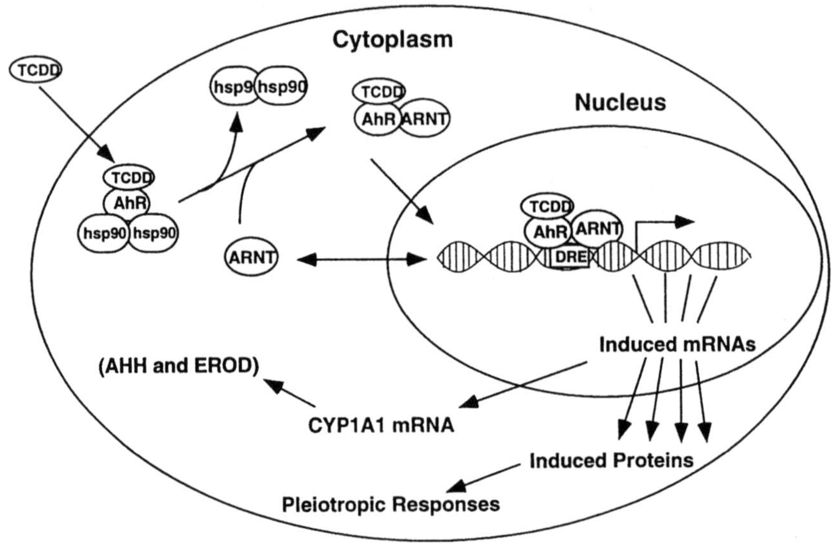

Figure 1. Molecular mechanism of induction of CYP1A1 gene expression by TCDD and related compounds.

Jaiswal, 1994). However, it is also evident that regulation of these genes is characterized by gene- and promoter-specific differences. For example, the glutathione S-transferase *Ya* gene contains a *cis*-acting XRE which is involved in Ah receptor-mediated induction of this gene (Rushmore et al., 1990). The constitutive factor C/EBPα shares a genomic binding site which overlaps with the XRE and the results suggest that cooperative interactions between the Ah receptor and C/EBPα regulate glutathione S-transferase *Ya* gene expression (Pimental et al., 1993). Other studies have reported that TCDD induces transforming growth factor α, interleukin-1β, and plasminogen activator inhibitor-2 gene expression in human keratinocytes; however, elevated expression of these genes is due to post-transcriptional processes (Choi et al., 1991; Sutter et al., 1991; Gaido et al., 1992). In addition, TCDD induces c-*fos* and c-*jun* protooncogene expression in wild-type and mutant mouse Hepa cells suggesting an Ah receptor-independent pathway (Puga et al., 1992).

Inhibition of Gene Expression

Several studies have reported that treatment of animals or cells with TCDD can result in decreased enzyme activities and/or decreased gene expression. For example, during the course of TCDD-induced porphyria, there is a marked decrease in uroporphyrinogen decarboxylase activity and this is accompanied by accumulation of higher carboxylated porphyrins (Kawanishi et al., 1983). Rats treated with high doses of TCDD (125 µg/kg) exhibit reduced hepatic phosphoenol pyruvate carboxy kinase (PEPCK) activity and mRNA levels; glucose-6-phosphatase and tryptophan 2,3-dioxygenase activities are also decreased (Stahl et al., 1993; Weber et al., 1991). It has been hypothesized that the lethal appetite suppression induced by TCDD is due to inhibition of gluconeogenesis and PEPCK gene expression. TCDD decreases a number of other activities including glucocorticoid, estrogen, progesterone and epidermal growth factor receptor binding in diverse cells/animals (Romkes et al., 1987; Harper et al., 1994; Hudson et al., 1985; Kärenlampi et al., 1983; Ryan et al., 1989, Madhukar et al., 1984, 1988; Newsted and Geisy, 1993; Astroff et al., 1990), rat uterine c-fos and epidermal growth factor (EGF) receptor mRNA levels, and transforming growth factor β mRNA levels in human keratinocytes (Astroff et al., 1990, 1991; Gaido et al., 1992). Thus, TCDD and related Ah receptor agonists both induce and decrease gene expression and enzyme activities; however, the role of these altered responses in development of TCDD-induced toxicity are unknown. Moreover, the molecular mechanisms of Ah receptor-mediated decreased gene expression have not been determined.

Inhibition of Estrogen (E2)-Induced Responses. Kociba and coworkers (1978) reported that dietary administration of TCDD to female Sprague-Dawley rats for up to 2 years resulted in decreased formation of benign uterine and mammary tumors which were observed at relatively high levels in control animals. Subsequent studies have demonstrated that TCDD inhibits mammary tumor growth in the athymic nude mouse model (Gierthy et al., 1993) and 7,12-dimethylbenzanthracene-induced mammary tumors in female Sprague-Dawley rats (Holcomb et al., 1994). Subsequent studies in the rodent model showed that TCDD and related compounds inhibit several estrogen-induced uterine responses including increased uterine wet weight, peroxidase activity, cytosolic and nuclear progesterone and estrogen receptor binding, EGF receptor binding, EGF receptor mRNA and c-fos mRNA levels (Romkes et al., 1987; Romkes and Safe, 1988; Astroff and Safe, 1988, 1990; Astroff et al., 1990, 1991; DeVito et al., 1992; Umbreit et al., 1988; Umbreit and Gallo, 1988; Gallo et al., 1986). In parallel studies, the antiestrogenicity of TCDD and related compounds has also been investigated in several human breast cancer cell lines. In estrogen receptor (ER)-positive cell lines, typified by MCF-7 and T47D cells, TCDD induces CYP1A1 gene expression and associated enzyme activities including benzo[a]pyrene hydroxylase, ethoxyresorufin *O*-deethylase, and several estradiol hydroxylases (Harris et al., 1990; Spink et al., 1990, 1992, 1994). These cell lines express the Ah receptor which forms a nuclear heterodimeric 6-7S complex. In contrast, ER-negative MDA-MB-231 cells are Ah-nonresponsive even though these cells form a ligand-induced nuclear Ah receptor complex (Harris et al., 1989). Recent studies reported that Ah-responsiveness is restored in MDA-MB-231 cells transiently transfected with a human ER expression plasmid and the results are consistent with cross-talk between the ER and Ah receptor in this cell line (Thomsen et al., 1994).

TCDD inhibits growth of MCF-7 and T47D cells (Biegel and Safe, 1990; Gierthy et al., 1987) but does not modulate proliferation of MDA-MB-231 cells. The antiestrogenic activity of structurally-diverse Ah receptor agonists has been extensively investigated in MCF-7 cells and inhibition of the following E2-induced responses have been reported: postconfluent focus production, secretion of tissue plasminogen activator activity, procathepsin D (52-kDa protein), cathepsin D (34-kDa protein), a 160-kDa protein, PR binding sites, glucose to lactate metabolism, pS2 protein levels, and PR, cathepsin D, ER and pS2 gene expression (Gierthy et al., 1987, 1993; Gierthy and Lincoln, 1988; Biegel and Safe, 1990; Krishnan and Safe, 1993; Krishnan et al., 1994; Wang et al., 1993; Harper et al., 1994; Zacharewski et al., 1994). For several of these responses, the structure-antiestrogenicity relationships for halogenated aromatic hydrocarbons paralleled their structure-Ah receptor binding and other structure-activity relationships (Krishnan and Safe, 1993; Spink et al., 1994). Gierthy and coworkers have proposed that the antiestrogenic activity of TCDD is due to induced oxidative metabolism of E2 and the resulting depletion of cellular hormone levels (Spink et al., 1990, 1992, 1994). Although induced metabolism of E2 may play a role in the antiestrogenic activity of Ah receptor agonists, the following results demonstrate that this response is independent of E2 metabolism.

(i) In rodents treated with an antiestrogenic dose of TCDD, circulating serum E2 levels are not significantly decreased (DeVito et al., 1992; Shiverick and Muther, 1982).

(ii) Different structural classes of Ah receptor agonists elicit antiestrogenic responses at concentrations which do not induce CYP1A1 (Biegel and Safe, 1990; Chaloupka et al., 1992; Liu et al., 1994; Harris et al., 1990).

(iii) TCDD rapidly inhibits some E2-induced responses such glucose to lactate metabolism (Narasimhan et al., 1991) and cathepsin D gene expression prior to induction of CYP1A1 (Krishnan et al., 1994).

Safe and coworkers (1991) proposed that inhibition of breast cancer cell growth may involve multiple pathways (Fig. 2) including:

(i) interaction of the nuclear Ah receptor complex with target sequences in E2-regulated genes, and/or
(ii) induction of gene products which directly or indirectly inhibit E2-induced gene transcription, downregulate the ER, or inhibit mitogen/growth factor (autocrine or paracrine)-induced gene transcription and cell growth.

TCDD and related compounds downregulate the ER (Harris et al., 1990; Wang et al., 1993) and inhibit growth factor-induced proliferation of breast cancer cells (Fernandez and Safe, 1992; Liu et al., 1992) and these responses are consistent with a direct or indirect role of induced gene products (M) in mediating these responses (Fig. 2).

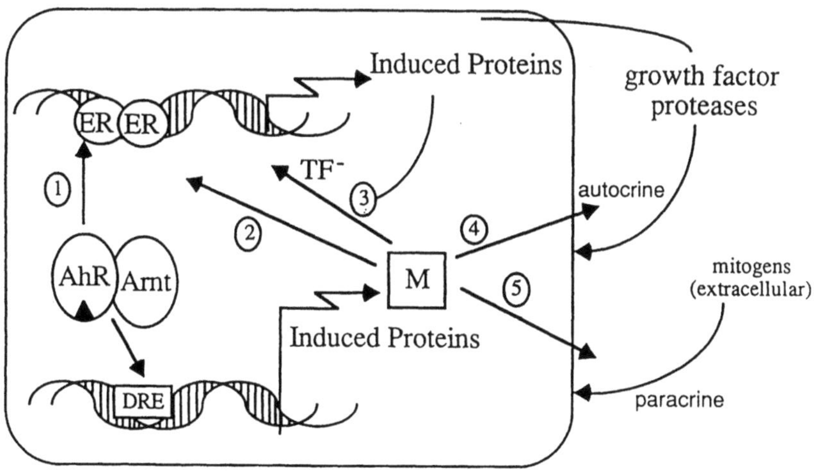

Figure 2. Ah receptor-mediated inhibition of estrogen-induced cell proliferation and gene expression (Safe et al., 1991). Pathways include direct interaction of the nuclear Ah receptor complex with E2-regulated genes (#1), induction of gene products which downregulate the ER (#2), inhibition of gene transcription (#3), and inhibition of autocrine- and paracrine-induced cell proliferation (#4, #5).

Recent studies in this laboratory have investigated the molecular biology of E2-induced cathepsin D gene transcription and inhibition of this response by Ah receptor agonists (Krishnan and Safe, 1993; Krishnan et al., 1994a,b). E2 induces cathepsin D gene expression and deletion analysis of the promoter region has shown that estrogen responsiveness requires the TATA box (-48/-45) and the -366/-13 5'-promoter sequence (Cavailles et al., 1993). Subsequent research in this laboratory identified an ERE half-site (-169/-165) and an Sp1 binding site (-199/-195) within the 5'-region (non-coding strand) of the cathepsin D gene (Fig. 3) (Krishnan et al., 1994b). A synthetic ER/Sp1 oligonucleotide duplex from the cathepsin D promoter was extensively investigated in gel electrophoretic mobility shift assays and the results confirmed the formation of an ER/Sp1 complex. Moreover, oligonucleotides mutated in either the ER and Sp1 genomic binding sites failed to form ER/Sp1 complexes with nuclear extracts from MCF-7 and HeLa cells. In parallel studies, the wild-type and mutant ER/Sp1 complexes were cloned into plasmids. The results of transient transfection studies in MCF-7 cells showed that E2-induced chloramphenicol acetyl transferase (CAT) activity in the plasmid containing the wild-type ER/Sp1 oligo whereas no induction response was observed with plasmids containing the mutant oligos.

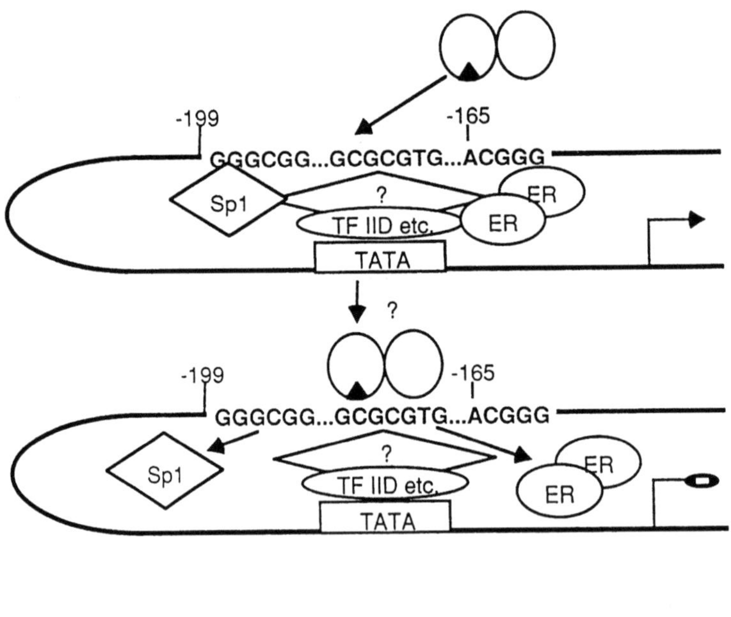

Figure 3. Mechanism of Ah receptor-mediated inhibition of estrogen-induced cathepsin D gene expression (Krishnan et al., 1994a).

E2-induced secretion of procathepsin D in MCF-7 cells is inhibited by diverse structural classes of Ah receptor agonists (Krishnan and Safe, 1993; Chaloupka et al., 1992) and recent studies in this laboratory have shown that TCDD inhibits E2-induced intracellular cathepsin D formation, cathepsin D mRNA levels and the rate of gene transcription (Krishnan et al., 1994a). Moreover, these inhibitory responses are observed within 0.5 to 2 hr after treatment of MCF-7 cells. Induction of a modulatory *trans*-acting factor (M) would probably not occur within 0.5 hr after treating the cells with TCDD suggesting that the inhibitory effects may be due to the nuclear Ah receptor complex. Subsequent analysis of the 5'-upstream DNA has shown that an imperfect DRE, GCGCGTG (-175/-181) is located between the ER and Sp1 genomic binding sites. *In vitro* gel electrophoretic mobility shift assays with the wild-type ER/Sp1 oligo showed that extracts from TCDD-treated cells do not form an ER/Sp1 complex; moreover, binding of the ER/Sp1 complex is abolished by addition of transformed cytosolic Ah receptor complex from MCF-7 cells. These results demonstrate a direct effect of the Ah receptor heterodimer on formation of the ER/Sp1 complex and this was confirmed in transient transfection assays which showed that TCDD inhibited E2-induced CAT activity mRNA levels. Moreover, an ER/Sp1 oligo mutated in the DRE sequence was estrogen-responsive in both the gel electrophoretic mobility shift and transient transfection assays. These studies thus define a mechanism for inhibition of E2-induced cathepsin D gene expression which involves direct interaction of the nuclear Ah receptor heterodimer with a genomic DRE situated between the *cis*-acting ER/Sp1 sequence (-199/-165). Zacharewski and coworkers have also identified a DRE sequence near the transcription start site of the pS2 gene which is responsible for inhibition of E2-induced pS2 gene expression (personal communication).

Inhibition of E2-induced cathepsin D and pS2 gene expression by direct interaction of the nuclear Ah receptor complex with a strategically-located DRE represents a new molecular model for understanding Ah receptor-mediated decreased gene expression.

ACKNOWLEDGEMENTS

The financial assistance of the National Institutes of Health (ES04176) and the Texas Agricultural Experiment Station is gratefully acknowledged. S. Safe is a Burroughs Wellcome Toxicology Scholar.

References

Asman DC, Takimoto K, Pitot HC, Dunn TJ and Lindahl R (1993) Organization and characterization of the rat class 3 aldehyde dehydrogenase gene. J Biol Chem 268:12530-12536

Astroff B, Rowlands C, Dickerson R and Safe S (1990) 2,3,7,8-Tetrachlorodibenzo-*p*-dioxin inhibition of 17β-estradiol-induced increases in rat uterine EGF receptor binding activity and gene expression. Mol Cell Endocrinol 72:247-252

Astroff B, Eldridge B and Safe S (1991) Inhibition of 17β-estradiol-induced and constitutive expression of the cellular protooncogene *c-fos* by 2,3,7,8-tetrachlorodibenzo-*p*-dioxin (TCDD) in the female uterus. Toxicol Lett 56:305-315

Astroff B and Safe S (1988) Comparative antiestrogenic activities of 2,3,7,8-tetrachlorodibenzo-*p*-dioxin and 6-methyl-1,3,8-trichlorodibenzofuran in the female rat. Toxicol Appl Pharmacol 95:435-443

Astroff B and Safe S (1990) 2,3,7,8-Tetrachlorodibenzo-*p*-dioxin as an antiestrogen: effect on rat uterine peroxidase activity. Biochem Pharmacol 39:485-488

Biegel L and Safe S (1990) Effects of 2,3,7,8-tetrachlorodibenzo-*p*-dioxin (TCDD) on cell growth and the secretion of the estrogen-induced 34-, 52- and 160-kDa proteins in human breast cancer cells. J Steroid Biochem Mol Biol 37:725-732

Bradfield CA and Poland A (1988) A competitive binding assay for 2,3,7,8-tetrachlorodibenzo-*p*-dioxin and related ligands of the Ah receptor. Mol Pharmacol 34:682-688

Bunce NJ, Landers JP and Safe S (1988) Kinetic models for association of 2,3,7,8-tetrachlorodibenzo-*p*-dioxin with the Ah receptor. Arch Biochem Biophys 267:384-397

Burbach KM, Poland AB and Bradfield CA (1992) Cloning of the Ah-receptor cDNA reveals a distinctive ligand-activated transcription factor. Proc Natl Acad Sci USA 89:8185-8189

Chaloupka K, Krishnan V and Safe S (1992) Polynuclear aromatic hydrocarbon carcinogens as antiestrogens in MCF-7 human breast cancer cells. Role of the Ah receptor. Carcinogenesis 13:2223-2239

Choi EJ, Toscano DG, Ryan JA, Riedel N and Toscano WA (1991) Dioxin induced transforming growth factor-α in human keratinocytes. J Biol Chem 266:9591-9597

DeVito MJ, Thomas T, Martin E, Umbreit TH and Gallo MA (1992) Antiestrogenic action of 2,3,7,8-tetrachlorodibenzo-*p*-dioxin: tissue-specific regulation of estrogen receptor in CD1 mice. Toxicol Appl Pharmacol 113:284-292

Dolwick KM, Schmidt JV, Carver LA, Swanson HI and Bradfield CA (1993) Cloning and expression of a human Ah receptor cDNA. Mol Pharmacol 44:911-917

Elferink CJ, Gasiewicz TA and Whitlock JP, Jr. (1990) Protein-DNA interactions at a dioxin-responsive enhancer. Evidence that the transformed Ah receptor is heteromeric. J Biol Chem 265:20708-20712

Ema M, Sogawa K, Watanabe N, Chujoh Y, Matsushita N, Gotoh O, Funae Y and Fujii-Kuriyama Y (1992) cDNA cloning and structure of the putative Ah receptor. Biochem Biophys Res Comm 184:246-253

Fernandez P and Safe S (1992) Growth inhibitory and antimitogenic activity of 2,3,7,8-tetrachlorodibenzo-*p*-dioxin (TCDD) in T47D human breast cancer cells. Toxicol Lett 61:185-197

Fiedler H (1993) Formation and sources of PCDDs/PCDFs. Organohalogen Compounds, Dioxin '93 11:221-228

Fujisawa-Sehara A, Sogawa K, Nishi C and Fujii-Kuriyama Y (1986) Regulatory DNA elements localized remotely upstream from the drug-metabolizing cytochrome P-450c gene. Nucleic Acids Res 14:1465-1477

Fujisawa-Sehara A, Yamane M and Fujii-Kuriyama Y (1988) A DNA-binding factor specific for xenobiotic responsive elements of P-450c gene exists as a cryptic form in cytoplasm: its possible translocation to nucleus. Proc Natl Acad Sci USA 85:5859-5863

Gaido KW, Maness SC, Leonard LS and Greenlee WF (1992) 2,3,7,8-Tetrachlorodibenzo-*p*-dioxin-dependent regulation of transforming growth factors-α and β_2 expression in a human keratinocyte cell line involves both transcriptional and post-transcriptional control. J Biol Chem 267:24591-24595

Gallo MA, Hesse EJ, MacDonald GJ and Umbreit TH (1986) Interactive effects of estradiol and 2,3,7,8-tetrachlorodibenzo-*p*-dioxin on hepatic cytochrome P-450 and mouse uterus. Toxicol Lett 32:123-132

Gasiewicz TA, Elferink CJ and Henry EC (1991) Characterization of multiple forms of the Ah receptor: recognition of a dioxin-responsive enhancer involves heteromer formation. Biochem 30:2909-2916

Gierthy JF and Lincoln DW (1988) Inhibition of postconfluent focus production in cultures of MCF-7 breast cancer cells by 2,3,7,8-tetrachlorodibenzo-*p*-dioxin. Breast Cancer Res 12:227-233

Gierthy JF, Lincoln DW, Gillespie MB, Seeger JI, Martinez HL, Dickerman HW and Kumar SA (1987) Suppression of estrogen-regulated extracellular plasminogen activator activity of MCF-7 cells by 2,3,7,8-tetrachlorodibenzo-*p*-dioxin. Cancer Res 47:6198-6203

Gierthy JF, Bennett JA, Bradley LM and Cutler DS (1993) Correlation of *in vitro* and *in vivo* growth suppression of MCF-7 human breast cancer by 2,3,7,8-tetrachlorodibenzo-*p*-dioxin. Cancer Res 53:3149-3153

Goldstein JA and Safe S (1989) Mechanism of action and structure-activity relationships for the chlorinated dibenzo-*p*-dioxins and related compounds. In: Kimbrough RD and Jensen AA (eds) Halogenated Biphenyls, Naphthalenes, Dibenzodioxins and Related Compounds. Elsevier-North Holland, Amsterdam, pp 239-293

Gonzalez FJ and Nebert DW (1985) Autoregulation plus upstream positive and negative control regions associated with transcriptional activation of the mouse cytochrome P_1-450 gene. Nucleic Acids Res 13:7269-7288

Hankinson O, Brooks BA, Weir-Brown KI, Huffman EC, Johnson BS, Nanthur J, Reyes H and Watson AJ (1991) Genetic and molecular analysis of the Ah receptor and of Cyp1a-1 gene expression. Biochimie 73:61-66

Harper N, Wang X, Liu H and Safe S (1994) Inhibition of estrogen-induced progesterone receptor in MCF-7 human breast cancer cells by aryl hydrocarbon (Ah) receptor agonists. Mol Cell Endocrinol (In Press)

Harris M, Zacharewski T, Astroff B and Safe S (1989) Partial antagonism of 2,3,7,8-tetrachlorodibenzo-*p*-dioxin-mediated induction of aryl hydrocarbon hydroxylase by 6-methyl-1,3,8-trichlorodibenzofuran : mechanistic studies. Mol Pharmacol 35:729-735

Harris M, Zacharewski T, Piskorska-Pliszczynska J, Rosengren R and Safe S (1990) Structure-dependent induction of aryl hydrocarbon hydroxylase activity in C57BL/6 mice by 2,3,7,8-tetrachlorodibenzo-*p*-dioxin and related congeners: mechanistic studies. Toxicol Appl Pharmacol 105:243-253

Holcomb M and Safe S (1994) Inhibition of 7,12-dimethylbenzanthracene-induced rat mammary tumor growth by 2,3,7,8-tetrachlorodibenzo-*p*-dioxin. Cancer Letters (In Press)

Hudson LG, Toscano WA,Jr. and Greenlee WF (1985) Regulation of epidermal growth factor binding in a human keratinocyte cell line by 2,3,7,8-tetrachlorodibenzo-*p*-dioxin. Toxicol Appl Pharmacol 77:251-259

Jaiswal AK (1994) Human NAD(P)H:quinone oxidoreductase: gene structure, activity and tissue-specific expression. J Biol Chem 269:14502-14508

Jones PB, Galeazzi DR, Fisher JM and Whitlock JP,Jr. (1985) Control of cytochrome P_1-450 gene expression by dioxin. Science 227:1499-1502

Jones PB, Durrin LK, Fisher JM and Whitlock JP,Jr. (1986) Control of gene expression by 2,3,7,8-tetrachlorodibenzo-*p*-dioxin: multiple dioxin-responsive domains 5'-ward of the cytochrome P_1-450 gene. J Biol Chem 261:6647-6650

Kawanishi S, Seki Y and Sano S (1983) Uroporphyrinogen decarboxylase: purification, properties, and inhibition by polychlorinated biphenyl isomers. J Biol Chem 258:4285-4292

Kärenlampi SO, Eisen HJ, Hankinson O and Nebert DW (1983) Effects of cytochrome P_1-450 inducers on the cell-surface receptors for epidermal growth factor, phorbol 12,13-dibutyrate, or insulin of cultured mouse hepatoma cells. J Biol Chem 258:10378-10383

Kimbrough RD (1985) Laboratory and human studies on polychlorinated biphenyls (PCBs) and related compounds. Environ Health Perspect 59:99-106

Kimbrough RD (1987) Human health effects of polychlorinated biphenyls (PCBs) and polybrominated biphenyls (PBBs). Annu Rev Pharmacol Toxicol 27:87-111

Kociba RJ, Keyes DG, Beger JE, Carreon RM, Wade CE, Dittenber DA, Kalnins RP, Frauson LE, Park CL, Barnard SD, Hummel RA and Humiston CG (1978) Results of a 2-year chronic toxicity and oncogenicity study of 2,3,7,8-tetrachlorodibenzo-*p*-dioxin (TCDD) in rats. Toxicol Appl Pharmacol 46:279-303

Krishnan V, Wang X, Ramamurthy P and Safe S (1994a) Effect of 2,3,7,8-tetrachlorodibenzo-*p*-dioxin (TCDD) in formation of estrogen-induced ER/Sp1 complexes on the cathepsin D promoter. Toxicologist 14:47

Krishnan V, Wang X and Safe S (1994b) ER/Sp1 complexes mediate estrogen-induced cathepsin D gene expression in MCF-7 human breast cancer cells. J Biol Chem 269:15912-15917

Krishnan V and Safe S (1993) Polychlorinated biphenyls (PCBs), dibenzo-*p*-dioxins (PCDDs) and dibenzofurans (PCDFs) as antiestrogens in MCF-7 human breast cancer cells: quantitative structure-activity relationships. Toxicol Appl Pharmacol 120:55-61

Landers JP and Bunce NJ (1991) The Ah receptor and the mechanism of dioxin toxicity. Biochem J 276:273-287

Liu H, Biegel L, Narasimhan TR, Rowlands C and Safe S (1992) Inhibition of insulin-like growth factor-I responses in MCF-7 cells by 2,3,7,8-tetrachlorodibenzo-*p*-dioxin and related compounds. Mol Cell Endocrinol 87:19-28

Liu H, Wormke M, Gentle AB, Bjeldanes LF and Safe S (1994) Indolo[3,2-b]carbazole exhibits both estrogenic and antiestrogenic activity in MCF-7 human breast cancer cells. Toxicologist 14:728

Madhukar BV, Brewster DW and Matsumura F (1984) Effects of *in vivo*-administered 2,3,7,8-tetrachlorodibenzo-*p*-dioxin on receptor binding of epidermal growth factor in the hepatic plasma membrane of rat, guinea pig, mouse, and hamster. Proc Natl Acad Sci USA 81:7407-7411

Madhukar BV, Ebner K, Matsumura F, Bombick DW, Brewster DW and Kawamoto T (1988) 2,3,7,8-Tetrachlorodibenzo-*p*-dioxin causes an increase in protein kinases associated with epidermal growth factor receptor in the hepatic plasma membrane. J Biochem Toxicol 3:261-277

N.I.H. (1982) Carcinogenesis Bioassay of 2,3,7,8-Tetrachlorodibenzo-*p*-dioxin (CAS No. 1746-01-6) in Osborne-Mendel Rats and B6C3F1 Mice (Gavage Study). U.S. National Institutes of Health, National Toxicology Program (NTP Tech. Report Series No. 209), Research Triangle Park, NC

Narasimhan TR, Safe S, Williams HJ and Scott AI (1991) Effects of 2,3,7,8-tetrachlorodibenzo-*p*-dioxin on 17β-estradiol-induced glucose metabolism in MCF-7 human breast cancer cells: ^{13}C-nuclear magnetic resonance studies. Mol Pharmacol 40:1029-1035

Nebert DW, Robinson JR, Niwa A, Kumaki K and Poland AP (1976) Genetic expression of aryl hydrocarbon hydroxylase activity in the mouse. J Cell Physiol 85:393-414

Neuhold LA, Shirayoshi Y, Ozato K, Jones JE and Nebert DW (1989) Regulation of mouse CYP1A1 gene expression by dioxin: requirement of two *cis*-acting elements during induction. Mol Cell Biol 9:2378-2386

Newsted JL and Giesy JP (1993) Effect of 2,3,7,8-tetrachlorodibenzo-*p*-dioxin (TCDD) on the epidermal growth factor receptor in hepatic plasma membranes of rainbow trout. Toxicol Appl Pharmacol 118:119-130

Okey AB, Riddick DS and Harper PA (1994) The Ah receptor: mediator of the toxicity of 2,3,7,8-tetrachlorodibenzo-*p*-dioxin (TCDD) and related compounds. Toxicol Lett 70:1-22

Pimental RA, Liang B, Yee GK, Wilhelmsson A, Poellinger L and Paulson KE (1993) Dioxin receptor and C/EBP regulate the function of the glutathione S-transferase Ya gene xenobiotic response element. Mol Cell Biol 13:4365-4373

Poland A, Glover E and Kende AS (1976) Stereospecific, high affinity binding of 2,3,7,8-tetrachlorodibenzo-*p*-dioxin by hepatic cytosol: evidence that the binding species is receptor for induction of aryl hydrocarbon hydroxylase. J Biol Chem 251:4936-4946

Poland A, Glover E and Taylor BA (1987) The murine Ah locus: a new allele and mapping to chromosome 12. Mol Pharmacol 32:471-478

Poland A and Glover E (1973) Chlorinated dibenzo-*p*-dioxins: potent inducer of δ-aminolevulinic acid synthetase and aryl hydrocarbon hydroxylase. II. A study of the structure-activity relationship. Mol Pharmacol 9:736-747

Poland A and Glover E (1975) Genetic expression of aryl hydrocarbon hydroxylase by 2,3,7,8-tetrachlorodibenzo-*p*-dioxin: evidence for a receptor mutation in genetically nonresponsive mice. Mol Pharmacol 11:389-398

Poland A and Glover E (1987) Variation in the molecular mass of the Ah receptor among vertebrate species and strains of rats. Biochem Biophys Res Comm 146:1439-1449

Poland A and Knutson JC (1982) 2,3,7,8-Tetrachlorodibenzo-*p*-dioxin and related halogenated aromatic hydrocarbons. Examinations of the mechanism of toxicity. Annu Rev Pharmacol Toxicol 22:517-554

Poland AP, Glover E, Robinson JR and Nebert DW (1975) Genetic expression of aryl hydrocarbon hydroxylase activity. Induction of monooxygenase activities and cytochrome, P-450 formation by 2,3,7,8-tetrachlorodibenzo-*p*-dioxin in mice generally 'nonresponsive' to other aromatic hydrocrbons. J Biol Chem 249:5599-5606

Puga A, Nebert DW and Carrier F (1992) Dioxin induces expression of c-*fos* and c-*jun* proto-oncogenes and a large increase in transcription factor AP-1. DNA Cell Biol 11:269-281

Quattrochi LC, Vu T and Tukey RH (1994) The human *CYP1A2* gene and induction by 3-methylcholanthrene: a region of DNA that supports Ah-receptor binding and promoter-specific induction. J Biol Chem 269:6949-6954

Rappe C (1992) Dietary exposure and human levels of PCDDs and PCDFs. Chemosphere 25:231-234

Rappe C (1993a) Sources of exposure, environmental concentrations and exposure assessment of PCDDs and PCDFs. Chemosphere 27:211-225

Rappe C (1993b) Environmental concentrations and ecotoxicological effects of PCDDs, PCDFs and related compounds. Organohalogen Compounds, Dioxin '93 12:163-170

Romkes M, Piskorska-Pliszczynska J and Safe S (1987) Effects of 2,3,7,8-tetrachlorodibenzo-*p*-dioxin on hepatic and uterine estrogen receptor levels in rats. Toxicol Appl Pharmacol 87:306-314

Romkes M and Safe S (1988) Comparative activities of 2,3,7,8-tetrachlorodibenzo-*p*-dioxin and progesterone on antiestrogens in the female rat uterus. Toxicol Appl Pharmacol 92:368-380

Rushmore TH, King RG, Paulson KE and Pickett CB (1990) Regulation of glutathione S-transferase Ya subunit gene expression: identification of a unique xenobiotic-responsive element controlling inductible expression by planar aromatic compounds. Proc Natl Acad Sci USA 87:3826-3830

Ryan RP, Sunahara GI, Lucier GW, Birnbaum LS and Nelson KG (1989) Decreased ligand binding to the hepatic glucocorticoid and epidermal growth factor receptors after 2,3,4,7,8-pentachlorodibenzofuran and 1,2,3,4,7,8-hexachlorodibenzofuran treatment of pregnant mice. Toxicol Appl Pharmacol 98:454-464

Safe S (1986) Comparative toxicology and mechanism of action of polychlorinated dibenzo-*p*-dioxins and dibenzofurans. Annu Rev Pharmacol Toxicol 26:371-399

Safe S (1990) Polychlorinated biphenyls (PCBs), dibenzo-*p*-dioxins (PCDDs), dibenzofurans (PCDFs) and related compounds: environmental and mechanistic considerations which support the development of toxic equivalency factors (TEFs). C R C Crit Rev Toxicol 21:51-88

Safe S, Astroff B, Harris M, Zacharewski T, Dickerson R, Romkes M and Biegel L (1991a) 2,3,7,8-Tetrachlorodibenzo-*p*-dioxin (TCDD) and related compounds as antiestrogens: characterization and mechanism of action. Pharmacol Toxicol 69:400-409

Safe S, Harris M, Biegel L and Zacharewski T (1991b) Mechanism of action of TCDD as an antiestrogen in transformed human breast cancer and rodent cell lines. In: Biological Basis for Risk Assessment of Dioxins and Related Compounds. Cold Spring Harbor Laboratory, Banbury Conference Proceedings #35, pp 367-377

Safe SH (1988) The aryl hydrocarbon (Ah) receptor. ISI Atlas of Science: Pharmacology 2:78-83

Schmidt JV, Carver LA and Bradfield CA (1993) Molecular characterization of the murine *Ahr* gene: organization, promoter analysis, and chromosomal assignment. J Biol Chem 268:22203-22209

Shiverick KT and Muther TF (1982) Effects of 2,3,7,8-tetrachlorodibenzo-*p*-dioxin on serum concentrations and the uterotrophic actions of exogenous estrone in rats. Toxicol Appl Pharmacol 65:170-176

Sogawa K, Fujisawa-Sehara A, Yamane M and Fuji-Kuriyama Y (1986) Location of regulatory elements responsible for drug induction in the rat cytochrome P-450c gene. Proc Natl Acad Sci USA 83:8044-8048

Spink DC, Lincoln DW, Dickerman HW and Gierthy JF (1990) 2,3,7,8-Tetrachlorodibenzo-*p*-dioxin causes an extensive alteration of 17β-estradiol metabolism in human breast cancer cells. Proc Natl Acad Sci USA 87:6917-6921

Spink DC, Eugster H-P, Lincoln DW,II, Schuetz JD, Schuetz EG, Johnson JA, Kaminsky LS and Gierthy JF (1992) 17β-Estradiol hydroxylation catalyzed by human cytochrome P4501A1: a comparison of the activities induced by 2,3,7,8-tetrachlorodibenzo-*p*-dioxin in MCF-7 cells with those from heterologous expression of the cDNA. Arch Biochem Biophys 293:342-348

Spink DC, Johnson JA, Connor SP, Aldous KM and Gierthy JF (1994) Stimulation of 17β-estradiol metabolism in MCF-7 cells by bromochloro- and chloromethyl-substituted dibenzo-*p*-dioxins and dibenzofurans: correlations with antiestrogenic activity. J Toxicol Environ Health 41:451-466

Stahl BU, Beer DG, Weber LWD and Rozman K (1993) Reduction of hepatic phosphoenolpyruvate carboxykinase (PEPCK) activity by 2,3,7,8-tetrachlorodibenzo-*p*-dioxin (TCDD) is due to decreased mRNA levels. Toxicology 79:81-95

Sutter TR, Guzman K, Dold KM and Greenlee WF (1991) Targets for dioxin: genes for plasminogen activator inhibitor-2 and interleukin-1β. Science 254:415-418

Thomsen JS, Wang X, Hines RN and Safe S (1994) Restoration of Ah responsiveness in MDA-MB-231 human breast cancer cells by transient expression of the estrogen receptor. Carcinogenesis 15:933-937

Tukey RH, Hannah RR, Negishi M, Nebert DW and Eisen HJ (1982) The Ah locus: correlation of intranuclear appearance of inducer-receptor complex with induction of cytochrome P-450 mRNA. Cell 31:275-284

Umbreit TH, Hesse EJ, MacDonald GJ and Gallo MA (1988) Effects of TCDD-estradiol interactions in three strains of mice. Toxicol Lett 40:1-9

Umbreit TH and Gallo MA (1988) Physiological implications of estrogen receptor modulation by 2,3,7,8-tetrachlorodibenzo-*p*-dioxin. Toxicol Lett 42:5-14

Wang X, Porter W, Krishnan V, Narasimhan TR and Safe S (1993) Mechanism of 2,3,7,8-tetrachlorodibenzo-*p*-dioxin (TCDD)-mediated decrease of the nuclear estrogen receptor in MCF-7 human breast cancer cells. Mol Cell Endocrinol 96:159-166

Weber LWD, Lebofsky M, Stahl BU, Gorski JR, Muzi G and Rozman K (1991) Reduced activities of key enzymes of gluconeogenesis as possible cause of acute toxicity of 2,3,7,8-tetrachlorodibenzo-*p*-dioxin (TCDD) in rats. Toxicology 66:133-144

Whitlock JP,Jr. (1986) The regulation of cytochrome P-450 gene expression. Annu Rev Pharmacol Toxicol 26:333-369

Whitlock JP,Jr. (1987) The regulation of gene expression of 2,3,7,8-tetrachlorodibenzo-*p*-dioxin. Pharmacol Rev 39:147-161

Whitlock JP,Jr. (1990) Genetic and molecular aspects of 2,3,7,8-tetrachlorodibenzo-*p*-dioxin action. Annu Rev Pharmacol Toxicol 30:251-277

Whitlock JP,Jr. (1993) Mechanistic aspects of dioxin action. Chem Res Toxicol 6:754-763

Zacharewski T, Bondy K, McDonell P and Wu ZF (1994) Antiestrogenic effects of 2,3,7,8-tetrachlorodibenzo-*p*-dioxin on 17β-estradiol-induced pS2 expression. Cancer Res 54:2707-2713

Inhibitors of the Sodium Pump: Toxins, then Drugs, and now hormones

J.-D. Horisberger

Institut de Pharmacologie et de Toxicologie, Université de Lausanne, Bugnon 27, CH-1005 Lausanne, Switzerland

The toxic properties of plants such as squills and the foxglove have been known since antiquity. These poisons induce emesis, diarrhea and death through cardiac arrhythmia. African hunters have extracted a toxin from the *Strophanthus* and used it as arrow poison. All these plants and many others contain glycosylated steroids that we now designate as cardiotonic steroids, because of their common property of inducing an increase of the strength of contraction of the myocardium. The therapeutic use of digitalis has first been described 200 years ago by Withering (1785) and since then digitalis derivatives have been extensively used for the treatment of congestive heart failure. The most commonly used compounds are digoxin for clinical purpose, and ouabain for laboratory experiments. Intoxication by digitalis-like compounds mainly occurs as the results of inappropriate dosage or because of accidental or suicidal acute intoxication. Accidental poisoning also occurs because of confusing cardiotonic steroid containing plants for herbal teas.

The cardiotonic steroid receptor

The target of cardiotonic steroid has first been disclosed in 1953 by Schatzman who observed inhibition of sodium and potassium transport across the erythrocyte membrane by ouabain. A few years later, this cation transport system, the Na,K-pump, has been identified by Skou as the membrane bound Na,K-ATPase (Skou 1957). The Na,K-ATPase is present in all animal cells and is responsible for the maintenance of the Na^+ and K^+ ion gradients across the cell membrane.

Structure and function of the Na,K-ATPase

The Na,K-ATPase is a member of the P-type ATPase family, a large group of ion motive ATPases that share a common general structure (Horisberger 1994). This family includes also the gastric proton pump, the various forms of the calcium ATPases, and H-ATPase and K-ATPases found in plant and bacteria, respectively. The Na,K-ATPase, like the gastric proton pump, consists of two subunits α and β. The α subunit, the largest one, contains the ATP-binding site and the elements responsible for the hydrolysis of ATP. It is therefore named the catalytic subunit. It contains also the binding site for the cardiotonic steroids.

The primary structure of the α subunit has been determined first by Shull et al. (1985) and Kawakami et al. (1985). It consists of slightly over 1000 amino acids. According to the best accepted model (Lingrel et al. 1994; Horisberger 1994) it exposes the main part of its mass to the intracellular side, while the extracellular portion is made of short loops connecting ten transmembrane segments. The smaller β subunit includes only one transmembrane segment.

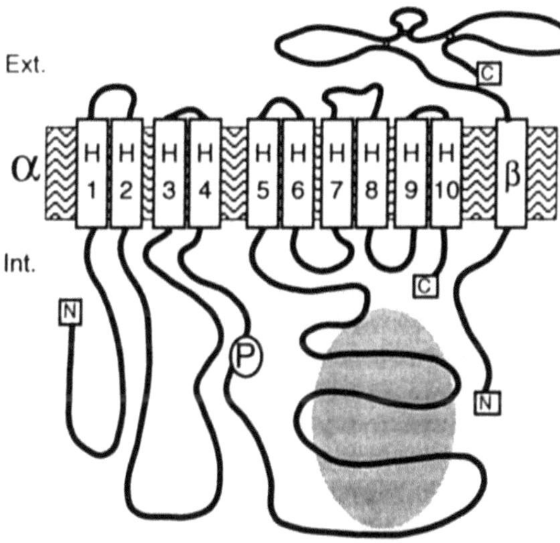

Figure 1. Two-dimensional scheme of the oligomeric structure and membrane topology of the Na,K-ATPase. The N- and C-terminal of the α and β chain are indicated. The circled P shows the position of the phosphorylation site. The region of the ATP binding site is indicated by the shaded area.

Organ differences in ouabain sensitivity have first suggested the existence of several isoforms. During the past ten years, molecular biology has allowed to identify 3 isoforms for the α subunit (see riew in Horisberger 1994). These isoforms have a specific tissue distribution and differ in their sensitivity to cardiotonic steroids (table 1).

Table 1. Isoforms of the α subunit of the Na,K-ATPase

Isoform	α1	α2	α3	references
Distribution	ubiquitous	nervous system skeletal muscle heart muscle	nervous system	Gick et al. 1993; Sweadner 1991
Ouabain sensitivity	(+) to ++	+++	++++ ?	Lingrel et al.1991; O'Brien et al. 1994
Control of expression	aldosterone (kidney)		?	Horisberger and Rossier 1992; Welling et al 1993 McDonough et al. 1992

With the few exceptions described in the table, physiological differences are not yet well demonstrated and the physiological rationale for the existence of different isoforms is still poorly understood.

Structure of the ouabain binding site

Ouabain and other membrane impermeant cardiotonic steroids act when added to the outside and therefore their binding site is expected to be formed by the short extracellular loops connecting the ten transmembrane segments. The first information concerning the structure of the binding site has come from comparison between isoforms with different ouabain sensitivity. A few species, the small rodents rat and mouse (Lingrel et al.1991), the toad *Bufo marinus* (Jaisser et al. 1992) and the Monarch butterfly (Holzinger et al. 1992) possess an ouabain-resistant enzyme. Comparison of the primary sequence of the α subunit in these species with that of ouabain-sensitive species or ouabain-sensitive isoforms in the same species revealed a few striking differences, namely the presence of charged amino acids at either end of the first (H1-H2) extracellular loop. Introduction of charged amino acids at these positions by site directed mutagenesis in ouabain-sensitive isoforms yielded ouabain-resistant mutants, demonstrating the essential role of this region in ouabain binding.

Generation of ouabain resistant mutants either by exposure to mutagenic agents (Canessa et al. 1992; Canessa et al. 1993) or by random mutagenesis applied to a defined segment of the protein (Schultheis et al. 1993) followed by selection of the ouabain-resistant mutant by growing the cells in ouabain containing media allowed to identify other sites involved in the binding of ouabain, namely the H1 segment and the H3-H4 and H7-H8 extracellular loops. Finally the systematic site directed mutagenesis of the H5-H6 extracellular loop allowed the identification of another amino acid essential for ouabain binding (Lingrel et al.1991; Feng and Lingrel 1994). Figure 2 summarizes the position of the amino acid involved in ouabain binding, i.e. the position at which site directed mutation results in an ouabain resistance phenotype.

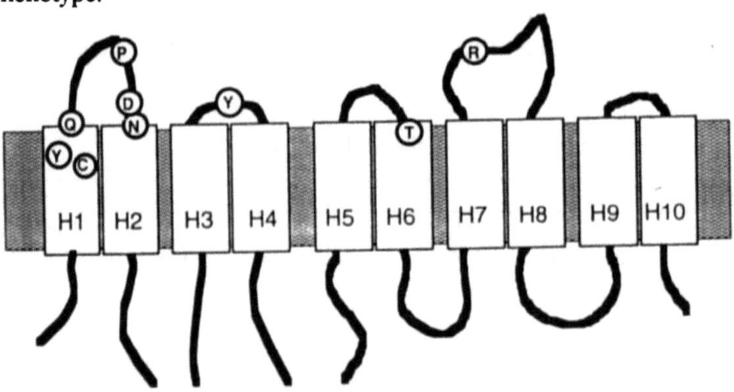

Figure 2. Amino acids of the α subunit of the Na,K-ATPase involved in the binding of ouabain. Only the membrane associated region of the protein is shown. Site directed mutagenesis at the circled positions induced an ouabain-resistant phenotype (Canessa et al. 1992, 1993; Lingrel et al.1991; Schultheis et al. 1993; Lingrel et al. 1994; Feng and Lingrel 1994)

In spite of these advances, the mechanism of ouabain binding is still poorly understood. First, although there is a about 20-fold difference in affinity between glycosylated steroids (like ouabain) and aglycone (like ouabagenin), none of the ouabain-resistance mutants show a significant alteration of this ratio, indicating that the binding site of the sugar moiety has not yet been identified (Canessa et al. 1992; O'Brien et al. 1993). Second, although a large number of compounds have been tested, we do not have an $\alpha 1$ specific inhibitor, namely a substance that would have a higher affinity for the $\alpha 1$ isoform that for the $\alpha 2/\alpha 3$ isoforms. The affinity seems to be determined by the structure of the receptor rather than by the structure of the ligand, which can be explained by the fact that the binding process includes one or several conformational changes of the protein (Forbush 1983; Horisberger 1994)

Mechanism of action of the cardiotonic steroids

How is the inhibition of the Na,K-ATPase related to the cardiotonic therapeutic effect and the multiple toxic effects of the digitalis ? The Na,K-ATPase transports actively sodium ions out of the cell and potassium ions into the cell. During one transport cycle it uses the energy provided by the hydrolysis of one molecule of ATP to exchange 3 Na^+ against 2 K^+. The gradients of Na^+ and K^+ across the cell membrane are essential for many functions of the cell. The cell membrane potential is maintained thanks to the K^+ gradient and K^+ channels. Na^+ and K^+ gradients are used for the control of cell volume. The sodium gradient is used to energize numerous secondary active coupled transport systems responsible for the maintenance of intracellular pH and calcium activity, the import of nutrients, etc. In excitable cells, the gradient of Na^+ and K^+ are especially important because they are responsible for the generation and the control of the action potential. In epithelia, the gradient of sodium established by the Na,K-pump drives the transepithelial transport, absorptive or secretory, of solute and water.

Considering the essential role of the Na,K-ATPase in the physiology of the cell it is not surprising that inhibition of the sodium pump by cardiotonic steroids will result in toxic effects in many organs, especially in those depending on the sensitive isoforms such as the heart and the central nervous system. However the cardiotonic steroids got their name because of their inotropic effect on the heart muscle. How is this effect explained ? During the myocardial action potential calcium channels are transiently open allowing the entry of calcium ion into the cytoplasm. The resulting increase in cytosolic free calcium triggers the release of a large quantity of calcium from the intracellular calcium store, the sarcoplasmic reticulum (SR). The strength of the mechanical contraction is proportional to the amount of calcium released. Partial inhibition of the Na,K-pump leads to a small increase of intracellular Na^+. The decreased transmembrane Na^+ gradient lowers the energy available for the sodium/calcium exchange process leading to a progressive accumulation of calcium in the cell and in the SR by the SR calcium pump and therefore to an release of larger amount of calcium at each action potential. This explains the increase of the muscle contraction following Na,K-pump inhibition. The other known cardiac effects of digitalis, the depressed conduction and the increase in automaticity, two effects responsible for the potentially lethal consequence of digitalis intoxication can be explained by alteration of the membrane potential and the ionic gradient.

The same mechanism that is responsible for the inotropic effect in the myocardium may be at work in many other cell type (Blaustein 1993) and explain for instance the vasoconstriction observed after Na,K-pump inhibition in isolated vessel preparation (Blaustein 1993; Poston et al. 1993). The vasoconstrictive effects are usually not obvious *in vivo*, probably because of physiologically antagonist effects through the autonomous nervous system. Indeed digitalis and related substance are known to affect cardiovascular function not only through their direct effect on the heart but also through inhibition of the sympathetic nerves and stimulation of the parasympathetic system. These effects are in part responsible for the slowing of the heart rate that is observed during digitalis treatment.

Although their mechanism is not understood in details, the numerous toxic effects on the nervous central system can easily be explained by alteration of the membrane potential or of the level of intracellular calcium and its relation to neurotransmitter secretion. In addition, most known mechanism of neurotransmitter reuptake by neurons and glial cells use the gradient of Na^+ and K^+ as energy source. Indeed inhibition of the Na,K-pump has been shown to increase excitatory amino acid toxicity (Brines and Robbins 1992). Therefore it is not surprising that dysfunction of the Na,K-pump has been hypothetically implicated in many type of central nervous system disorders, from epilepsy to depression (Lees 1991). One has to admit however, that there are no direct proofs of these hypothesis for any human disease. One genetic dysfunction of the Na,K-pump in *Drosophila*, the so-called « bang » mutation that is associated with a lowered expression of Na,K-pump in the nervous system of the fly results in a syndrome that is reproducible by injection of ouabain: the affected *Drosophila* undergo a temporary paralysis after exposure to a mechanical shock stimulus (Schubiger et al. 1994).

Finally inhibition of the Na,K-pump should results in a reduction of transepithelial sodium and fluid reabsorbtion, resulting in a diuretic effect. This effect is however difficult to observe with ouabain *in vivo* because of the relatively low sensitivity of the renal isoform.

Exogenous and Endogenous Sodium pump inhibitors

Considering the very highly conserved characteristic of the Na,K-pump for ouabain binding, at least in the $\alpha2$ and $\alpha3$ isoforms, it was tempting to speculate that this property of corresponded to a physiological role. In addition, the existence of a physiological sodium pump inhibitor with diuretic and vasoconstrictive properties has been postulated since a long time in a physiopathological model of chronic hypertension (Blaustein 1983; Blaustein 1994). According to this model a defect in renal sodium handling would lead to chronic sodium retention. This in turn would stimulate the production of a natriuretic hormone leading to normalization of the extracellular fluid volume but, because of the vasoconstrictive effect of sodium pump inhibition, at the expense of an increase in vascular tone leading to chronic arterial hypertension. Because of the physiological and pathophysiological importance of these questions the search for circulating Na,K-pump inhibitors has been a very active field of research since many years.

In 1991, the groups of Hamlyn and Blaustein have published a series of papers reporting the results of their extensive work searching for a circulating sodium pump inhibitor in man (Ludens et al. 1991; Mathews et al. 1991; Harris et al. 1991; Hamlyn et al. 1991). Starting from a large amount of plasma of volume expanded patients and through a series of purification steps including an affinity column on sheep kidney Na,K-ATPase, they were able to isolate 12 μg of Na,K-pump inhibitor as a pure substance. This substance could not be distinguished from ouabain by chromatographic and mass spectrometric methods. In addition it bound to the Na,K-ATPase and inhibited rubidium transport with the same kinetic parameters. Finally this « endo-ouabain » had an inotropic effect when applied to the guinea-pig heart. Several observations suggest that this compound is indeed endogenous (Ludens et al. 1992; Boulanger et al. 1993). First, its plasma level persisted in patient under total parenteral nutrition, i.e. with zero controlled input of cardiotonic steroid. Second the plasma level decreased after adrenalectomy, and most convincingly, a significant increase in concentration between arterial blood and blood in the adrenal vein could be demonstrated.

However, several groups have presented results suggesting other origins than the adrenal glands, in particular the hypothalamus (Tymiak et al. 1993; Sancho et al. 1993) and ouabain-like activity could be stimulated by dietary sodium in adrenalectomized rats (Leenen et al. 1993). Another difficulty comes from the presence of numerous cardiotonic steroid substances in food. For instance Longerich et al (1993) have determined the content in digoxin-like substances in teas brewed from freely available herbal teas. In a number of cases they observed large amounts (up to several mg of digoxin equivalent per cup) of substances reacting with anti-digoxin antibodies or competing with ouabain for binding to Na,K-ATPase. Trying to dissociate endogenous from exogenous circulating Na,K-pump inhibitors, Tamura et al. (1994) observed that several substances disappeared from urine when rat were fed a synthetic diet devoided of any cardiotonic steroids. They were able to show the presence of these substances in the normal rat chow. However the urinary excretion of another, ouabain-like, substance did not decrease after one week of synthetic diet and could not be found in the normal chow. Although it is obvious that there are circulating inhibitors of the Na,K-pump from exogenous origin, it is not yet clear if the ouabain-like substance is just stored in the organism, if it is a metabolite of one or the other of the exogenous cardiotonic steroids, or if it is entirely synthesized in the body. Although higher levels of ouabain-like substances have been observed after volume expansion (Leenen et al. 1993; Ludens et al. 1993) or in physiopathological conditions such as heart failure (Gottlieb et al. 1992; Blaustein 1993) the regulation of the production or release of this putative endogenous Na,K-pump regulator is still very poorly understood.

Whether or not an endogenous cardiotonic steroid is used for the physiological regulation of one or several isoforms of the Na,K-ATPase, the presence in food of several inhibitors of the Na,K-pump poses interesting toxicological questions. The adverse effects of an acute intoxication with digoxin or any of the other compounds used in human therapeutics are very well known. However the effects of chronic exposure to ouabain-like compounds have not been investigated in great details. Experimental works has demonstrated that chronic administration of ouabain induces an upregulation of the Na,K-ATPase in many tissues (Li et al. 1993). The long term consequences of this change are not known. On the other hand, and in accordance with the sodium-pump-inhibitor hypothesis for the physiopathology of chronic hypertension, chronic administration of low doses of ouabain to rats leads to the development of arterial hypertension (Yuan et al. 1993). Considering the large difference in ouabain sensitivity between human and rat it is difficult to evaluate the relevance of these findings to human physiopathology.

Conclusions

Although the presence of circulating Na,K-pump inhibitors have been well demonstrated in man and animals many problems remains to be solved concerning the exact nature, the exogenous or endogenous origin and the potential physiological and pathophysiological roles of these substances and, in the first place, several fundamental questions about the specific functional characteristics and the role of ouabain-sensitive and ouabain-resistant isoforms of the Na,K-ATPase will have to be answered. The role of common exposure to sodium pump inhibitors in the pathogenesis disease of major public health importance such as arterial hypertension or central nervous system diseases deserves to be carefully evaluated.

References

Blaustein MP (1984) Sodium transport and hypertension. Where are we going? Hypertension 6:445-453

Blaustein MP (1993) Physiological effects of endogenous ouabain: Control of intracellular Ca2+ stores and cell responsiveness. Am J Physiol Cell Physiol 264:C1367-C1387

Boulanger BR, Lilly MP, Hamlyn JM, Laredo J, Shurtleff D, Gann DS.(1993) Ouabain is secreted by the adrenal gland in awake dogs. Am J Physiol Endocrinol Metab 264:E413-E419

Brines ML, Robbins RJ.(1992) Inhibition of a2/a3 sodium pump isoforms potentiates glutamate neurotoxicity. Brain Res 591:94-102

Canessa C, Horisberger J-D, Louvard D, Rossier BC.(1992) Mutation of a cysteine in the first transmembrane segment of Na-K-ATPase a subunit confers ouabain resistance. EMBO J 11:1681-1687

Canessa C, Horisberger J-D, Rossier BC.(1993) Mutation of a tyrosine in the H3-H4 ectodomain of Na,K-ATPase confers ouabain resistance. J Biol Chem 268:17722-17726

Feng J, Lingrel JB.(1994) Analysis of amino acid residues in the H5-H6 transmembrane and extracellular domains of Na,K-ATPase a subunit identifies threonine 797 as a determinant of ouabain sensitivity. Biochemistry 33:4218-4224

Forbush BI (1983) Cardiotonic steroid binding to Na,K-ATPase. Current Topics in Membranes and Transport 19:167-201

Gick GG, Hatala MA, Chon D, Ismail-Beigi F.(1993) Na,K-ATPase in several tissues of the rat: Tissue-specific expression of subunit mRNAs and enzyme activity. J Membrane Biol 131:229-236

Gottlieb SS, Rogowski AC, Weinberg M, Krichten CM, Hamilton BP, Hamlyn JM.(1992) Elevated concentrations of endogenous ouabain in patients with congestive heart failure. Circulation 86:420-425

Hamlyn JM, Blaustein MP, Bova S et al (1991) Identification and characterization of a ouabain-like compound from human plasma. Proc Natl Acad Sci USA 88:6259-6263

Harris DW, Clark MA, Fisher JF et al (1991) Development of an immunoassay for endogenous digitalislike factor. Hypertension 17:936-943

Holzinger F, Frick C, Wink M.(1992) Molecular basis for the insensitivity of the Monarch (*Danaus plexippus*) to cardiac glycosides. FEBS Lett 314:477-480

Horisberger J-D (1994) The Na,K-ATPase: Structure-Function Relationship. Austin: Landes,

Horisberger J-D, Rossier BC.(1992) Aldosterone regulation of gene transcription leading to control of ion transport. Hypertension 19:221-227

Jaisser F, Canessa C, Horisberger J-D, Rossier BC.(1992) Primary sequence and functional expression of a novel ouabain-resistant Na,K-ATPase. J Biol Chem 267:16895-16903

Kawakami K, Noguchi S, Noda M et al (1985) Primary structure of the alpha-subunit of *Torpedo californica* ($Na^+ + K^+$)ATPase deduced from cDNA sequence. Nature 316:733-736

Leenen FHH, Harmsen E, Yu H, Yuan B.(1993) Dietary sodium stimulates ouabainlike activity in adrenalectomized spontaneously hypertensive rats. Am J Physiol Heart Circ Physiol 265:H421-H424

Lees GJ (1991) Inhibition of sodium-potassium-ATPase: A potentially ubiquitous mechanism contributing to central nervous system neuropathology. Brain Res Rev 16:283-300

Li PWC, Ho CS, Swaminathan R.(1993) The chronic effects of long-term digoxin administration on Na^+/K^+-ATPase activity in rat tissues. Int J Cardiol 40:95-100

Lingrel, Orlowski, Price, Pathak.(1991) Regulation of the a-subunit genes of the Na,K-ATPase and determinant of cardiac glycoside sensitivity. In: Kaplan JH, De Weer P, eds. The Sodium Pump: Structure, Mechanism, and Regulation. New York: Rockefeller University Press, 1-16.

Lingrel JB, Van Huysse J, O'Brien W, Jewell-Motz E, Askew R, Schultheis P.(1994) Structure-function studies of the Na,K-ATPase. Kidney Int 45 Suppl. 44:S32-S39

Longerich L, Johnson E, Gault MH.(1993) Digoxin-like factors in herbal teas. Clin Invest Med 16:210-218

Ludens JH, Clark MA, DuCharme DW et al (1991) Purification of an endogenous digitalislike factor from human plasma for structural analysis. Hypertension 17:923-929

Ludens JH, Clark MA, Kolbasa KP, Hamlyn JM.(1993) Digitalis-like factor and ouabain-like compound in plasma of volume-expanded dogs. J Cardiovasc Pharmacol 22 Suppl. 2:S38-S41

Ludens JH, Clark MA, Robinson FG, DuCharme DW.(1992) Rat adrenal cortex is a source of a circulating ouabainlike compound. Hypertension 19:721-724

Mathews WR, DuCharme DW, Hamlyn JM et al (1991) Mass spectral characterization of an endogenous digitalislike factor from human plasma. Hypertension 17:930-935

McDonough AA, Azuma KK, Lescale-Matys L et al (1992) Physiologic rationale for multiple sodium pump isoforms. Differential regulation of a1 vs a2 by ionic stimuli. Ann NY Acad Sci 671:156-169

O'Brien WJ, Lingrel JB, Wallick ET.(1994) Ouabain binding kinetics of the rat alpha two and alpha three isoforms of the sodium-potassium adenosine triphosphate. Arch Biochem Biophys 310:32-39

O'Brien WJ, Wallick ET, Lingrel JB.(1993) Amino acid residues of the Na,K-ATPase involved in ouabain sensitivity do not bind the sugar moiety of cardiac glycosides. J Biol Chem 268:7707-7712

Poston L, Woolfson RG, Graves JE.(1993) Effects of sodium-transport inhibition in human resistance arteries. J Cardiovasc Pharmacol 22 Suppl. 2:S1-S3

Sancho JM, Garcia-Martin E, Garcia-Robles R et al (1993) Properties of the purified hypothalamic pituitary NA/K-ATPase inhibitor. J Cardiovasc Pharmacol 22 Suppl. 2:S32-S34

Schatzmann HJ (1953) Herzglycoside als Hemmstoffe für den aktiven Kalium und Natrium Transport durch die Erythrocyten-membran. Helv Physiol Pharmacol Acta 11:346-354

Schubiger M, Feng Y, Fambrough DM, Palka J.(1994) A mutation of the Drosophila sodium pump a subunit gene results in bang-sensitive paralysis. Neuron 12:373-381

Schultheis PJ, Wallick ET, Lingrel JB.(1993) Kinetic analysis of ouabain binding to native and mutated forms of Na,K-ATPase and identification of a new region involved in cardiac glycoside interaction. J Biol Chem 268:22686-22694

Shull GE, Schwartz A, Lingrel JB.(1985) Amino-acid sequence of the catalytic subunit of the $(Na^{+}+K^{+})$ATPase deduced from a complementary DNA. Nature 316:691-695

Skou JC (1957) The influence of some cations on an adenosinetriphosphatase from peripheral nerves. Biochim Biophys Acta 23:394-401

Sweadner (1991) Overview: Subunit diversity in the Na,K-ATPase. In: Kaplan JH, De Weer P, eds. The Sodium Pump: Structure, Function and Regulation. New York: Rockefeller University Press, 63-76.

Tamura M, Harris TM, Phillips D et al (1994) Identification of two cardiac glycosides as Na^{+}-pump inhibitors in rat urine and diet. J Biol Chem 269:11972-11979

Tymiak AA, Norman JA, Bolgar M et al (1993) Physicochemical characterization of a ouabain isomer isolated from bovine hypothalamus. Proc Natl Acad Sci USA 90:8189-8193

Welling PA, Caplan M, Sutters M, Giebisch G.(1993) Aldosterone-mediated Na/K-ATPase expression is a1 isoform specific in the renal cortical collecting duct. J Biol Chem 268:23469-23476

Withering W (1785) An Account of the Foxglove and Some ot its Medical Uses. London: Robinson,

Yuan CM, Manunta P, Hamlyn JM et al (1993) Long-term ouabain administration produces hypertension in rats. Hypertension 22:178-187

Rodent Responses To Peroxisome Proliferators - A Receptor-Mediated Phenomenon?

J.D. Tugwood
ZENECA Central Toxicology Laboratory, Alderley Park, Macclesfield, Cheshire. SK10 4TJ. U.K.

The Effects Of Peroxisome Proliferators In Rodents

The phenomenon of rodent peroxisome proliferation was first observed some years ago when the hypolipidaemic drug clofibrate was administered to rats. Since then, a wide variety of structurally diverse and industrially important compounds have been shown to have similar effects; these compounds include other hypolipidaemic drugs, leukotriene antagonists, herbicides and plasticisers (for review, see Reddy & Lalwani, 1982). In addition to increases in size and number of hepatic peroxisomes, peroxisome proliferators (PPs) in susceptible species also bring about induction of hepatic peroxisomal enzymes, and cause increases in S-phase and ultimately carcinogenesis of the liver (Reddy *et al.*, 1980), although the potency of these compounds differs considerably in terms of their carcinogenicity. This rodent hepatocarcinogenesis is important in the context of human risk assessment, since humans are exposed to many PPs both environmentally and otherwise. PPs are termed non-genotoxic carcinogens since they are almost exclusively negative in routine genotoxicity assays. This raises the question as to the mechanism of carcinogenicity of these compounds. It has been proposed that the generation of hydrogen peroxide through increase in peroxisomal fatty acid β-oxidation causes DNA damage indirectly through oxidative stress. Perhaps more likely, however is the suggestion that PPs, by inducing DNA replication, are able to 'fix' spontaneous and environmentally-induced mutations which eventually give rise to tumours (Cattley *et al.*, 1989).

While inducing these pleiotropic effects in rats and mice, PPs have essentially no effect on the guinea pig, larger mammals such as the dog, or on non-human primates. Epidemiological evidence and limited work with human hepatocytes *in vitro* suggest that humans also do not respond to PPs. The molecular basis of these marked species differences in responses is not yet understood.

Cloning The Peroxisome Proliferator Activated Receptor (PPAR)

The properties of PPs are able to give some clues as to the possible molecular mechanism by which these compounds act. Firstly, rat liver cytosol contains a protein which binds to these compounds (Lalwani *et al.*, 1983). Furthermore, the administration of PPs results in the increase in activity of a number of enzymes, namely those involved in the peroxisomal β-oxidation of fatty acids (Lazarow & de Duve, 1976), and members of the cytochrome P450 IVA class (Orton & Parker, 1982). Importantly, this increase in enzyme activity is brought about by regulation at the level of transcription (Hardwick *et al.*, 1987). These observations imply that the mechanism by which PPs act may be similar to that of steroid hormones, i.e. via the activation of an intracellular ligand-activated transcription factor. With this *in mind*, a cDNA library from mouse liver was screened using a probe derived from the combined nucleotide sequences of several such nuclear receptors. Of the novel receptors identified by this approach, one was shown to be transcriptionally activated by PPs in an *in vitro* assay (Issemann & Green, 1990) This activation was detected using a structurally diverse range of PPs (Fig.1), and this novel nuclear receptor was termed the Peroxisome Proliferator Activated Receptor (PPAR).

Transcriptional Regulation Of Acyl CoA Oxidase By PPAR

It was important to establish a link between activation of PPAR *in vitro*, and the observed *in vivo* effects of PPs. In keeping with other nuclear receptors, it was anticipated that PPAR would regulate gene expression by binding specifically to DNA in the promoter region of responsive genes. Furthermore, the amino acid sequence in the region responsible for DNA binding suggested that the DNA element recognised by PPAR would be similar to that bound by a subset of nuclear receptors including those for retinoic acid and thyroid hormone (Umesono & Evans, 1989). This element is a direct repeat of the sequence AGGTCA. The acyl CoA oxidase (ACO) gene promoter was analysed for the presence of such a sequence; this gene is important since it is the rate-limiting enzyme of the peroxisomal β-oxidation pathway. Such elements are present in the ACO promoter (Osumi *et al.*, 1987), and a combination of DNA binding and *in vitro* activation experiments have demonstrated that one such element, consisting of the direct repeat sequence separated by a single nucleotide, can mediate transcriptional regulation of the ACO gene by PPAR (Tugwood *et al.*, 1992). This DNA sequence, termed a peroxisome proliferator response element (PPRE), behaves as a transcriptional enhancer since it regulates transcription in a position and orientation-independant fashion (Fig.2).

Nuclear receptors for retinoic acid, thyroid hormone and vitamin D3 also recognise a direct repeat of the AGGTCA sequence, and the relative spacing of these repeat elements determines the specificity for their cognate receptor (Umesono *et al.*, 1991). It has been shown that PPAR interacts specifically with repeat elements spaced by one nucleotide ('DR-1'), and that spacings of 0bp or 2-5bp are not PP response elements [I.Issemann, unpublished results]. Interestingly, this specificity is shared by the retinoid X receptor (RXR) (Kliewer *et al.*, 1992b).

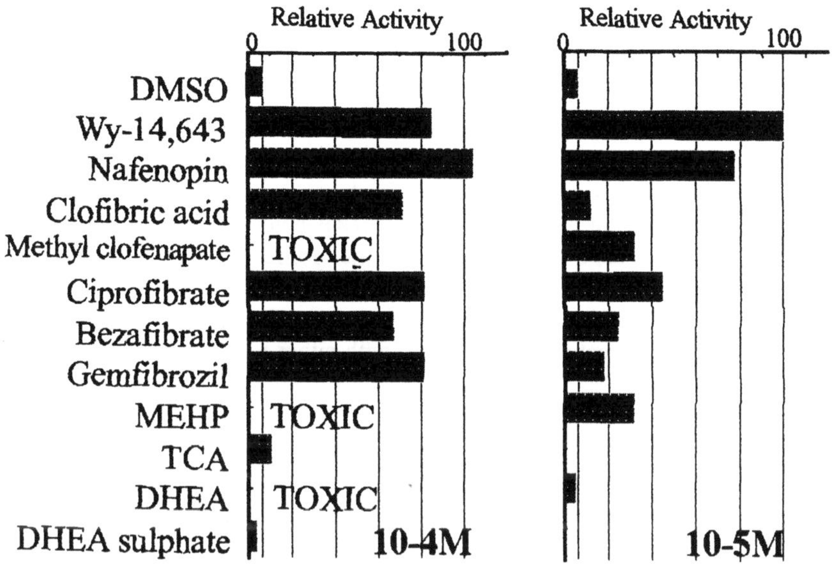

Fig.1. Activation of mouse PPAR-alpha by peroxisome proliferating drugs.
The ability of PP drugs to activate mPPAR-alpha in an *in vitro* activation assay was determined.. A mPPAR-alpha expression plasmid was transfected into mouse hepatoma cells along with a PPRE-containing reporter plasmid, and the cells were treated with drugs at the above concentrations. Cell extracts were assayed for reporter gene activity, and normalised to 10-5M Wy-14,643 = 100%.

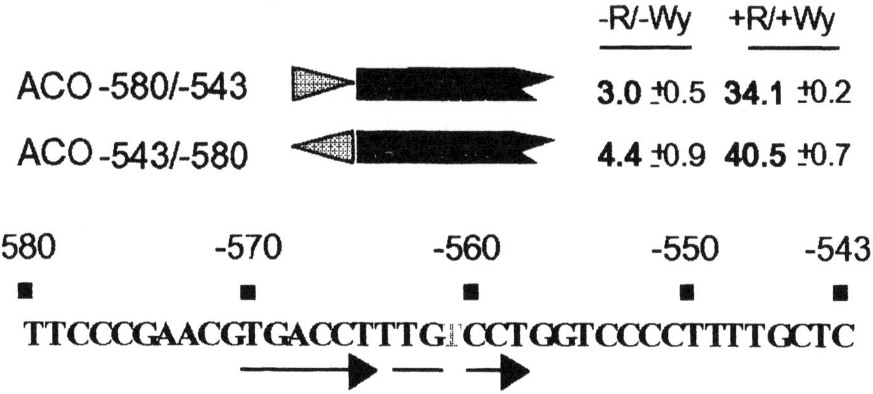

Fig.2. The Acyl CoA Oxidase PPRE Behaves As An Enhancer Element.
A DNA fragment from the ACO promoter (-580 to -543bp) was inserted into globin-CAT reporter plasmids in both orientations and transfected into mouse hepatoma cells along with a mPPAR-alpha expression vector and in the presence of WY-14,643 (+R/+Wy). CAT enzyme activity was measured , and expressed as percentage of total chloramphenicol acetylated. The sequence of this fragment is shown below, with arrows depicting TGACCT-like repeats.

PPAR Activates Transcription By Forming Heterodimers With RXR

Since RXR forms DNA-binding heterodimers with other nuclear receptors recognising a direct repeat element (Kliewer *et al*, 1992a), it seemed likely that the same would be true for PPAR. Indeed, this has been shown to be the case - the ACO direct repeat element is bound by PPAR/RXR heterodimers and the ligand for RXR, 9-*cis* retinoic acid, enhances peroxisome proliferator activity (Gearing *et al.*, 1993; Keller *et al.*, 1993). This activity is specific for DR-1 elements, and when both PPAR and RXR are co-expressed in the same cells in an *in vitro* assay a greater transcriptional effect is observed than with either receptor expressed alone (Fig.3).

Direct Transcriptional Regulation Of Cytochrome P450IVA1 By PPAR

Of the enzymes that are regulated by peroxisome proliferators *in vivo*, cytochrome P450IVA1 (CYP4A1) expression is induced the most rapidly, increases in rate of transcription being detectable within 30 minutes as opposed to a several hour time course for the induction of the peroxisomal enzymes (Bell *et al.*, 1991). Furthermore, since CYP4A1 is induced so early, it is conceivable that activity of this enzyme is required for the induction of 'later' PP-responsive genes. Thus, an 'alternative' scenario of the role of PPAR in liver enzyme induction can be postulated, which depends on CYP4A1 activity to generate a proximate ligand which would activate PPAR (Fig.4). Therefore, to distinguish between these two alternative models of enzyme induction it is important to determine whether or not CYP4A1 is directly transcriptionally regulated by PPAR.

In our laboratory, we have recently addressed this question by cloning large regions of CYP4A1 flanking DNA, and using functional assays to search for putative PPREs (T.C. Aldridge, J.D. Tugwood and S. Green, manuscript in preparation). Briefly, we have identified a DNA sequence approximately 4kbp 5' of the CYP4A1 transcription initiation site that is able to interact with PPAR/RXR heterodimers and confer responsiveness to PPs in *in vitro* assays. Unlike the ACO PPRE, this regulatory region is structurally complex in that it contains two elements with similarity to the DR-1 sequence; one of these (element I) has ten out of twelve bases identical to the 'TGACCTnTGACCT' consensus, and the other (element II) has nine out of twelve identical (Fig.5). Interestingly, the results show that element II confers responsiveness to PPAR/RXR heterodimers whereas element I does not; furthermore, element II is bound by PPAR/RXR heterodimers and element I is not. There is tentative evidence to suggest that element I mediates transcriptional control of CYP4A1 through binding of RXR plus an as yet unidentified nuclear factor - further experiments are in progress to clarify this.

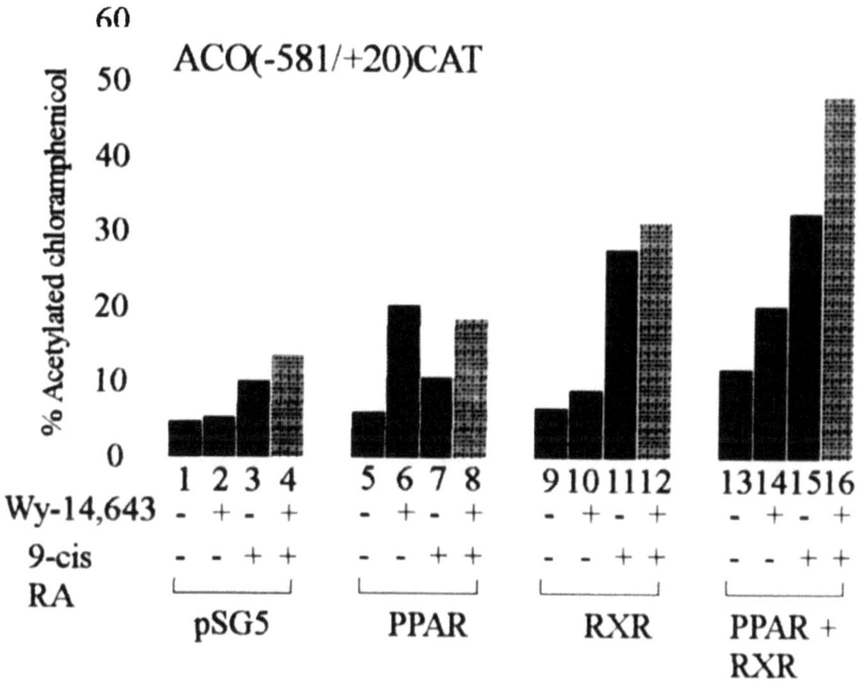

Fig.3. Interaction Of PPAR-alpha And RXR In Transcriptional Activation.

The reporter plasmid pACO(-581/+20)CAT, containing a PPRE, was transfected into mouse hepatoma cells either with or without PPAR and RXR expression plasmids, and with or without ligands as shown. Cell extracts were assayed for CAT activity.

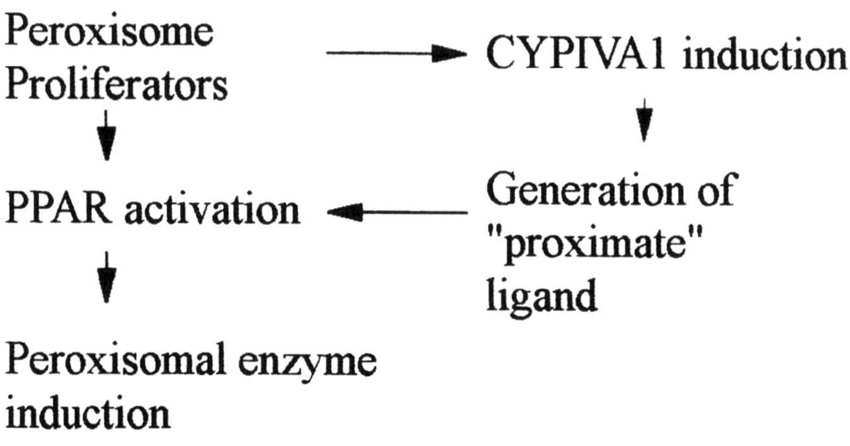

Fig.4. Alternative Hypotheses For PPAR Activation

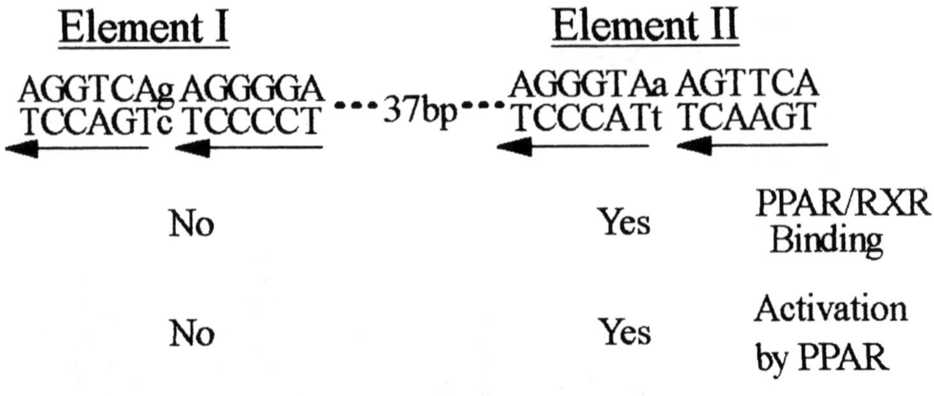

Fig.5. Structure Of The Rat Cytochrome P450IVA1 PPRE
The abilities of elements I and II to function in *in vitro* DNA binding assays and in activation assays is indicated.

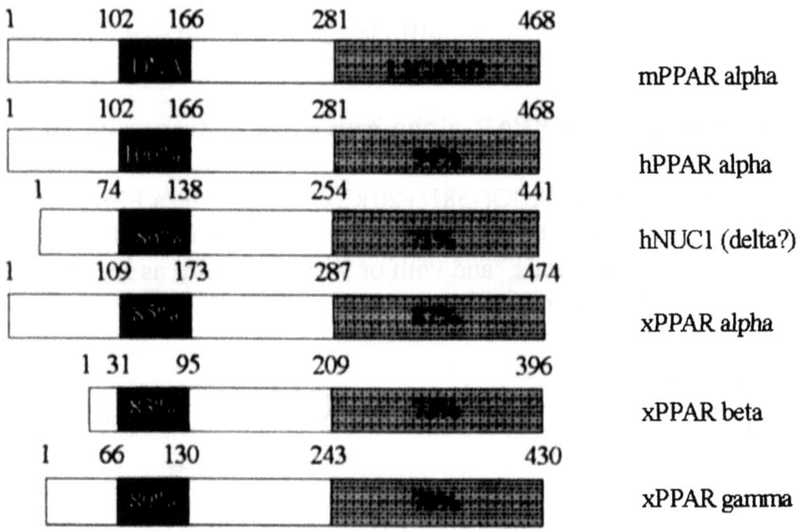

Fig.6. The Peroxisome Proliferator Activated Receptor Family.
Alignment of known PPARs with mouse PPAR-alpha. The percentages indicate amino acid identity, and the figures above the receptors amino acid positions.

PPARs Are A Complex Multigene Family

Following the isolation of PPAR from mouse liver, it quickly became apparent that rather than being a single receptor PPAR was a member of a multigene family that is becoming increasingly complex. Three receptors related to mouse PPAR have been identified in *Xenopus*, and these have been termed xPPARα, β and γ (Dreyer *et al.*, 1992). The prototype mouse receptor is termed mPPARα by this convention. PPARs have now been identified in rat and hamster, and perhaps more surprisingly in humans. A human homologue of mPPARα has been cloned from liver, and shown to be activated by the peroxisome proliferator Wy-14,643 (Sher *et al.*, 1993). This observation is intriguing given that humans are considered 'non-responsive', and it will be interesting to discover the precise mechanism which accounts for the apparent lack of response in man. An additional PPAR subtype, termed NUC1, has been isolated from human osteosarcoma cells which is also able to be activated by Wy-14,643 in an *in vitro* assay (Schmidt *et al.*, 1992). This receptor is related to but distinct from the previously isolated PPARα, β and γ subtypes, and this suggests at least four classes of PPARs exist. The relationship between PPARs thus far isolated is depicted in Fig.6.

PPARs can be Activated By Fatty Acids

An obvious question that arises, given the usual specificity of receptor/ligand interactions, is how a diverse collection of compounds such as PPs can all activate the same class of receptors. One possible clue to this has come from the observation that certain fatty acids are able to activate PPARs - this was first observed in *in vitro* assays using a chimaera between the rat PPARα and the glucocorticoid receptor (Göttlicher *et al.*, 1992). This suggests that fatty acids may be the natural activating ligands for PPARs, and that PPs activate by influencing the intracellular concentration of these compounds. One mechanism by which this may occur is via the liver fatty acid binding protein (l-FABP). This protein is thought to be involved in intracellular fatty acid transport, and interestingly some PPs are able to displace fatty acids from l-FABP (Cannon & Eacho, 1991). This has an interesting parallel with the action of retinoic acid (RA), in that intracellular concentrations of RA are regulated by the cellular retinoic acid binding protein (CRABP). Thus the activation of retinoic acid receptors by RA is influenced by the concentration and disribution of CRABP.

It is clear that the study of PPARs and their activation is likely to generate information pertinent to a number of research areas, including how receptors interact with ligands, mechanisms of transcriptional activation, and the assessment of human risk from toxic compounds. This latter area is of particular interest given the marked species differences in responses to PPARs. What is the natural role of human PPARs, and why are they apparently unable to mediate the pleiotropic toxic effects of this class of compound *in vivo* and *in vitro*? The answers to these questions will profoundly influence the assessment of human risk from peroxisome proliferators.

References

Bell, D.R., Bars, R.G., Gibson, G.G. and Elcombe, C.R. (1991). Localisation and differential induction of cytochrome P450IVA and acyl CoA oxidase in rat liver. Biochem. J. 275:247-252.

Cannon, J.R. and Eacho, P.I. (1991). Interaction of LY171883 and other peroxisome proliferators with fatty-acid-binding protein isolated from rat liver. Biochem. J. 280:387-391.

Cattley, R.C., Marsman, D.S. and Popp, J.A. (1989). Failure of the peroxisome proliferator Wy-14,643 to initiate growth-selectable foci in rat liver. Toxicol. 56:1-7.

Dreyer, C. Krey, G. Kellor, H., Givel, F., Helftenbein, G. and Wahli, W. (1992). Control of the peroxisomal β-oxidation pathway by a novel family of nuclear hormone receptors. Cell 68:879-887.

Gearing, K.L., Göttlicher, M, Teboul, M., Widmark, E. and Gustafsson, J-Å. (1993). Interaction of the peroxisome-proliferator-activated receptor and retinoid X receptor. Proc. Natl. Acad. Sci. USA 90:1440-1444.

Göttlicher, M, Widmark, E., Li Q. and Gustafsson, J-Å. (1992). Fatty acids activate a chimera of the clofibric acid-activated receptor and the glucocorticoid receptor. Proc. Natl. Acad. Sci. USA 89:4653-4657.

Hardwick, J.P., Song, B.J., Huberman, E. and Gonzalez, F.J. (1987). Isolation, complementary DNA sequence, and regulation of rat hepatic lauric acid ω-oxidation (cytochrome P-450). J. Biol. Chem. 262:801-810.

Issemann, I. and Green, S. (1990). Activation of a member of the steroid hormone receptor superfamily by peroxisome proliferators. Nature 347:645-650.

Keller, H., Dreyer, C. Medin, J., Mahfoudi, A., Ozato, K. and Wahli, W. (1993). Fatty acids and retinoids control lipid metabolism through activation of peroxisome proliferator-activated receptor—retinoid X receptor heterodimers. Proc. Natl. Acad. Sci. USA 90:2160-2164.

Kliewer, S.A., Umesono, K., Mangelsdorf, D.J. and Evans, R.M. (1992a). Retinoid X receptor interacts with nuclear receptors in retinoic acid, thyroid hormone and vitamin D_3 signalling. Nature 355:446-449.

Kliewer, S.A., Umesono, K., Noonan, D.J., Heyman, R.A. and Evans, R.M. (1992b). Convergence of 9-*cis* retinoic acid and peroxisome proliferator signalling pathways through heterodimer formation of their receptors. Nature 358:771-774.

Lalwani, N.D., Fahl, W.E. and Reddy, J.K. (1983). Detection of a nafenopin binding protein in rat liver cytosol associated with the induction of peroxisome proliferation by hypolipidemic compounds. Biochem. Biophys. Res. Comm. 116:388-393.

Lazarow, P.B. and de Duve, C. (1976). A fatty acyl-CoA oxidizing system in rat liver peroxisomes, enhancement by clofibrate, a hypolipidemic drug. Proc. Natl. Acad. Sci. USA 73:2043-2046.

Orton, T.C. and Parker, G.L. (1982). The effect of hypolipidaemic agents on the hepatic microsomal drug-metabolising enzyme system of the rat. Drug Metab. Dispos. 10:110-115.

Osumi, T., Ishii, N., Miyazawa, S. and Hashimoto, T. J. (1987). Isolation and structural characterization of the rat acyl-CoA oxidase gene. Biol. Chem. 262:8138-8143.

Reddy, J.K., Azarnoff, D.L. and Hignite, C.E. (1980). Hypolipidemic hepatic peroxisome proliferators form a novel class of carcinogens. Nature 283:397-398.

Reddy, J.K. and Lalwani, N.D. (1983). Carcinogenesis by hepatic peroxisome proliferators: evaluation of the risk of hypolipidemic drugs and industrial plasticisers to humans. CRC Crit. Rev. Toxicol. 12:1-58.

Schmidt, A., Endo, N., Rutledge, S.J., Vogel, R., Shinar, D. and Rodan, G.A. (1992). Identification of a new member of the steroid hormone receptor superfamily that is activated by a peroxisome proliferator and fatty acids. Mol. Endocrinol. 6:1634-1641.

Sher, T., Yi, H-F., McBride, O.W. and Gonzalez, F.J. (1993). cDNA cloning, chromosomal mapping, and functional characterization of the human peroxisome proliferator activated receptor. Biochemistry 32:5598-5604.

Tugwood, J.D., Issemann, I., Anderson, R.G., Bundell, K.R., McPheat, W.L. and Green, S. (1992). The mouse peroxisome proliferator activated receptor recognizes a response element in the 5' flanking region of the rat acyl CoA oxidase gene. EMBO J. 11:433-439.

Umesono, K. and Evans, R.M. (1989). Determinants of target gene specificity for steroid/thyroid hormone receptors. Cell 57:1139-1146.

Umesono, K., Murakami, K.K., Thompson, C.C. and Evans, R.M. (1991). Direct repeats as selective response elements for the thyroid hormone, retinoic acid, and vitamin D_3 receptors. Cell 65:1255-1266.

Role of Second Messengers in Mineral Particle-Induced Production of Reactive Oxygen Species by Phagocytes

K. Savolainen[*,1] *and M. Ruotsalainen*[*]
[*]Department of Toxicology, National Public Health Institute, and [1]Department of Environmental Sciences, University of Kuopio, Kuopio, Finland.

Exposure to asbestos has been associated for a number of years with pulmonary diseases such as pulmonary fibrosis and carcinoma. Occupational exposure to chrysotile has been a frequent cause for lung carcinoma. More recently, also an association between exposure to quartz dust and increased risk of lung carcinoma has been found (Koskela et al. 1987). Exposure to both asbestos and quartz is associated also with an increased risk of lung fibrosis.

Development of pulmonary inflammation leading to fibrosis is often associated with exposure to mineral particles. Inhalation of mineral particles results in their uptake by alveolar macrophages (Doelman et al. 1990). Phagocytized particles may induce inflammatory reaction in the lung during which chemotactic factors are released from macrophages. Subsequently, infiltration of leukocytes to the site of inflammation takes place upon their stimulation by macrophage- or microbe-derived chemotactic compounds (Lugano et al. 1982). Prolonged exposure to fibrous material also causes accumulation of activated phagocytic cells, i.e. neutrophils, eosinophils and basophils, in the lungs, often with pulmonary fibrosis as the end result. Mineral dusts are also effective activators of alveolar macrophages (Doelman et al. 1990) whose activation causes the release of proteolytic enzymes and reactive oxygen metabolites (ROM) such as superoxide anion radical, hydrogen peroxide, hydroxyl radical, and singlet oxygen (Adamson and Bowden 1984).

Even if quartz- and asbestos-induced pulmonary diseases have been well characterized, little is known of their pathophysiological mechanisms. A number of studies suggest, however, that mineral particles stimulate phagocytic cells to produce ROM (Mossman and Marsh 1989). The role of oxygen radicals, superoxide anion and hydroxyl radical, in these diseases seems possible (Halliwell and Gutteridge 1984, Seifert and Schultz 1991). In fact, it is the inability of phagocytic cells to destroy mineral particles that leads to pathologically persistent production of ROM, often called frustrating phagocytosis. When excessive amounts of ROM are produced, they may damage surrounding cells and tissue by attacking biological macromolecules (Doelman et al. 1990, Adachi et al. 1992).

Recent findings suggest that nitric oxide (NO), also a radical species produced by phagocytic cells, may have a role in the cytotoxicity induced by activation of immunological cells (Barnes and Belvisi 1993, Moncada and Higgs 1993). The role of NO in the activation of phagocytic cells by mineral particles will also be briefly discussed in this context. The general aim is to review data on the possible role of receptor-mediated events in mineral particle-induced stimulation of phagocytes to produce ROM.

Phagocytic Cells, Mineral Particles and Lung Pathology

Mineral particles induce inflammatory reactions in the lung. Upon stimulation by chemotactic factors released by macrophages, and from destroyed bacteria, leukocytes infiltrate from the blood stream into the pulmonary tissue. Asbestos is also involved in the production of chemotactic factors (Yano 1988). A number of soluble and particulate stimuli activate alveolar macrophages (Roney and Holian 1989) and human polymorphonuclear leukocytes (PMNL) to produce ROM (Tuomala et al. 1993a, b). In fact, sustained production of ROM by immunological cells upon their stimulation by mineral particles is involved in several pathological lung conditions (Mossman et al. 1983). Driscoll and coworkers (1990) found an association between increased quartz dust retention, lung injury, activation of alveolar macrophage fibronectin release, and the development of fibrosis in rat lung. Animal studies are in a good agreement with human studies. Prolonged occupational exposure to asbestos or quartz has been implicated in pulmonary fibrosis (Mossman et al. 1983) and in the increased risk of lung cancer in humans (Troll and Wiesner 1985, Koskela et al. 1987). Prolonged exposure to chrysotile asbestos causes also mesothelioma (Mossman and Gee 1989).

There is evidence that mineral particle-induced cell activation may be partially a receptor-mediated phenomenon (Korchak et al. 1988a, Tuomala et al. 1992, Ruotsalainen and Savolainen 1994). Chrysotile asbestos stimulates the production of superoxide anion in guinea pig alveolar macrophages by increasing the turnover of phosphatidic acid (PA) (Roney and Holian 1989). The change in PA turnover reflects activation of inositol lipid signalling system because hydrolysis of phosphatidylinositol-4,5-bisphosphate (PIP_2) yields diacylglycerol (DG). DG is, in turn, metabolized to PA (Berridge and Irvine 1989, Putney 1993). Tuomala and coworkers (1992) have shown that facilitated inositol lipid signalling and increased production of another PIP_2-derived second messenger, inositol-1,4,5-trisphosphate ($InsP_3$) precedes the production of ROM in PMNL by quartz. Facilitation of inositol lipid signalling also precedes the elevation of levels of free intracellular Ca^{2+} ($[Ca^{2+}]_i$), because $InsP_3$ releases Ca^{2+} from intracellular stores. DG activates, in turn, PKC (Nishizuka 1986, Berridge and Irvine 1989, Seifert and Schultz 1991).

The mechanisms whereby mineral particles activate phagocytic cells are poorly known whereas those of soluble agonists have been better clarified. Therefore, mechanisms of phagocytic cell activation by soluble stimuli will be discussed first, and used as a background information for reviewing activation mechanisms of phagocytic cells by mineral particles.

Chemotactic Peptide fMLP-Induced Phagocyte Activation

A chemotactic peptide, fMLP, released from bacterial membrane, causes the invasion of leukocytes to the site of fMLP release. fMLP is also a standard stimulant for mechanisms of receptor-mediated cell activation. fMLP causes the aggregation of PMNL, depolarization of the cells, and stimulation of respiratory burst, i.e. increased O_2 consumption and increased activity of hexose monophosphate shunt (Holian and Stickle 1985). Binding of fMLP to its cell surface receptor activates G-protein and stimulates G-protein-coupled phospholipase C (PLC) thereby initiating a bifurcating cell signalling pathway that consists of inositol lipid-derived second messengers. Membrane PIP_2 is hydrolyzed by PLC to active second messengers, $InsP_3$ and DG (Berridge and Irvine 1984, Korchak et al. 1988a,b, Baggiolini et al. 1993). Obviously $InsP_3$-induced release of Ca^{2+} from intracellular stores (Putney 1993), and DG-induced stimulation of protein kinase (PKC) (Nishizuka 1986) are essential for fMLP-induced leukocyte activation. Subsequently, PKC phosphorylates plasma membrane NADPH oxidase which then catalyzes the reduction of oxygen to superoxide anion (Cox et al. 1985, Sadler and Badwey 1988, Baggiolini and Wyman 1990, Seifert and Schultz 1991). This conclusion agrees well with the observation that oxidative burst correlates with the amount of PKC in the cells (Salamino et al. 1991).

Phorbol Myristate Acetate-Induced Phagocyte Activation

Phorbol myristate acetate (PMA), a tumor promoter, is a direct activator of PKC that is able to by-pass receptor stimulation (Nishizuka 1986). PKC has a role in receptor-mediated and G-protein coupled inositol lipid signalling pathway as one of the regulatory enzymes of cellular functions and activation (Lad et al. 1992, Putney 1993). Because of the ability of PMA to directly stimulate PKC, PMA has been widely used to explore independent roles of PKC in various types of activation of phagocytic cells. An important feature of DG, the other key messenger in the inositol lipid signalling pathway in addition of $InsP_3$ (Putney 1993) is that it induces a transient translocation, and subsequent activation of PKC, into the membrane and concomitantly causes the activation of NADPH oxidase. The end result is increased ROM production. In contrast to DG, PMA induces a permanent translocation of PKC into the membrane. This difference in effects between DG and PMA may explain many of the different effects in phagocytic cells during their receptor-mediated or direct stimulation of PKC. For example, Chabot et al. (1992) suggest that fMLP stimulates DG production through phospholipase D (PLD), whereas PMA-induced DG generation may involve several metabolic pathways including PLD activation, and activation of PA phosphohydrolase.

Role of Protein Kinase C in Phagocyte Activation

Several recent observations emphasize the role of cellular lipid metabolism and subsequent activation of PKC in agonist-induced production of ROM in phagocytic cells. Umei et al. (1993) reported that priming of neutrophils with PMA, but not with fMLP, prior to their stimulation with the same compounds causes the translocation of PKC activity into the membrane which is a prerequisite for the enzyme activity, subsequent phosphorylation of NADPH oxidase, and ROM production. Ding and Badwey (1993a,b) have shown in guinea pig neutrophils that they contain a battery of protein kinases that are rapidly activated upon stimulation by fMLP. The activity was maximal within 30 sec with this stimulus, and returned to basal level within 3 min. This is consistent with the time-course of fMLP-induced increase in inositol lipid signalling (Korchak et al. 1988a,b), elevation of $[Ca^{2+}]_i$, and the increase of the production of ROM (Tuomala et al. 1992, 1993b). In contrast, stimulation of guinea pig neutrophils with PMA resulted in a diminution of these kinase activities. However, stimulation of PKC with PMA obviously stimulates another battery of protein kinases, possible various PKC (Nishizuka 1986) isoenzymes because this stimulation induces a strong oxidative burst in PMNL that peaks between 9 and 16 min after to the stimulation. Moreover, these responses can be almost completely inhibited with a novel PKC inhibitor, Ro 31-7549 (Twomey et al. 1990, Ruotsalainen and Savolainen 1994).

The complexity of the signalling pathway leading to the activation of phagocytic cells is further emphasized by the observation of Kessels et al. (1993) who suggested that PKC activation is not involved in fMLP-induced activation of PLD that occurs simultaneously with oxidative burst in PMNL. The activation of PLD has been shown to result in the production of PA and to play an important role in the stimulation of PMNL respiratory burst. In fact, PA derived from phosphatidylcholine has been shown to induce respiratory burst in electropermeabilized human neutrophils by acting on a downstream step of PKC. This is most likely due to direct activation of NADPH oxidase which increases the production of superoxide anion (Mitsuyama et al. 1993). This emphasizes the complexity of PMNL stimulation because PA is further metabolized to DG by PA phosphohydrolase resulting in PKC activation. In some cases, the activation of NADPH oxidase seems to be independent of PKC activation by DG.

Activation of PKC occurs in concert with the activation of NADPH oxidase, responsible for fMLP-induced rapid and transient oxidative burst. PKC inhibitors staurosporine and cpg41251 inhibit PMA- and fMLP-induced oxidative burst. PKC inhibitor staurosporine amplified fMLP-induced activation of PLD and oxidative burst indicating that PKC may have a modulating role in PLD activation responsible for PA-mediated respiratory burst in PMNL (see Kessels et al. 1993). Thus, even if PKC seems to play a key-role in PMA-induced activation of NADPH oxidase and subsequent oxidative burst, this may not necessarily be the case in fMLP-induced receptor-mediated activation of PMNL. Details of phagocyte transmembrane signalling pathways are shown in Figure 1.

Role of Nitric Oxide in Phagocyte Activation

An important cellular messenger in addition to the elements of cell signalling pathways discussed above is nitric oxide (NO). NO is generated by mammalian cells through a hitherto unrecognized metabolic route by NO synthase which by using L-arginine as a substrate produces citrulline and NO. Actions of NO are mediated through the activation of soluble guanylate cyclase and the subsequent increase in the concentration of cyclic guanosine monophosphate (cGMP) in target cells. NO is produced in large quantities also during host defence and immunologic reactions, and is generated by macrophages PMNL. The stimulation of peritoneal macrophages by PMA also increases the production of NO. Production of NO in macrophages and PMNL from L-arginine is responsible e.g. for the cytotoxicity of these cells against tumor cells and bacteria. The biochemical basis for the cytotoxicity induced by NO depends on the combination of NO with iron-containing moieties in key enzymes in the respiratory chain and of the synthesis of DNA (see Moncada and Higgs 1993). There are data to suggest that NO may also play a part both in acute and chronic inflammation, and that NO may also play a part in tissue damage. It may not only be cytotoxic to invading particles but to cells that produce it and for neighboring cells. Moreover, even if NO itself is a toxic radical species, it may also react with oxygen-derived radicals to generate molecules, such a peroxynitrites, that amplify NO toxicity (Moncada and Higgs 1993). It is also possible that NO may play a role in mineral particle-induced cell activation and subsequent cellular injury.

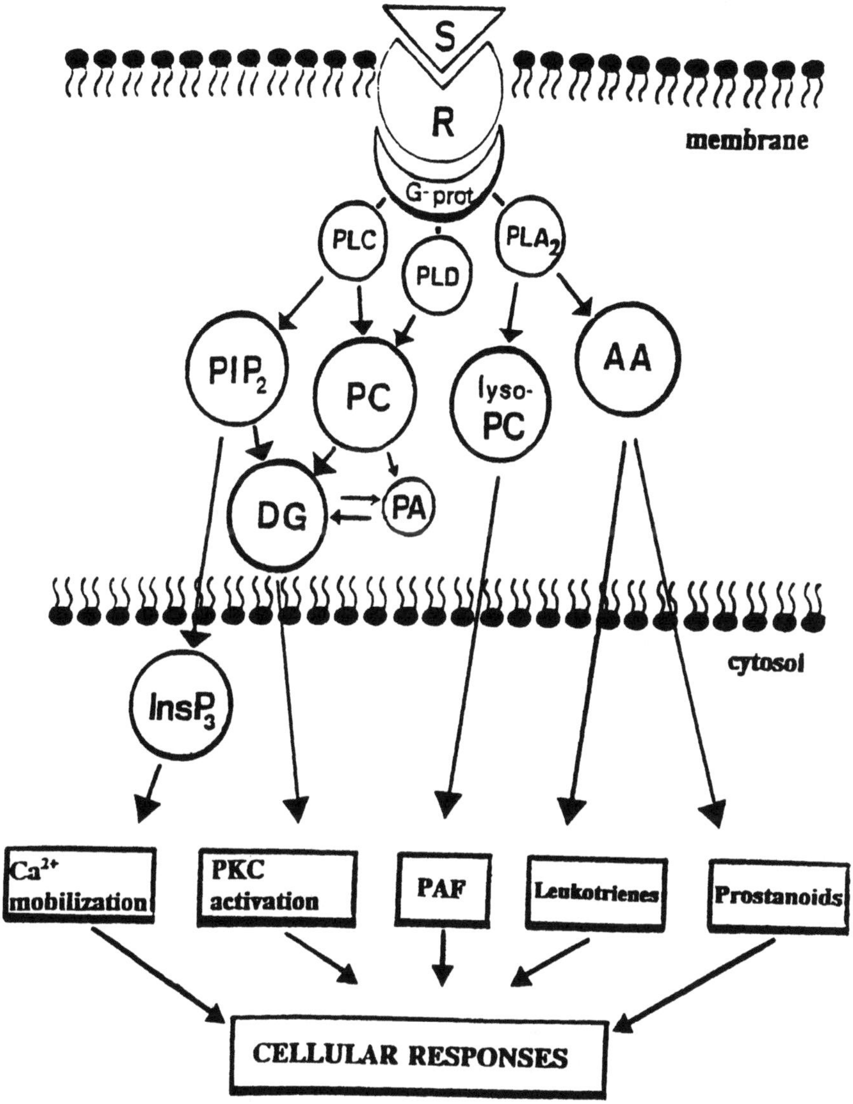

Fig. 1. Summary of phagocyte transmembrane signalling systems. Recognition of an agonist (e.g. fMLP, mineral particle) cause receptor-mediated activation of an effector enzyme, phospholipase C (PLC), through G-protein activation that causes hydrolysis of membrane phosphatidylinositol-4,5-bisphosphate (PIP_2). Hydrolysis of PIP_2 generates two second messengers, inositol-1,4,5-trisphosphate ($InsP_3$) and diacylglycerol (DG). $InsP_3$ mobilizes Ca^{2+} from intracellular stores and DG activates a phospholipid-dependent protein kinase C (PKC). Hydrolysis of lysophosphatidylcholine (lyso-PC) by phosphalipase A_2 (PLA_2) leads to generation of platelet activating factor (PAF). DG-kinase phosphorylates DG to phosphatidic acid (PA), and PA-phosphohydrolase forms DG from PA. Arachidonic acid (AA) is converted via the cyclo-oxygenase and lipo-oxygenase pathways to leukotrienes and prostanoids.

Mineral Particle-Induced Activation of Phagocytic Cells

The ability of asbestos and quartz to increase the production of ROM in human PMNL and macrophages (Hedenborg and Klockars 1987,1989, Klockars et al. 1990, Nyberg and Klockars 1990a,b,c) has been convincingly demonstrated. There are also data to show that both quartz and chrysotile asbestos generate a strong oxidative burst in rat and guinea pig alveolar macrophages (Roney and Holian 1989, Chen et al. 1991). The correlation between fiber-induced pulmonary fibrosis and bronchogenic carcinoma (Mossman et al. 1983, Mossman and Marsh 1989) as well as pleural mesothelioma has been well demonstrated (Doelman et al. 1990). The role of ROM is this pathology has received increasing amount of support from recent studies (Mossman and Gee 1989). It is clear that ROM production is a part of normal functions of phagocytic cells, and amplification of ROM production by mineral particles has proved a tempting hypothesis which has received support from a number of recent studies (Roney and Holian 1989, Tuomala et al. 1992, 1993a,b, Ruotsalainen and Savolainen 1994).

The effects of mineral fibers resemble in many ways those induced by soluble stimuli whose mechanisms have been better clarified (Lad et al. 1992, Baggiolini et al. 1993). Perhaps the most striking new feature in mineral particle-induced phagocytic cell activation and ROM production is that the involvement of G-protein-coupled recognition sites for mineral particles, subsequent activation of inositol lipid signalling pathway, $[Ca^{2+}]_i$, and the activity of PKC may play an important role in mineral particle-induced activation of immunological cells.

Mineral Particles and Second Messengers

Korchak et al. (1988a,b) provided evidence that Ca^{2+}-mobilizing ligands, fMLP and concanavalin A, activate PMNL to produce ROM by stimulating inositol lipid signalling and elevating $[Ca^{2+}]_i$. In an important paper Meshulam et al. (1988) showed that particulate stimuli, notably serum opsonized and non-opsonized zymosan, elicited oxidative burst in PMNL. Increased production of $InsP_3$ and elevation of $[Ca^{2+}]_i$ preceded this particle-induced production of ROM. The kinetics of these events were such that they clearly proposed for the first time an association between the oxidative burst, and the production of inositol lipid second messengers. In the same paper, also Candida albicans hyphae induced similar effects in PMNL. Some of these observation lead several investigators to hypothesize that also mineral particle-induced cell activation may be a receptor-mediated phenomenon.

Roney and Holian (1989) induced excessive production of ROM in guinea pig alveolar macrophages by exposing them to chrysotile asbestos.

Increased turnover of PA, a metabolite of inositol lipid-derived DG (Berridge and Irvine 1989), preceded chrysotile-induced production of ROM in these cells. Increased PA production was followed by elevations of $[Ca^{2+}]_i$, most likely because it more generally reflect the activity of inositol signalling pathway, a part of which is the production of Ca^{2+}-releasing $InsP_3$. Moreover, chrysotile-induced ROM production could be inhibited with a non-specific PKC inhibitor, staurosporine, and partially blocked with pertussis toxin that inhibits a stimulatory G-protein coupled to PLC (Roney and Holian 1989). Therefore, these results provide evidence that chrysotile induce ROM production in guinea pig alveolar macrophages takes place by acting through a receptor-coupled and second messenger-mediated mechanism in which a G-protein and PKC may have at least a contributing role.

Another important piece of information suggesting the role for second messengers in mineral fiber-induced ROM production in phagocytic cells has derived from studies of Tuomala et al. (1992) who demonstrated in PMNL that quartz increased the production of $InsP_3$ prior to the increased production of ROM. In another study, Tuomala et al. (1993b) showed that quartz also dose-dependently elevates $[Ca^{2+}]_i$ in stimulated PMNL. The time-course of the quartz-induced $[Ca^{2+}]_i$ elevation was such that it was clear that also the changes in $[Ca^{2+}]_i$ preceded increased ROM production. This is consistent with the assumption that quartz may induce the production of ROM through a mechanism which requires both the generation of inositol lipid second messengers and subsequent elevation of $[Ca^{2+}]_i$. It was notable that the quartz-induced ROM production, both in terms of time-course and dose-effect, resembled closely those induced by fMLP and PMA. This is consistent with the view that receptor-mediated mechanisms and PKC may be involved in quartz-induced PMNL activation. In the same study, chrysotile had effects on $[Ca^{2+}]_i$ and ROM production that closely resembled the effects of quartz.

Mineral Particles, G-proteins, Calcium and Cell Damage

To further explore the mechanisms whereby quartz induces ROM production in PMNL, Tuomala et al. (1993a) studied the effects of repeated doses of quartz, and interactions of quartz with fMLP on $[Ca^{2+}]_i$ and ROM production. These studies revealed that a single dose of quartz dose-dependently elevated $[Ca^{2+}]_i$ and ROM production, whereas such effect was not seen after repeated doses of quartz. On the other hand, quartz administered prior to fMLP did not modify fMLP-induced $[Ca^{2+}]_i$, but attenuated fMLP-induced ROM production. Priming of PMNL with fMLP prior to quartz did not affect quartz-induced $[Ca^{2+}]_i$ changes but quartz-induced ROM production was greatly amplified. These studies provide evidence that both quartz and fMLP increase the production of ROM in PMNL by mechanisms in which $[Ca^{2+}]_i$ may have a key-role. It seems possible that both $[Ca^{2+}]_i$-dependent and -independent mechanisms may account to quartz-induced ROM production in PMNL. Moreover, Ruotsalainen and Savolainen (1994) have shown that a novel PKC inhibitor, Ro 31-7549 (Twomey et al. 1990), almost completely blocks both quartz- and chrysotile-induced production of ROM in PMNL suggesting a role for PKC in mineral particle-induced oxidative burst.

Pretreatment of cells with cholera toxin, but not with pertussis toxin, partially inhibited quartz- and chrysotile-induced production of ROM thereby providing evidence that a cholera but not a pertussis toxin sensitive G protein may be somehow involved in quartz- and chrysotile-induced activation of PMNL. These observations were further substantiated by the finding that a stable GTP analogue, GTPγS, amplified quartz- but not chrysotile-induced production of ROM (Ruotsalainen and Savolainen, unpublished).

Ruotsalainen and coworkers (unpublished) have shown that both extra- and intracellular Ca^{2+} are important for quartz- and chrysotile-induced activation of PMNL. A calcium channel blocker verapamil attenuated both quartz- and chrysotile-induced elevations of Ca^{2+} suggesting that the influx of Ca^{2+} through verapamil-sensitive Ca^{2+} channels may play a role in mineral particle-induced activation of PMNL. Moreover, the steepness of quartz- and chrysotile-induced elevation of $[Ca^{2+}]_i$ was dose-dependently decreased by verapamil indicating that verapamil inhibited the influx of extracellular Ca^{2+} into the cell. These conclusions were, in fact, confirmed by similar findings when extracellular Ca^{2+} was chelated with EGTA. These data suggest that a release of Ca^{2+} from intracellular stores is the initial event in mineral particle-induced activation of PMNL. Thus, even if both intra- and extracellular Ca^{2+} are essential for PMNL activation by mineral particles, the release of Ca^{2+} from intracellular stores seem to be essential for the initiation of the activation which is then amplified by slower influx of extracellular calcium, possibly due a co-operative action of elevated $[Ca^{2+}]_i$ on membrane calcium channels (von Tscharner et al. 1986, Putney 1993).

Chen et al. (1991) have shown that quartz induces a dose-dependent increase in $[Ca^{2+}]_i$ in rat alveolar macrophages providing evidence for the key-role of Ca^{2+} in mineral particle-induced cell activation. Moreover, chelation of extracellular calcium by EGTA partially inhibited quartz-induced $[Ca^{2+}]_i$ elevation, lactatedehydrogenase release, and cellular damage (Chen et al. 1991). They concluded that elevation of $[Ca^{2+}]_i$ is a very early event in quartz-induced macrophage damage subsequent to excessive cell stimulation by mineral particles. These results have been in essence confirmed by the findings of Rojanasakul et al. (1993) who showed that quartz produces damage in rat alveolar macrophages, associated with increased levels of $[Ca^{2+}]_i$, and partially blocked by a calcium channel blocker nifedipine, or by chelation of extracellular calcium by EGTA.

Mineral Particles and Nitric Oxide

The role of NO in mineral particle-induced activation of phagocytic cells has not yet been fully clarified. Some studies have been carried out with particulate stimuli, such as serum-opsonized zymosan (Wright et al. 1989). Recently, Ruotsalainen et al., (unpublished) stimulated murine macrophage cell line RAW 264.7 with interferon-γ (INF), chrysotile, quartz, or combinations of INF with chrysotile or quartz. In these studies both quartz and chrysotile seemed slightly to stimulate inducible NO-synthase which was, however, modest as compared to INF-induced stimulation of NO production. INF was remarkably cytotoxic to the macrophages, and both quartz and chrysotile further amplified INF-induced cytotoxicity even if both were inactive alone. Thus, even if quartz and chrysotile greatly increased INF-cytotoxicity, and both slightly stimulated inducible NO-synthase, this INF-mineral particle interaction was most likely not due to production of NO.

Conclusions

Together these data provide evidence that activation of phagocytic cells to produce ROM may at least partially be mediated through receptor-coupled mechanisms. Involvement of a cholera-sensitive G-protein, possibly linked with a cell membrane structure recognizing mineral particles, requires further clarification. Quartz or chrysotile are probably capable of increasing the production of inositol lipid second messengers and concomitantly elevating levels of $[Ca^{2+}]_i$. Elevations of $[Ca^{2+}]_i$ are probably prerequisites for mineral particle-induced activation of phagocytic cells to produce ROM. Intra- and extracellular calcium are important for phagocytic cell activation, intracellular calcium having a role of an initiator, and extracellular calcium one as an amplifier of cellular responses. PKC probably plays a key-role in the activation of phagocytes to produce ROM. Production of NO may be involved in the cascade of event leading to phagocytic cell activation; it may modify mineral particle-induced responses of phagocytic cells.

Acknowledgements

The authors wish to thank Ms. Leena Heikkinen and Ms. Liisa Kujanpää for skillful technical assistance, Ms. Heli Leinonen for careful typing of the manuscript, and the Finnish Work Environment Fund and The Academy of Finland for financial support.

References

Adachi S, Kawamura K, Yoshida S, Takemoto K (1992) Oxidative damage on DNA induced by asbestos and man-made fibres in vitro. Int Arch Occup Health 63:553-557

Adamson IY, Bowden DH (1984) Relationship of alveolar epithelial injury and repair to induction of pulmonary fibrosis. Am J Pathol 130:377-383

Baggiolini M, Boulay F, Badwey JA, Curnutte JT (1993) Activation of neutrophil leukocytes: chemoattractant receptors and respiratory burst. FASEB J 7:1004-1010

Baggiolini M, Wymann P (1990) Turning on the respiratory burst. Trends Biol Sci 15:69-72

Barnes PJ, Belvisi MG (1993) Nitric oxide and lung disease. Thorax 48:1034-1043

Berridge MJ, Irvine RF (1984) Inositol trisphosphate, a novel second messenger in cellular signal transduction. Nature 312:315-321

Berridge MJ, Irvine RF (1989) Inositol phosphates and cell signalling. Nature 341:197-205

Chabot MC, McPhail C, Wykle RL, Kennerly AA, McCall CE (1992) Comparison of diglyceride production from choline-containing phosphoglycerides in human neutrophils stimulated with N-formylmethionyl-leucyphenylalanine, ionophore A23187 or phorbol 12-myristate 13-acetate. Biochem J 286:693-699

Chen J, Armstrong LC, Liu S, Gerriets JE, Last JA (1991) Silica increases cytosolic free calcium ion concentration of alveolar macrophages in vitro. Toxicol Appl Pharmacol 111:211-220

Cox JA, Jeng AY, Sharkey NA, Blumberg PM, Tauber AI (1985) Activation of the human neutrophil nicotinamide adenine dinucleotide phosphate (NADPH)-oxidase by protein kinase C. J Clin Invest 76:1932-1938

Ding J, Badwey JA (1993a) Neutrophils stimulated with a chemotactic peptide or a phorbol ester exhibit different alterations in the activities of a battery of protein kinases. J Biol Chem 268:5234-5240

Ding J, Badwey JA (1993b) Stimulation of neutrophils with a chemoattractant activates several novel protein kinases that can catalyze the phosphorylation of peptides derived from the 47-kDa protein component of the phagocyte oxidase and myristoylated alanine-rich C kinase substrate. J Biol Chem 268:17326-17333

Doelman CJA, Leurs R, Oosterom WC, Bast A (1990) Mineral dust exposure and free radical-mediated lung damage. Exp Lung Res 16:41-55

Driscoll KE, Maurer JK, Lindenschmidt RC, Romberger D, Rennard SI, Crosby L (1990) Respiratory tract responses to dust: Relationship between dust burden, lung injury, alveolar macrophage fibronectin release, and the development of pulmonary fibrosis. Toxicol Appl

Pharmacol 106:88-101
Halliwell B, Gutteridge JMC (1984) Oxygen toxicity, oxygen radicals, transition metals and diseases. Biochem J 219:1-14
Hedenborg M, Klockars M (1987) Production of reactive oxygen metabolites induced by asbestos fibres in human polymorphonuclear leucocytes. J Clin Pathol 40:1189-1193
Hedenborg M, Klockars M (1989) Quartz-dust-induced production of reactive oxygen metabolites by human granulocytes. J Lung 167:23-32
Holian A, Stickle D (1985) Calcium regulation of phosphatidyl inositol turnover in macrophage activation by formyl peptides. J Cell Physiol 123:39-45
Kessels GC, Krause K-H, Verhoeven AJ (1993) Protein kinase C activity is not involved in N-formylmethionyl-leucyl-phenylalanine-induced phospholipase D activation in human neutrophils, but in essential for concomitant NADPH oxidase activation: studies with a staurosporine analogue with improved selectivity for protein kinase C. Biochem J 292:781-785
Klockars M, Hedenborg M, Vanhala E (1990) Effect of two particle surface-modifying agents, polyvinylpyridine-N-oxide and carboxymethylcellulose, on the quartz and asbestos mineral fiber-induced production of reactive oxygen metabolites by human polymorphonuclear leukocytes. Arch Environ Health 45:8-14
Korchak HM, Vosshall LB, Zagon G, Ljubich P, Rich AM, Weissmann G (1988a) Activation of the neutrophil by calcium-mobilizing ligands. J Biol Chem 263:11090-11097
Korchak HM, Vosshall LB, Zagon G, Ljubich P, Rich AM, Weissmann G (1988b) Activation of the neutrophil by calcium-mobilizing ligands II. Correlations of calcium, diacylglycerol, and phosphatidic acid generation with superoxide anion generation. J Biol Chem 263:11098-11105
Koskela R-S, Klockars M, Järvinen E, Kolari PJ, Rossi A (1987) Cancer mortality of granite workers. Scand J Work Environ Health 13:26-31
Lad PM, Kaptein JS, Lin C-KE, Kalunta CI, Scott SJ, Gu DG (1992) G-proteins and the role of second messengers in the regulation of the human neutrophil. Immunol Ser 57:107-136
Lugano EM, Dauber JH, Daniele RP (1982) Acute experimental silicosis. Lung morphology, histology, and macrophage chemotaxin secretion. Am J Pathol 109:27-36
Meshulam T, Diamond RD, Lyman CA, Wysong DR, Melnick DA (1988) Temporal association of calcium mobilization, inositol trisphosphate generation, and superoxide anion release by human neutrophils activated by serum opsonized and nonopsonized particulate stimuli. Biochem Biophys Res Commun 150:532-539
Mitsuyama T, Takeshige K, Minakami S (1993) Phosphatidic acids induces the

respiratory burst of electropermeabilized human neutrophils by acting on a downstream step of protein kinase C. FEBS Lett 328:67-70
Moncada S, Higgs A (1993) The L-argine-nitric oxide pathway. New Engl J Med 329:2002-2012
Mossman BT, Gee JBL (1989) Asbestos-related diseases. New Engl J Med 320:1921
Mossman BT, Light W, Wei E (1983) Asbestos: Mechanisms of toxicity and carcinogenicity in the respiratory tract. Ann Rev Pharmacol Toxicol 23:595-615
Mossman BT, March JP (1989) Evidence supporting a role for active oxygen species in asbestos induced toxicity and lung disease. Env Health Perspect 81:91-94
Nishizuka Y (1986) The role of protein kinase C in cell surface signal transduction and tumor promotion. Science 233:305-312
Nyberg P, Klockars M (1990a) Measurement of reactive oxygen metabolites produced by human monocyte-derived macrophages exposed to mineral dusts. Int J Exp Path 71:537-544
Nyberg P, Klockars M (1990b) Quartz-induced production of reactive oxygen metabolites by activated human monocyte-derived macrophages. APMIS 98:823-827
Nyberg P, Klockars M (1990c) Effect of immunoglobulins on mineral dust-induced production of reactive oxygen metabolites by human macrophages. Inflammation 14:621-629
Putney JW Jr (1993) Excitement about calcium signalling in inexcitable cells. Science 262:676-678
Rojanasakul Y, Wang L, Malanga CJ, Ma JYC, Banks DE, Ma JKH (1993) Altered calcium homeostasis and cell injury in silica-exposed alveolar macrophages. J Cell Phys 154:310-316
Roney PL, Holian A (1989) Possible mechanism of chrysotile asbestos-stimulated superoxide anion production in guinea pig alveolar macrophages. Toxicol Appl Pharmacol 100:132-144
Ruotsalainen M, Savolainen K (1994) Effects of a protein kinase C inhibitor, Ro 31-7549, on the activation of human leukocytes by particulate stimuli. Hum Exper Toxicol (in Press)
Sadler KL, Badwey JA (1988) Second messengers involved in superoxide production byneutrophils. Hemat Oncol Clin North Am 2:185-200
Salamino F, Sparatore B, De Tullio, Mengotti P, Melloni E, Pontremoli S (1991) Respiratory burst in activated neutrophils is directly correlated to the intracellular level of protein kinase C. Eur J Biochem 200:573-577
Seifert R, Schultz G (1991) The superoxide-forming NADPH oxidase of phagocytes: an enzyme system regulated by multiple mechanism. Reviews of Physiology, Biochemistry, and Pharmacology 117, Springer-Verlag, Heidelberg, p. 19

Troll W, Wiesner R (1985) The Role of Oxygen Radicals as a Possible Mechanism of Tumor Promotion. Annu Rev Pharmacol Toxicol 25:509-528

von Tscharner V, Prod'hom B, Baggiolini M, Reuter H (1986) Ion channels in human neutrophils activated by a rise in free cytosolic calcium concentration. Nature 324:369-372

Tuomala M, Hirvonen M-R, Holopainen M, Savolainen K (1993a) Stimulation of human poly morphonuclear leukocytes by consecutive doses of quartz and interactions of quartz with fMLP. Toxicol Appl Pharmacol 118:224-232

Tuomala MH, Hirvonen M-R, Savolainen KM (1992) Production of inositol phosphates and reactive oxygen metabolites in quartz-dust-stimulated human polymorphonuclear leukocytes. FEBS Lett 296:57-60

Tuomala M, Hirvonen M-R, Savolainen KM (1993b) Changes in free intracellular calcium and production of reactive oxygen metabolites in human leukocytes by soluble and particulate stimuli. Toxicology 80:71-82

Twomey B. Muid RE, Nixon JS, Sedwick AD, Wilkonson SE, Dale MM (1990) The effect of new potent selective inhibitors of protein kinase C on the neutrophil respiratory burst. Biochem Biophys Res Commun 171:1087-1092

Umei T, Ohhara N, Okamura S, Harada M, Nakao M, Shirai T, Niho Y (1993) Activation of neutrophils NADPH oxidase by PMA: cytosol activity is translocated in phorbol-primed neutrophils. Int J Biochem 25:631-633

Wright CD, Mülsch A, Busse R, Osswal H (1989) Generation of nitric oxide by human neutrophils. Biochem Biophys Res Commun 160:813-819

Yano E (1988) Mineral fiber-induced malondialdehyde formation and effect of oxidant scavengers in phagocytic cells. Int Arch Occup Environ Health 61:19-23

Estrogenic Effects of Some Xenobiotics

M. Vargová, M. Gajdová, J., Jakubovský, and L. Wsólová
Institute of Preventive and Clinical medicine, Limbová 4, 833 01 Bratislava, Slovak Republic

While investigating the estrogenic properties of the pesticide Fosmet administered to neonatal rats we found that a 10% emulsion concentrate of Tween 80 can increase the mean relative weight of the uterus and cause changes in the endometrium that are characteristic of the action of estrogens (Gajdová et al., 1988). In the rats exposed to Tween 80 we found a significant prolongation of the oestrous cycle and persistent vaginal estrus in adulthood (Vargová et al., 1990). Similar changes after the application of some substances with estrogenic effect were described by Bulger and Kupfer (1985).

The results of the examination of the controls indicated that the estrogenic effects that were observed may be caused by the presence of the emulsifier. We decided, therefore, to analyze in detail the possible estrogenic effect of Tween 80 and Polyethyleneglycol 300. In order to do this, we observed the onset of vaginal opening and the length of the estrous cycle, as well as its cytologic characterization, in rats treated neonatally. Female rat sexual organs were examined histologically.

MATERIAL AND METHODS

Tested substances: Tween 80 (Merk, Schuchard, Germany) and Polyethyleneglycol 300 (PEG 300) delivered by Aldrich chemie, Steinheim, Germany.

Methods: Mature rats were mated (two females to one male) and the female rats from the resultant litters were injected i.p. on days 4, 5, 6, 7 after birth with 1,5 or 10% aqueous solution of Tween 80, and PEG 300 at a volume 0.1 ml/rat. Control rats were injected with aqua pro injectione and positive controls were given 59 ug of Agostilben (diethylstilbestrol-DES), Biotika, Slovenska Lupča. In the group 1 the uteri were taken on day 7 for histological examination. In the remaining rats the onset of vaginal opening was further observed. The estrus cycle was examined at wk 10, 14 and 18 of life. Vaginal smears were obtained daily between 9,00 and 10,00 hr for 14 days; tap water at room temperature was introduced into the vagina using an eye dropper and the vaginal content was removed. One drop of the content was put on an ethanol - washed microscope slide, left to dry and fixed with methanol for 5 min. Staining was performed according to Pekárek (Luksch, 1953) and the slides were examined microscopically and classified as to the stage of the estrous cycle according to the criteria of Allen (1922). At 5 months of age, the rats were killed by cervical dislocation and the uterus, ovaries, adrenal glands and pituitary gland were removed. The organs were weighed and then fixed with 10% buffered formol, and embedded in paraffin. Sections from paraffin blocks were stained with haematoxylin and eosin.

The conditions of the experiment: The rats were kept under controlled conditions, with a 12-hr light/dark cycle, and at 20-22^0 C and 50-60 % relative humidity. Food and water were available ad libitum.
Statistical evaluation: The data were evaluated by analysis of variance and least square difference - test.

RESULTS

None of the tested substances affected the growth and viability of neonatal female rats. A statistically significant (P< 0,01) decrease in body weight was found in adult rats that had been given 1% Tween 80, PEG 1,5, 10 % or DES in comparison with the untreated control. All the rats treated with Tween 80 had significantly advanced vaginal opening in comparison with untreated and negative controls (Fig.1). PEG did not influence significantly the time of vaginal opening (Fig. 2). Significant prolongation of the estrus cycle and induction of persistent vaginal estrus were found in groups treated with Tween 80 (Fig. 3).

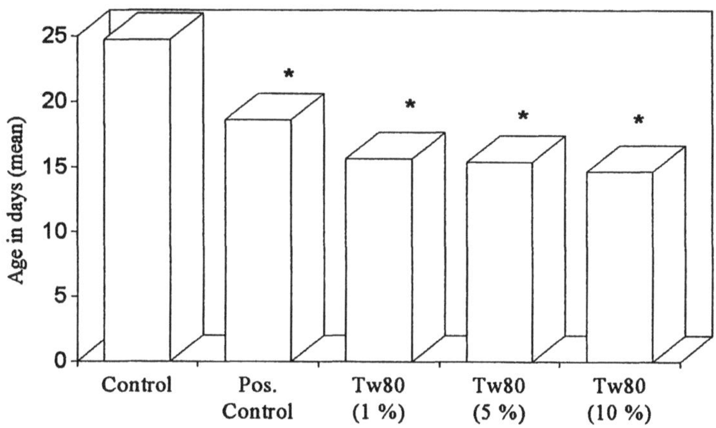

Fig. 1 :Time of vaginal opening in female rats neonatally treated with Tween 80. Values marked with asterix differ significantly (analysis of variance and LSD - test) from the corresponding value for control (* = P < 0,05)

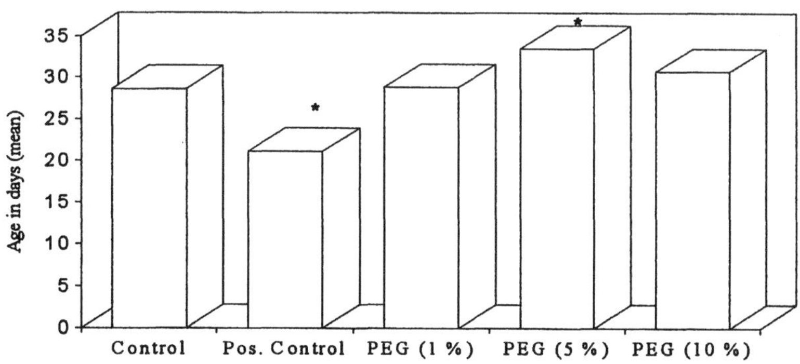

Fig. 2 : Time of vaginal opening in female rats neonatally treated with PEG

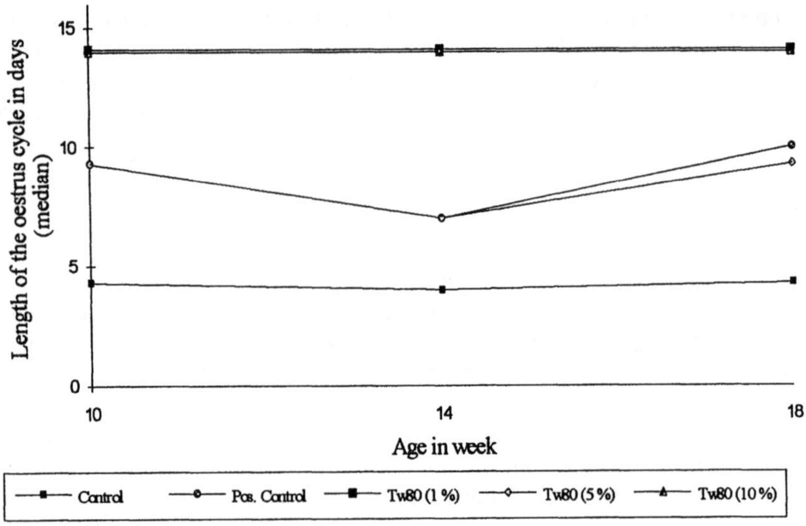

Fig. 3 : Length of the oestrus cycle in rats treated neonatally with Tween 80 (median)

In the positive control and in all groups of rats that were given Tween 80, cytological investigation revealed a shortening of diestrus and earlier onset of prestrous, and the prolongation of the end of proestrus and the beginning of metaestrus. In most rats the late phases of metaestrus were absent. PEG 300 did not influence significantly the length of estrus cycle (Fig 4).

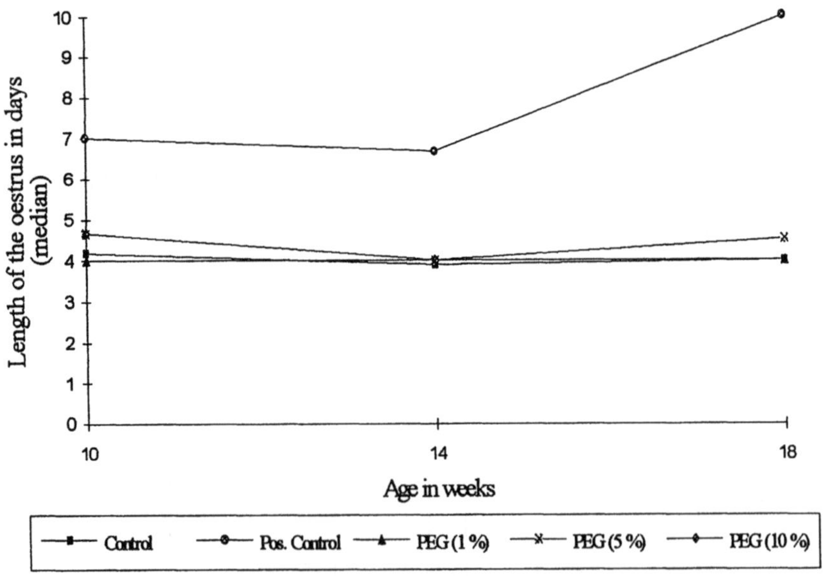

Fig 4 : Length of the oestrus cycle in rats treated neonatally with PEG (median)

Histopatological investigation revealed planocellular metaplasia in the epithelium lining of the uterus lumen and in the endometrial glands of the group given Tween 80 % and 1,5 and 10 % PEG 300.

DISCUSSION

Tween 80 is a classical non-ionic surface-active detergent. It is widely used as an additive in pharmaceuticals and in the food industry. Some of the Tweens are considered to be carcinogens or cocarcinogens (Lewis, 1979).

In our study, we found that 4-day administration of Tween 80 to female rats during the period crucial for the development and function of reproductive organs accelerated the maturation of these organs. We found significant prolongation of the estrous cycle and induction of persistent vaginal estrus. The persistence of vaginal estrus after the perinatal exposure of female rats to potential estrogens used to be ascribed to estrogenic effects (Barraclough, 1961; Gorski, 1971).

The most apparent histological finding was planocellular metaplasia of the endometrial epithelium of the uterus in the group given 10% Tween 80 and 1, 5 and 10% PEG 300. The changes were in many ways similar to those resulting from the administration of DES to the positive controls, and identical to the changes seen in the uterus and ovaries after chronic endocrinal stimulation by oestrogens (Heywood and Wadsworth, 1980; Highman *et al.*, 1977).

CONCLUSION

4-days administration of Tween 80 to female rats during the period crucial for the development and function of reproductive organs accelerated maturation, prolonged the estrus cycle, and induced persistend vaginal estrus. Polyethylenglycol 300 did not influence the maturation of reproductive organs so remarkable as Tween 80 did. Squamous cell metaplasia of the epithelial lining of the uterus and cytological changes found in the uterus after administration of 10% Tween 80 and 1, 5 and 10 % PEG 300 were indicative of chronic estrogenic stimulation.

REFERENCES

Allen E. (1922) The oestrous cycle of the mouse. American Journal of Anatomy 30, 297-350.

Barraclough C.A.(1961) Production of anovulatory, sterile rats by single injection of testosteron propinate. Endocrinology 68, 62-67.

Bulger W. H. and Kupfer D. (1985) Estrogenic activity of pesticides and xenobiotics on uterus and male reproductive tract. In Endocrine Toxicology. Edited J.A.Thomas and K. S. Korach. pp. 1-33. Raven Press, New York.

Gajdová M., Vargová M., Jakubovský J., Grunt J., Války J. and Galbavý Š., (1988) Estrogenic effects of fosmet on neonatal rat uterus. Bratislavské Lekárske Listy 89, 843-847.

Gorski R. A. (1971) Gonadal hormones and perinatal development of neuroen-docrine function. In Frontiers in Neuroendocrinology. Edited by L. Martini and W.F. Ganong. pp. 237-290. Oxford University Press, New York.

Heywood R. and Wadsworht P.F. (1980) Experimental toxicology of estrogens. Pharmacology and Therapeutics 8, 125-142.

Highman B., Norwel M. J. and Schnellenberger T. E. (1977) Pathological changes in female C3H mice continuously fed diets containing diethylstilboestrol or 17 estradiol. Journal of Environmental and Pathological Toxicology 1, 1-30.

Lewis R. j., Sr (1979) Registry of Toxic Effects of Chemical Substances. 1978 Ed. DHEW/NIOSH Publ. No. 79-100. Cincinnati, OH.

Luksch F. (1953) Úvod do poševní cytológie. Státní zdravotní nakladatelství p.27. (In Czech).

Vargová M. et al. (1990) Study of toxic and genotoxic effects of selected chemicals on the organism. Final Report on the Task State Plan of Technical Development N-05-535-863-03, p. 62-89, ÚPKM, Bratislava.

In Vitro tests in regulatory toxicology

(Chair: Prof. Dr. M. Balls, Italy)

In Vitro Methods in Regulatory Toxicology: The Crucial Significance of Validation

M. Balls
European Centre for the Validation of Alternative Methods (ECVAM),
JRC Environment Institute, 21020 Ispra (VA), Italy

Regulatory toxicity testing in animals raises a number of scientific, humanitarian, legislative, practical and economic questions and conflicts. For example, while the use of animal procedures in identifying the potential hazards represented by chemicals and products of many kinds is required by various specific and general laws and regulations, there are other laws of no less importance (e.g. European Union [EU] *Directive 86/609/EEC* [Anon, 1986]), which are intended to protect laboratory animals from unnecessary pain, suffering, distress or lasting harm. They stipulate that the use of laboratory animals in potentially harmful procedures should only be permitted when such use can be justified as necessary on strong scientific grounds. However, the scientific basis of many animal tests themselves and, in particular, of the ways whereby the data they provide are currently applied in human risk assessment, is weak. In addition, the present dependence of hazard prediction on animal tests requires considerable human and economic resources, as a result of which only a very small proportion of the chemicals which might threaten the wellbeing of humans, or be hazardous to the environment in general, have been subjected to anything approaching a scientifically-satisfactory evaluation.

Non-animal tests and testing strategies can offer solutions to many of the problems caused by the current over-reliance of regulatory hazard prediction and risk assessment on the routine, check-list application of animal tests. They can offer a more-mechanistic basis for understanding toxic effects at the systemic, organ, cellular and molecular levels. When human cells and tissues are used, or modelling is based on human exposure and responses, the problem of species differences, which severely limits the relevance of animal tests, can be avoided. The change of emphasis from *in vivo* to *in vitro* approaches must therefore also embrace the increasing use of properly-obtained and safely-maintained *human*, rather than *animal*, cells and tissues (Rogiers *et al.*, 1993; Hawksworth, 1994). Well-designed *in vitro* studies can also assist with the selection of the most-appropriate laboratory animal species for any subsequent, scientifically necessary (and therefore morally justifiable) use as models for man or other species.

Non-animal methods do not only offer routes of escape from many of the scientific, ethical and legal conflicts inherent in current practices. They can also offer the possibility of more-manageable, less time-consuming and less expensive testing strategies, so that the practical and economic limitations which currently prevent the adequate testing of many chemicals and products can be overcome.

Nevertheless, despite the obvious benefits which would accrue from the orderly development and successful application of non-animal tests and testing strategies, various barriers to their acceptance must be dealt with before they can fully contribute to the transformation of hazard prediction and risk assessment from a poorly-practised, pseudo-scientific art into an effective and modern scientific discipline. Clark (1994) considered that the most important of these barriers are the validation barrier and the scientific barrier, but I would also like to emphasise another barrier - that represented by the inadequacy of the available *in vivo* toxicity data base.

The Validation Barrier

Validation has been defined as *the process by which the reliability and relevance of a procedure are established for a specific purpose* (Balls *et al.*, 1990a). It is now widely agreed that formal validation studies should be preceded by proper test development (including the production of a protocol or standard operating procedure), and that an interlaboratory trial involving appropriate sets of coded chemicals (of known and different degrees of *in vivo* toxicity) is an essential part of the process. Only when the outcome of a successful validation study has been subjected to independent assessment should the formal acceptance of a new test, test battery or testing strategy into regulatory practice be proposed and considered (Balls *et al.*, 1990b).

Although there are various forms of validation (Balls, 1992), some of which are more useful than others, the process should primarily be seen as a scientific activity and should be the responsibility of scientists who are committed to new test development, and of those toxicologists and regulators who are prepared to assist in the orderly and sensible incorporation of new methodology into regulatory toxicology.

Up to now, the validation process has not been applied with notable success. This is partly because it is much more difficult than we had imagined, and partly because of a misunderstanding of why animal tests are conducted in the first place and therefore, how they should be replaced. Animal testing is performed for two main reasons and is of two main types:

Type A. in compliance with (or in perceived compliance with) the *specific requirements* of regulatory authorities for toxicity test data; or

Type B. in compliance with the (common law-based or product liability-based) need for *safety assessment*, as interpreted by manufacturing and/or retail companies themselves.

Type A testing involves industrial chemicals produced on any significant scale, certain types of cosmetic ingredients (e.g. preservatives, colouring agents, hair dyes, and UV filters, as specified in the EU by the Cosmetics Directive, *Directive 76/768/EEC* (Anon, 1976), and certain kinds of products (e.g. medicines, pesticides). Type B testing, not being "required" by specific legislation, tends to be a matter of decision for companies themselves. That is why many cosmetic companies have declared that they have stopped animal testing. What they have done is to stop doing what they were never specifically required to do in the first place, and they may now use *in vitro* tests instead, often followed by human volunteer studies, as a basis for safety assurance. This has led to a significant, and welcome, reduction in the testing of cosmetic formulations, but has had no impact, of course, on the legally required testing of industrial chemicals, certain cosmetic ingredients, medicines, pesticides, etc. This is an important point of emphasis, because, although it is vital that cosmetic companies produce consumer products (shampoos, etc.) which are as safe as possible, there are no specific regulations, at least in Europe, which make mandatory their toxicity testing in animals.

The distinction between regulatory and non-regulatory animal testing has important implications in terms of the way we think about, and therefore conduct, validation studies, which are, as a consequence, of two main types:

Type A. validation of the relevance and reliability of methods, batteries or testing strategies which could replace, refine or reduce animal testing for *regulatory* purposes; and

Type B. validation of the relevance and reliability of non-animal procedures for their general use for *non-regulatory* purposes.

This distinction has two further consequences. Firstly, it leads to *in-house validation*, in which a company evaluates a testing strategy for its own purposes (i.e. for Type B testing, perhaps also in the hope of contributing to the development of alternatives for use in Type A testing). This often involves the use of chemicals and formulations which are the property of a company, for which it also has (and owns) animal test data, details of which it is not prepared to make public. Many companies including Beiersdorf, Henkel, L'Oréal, Procter & Gamble, Unilever and Zeneca, have devised in-house procedures for use as screens before a decision is taken about the need for the animal testing of new chemicals or about the comparative acceptability of a new product in terms of consumer safety. Secondly, such in-house validation does not replace (although it can usefully precede) *scientific validation*, in which the names and structures/contents of the chemicals and formulations used must be freely available, as must the animal toxicity test data and the protocols of the tests or testing strategies undergoing validation, and in which the results and conclusions of validation studies must be published in detail in the peer-review literature. Successful scientific validation is (and should always be) an essential and unavoidable prerequisite to the acceptance of alternatives for legally required testing (i.e. for regulatory testing, Type A).

The Scientific Barrier

Clark (1994) emphasised that alternative methods "must be based on good science and must not make extravagant claims that run counter to common sense". It is not difficult to agree with that sentiment, but there is much controversy about just how "good" the scientific basis of *in vitro* toxicity tests ought to be or could practically be expected to be (e.g. Balls, 1992; Flint, 1992).

Flint (1994) has proposed an approach to validation which differs from the formal process envisaged by Balls *et al.* (1990a), and which he claims has a greater inbuilt flexibility. He proposes three main steps:

Step 1. An examination of the proposed test system itself, to establish precisely what toxicity is being measured and whether the cells and endpoints selected are appropriate.

Step 2. An evaluation with model positive and negative controls, followed by extensive interlaboratory studies with chemicals of unknown activity, whose predicted toxicity is later confirmed by additional *in vivo* studies.

Step 3. Use of the test system to investigate the mechanisms of toxicity of a variety of agents, which would reveal the limitations of the system and might suggest ways of adapting it to avoid these limitations or suggest the need for a new type of test.

At some point, he argues, the performance of the new test will have been sufficiently optimised to justify its recommendation as an accepted method for regulatory purposes.

Ideally, of course, there *should* be sufficient understanding of all the toxic mechanisms of action *in vivo* for appropriate *in vitro* systems and endpoints to be selected. However, while this might be possible with some pharmaceuticals, it is very unlikely that such an understanding could be provided in the foreseeable future for the vast majority of the other types of chemicals for which hazard prediction is necessary as a basis for risk assessment. Recognition of this truth has led to support for the *correlative* approach, where correlation is defined as similarity based on a functional relationship between two variable quantities connected by a simple formula, which may or may not have a mechanistic relevance (Balls & Fentem, 1992). The lack of an established mechanistic connection between the systems providing two sets of data does not necessarily mean that such a connection does not exist, but merely that it is not (yet) known or understood. Thus, the mechanistic and correlative approaches are not mutually exclusive, and which one is appropriate depends on the specific circumstances in question.

The *In Vivo* Data Barrier

Whether a new, non-animal test is based on a mechanistic understanding of toxic effects and/or responses or on the correlative approach, evaluation of its relevance and reliability must be based on its ability to provide information which could usefully and safely be incorporated into the hazard identification and risk assessment process. This, in turn, necessarily means comparing the information it provides with information gained in conventional animal toxicity tests and/or of knowledge gained following human exposure. This invevitably leads to serious difficulties, whether the approach to validation is that spelled out by Balls *et al.* (1990) or that favoured by Flint (1994), because:

1. The relevance and reliability of regulatory animal toxicity tests themselves have not been satisfactorily established.
2. Many of the data produced in such tests were produced before Good Laboratory Practice was introduced in an attempt to improve their quality.
3. It is likely that many tests are still so poorly designed that they may lead to the wrong conclusions (Festing, 1994).
4. It is not clear whether, in the words of Roe (1994), the historical *in vivo* test data base represents "valuable treasure or worthless trash", although, in general, the latter is more likely to be the case than the former.
5. Reliable human data are very hard to obtain.

It is therefore proving very difficult to find test materials in sufficient numbers, of sufficient variety, and sufficiently representative of the range of types and levels of toxicity required for validation, despite the efforts of organisations such as ECETOC (Bagley *et al.*, 1992). This is partly because many of the animal tests do not provide data which can satisfy a reasonable set of quality criteria, and partly because many of the data are the property of commercial companies which are unwilling to make them available. However, there is also another problem, the use of non-animal screens and limit tests can mean that tests are being less and less conducted in ways which will lead to severe toxic effects.

For these reasons, it is questionable whether *in vivo* data should ever be accepted as the "gold standard" which the developers of non-animal tests have to strive to attain, except where an adequate mechanistic understanding of events is available.

There is yet another problem - even if animal data were available, their relevance to humans would be limited by species differences and by various biokinetic factors. Thus, an *in vitro* test which gave the same results as an animal test could be no better than the test for which a replacement was being sought.

This means that those who are involved in the validation of the new alternative procedures are faced with a seemingly insuperable problem. We can establish reproducibility, but how are we to establish relevance?

The Way Forward

Overcoming the validation barrier to the acceptance of *in vitro* methods and other alternatives to animal tests will require that the validation process itself is kept under constant review, and the useful contributions being made, for example, by Barile & Ekwall (1993), Goldberg *et al.* (1993) and Flint (1994) are to be welcomed. However successful validation requires that the scientific barrier and the *in vivo* data barrier can also be surmounted. Greater investment is needed in the science of toxicology, so that tests with a sound mechanistic basis can progressively become more feasible (Roberfroid, 1994). Providing reliable reference data for the evaluation of the new methods is even more difficult, and there seems to be no alternative to the setting up of international reference collections of chemicals carefully selected for their relevance to specific toxicological endpoints, carefully reviewed and classified by experts, and made available to individuals and agencies involved in validation studies, as suggested by Purchase (1990).

However, what is needed most of all is a revolution in thinking about the purpose and practice of regulatory toxicology, with a recognition of the inadequacy of both the data on toxic hazard provided by the current animal tests and of the ways in which they are used in risk assessment. This, in turn, will require a change of heart among many of the regulators themselves. Progress has been disappointing so far, as McLean (1991) said in his concluding remarks at the end of a meeting held to discuss the Second Report of the FRAME Toxicity Committee (Balls *et al.*, 1991): "The idea that regulators should still demand out-of-date, useless, scientifically-invalid, hopeless studies, which involve animals, is a scandal at the political level, which is to a considerable extent beyond our power [as toxicologists]. All we can do is to say, as forcefully as possible, that we do not want animal toxicology studies which are excessive, unnecessary or repetitive."

There are signs that the situation could improve. For example, an EC-funded workshop on the regulatory acceptance of validated non-animal toxicity test procedures, at which the majority of the participants were regulators (Balls *et al.*, 1990b), concluded that: "All conceivable and practical steps should be taken to make the formal acceptance and incorporation of non-animal toxicity test procedures into regulatory practice as smooth and rapid a process as possible."

Meanwhile, an Inter-agency Regulatory Alternatives Group (IRAG) is active in the United States, the OECD has modified its guideline updating procedures to allow for the incorporation of validated non-animal test procedures, and regulators in the EU recognise the need for new methods for risk assessment based on safety under conditions of use, rather than merely on testing to toxic levels in animals (Bass & Hildebrandt, 1994).

It is also essential that the laws which control animal experimentation are rigorously applied, so that animal suffering is minimised in regulatory toxicity testing and the smallest possible numbers of animals are used. To this end, the Animal Procedures Committee, which advises the British Government on these matters, has just produced a report containing a number of important recommendations (Anon, 1994). Let us hope that they will be accepted and acted upon.

References

Anon (1976) Council Directive 76/768/EEC of 27 July 1976 on the approximation of the laws of the Member States relating to cosmetic products. Off J Eur Comm L262:169

Anon (1986) Council Directive 86/609/EEC of 24 November 1986 on the approximation of laws, regulations and administrative provisions of the Member States regarding the protection of animals used for experimental and other scientific purposes. Off J Eur Comm L358:1

Anon (1994) Animal Procedures Committee: report to the Home Secretary on regulatory toxicity testing. Home Office, London, 20 pp.

Bagley DM, Botham PA, Gardner JR, Holland G, Kreiling R, Lewis RW, Stringer DA, Walker AP (1992) Eye irritation: reference chemicals data bank. Toxic in Vitro 6:487

Balls M (1992) *In vitro* test validation: high hurdling, but *not* pole vaulting. ATLA 20:355

Balls M, Fentem JH (1992) The use of basal cytotoxicity and target organ toxicity tests in hazard identification and risk assessment. 20:368

Balls M, Blaauboer B, Brusick D, Frazier J, Lamb D, Pemberton M, Reinhardt C, Roberfroid M, Rosenkranz H, Schmid B, Spielmann H, Stammati A-L, Walum E (1990a) Report and recommendations of the CAAT/ERGATT workshop on the validation of toxicity test procedures. ATLA 18:313

Balls M, Botham P, Cordier A, Fumero S, Kayser D, Koëter H, Koundakjian P, Gunnar Lindquist N, Meyer O, Pioda L, Reinhardt C, Rozemond H, Smyrniotis T, Spielman H, van Looy H, van der Venne M-T, Walum E (1990b) Report and recommendations of an internatinal workshop on promotion of the regulatory acceptance of validated non-animal toxicity test procedures. ATLA 18:339

Balls M, Bridges J, Southee J (eds) (1991) Animals and alternatives in toxicology: present status and future prospects. Macmillan, London, and VCH Publishers, New York, 390 pp.

Barile FA, Ekwall B (1993) Standardization and validation. In: Barile FA (ed) Introduction to *in vitro* cytotoxicology. CRC Press, Boca Raton, pp. 189-208

Bass R, Hildebrandt AG (1994) New methods for risk assessment of drugs. In: Fracchia GN (ed) European medicines research - perspectives in pharmacotoxicology and pharmacovigilance. IOS Press, Amsterdam, pp. 22-29

Clark DG (1994) Barriers to the acceptance of *in vitro* alternatives. Toxic in Vitro 8:in press

Festing MFW (1994) Reduction of animal use: experimental design and quality of experiments. Lab Animals 28:212

Flint OP (1992) *In vitro* test validation: a house built on sand. ATLA 20:196

Flint O (1994) Rational programmes for the assessment of drug toxicity. In: Lees P (ed) Proceedings of the 6th Internatinal Congress of the European Association for Veterinary Pharmacology and Toxicology. Blackwell, Oxford, pp. 9-12

Goldberg AM, Frazier JM, Brusick D, Dickens MS, Flint O, Gettings SD, Hill RN, Lipnick RL, Renskers KJ, Bradlaw JA, Scala RA, Veronesi B, Green S, Wilcox NL, Curren RD (1993) Framework for validation and implementation of in vitro toxicity tests: report of the Validation and Technology Transfer Committee of the Johns Hopkins Center for Alternatives to Animal Testing. J Am Coll Toxic 12:23

Hawskworth GM (1994) Advantages and disadvantages of using human cells for pharmacological and toxicologial studies. Hum Exp Toxic 13:568

McLean A (1994) Concluding remarks. In: Balls M, Bridges J, Southee J (eds) Animals and alternatives in toxicology - present status and future prospects. Macmillan, London, and VCH Publishers, New York, pp. 341-344

Purchase IFH (1990) An international reference chemical data bank would accelerate the development, validation and regulatory acceptance of alternative toxicity tests. ATLA 18:345

Roberfroid M (1994) Alternatives in safety testing: progress or uselessness? ATLA 22:in press

Roe FJC (1994) Historical histopathological control data for laboratory rodents: valuable treasure or worthless trash? Lab Animals 28:148

Rogiers V, Sonck W, Shephard E, Vercruysse A (eds) (1993) Human cells in *in vitro* pharmaco-toxicology - presents status within Europe. VUB Press, Brussels, 388 pp.

Scientific, ethical and legal aspects of the acceptance of *in vitro* methods in regulatory toxicology

E. Walum[1,2], B. Forsman[3] and T. Stagh[4]
[1]Department of Neurochemistry and Neurotoxicology, Stockholm University, S-106 91 STOCKHOLM and [2]Pharmacia AB, BioScience Center, S-112 87 STOCKHOLM; [3]Centre for Research Ethics, University of Göteborg, Brogatan 4, S-413 01 GÖTEBORG; [4]National Board For Laboratory Animals, Ministry of Agriculture, S-103 33 STOCKHOLM, Sweden.

Society demands greater safety in the manufacture, distribution and use of chemicals and greater protection of the environment. At the same time currently used, animal based, methods for the identification of toxic chemicals and estimation of risk are often considered scientifically and ethically inadequate. This problem may be, at least in part, solved by the introduction of *in vitro* methods in regulatory toxicology (Balls et al. 1991). *In vitro* methods provide means for the development of test procedures that employ human cells and tissues, that make mechanism based risk estimations readily available and that reduces animal strain and suffering. However, the reliability and relevance of such new procedures have to be established through validation studies before they can be accepted.

At an international workshop, held in Greece in 1990, representatives for academia, industry and regulatory authorities discussed validation, the dissemination of information, independent assessment of the relevance and usefulness of validated alternative test procedures, and promotion of their incorporation into regulatory practice. The document summarising the results of that workshop has become known as the Vouliagmeni Report (Balls et al. 1990). In the present paper we discuss certain aspects on validation and risk estimation and how ethical and legal considerations may influence the acceptance of *in vitro* methods in regulatory toxicology.

Validation

The aim of validation studies is to identify procedures that are capable of performing according to certain demands: the methods must have been proven to be reproducible, reliable and transferable (Balls and Clothier 1987). It must be possible to extrapolate the data from the tests to an *in vivo* (human) situation and to integrate them into risk studies. Validation is thus the process by which the relevance and reliability of a test method are established for a particular purpose

(Balls et al. 1990). Most toxicologists agree on that validation is a necessary step in the transfer of test from development to regulatory application. However, there is no complete consensus on how validation studies should be performed and an extensive literature describes different opinions. This discussion was recently reviewed in connection to the Third International Conference on Practical *In vitro* Toxicology (Walum et al. 1994).

At a satellite symposium, *In vitro* Neurotoxicology, to the XIIth International Congress of Pharmacology validation was defined as "the rate limiting step between the development of tests and their application and acceptance" (Green, 1994). When pondering upon this statement, one has to wonder what might determine the rate constant in the transition. It is evident the transition rate will be low if tests are validated and rejected because they are not capable of performing according to the demands. If a validation study is designed to include a small number of "ready" developed tests this is likely to happen. For every test procedure that is validated and has to re-enter test development (Walum et al. 1994) the overall rate by which tests for a certain purpose are accepted for regulatory use is reduced. Consequently, a more efficient mode of progress appears to be an initial validation of endpoints, and not of complete test procedures. It is of course more important to find out, as a first step, how predictive a certain endpoint is for a particular type of toxicity, than how that endpoint should be best determined; i.e. relevance should be determined before reliability (Ekwall 1992a). Another crucial factor determining the rate of acceptance is the number of tests that enters a particular validation study. A validation process, aided by computer based statistics, where a large number of tests is introduced will more rapidly result in the selection of a desired number of high quality tests than a procedure where a few, or even single , tests are validated in succession. Thus, a parallel approach to the introduction of methods in a validation study is preferable, in terms of efficiency, to a serial one (Ekwall 1992a). These principles are among the corner-stones of the MEIC-programme (Bondesson 1989), in which the relevance of *in vitro* cytotoxicity determinations for the prediction of human acute lethal and sublethal toxicity, subacute toxicity, local toxicity and organ-specific toxicity is investigated (Walum et al. 1994). Furthermore, the risk of reaching the "brick-wall-situation", where a test is completely resistant to further development due to the establishment of a standard protocol, is reduced if the actual test procedure is decided on as late in the development-validation-acceptance process as possible.

Risk assessment

In the determination of risk, the actual exposure is related to the exposure known to be required for the appearance of a particular injury. Thus, if the concentration of a chemical present in the blood over certain time, the magnitude of the main distributing factors for that chemical and the value for the critical toxic concentration in a certain organ is known, it is possible to estimate risk (Walum et al. 1993). *In vitro* systems are powerful tools for the determination of critical toxic concentrations, since they are well suited for mechanistic studies. When the most sensitive cell type and cell physiological or biochemical parameter has been identified by means of a stepwise *in vitro* test procedure (Atterwill 1989, Walum 1992a) the concentration-effect relation can be carefully determined and the critical concentration thus obtained. Toxicokinetic data may be derived from several sources, e.g. human exposures, animal experiments, *in vitro* models and computer based physiological and chemical models, and can be used to extrapolate the critical toxic concentration to a critical blood concentration. By comparing this blood concentration to that expected or known to occur in a population under specified conditions, it should be possible to estimate the risk. In fact, these ideas are tested in a multicentre project (ECITTS, Blaauboer et al. 1994) supported by the Swedish National Board for Laboratory Animals (CFN) and the European Centre for Validation of Alternative Methods (ECVAM).

Several authors have advocated the use of *in vitro* models in risk assessment (Castell 1990, Ekwall 1992b, Balls and Fentem 1992, Zbinden 1992) and it is often argued that the possibility of using human material for toxicological studies is one of the major advantages of *in vitro* methods. At a recent EU-conference on Human Cells in *In vitro* Pharmaco-Toxicology it was stated that "There is an obvious need, scientifically and ethically, to implement human tissues and cells in the study of effects of xenobiotics. The incorporation of such model systems would imply important improvements in toxicological risk evaluation" (Blauboer 1993). It is our opinion that the possible use of human *in vitro* systems in mechanism based risk assessments represents the best scientific argument for the incorporation of *in vitro* methods in regulatory toxicology. Accordingly, we would strongly endorse validation studies, which address the question of how results from *in vitro* test procedures should be converted into risk determinations (Walum 1992b).

Ethical aspects

It seems to be self-evident that *in vitro* methods entail ethical advantages. But let us first examine possible disadvantages with those methods. The issue has then to be divided into one question of general (including human) ethics and one of animal ethics.

Concerning general ethics, the main possible disadvantage is that *in vitro* methods might not fulfil the demands of being scientifically adequate. However, there is a disadvantage only if the *in vitro* methods are more poor than animal *in vivo* methods are. Of course, there is an ethical problem if both kinds of methods are equally poor. In both cases, no safety can be secured. Most likely, different *in vitro* methods are of different scientific quality, and that is what validation programmes have to determine. Consequently, validation has a general ethical dimension.

Concerning animal ethics, there are two possible ethical problems connected to *in vitro* methods, one of which can possibly be regarded as a disadvantage. The first problem is that animals often are killed in order to harvest cells or tissue for the experiment. Still, we do not regard the mere killing of non-human animals as a significant ethical problem. But, of course, it is important how the animals are treated before they are killed.

The second problem is more serious. That relates to the use of serum in the medium of cell cultures. Serum, notably bovine and horse serum, sometimes comes from abused animals. Some scientists using whole animals in their experiments have pointed out the use of serum as a disadvantage of *in vitro* methods, and certainly it could be a disadvantage if the animals (cows, calves and horses) are seriously ill-treated and used for serum production during long periods. Nevertheless, the animal-centred disadvantages can be handled. Under no circumstances is it necessary to treat animals for cell or serum production badly. The adequate treatment of animals may perhaps increase the cost of the final products, but that is not a valid argument for lower animal maintenance standards. We believe that if scientists become aware of the problems and make their opinions clear, those problems will be easy to solve.

Now let us look at the possible ethical advantages of *in vitro* methods. Also in this case the issue has to be divided into general (including human) ethics and animal ethics. Concerning general aspects, the simplicity and rapidity of *in vitro* test methods make it possible to get more substances tested for safety. As long as regulatory bodies have to resort to animal *in vivo* methods, there is not enough capacity to test most chemicals, let alone combinations of various substances and products. *In vitro* methods reasonably improve the capacity to get pure chemicals and combinations of substances tested. The enlarged testing capacity is an ethical advantage with *in vitro* methods compared to *in vivo* methods, provided that we do not return to pre-technical era.

Regarding aspects of animal ethics, the great advantage is, of course, that tissues and cells are not sensitive beings. No painful procedures are thus conducted in the test situation.

On the whole, ethical advantages overrule by far the disadvantages, provided that *in vitro* methods are not scientifically less relevant or less reliable than animal tests are. Finally, ethics should play an important part in the enterprise of science as well as in politics. "The purpose of ethics is to overcome our psychological limitations", as a Danish philosopher has put it. Using ethics in the choice of research and test methods would make it a rational choice, unlike resorting to feelings or market demands only.

Legal aspects

The Council of Europe Convention on the protection of Vertebrate Animals Used for Experimental and Other Scientific Purposes lays down, in the preamble and articles, that Member States are to develop alternatives to animal experiments and that animal experiments must not be undertaken if adequate scientific methods that do not involve the use of animals are available. The provisions of EC Directive 1986/609 are basically similar. With reference to these two documents, one has to point out that most European countries are now members of the Council of Europe and many of them are also, or may soon become, members of the European Union.

The Convention and the EC Directive also contain a number of articles with practical provisions on animal management. These provisions, which are directly binding on Member States, cover matters such as cage sizes, breeding facilities, statistics and training. Unlike the concrete animal protection rules in the convention and the EC Directive, the provisions on alternatives to animal experiments take the form of overall goals. They may be regarded as declarations of intent to strengthen animal protection: no animals should be used where non-animal methods are available. However, there is every reason to believe that this objective, which also lies at the heart of national legislation in many countries, has made a substantial impact an the attitudes of scientists and general public. Generally speaking, however, international organisations such as the Council of Europe and the European Union have neither the expertise nor the means to appraise the toxicological test methods that are being developed constantly. Nevertheless, agreements or policy decisions by international organisations and policy declarations in national legislation are a major force in promoting the development of alternatives to animal experiments, in particular in toxicology.

The practical development of new methods is driven by the policies of various international bodies in the field of product safety, which in their turn shape the course of national testing activities. Mention should be made in this connection of the innovative work done by the OECD over the years in reducing the adverse effects of test methods in those cases where the use of animals is still necessary for toxicological purposes. Limitation of the use of laboratory animals is also the approach adopted by the specialised bodies that formulate guidelines for laboratory experiments, such as the Council of Europe's Pharmacopoeia Commission and the European Union's European Medicine Agency (EMA).

The task of drafting detailed rules on test methods has in the last 10-15 years increasingly been assumed by international bodies on the basis of co-operation between many different countries. Methods developed by individual countries are insignificant by comparison. Decisions by the relevant international bodies are essential for a method to have any chance of being recognised for registration purposes on a international scale. The national agencies responsible for the use of chemicals, drugs and food additives have a say in the methods authorised for use as representatives of these organisations. In order to achieve success in the acceptance of *in vitro* methods, scientists should endeavour to maintain close contacts with the relevant agencies and international bodies. Such contacts promote exchanges of information and facilitate adjustments of existing guidelines to recent developments and validations.

Today *in vitro* methods are widely used in connection with product development and the screening of new substances. However, this progress is due to the fact that these methods are faster, more accurate, simpler and cheaper than *in vivo* methods rather than to a new ethical attitude towards the use of animals. This trend can be expected to continue; from a scientists point of view the search for more effective, improved *in vitro* methods must go on. The development of such methods is not just only affected by scientific criteria, but to some extent also by political considerations. Very often, political decisions are influenced by emotions and subjective convictions about what is right and proper rather than by reality. It is therefore important for scientists to emphasise that decisions to restrict animal experiments are only relevant if the substitutes represent equivalent or improved methods.

Conclusions

With respect to scientific as well as ethical and legal aspects it is the quality of a new test procedure that determines if it is going to be accepted for regulatory purposes or not. The quality of a method is determined through validation studies and we argue in favour of the idea that relevance should be determined before reliability. The object of performing risk assessment should be integrated in all test development and validation projects. Furthermore, the use of ethics in the selection of test methods make the choice a rational one. The responsibility for and the main influence on the regulatory process governing the authorisation of toxicological methods lies with international bodies in various fields and the relevant national agencies. This fact should always be borne in mind when new test methods are developed.

References

Atterwill CK (1989) Brain reaggregate cultures in neurotoxicological investigations: studies with cholinergic neurotoxins. ATLA 16:221-230

Balls M, Clothier R (1987) Validation of alternative toxicity test systems: lessons learned and to be learned. Molecul Toxicol 1:547-559

Balls M, Botham P, Cordier A, Fumero S, Kayser D, Koeter H, Koundakjian P, Lindquist NG, Meyer O, Pioda L, Reinhardt C, Rozemond H, Smyrniotis T, Spielmann H, Van Looy H, van der Venne M-T, Walum E (1990) Report and recomendations of an international workshop on promotion of the regulatory acceptance of validated non-animal toxicity test procedures. ATLA 18:339-344

Balls M, Bridges J, Southee (1991) Animals and Alternatives in Toxicology. Present Status and Future Prospects. Macmillan Press, London, pp 390.

Balls M, Fentem JH (1992) The use of basal cytotoxicity and target organ toxicity tests in hazard identification and risk assessment. ATLA 20:368-388

Blauboer B (1993) The significance and applicability of human tissues and cells as *in vitro* models in toxicology. In: Rogiers V, Sonck W, Shophard E, Vercruysse A (eds) Human cells in *in vitro* pharmaco-toxicology. VUBpress, Brussels, pp 1-9

Blauboer B, Balls M, Bianchi V, Bolcsfoldi G, Guillouzo A, Moore GA, Odland L, Reinhardt CA, Spielmann H, Walum E (1994) The ECITTS integrated toxicity testing scheme: the application of *in vitro* test systems to the hazard assessment of chemicals

Castell JV (1990) Remarks of the winner of the EFPIA award. In: Medicines research and animal experimentation: further steps to a solution. Aulendorf, Germanty, pp 17-19

Ekwall B (1992a) Validation of *in vitro* cytotoxicity tests. In: Castell JV, Gómez-Lechón M (eds) *In vitro* alternatives to animal pharmaco-toxicology. Farmaindustria, Madrid, pp 361-390

Ekwall B (1992b) Basal cytotoxicity data (BC-data) in human risk assessment. In: Miura T, Soma Y, Tohyama C (eds) Proceedings of the workshop on risk assessment and risk management of toxic chemicals. National Institute of Environmental Studies, Japan, pp137-142

Green S (1994) Validation of *in vitro* tests. Oral presentation at In-Vitro Neurotoxicology, Satellite Symposium of XIIth International Congress of Pharmacology, July 29-Aug 1, Val Morin, Québec, Canada.

Walum E (1992a) Studies of neurotoxicity in cellular models. In: Castell JV, Gómez-Lechón M (eds) *In vitro* alternatives to animal pharmaco-toxicology. Farmindustria, Madrid, pp 205-237

Walum E (1992b) *In vitro* test validation: should it obey laws, recognise uncertainty principles or follow scientific principles? ATLA 20 :502-503

Walum E, Nordin M, Beckman M, Odland L (1993) Cellular methods for identification of neurotoxic chemicals and estimation of neurotoxicological risk. Toxic *in vitro* 7:321-326

Walum E, Clemedson C, Ekwall B (1994) Principles for the validation of *in vitro* toxicology test methods. Toxic *in vitro*, in press

Zbinden G (1992) Development of *in vitro* toxicology. In: Jolles G, Cordier A (eds) *In vitro* methods in toxicology. Academic press, London, pp3-19

The Validation and Use of In Vitro Teratogenicity Tests

R. Bechter

Drug Safety, Sandoz Pharma Ltd., CH-4002 Basle, Switzerland

Introduction

Developmental toxicology in a pharmaceutical industry setting is concerned with four major tasks during the developmental process of new drugs. These consist of 1) safety evaluations prior to the clinical testing of a given compound in women, 2) submission of data which allow risk assessment as prerequisite for drug registration, 3) mechanistic investigations into adverse effects, and 4) prescreening/screening of new compounds for their embryotoxic/teratogenic potential as a basis for the promotion for further development. Safety evaluation and risk assessment can at present only be done based on data collected from animal experiments (Kimmel 1990; Koeter 1993; Palmer 1993) and the studies required are well defined by regulatory guidelines on an international level. To fulfill all these requirements, a large number of animals are used (Manson 1991). Recent attempts to harmonize the regulations for drug testing for reproductive toxicity, in order to reduce the numbers of animals needed for world wide registrations, have been very successful. They led to an ICH harmonized tripartite guideline on detection of toxicity to reproduction of medicinal products in 1993. However, testing for developmental toxicity is still very costly and time consuming. It is, therefore, of major interest to perform these labor intensive and expensive studies with compounds which have a high probability to prove non-toxic to the developing embryo. Quick and reliable short term tests performed in the early phase of drug development can save resources and reduce the time of drug development for successful candidates.

Developmental toxicity of a compound may become manifest as an intrauterine growth retardation and/or death, structural anomalies, effects on the offsprings' functional development (Gras and Mondain 1981) including very late serious health problems (Herbst et al. 1979). These adverse effects can be observed in the presence or absence of maternal toxicity. General and organ specific toxicity of a compound, therefore, should be considered in the risk assessment process. The possible mechanisms leading to developmental toxicity are manyfold and still unclear. Attempts to elucidate these mechanisms are based on processes which are prerequisites for normal embryonic development. In vitro test systems either measure one of these processes as endpoints or an overall adverse effect (table 1). In cases of abnormal outcome in animal studies, in vitro models provide valuable information on the compounds' embryotoxic mechanisms. The exclusion of maternal factors such as metabolism or placental hemodynamics, or testing of metabolites for their direct action on the processes of development can help to identify the embryotoxic principle of the compound under study. For the selection of drug candidates it would seem important to assess the compound's potential to produce embryotoxic effects as well as to determine the potency in relation to other structurally related candidates to predict in vivo outcome. In vitro models for developmental toxicity which have been proposed during the last decades have been extensively reviewed in recent years (Freeman and Brown 1987; Faustmann 1988; ECETOC 1989; Daston and D'Amato 1989; Welsch 1990; Whittaker and Faustmann 1994). In this paper, the most promising in vitro test systems which allow to evaluate certain aspects of developmental toxicity for the purpose of drug selection in pharmaceutical industry are described and critically evaluated.

Requirements for a Screening System in Drug Development

It is generally accepted that in vitro systems for developmental toxicity screening should be predictive for human outcome. They do not necessarily have to predict the outcome of animal testing (Koeter 1993). This is especially important for the screening of chemicals for which no animal testing is planned or has been done before they are released into the environment (Peters and Piersma 1990). For pharmaceuticals, safety evaluation and risk assessment prior to human use are strictly regulated and based on preclinical animal data. This holds true also for reproductive toxicity. A new drug (exception life saving indications) will not be administered to women of childbearing potential with the appropriate safety margins or the appropriate safety precautions if it has not proven to be non-toxic to intrauterine development in animal models. A developmentally toxic compound with a (false) negative in vitro result will be identified during further in vivo

Table 1: Proposed in vitro and short term test systems for developmental toxicity

Test system	Endpoint(s)	References
Cell lines		
Mouse ovary tumor cell (MOT)	Attachment	Braun et al. 1979, 1981, 1982
Human embryonic palatal mesenchyme cell (HEPM)	Proliferation	Pratt et al. 1982; Pratt and Willis 1985
Neuroblastoma cell	Differentiation	Mummery et al. 1984, 1993
Embryonic stem cell	Differentiation Proliferation Toxicity	Laschinski et al. 1991; Newall and Beedles 1993; Mummery et al. 1993
Pox virus	Expression	Keller and Smith 1982
V79 cell	Communication	Trosko et al. 1982
Primary cell cultures		
Drosophila embryonic cell	Differentiation Toxicity	Bournias-Vardiabasis et al. 1983; Bournias-Vardiabasis 1990
Neural tissue cell	Differentiation Proliferation Toxicity	Flint and Orton 1984; Seifert 1989; Langille et al. 1989; Daston et al. 1991
Limb bud cell	Differentiation Proliferation Toxicity	Hassell and Horigan 1982; Flint and Orton 1984; Kistler 1987; Wiger et al. 1988; Renault et al. 1989
Brain cell aggregate	Differentiation Proliferation Toxicity	Honegger et al. 1979; Honegger and Werffeli 1988; Honegger and Schilter 1992, 1993; Kucera et al. 1993; Schmid et al. 1993

Table 1: continued

Test system	Endpoint(s)	References
Organ cultures		
Limb bud	Differentiation Proliferation Toxicity	Aydelotte and Kochhar 1972; Neubert et al. 1974a, 1974b; Kochhar 1975; Review in Barrach and Neubert 1980
Palatal shelves	Differentiation Closure	Abbott and Buckalew 1992; Al-Obaidi et al. 1994
Other	Differentiation Proliferation Toxicity	Reviews by Brown and Fabro 1982; Neubert 1982; Faustmann 1988; Whittaker and Faustmann 1994
Invertebrate		
Planaria	Regeneration Lethality	Best and Morita 1982
Hydra	Reaggregation Lethality	Johnson 1980; Johnson et al. 1982
Sea Urchin embryo	Differentiation Lethality	Estus and Blumer 1989
Shrimp embryo	Differentiation Lethality	Sleet and Brendel 1983, 1984, 1985; Sleet 1992
Cricket embryo	Differentiation Lethality	Walton 1983
Drosophila embryo	Differentiation Lethality	Schuler et al. 1982, 1985; Ranganathan et al. 1987

Table 1: continued

Test system	Endpoint(s)	References
Vertebrate embryos		
Chicken	Differentiation Retardation Lethality	Jelinek 1982; Kucera and Burnand 1987, 1988; Kucera et al. 1993; Schmid et al. 1993
Fish	Differentiation Lethality	Birge et al. 1983; Herrmann 1933
Frog (FETAX)	Differentiation Lethality	Dumont et al. 1983; Dawson and Bantle 1987; Fort et al. 1988
Mammalian embryos		
Rodent	Differentiation Retardation Lethality	New 1978; Fantel 1982; Schmid 1985; Sadler et al. 1982; Klug et al. 1985; Bechter and Schmid 1987
Non-rodent	Differentiation Retardation Lethality	Naya et al. 1991; Ninomiya et al. 1993

testing in animal studies (Williams 1994). However, loosing a novel and efficient therapy principle due to false positive in vitro data cannot be accepted. Therefore, in drug development, a screening model should be optimized to have a high specificity.

The majority of compounds developed for human therapy are, unfortunately for the in vitro toxicologist, lipophilic. A reasonable concentration of the test compound needs to be achieved in the aqueous medium of the in vitro test or appropriate delivery systems have to be available (Kitchin and Ebron 1984). In addition, the small amount of compound which is usually synthesized at early stages of drug development will limit the palette of suitable models.

In drug screening often a series of structurally related follow-up compounds is tested. In most of these investigated classes of compounds a fair amount of pharmacological data has been produced at the time of toxicological screening and some information on kinetics, metabolism and potentially adverse effects is usually available from one or the other compound of this class. In addition, results from animal studies and/or human data is available for one or several congeners. Testing of a new chemical is very rarely a 'blind' activity under these circumstances. The in vitro test has to produce not only unspecific 'yes' or 'no' answers but also a rating as concerns the compounds potencies to adversely affect normal development has to be facilitated. In addition, facilitation of drug metabolism in vitro might be desirable.

For the same reasons, an in vitro system might be chosen for its properties of representing a particular mode or mechanism relevant for the potential developmental toxicity of a given compound class (Kistler 1987; Kistler et al. 1990). The same in vitro test might be of very low predictive value for another set of by themselves related chemicals since they do not act via disturbances of the same specific developmental processes (Tsuchiya et al. 1991). Large validation studies with a wide variety of chemicals reported in the scientific literature as being in vivo teratogenic or non-teratogenic, therefore, are of minor importance for the employment of an in vitro screening tool in drug development. In addition, in vivo animal data published often lack quality to be used to validate in vitro data against. Also, the selection of chemicals to be tested as part of a validation study is often biased to prove the suitability of a proposed test system as a screening test (Palmer 1993). However, there is a need to move the most promising test systems further along the validation pathway with well designed and controlled (multi-laboratory) studies.

Mammalian Postimplantation Whole Embryo Culture

Mammalian postimplantation embryos have been cultured successfully for the last approximately 20 years over a wide range of stages of organogenesis (Cockroft 1976; New 1978; Fujinaga and Baden 1992; Barber et al. 1993; Ninomiya et al. 1993). The culture of embryos from the head fold/early somite stage onward has been proposed as teratogenicity screening systems in the early eighties, using the mouse (Sadler et al. 1982, 1985) or the rat (Schmid 1985). The culture of rabbit embryos has recently been optimized (Ninomiya et al. 1993). In principle, for testing of compounds in the rat and mouse, head fold or early somite stage embryos are dissected free from maternal tissue and parietal yolk sac and Reichert's membrane are removed leaving the visceral yolk sac and ectoplacental cone intact. The conceptus is cultured in medium under defined gassing conditions for 24 to 48 hours usually in a roller bottle system. A variety of media have been used with either species. Rat serum which can be diluted up to 50 % with composition salt solutions works well for the rat (Bechter et al. 1991; Terlouw et al. 1993) and mouse (Sadler 1979), but also serum from cows (Klug et al. 1985), monkeys (Klein et al. 1982) and humans (Chatot et al. 1980; Van Maele-Fabry et al. 1993a, Abir et al. 1993) allows satisfactory embryonic growth. Serum from rabbits diluted up to 40 % has been shown to support rabbit embryonic growth in vitro (Ninomiya et al. 1993). The test compound under investigation can be added to the culture during defined culture periods (Sadler et al. 1988; Stahlmann et al. 1993) or the entire culture length. Drug metabolism can be mimicked in various ways. A liver homogenate (S9) and appropriate cofactors have been added to the culture medium using liver from different sources like rat (Fantel et al. 1979; Kitchin et al. 1981) or human (Zhao et al. 1993). Oglesby et al. (1992) have cocultured hepatocytes during the culture period producing metabolites concurrently in the culture medium. Sequential hepatocyte/whole embryo culture, which allow to study the kinetics of toxification and detoxification in vitro have been reported (Bechter et al. 1989). Using serum as culture medium from animals retreated with the test compound also allows to metabolically activate the compound and study the effects of its metabolites on embryonic development in vitro (Schmid et al. 1982).

At the end of the culture period, a number of endpoints are measured. Effects on the development of the visceral yolk sac vascularization and circulation, on hematopoiesis, embryonic growth (e.g. size, protein and DNA content) and differentiation (number of somites, morphological score) as well as dysmorphogenic effects are determined (Sadler et al. 1982; Van Maele-Fabry et al. 1993b; Schmid 1985; Brown and Fabro 1981; Bechter and Brouillard 1988). The data interpretation considers adverse effects on yolk sac development, embryonic growth and differentiation as well as specific dysmorphogenesis (figure 1).

Validation studies (table 2) have been published by Schmid et al. 1983 and Cicurel and Schmid 1988a, 1988b. An interlaboratory validation study is currently ongoing (Van Maele-Fabry et al. 1991; Govers et al. 1988; Piersma et al. 1994). Results of an intersystem validation study have been published by Kucera et al. 1993. Results of 6 pairs of compounds were compared in the cultured chick and rat embryo and in a brain cell aggregate culture. We reported excellent comparison between in vivo and in vitro data for a series of retinoids (Bechter et al. 1992). In an internal prospective validation study, ongoing for the last 10 years, with so far 5 in vivo teratogenic and 49 in vivo non-teratogenic compounds, the specificity of the test system was 90%, the sensitivity 60% and over all accuracy 87% (Bechter 1994, unpublished results).

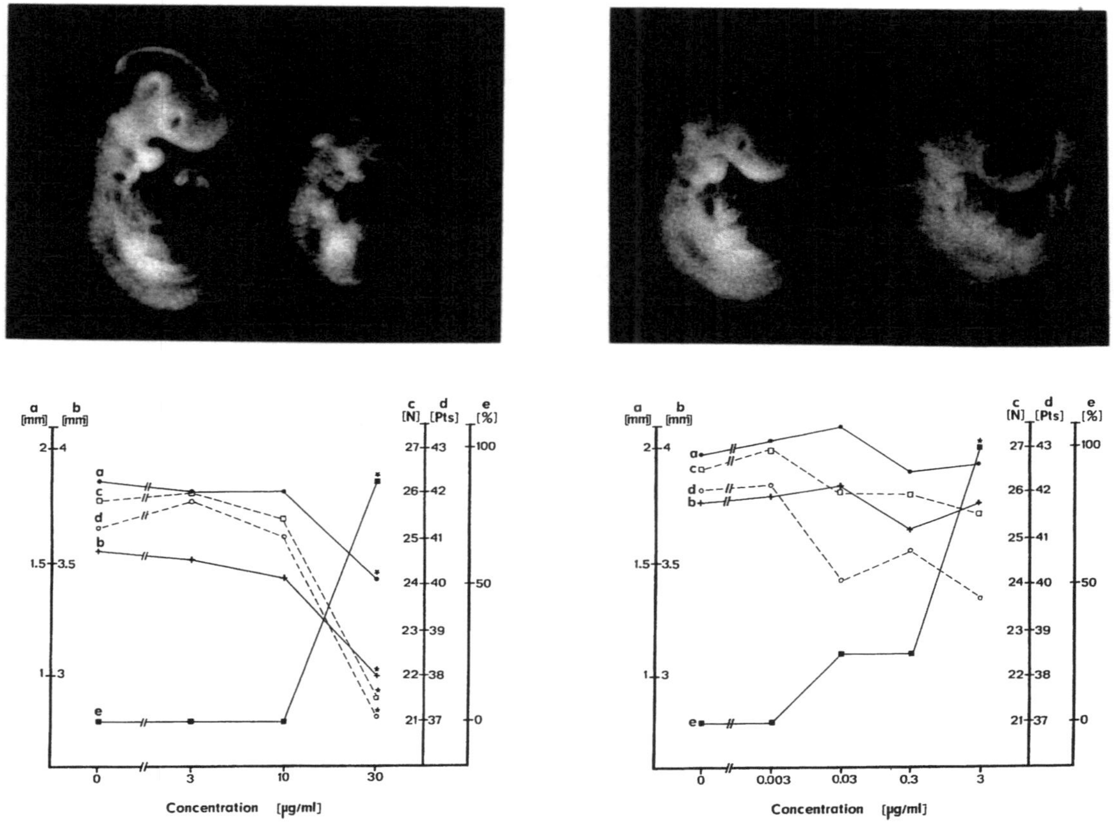

Fig. 1 Typical results of whole embryo culture experiments with a non-teratogen (left panel) and a teratogen (right panel). *a* crown-rump length, *b* head length, *c* somite numbers, *d* morphological score points, *e* frequency of dysmorphogenic embryos.

Table 2: Predictability of test systems proposed for developmental toxicity screening for which validation data are reported

Test system	Compounds tested	Metabolism possible	Sensitivity	Specificity	Accuracy
Mouse ovary tumor cell (MOT)	102 (74/28)[a]	yes	81	75	79
Human embryonic palatal mesenchyme cell (HEPM)	55 (35/20)	yes	66	60	64
Neuroblastoma cell	54 (36/18)	no	86	78	83
Embryonic stem cell	44[b] (21/13)	yes[c]	67	78	71
Pox virus	51 (42/9)	no	79	89	80
Drosophila embryonic cell	99 (61/38)	yes	71	82	75
Limb bud cell	46[d] (27/19)	yes	93	89	91
	27[e] (22/5)		86	100	89
	25[f] (15/10)		73	80	76
	69[g] (44/25)		48-66	56-79	58-62
	122[h] (85/37)		45	86	57
	38[i] (5/33)		40	58	55

Table 2: continued

Test system	Compounds tested	Metabolism possible	Sensitivity	Specificity	Accuracy
Drosophilla embryo	32	yes	77	80	78
Chicken embryo (CHEST[k])	130	no	oversensitive (lethality vs dysmorphogenesis)		
Chicken embryo (Modified)	12[l] (2/9)	no	100	0	82
Frog embryo (FETAX)		yes	85	80	93
Rodent whole embryo	27[m] (17/10)	yes	100	80	93
	12[l] (2/9)		89	50	82
	8[n] (4/4)		50-100	100	75-100
	54[o] (5/49)		60	90	87

[a] In vivo teratogens/in vivo non-teratogens. [b] Combined from Newall and Beedles 1993/ Laschinski et al. 1991. [c] Cells are metabolically active by themselves (Newall and Beedles 1993). [d] Combined with midbrain cell micromass (Flint and Orton 1984). [e] Guntakatta et al. 1984. [f] Combined with midbrain cell micromass (Uphill et al. 1990). [g] Combined with midbrain cell micromass, participation of 3 Laboratories (Koelman et al. 1991). [h] Renault 1991. [i] Internal validation (Bechter 1994, data not published). [k] (Chest I+II) Jelinek 1982. [l] Interlab/intersystem validation Kucera et al 1993. [m] Cicurel and Schmid 1988. [n] Interlab validation (Govers et al. 1988 and Piersma et al. 1994). [0] Internal prospective validation (Bechter 1994, data not published).

The rodent whole embryo culture system is a developed in vitro test system for the detection of potentially teratogenic compounds and for the elucidation of mechanisms of teratogenicity. It is used in many laboratories and has proven to be a valuable tool. It allows to detect dysmorphogenesis in many organ primordia and to distinguish between specific dysmorphogenic effects and adverse effects on the embryos' general retardation of growth and differentiation. In addition, it facilitates the rating of the potencies of structurally related compounds. Also advantageous is the small amounts of test chemicals which are needed for valid test results. It allows the mimicking of drug metabolism by the possibility to use a variety of different drug activating systems. Test compound and/or metabolite concentrations in culture medium and conceptus tissue can easily be monitored. However, the system is relatively complex, it covers only a part of organogenesis, and high technical skill is required. Also, whole embryo culture still requires pregnant animals as the source of embryos to be cultured. The performance of this test can be costly in a setting where animals which are used for the production of serum are not easily available (e.g. old male breeders, veterinary school with cows).

Limb Bud Cell Culture

High density cultures of mouse or chick limb bud cells have been proposed as teratogenicity screening systems by Wilk et al. (1980); Hassell and Horigan (1982) and Guntakatta et al. (1984). Flint and Orton (1984) and Kistler (1985) used cells from rat limb buds and Renault et al. (1993) modified the system to be used with rabbit limb bud cells. In principle, prechondrogenic mesenchymal limb bud cells previously dissociated by trypsin treatment are seeded as a droplet at high density (micromass) in plastic dishes, using 35 mm dishes (Flint and Orton 1984), 24 well plates (Kistler 1985) or 96 well plates (Paulsen and Solursh 1988, Renault et al. 1989). The cells are allowed to adhere for a few hours and are then flooded with the culture medium. The cells proliferate and undergo differentiation into chondrocytes after 5 to 6 days in culture. The rate of proliferation and differentiation into cartilaginous nodules can easily be measured quantitatively by simple staining methods with subsequent spectrophotometry and fully automated image analysis (Renault et al. 1989) or by radiochemistry (Guntakatta et al. 1984). The test compound is added to the cultures in different concentrations. For best results, the addition should be preceded by a 24 hour incubation period without test article. At the end of the 6 day culture period the test compound concentrations leading to a 50% inhibition (IC50, compared to a control culture without test compound) of cell proliferation and differentiation are determined. Comparison of the IC50 values for proliferation and differentiation allow an

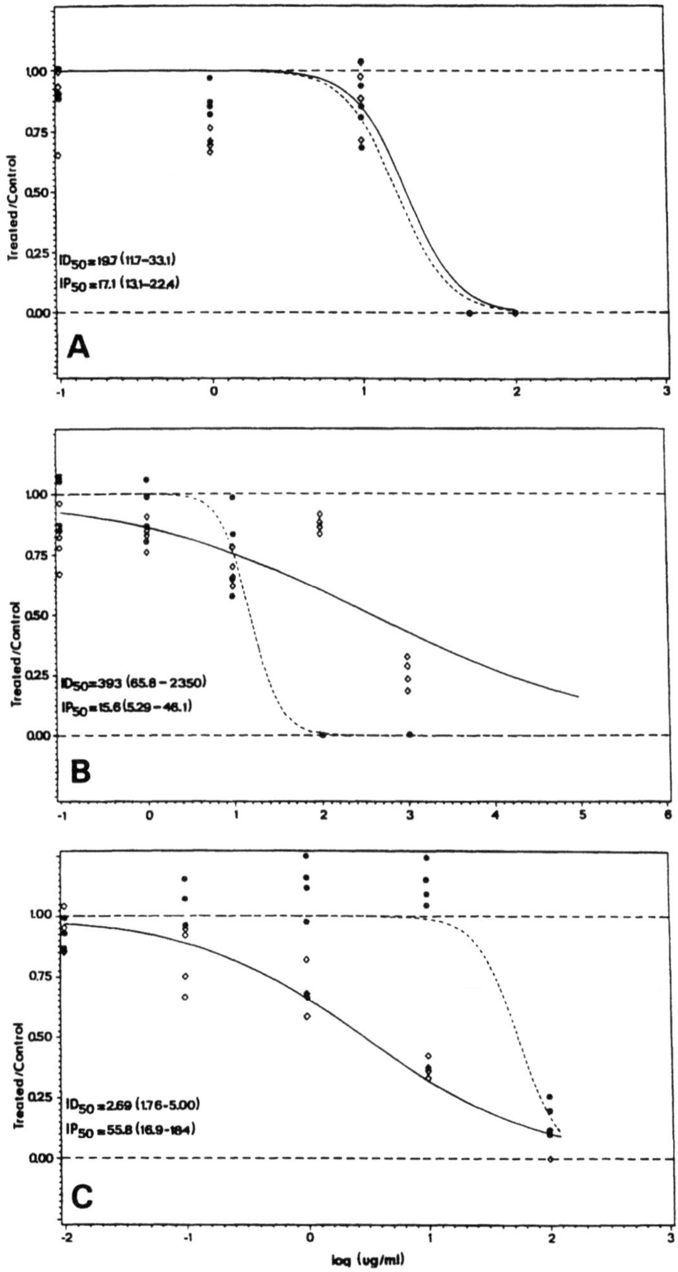

Fig. 2 Typical results of limb bud cell culture experiments with a non-teratogen *(A)*, and compounds which specifically inhibited proliferation *(B)* or differentiation *(C)*. IP_{50}:concentration inhibiting cell proliferation (dotted line) by 50%. ID_{50}:concentration inhibiting cell differentiation (solid line) by 50%.

estimate of the teratogenic potential of the test compound by applying the 2-fold rule (figure 2, Renault et al. 1989). In addition, and probably more meaningful, this test allows to determine the potency of a compound to interfere with these developmental processes and allows to rate chemically related compounds concerning their potency to inhibit cell differentiation in vitro. The interpretation of the results is somewhat arbitrary and different rules for cut off concentrations have been used (e.g. Koelman et al. 1991). Attempts to add metabolic activating systems have been successfully undertaken (Guntakatta et al. 1984; Flint and Orton 1984; Ghaida and Merker 1992).

Flint and Orton (1984) combined the micromass culture of limb bud cells with a high density culture of embryonic mid brain cells which differentiate under appropriate culture conditions into neuronal cells to test 27 teratogens and 19 non-teratogens in a blind trial (table 2). They reported 93% sensitivity, 89% specificity and 91% overall accuracy. Guntakatta et al. (1984) reported a sensitivity of 100% and a specificity of 86% with 22 teratogens and 5 non-teratogens tested. The useful application of the culture system in the selection of drug candidates was demonstrated with a series of antifungal triazoles (Flint and Boyle 1985) and retinoids containing a carboxylic endgroup (Kistler 1985, 1987; Kistler et al. 1990). Renault et al. (1989) reported a high specificity (100%), 61% sensitivity and an overall accuracy of 75% with 33 in vivo teratogens and 18 in vivo non-teratogens. In our laboratory, we tested 5 in vivo teratogenic and 33 in vivo non-teratogenic developmental compounds with a specificity of 58%, a sensitivity of 40% and an overall accuracy of 55% (Bechter 1994, unpublished results). Preliminary conclusions from a recent multicenter validation study (Koelman et al. 1991) including 69 compounds were that teratogenicity of potent teratogens could be predicted with sufficient probability (48-66% sensitivity) and that the in vitro micromass test would have its greatest value when applied for selection purposes within a specific chemical class. Validation studies including retinoids (Kistler 1987; Kistler et al. 1990) and a series of structurally related herbicides (Tsuchiya et al. 1991) nicely show that the predictive value of this test system for in vivo embryotoxicity depends on the chemical class of the compounds investigated.

The micromass test is a simple system which allows the screening of a large number of compounds in a short time, especially if it is fully automated. The amount of test chemical needed is small and only a few animals are used for the preparation of the cell cultures. However, its predictiveness for certain classes of chemicals is limited and the rules applied for the interpretation of the test results are critical.

Embryonic Stem Cell Culture

A pluripotent embryonic stem cell line was first derived from the inner cell mass of mouse blastocysts by Robertson (1986). Embryonic stem cells can be grown and kept in an undifferentiated state for many passages on an embryonic fibroblast feeder layer (Laschinski et al. 1991) or in the presence of leucemia inhibiting factor (LIF) in the culture medium (Newall and Beedles 1993). Cells can be washed, harvested by trypsinization and seeded as microdroplets in multiwell plates in a similar fashion as the limb bud cells discussed above. Cultured without feeder cells or in medium without LIF they differentiate within 7 days in culture into embryonic endoderm which is morphologically different from the embryonic stem cells. A test compound might interfere with cell growth and/or differentiation leading to cytotoxicity and a reduction of differentiated embryonic stem cells. Laschinski et al. (1991) cultured fibroblasts in parallel with the embryonic stem cells to test the same chemical under investigation. The IC10 and IC50 values for fibroblast cytotoxicity (as determined by the MTT-assay) and for embryonic stem cell cytotoxicity were determined. A possible teratogen may be identified by a higher sensitivity of embryonic stem cells compared to fibroblasts (difference of IC50's greater than one order of magnitude) or by a different shape of the concentration-response curve for the two cell types. The authors tested a series of compounds for which metabolic activation for teratogenic activity was not required. Positive results were mainly restricted to antineoplastic agents. In vivo non-teratogenic agents were identified negative by the test system. However, a number of compounds were false negative in embryonic stem cells. Newall and Beedles (1993) modified the stem-cell test using the protocol developed for the limb bud cell culture by Flint and Orton (1984). Cytotoxicity in the embryonic stem cells was determined with the MTT-assay and the extent of differentiation was evaluated in the same cells by measuring the area stained by Nile blue. Potentially teratogenic compounds were identified by a specific inhibition of differentiation (ratio greater than 2 of the IC50's for the two endpoints). A number of 25 compounds have been tested in a blind trial. The stem cell test predicted teratogenicity with similar sensitivity and specificity as other in vitro test systems (table 2). Furthermore, it appears that the test identified teratogens requiring activation without an additional metabolizing system.

The embryonic stem cell test seems to be a promising system, especially for screening of drug candidates. It is similarly simple as the micromass test, it is quick, cheap and only requires small amounts of chemical. In addition, no animals are used and it can be set up whenever needed by thawing frozen portions of stored cells. However, the stem cell culture needs to be further tested for its suitability as a reliable screening tool for new compounds in drug development.

Conclusions

In this paper, three in vitro teratogenicity screening systems are described. The first two were chosen because they are used in our laboratory for the practical purpose of compound selection. In addition, both systems have been validated by different laboratories with a wide variety of compounds and compound classes over a long period of time. The third system, in the authors opinion, seems to be a promising tool for the same purpose.of drug selection. This subjective choice does not intend to reject other in vitro or short term tests for embryotoxicity. Each of these systems may have its merits for particular purposes. However, it seems important that the chosen test provides a relevant answer to the problem in question, whether employed for screening, testing or investigating the mechanisms of adverse effects. In addition, the investigator needs to be aware of the different shortcomings associated with all of these tests (including in vivo animal testing).and the limits of data interpretation. Furthermore, a test system might work very well in one laboratory (but not in others) for a particular purpose, a particular class of compounds or might very closely mimic a mechanism. To validate such a system for their particular application in their own laboratory has to be the responsibility of the investigators themselves before reliable data can be created.

Acknowledgements

The author wishes to thank R. Chu for her excellent secretarial help and for deceiving the computer to bring the manuscript into a camera ready form, N. Krafft and P. Stirnimann for technical assistance and Drs. K. Bartmann and E. Koch for critically reviewing the contents.

References

Abbott BD, Buckalew AR (1992) Embryonic palatal responses to teratogens in serum-free organ culture. Teratology 45:369-382

Abir R, Ornoy A, Hur HB, Jaffe P, Pinus H (1993) IgG exchange as a means of partial correction of anomalies in rat embryos in vitro, induced by sera from women with recurrent abortion. Toxic in Vitro 7:817-826

Al-Obaidi N, Kastner U, Merker HJ, Klug S (1994) Development of a suspension organ culture of the fetal rat palate. Submitted 1994

Aydelotte MB, Kochhar DM (1972) Development of mouse limb buds in organ culture: chondrogenesis in the presence of a proline analog, L-azetidine-2-carboxylic acid. Developmental Biology 28:191-201

Barber CV, Carda MB, Fantel AG (1993) A new technique for culturing rat embryos between gestation days 14 and 15. Toxic in Vitro 7:695-700

Barrach HJ, Neubert D (1980) Significance of organ culture techniques for evaluation of prenatal toxicity. Arch Toxicol 45:161-187

Bechter R, Schmid B, Mayer FK (1986) Teratogenic potential of antimycotic drugs evaluated in the whole-embryo culture system. Fd Chem Toxic 24:641-642

Bechter R, Schmid BP (1987) Teratogenicity in vitro - a comparative study of four antimycotic drugs using the whole-embryo culture system. Toxic in Vitro 1:11-15

Bechter R, Brouillard JF (1988) The effects of different chemical forms of a test compound on embryotoxicity, distribution and metabolism in vitro. Toxic in Vitro 2:181-188

Bechter R, Bouis P, Fischer V (1989) Primary hepatocyte culture as an activating system for xenobiotica tested in the rat whole embryo in vitro. In: Goldberg AM, Principe ML (eds) Alternative methods in toxicology (vol 7) - in vitro toxicology new directions. Mary Ann Liebert, Inc, New York, USA, pp 313-326

Bechter R, Terlouw GDC, Lee QP, Juchau MR (1991) Effects of QA 208-199 and its metabolite 209-668 on embryonic development in vitro after microinjection into the exocoelomic space or into the amniotic cavity of cultured rat conceptuses. Teratogenesis, Carcinogenesis, and Mutagenesis 11:185-194

Bechter R, Terlouw GDC, Tsuchiya M, Tsuchiya T, Kistler A (1992) Teratogenicity of arotinoids (retinoids) in the rat whole embryo culture. Arch Toxicol 66:193-197

Best JB, Morita M (1982) Planarians as a model system for in vitro teratogenesis studies. Teratogenesis, Carcinogenesis, and Mutagenesis 2:277-291

Birge WJ, Black JA, Westerman AG, Ramey BA (1983) Fish and amphibian embryos - a model system for evaluating teratogenicity. Fundamental and Applied Toxicology 3:237-242

Bournias-Vardiabasis N (1990) Drosophila melanogaster embryo cultures: an in vitro teratogen assay. ATLA 18:291-300

Bournias-Vardiabasis N, Teplitz RL, Chernoff GF, Seecof RL (1983) Detection of teratogens in the drosophila embryonic cell culture test: assay of 100 chemicals. Teratology 28:109-122

Braun AG, Emerson DJ, Nichinson BB (1979) Teratogenic drugs inhibit tumour cell attachment to lectin-coated surfaces. Nature 282:507-509

Braun AG, Dailey JP (1981) Thalidomide metabolite inhibits tumor cell attachment to concanavalin a coated surfaces. Biochemical and Biophysical Research Communications 98:1029-1034

Braun AG, Buckner CA, Emerson DJ, Nichinson BB (1982) Quantitative correspondence between the in vivo and in vitro activity of teratogenic agents. Proc Natl Acad Sci USA 79:2056-2060

Brown NA, Fabro S (1981) Quantitation of rat embryonic development in vitro: a morphological scoring system. Teratology 24:65-78

Brown NA, Fabro SE (1982) The in vitro approach to teratogenicity testing. In: Snell K (ed) Developmental Toxicology. Croom Helm, London, England, pp 31-57

Chatot CL, Klein NW, Piatek J, Pierro LJ (1980) Successful culture of rat embryos in human serum: use in the detection of teratogens. Science 207:1471-1473

Cicurel L, Schmid BP (1988a) Post-implantation embryo culture: validation with selected compounds for teratogenicity testing. Xenobiotica 18:617-624

Cicurel L, Schmid BP (1988b) Postimplantation embryo culture for the assessment of the teratogenic potential and potency of compounds. Experientia 44:833-840

Cockroft (1976) Comparison of in vitro and in vivo development of rat foetuses. Developmental Biology 48:163-172

Daston GP, D'Amato RA (1989) In vitro techniques in teratology. Toxicology and Industrial Health 5:555-585

Daston GP, Rogers JM, Versteeg DJ, Sabourin TD, Baines D, Marsh SS (1991) Interspecies comparisons of A/D ratios: A/D ratios are not constant across species. Fundamental and Applied Toxicology 17:696-722

Dawson DA, Bantle JA (1987) Development of a reconstituted water medium and preliminary validation of the frog embryo teratogenesis assay - xenopus (FETAX). Journal of Applied Toxicology 7:237-244

Dumont JN, Schultz TW, Buchanan MV, Kao GL (1983) Frog embryo teratogenesis assay: xenopus (FETAX) - a short-term assay applicable to complex environmental mixtures. Environmental Science Research. (Series) New York 27:393-405

European Chemical Industry Ecology & Toxicology Centre (ECETOC) (1989) Alternative approaches for the assessment of reproductive toxicity (with emphasis on embryotoxicity/teratogenicity) Monograph No 12

Estus S, Blumer JL (1989) Role of microtubule assembly in phenytoin teratogenic action in the sea urchin (arbacia punctulata) embryo. Molecular Pharmacology 36:708-715

Fantel AG, Greenaway JC, Juchau MR, Shepard TH (1979) Teratogenic biactivation of cyclophosphamide in vitro. Life Sciences 25:67-72

Fantel AG (1982) Culture of whole rodent embryos in teratogen screening. Teratogenesis, Carcinogenesis, and Mutagenesis 2:231-242

Faustmann EM (1988) Short-term tests for teratogens. Mutation Research 205:355-384

Flint OP, Boyle FT (1985) An in vitro test for teratogens: its application in the selection of non-teratogenic triazole antifungals. In: Homburger F (ed) Concepts in Toxicology. Karger, Basel, Switzerland, vol 3, pp 29-35

Flint OP, Orton TC (1984) An in vitro assay for teratogens with cultures of rat embryo midbrain and limb bud cells. Toxicology and Applied Pharmacology 76:383-395

Fort DJ, Dawson DA, Bantle JA (1988) Development of a metabolic activation system for the frog embryo teratogenesis assay: xenopus (FETAX). Teratogenesis, Carcinogenesis, and Mutagenesis 8:251-263

Freeman SJ, Brown NA (1987) Sub-mammalian and sub-vertebrate models in teratogenicity screening. In: Atterwill CK, Steele CE (eds) In vitro methods in toxicology. Cambridge University Press, Cambridge, UK, pp 391-409

Fujinaga M, Baden JM (1992) Variation in development of rat embryos at the presomite period. Teratology 45:661-670

Ghaida J, Merker HJ (1992) Effects of cyclophosphamide and acrolein in organoid cultures of mouse limb bud cells grown in the presence of adult rat hepatocytes. Toxic in Vitro 6:27-40

Govers MJAP, Peters PWJ, Piersma AH, Verhoef A, Schmid BP, Attenon J, Van Maele-Fabry G, Picard JJ, Stadler J, Verseil C, Bechter R (1988) Interlaboratory validation of rodent postimplantation embryo culture as a screening test for teratogenic compounds. Presented at the European Teratology Society Meeting

Gras G, Mondain J (1981) Pollution des produits de la peche par le mercure et le methylmercure. Toxicol Eur Res 3:243-59

Guntakatta M, Matthews EJ, Rundell JO (1984) Development of a mouse embryo limb bud cell culture system for the estimation of chemical teratogenic potential. Teratogenesis, Carcinogenesis, and Mutagenesis 4:349-364

Hassell JR, Horigan EA (1982) Chondrogenesis: a model developmental system for measuring teratogenic potential of compounds. Teratogenesis, Carcinogenesis, and Mutagenesis 2:325-331

Herbst AL, Scully RE, Robboy SJ (1979) Prenatal diethylstilbestrol exposure and human genital tract abnormalities. Natl Cancer Inst Monogr 51:25-35

Herrmann K (1993) Effects of the anticonvulsant drug valproic acid and related substances on the early development of the zebrafish (brachydanio rerio). Toxic. in Vitro 7:41-54

Honegger P, Schilter B (1992) Serum-free aggregate cultures of fetal rat brain and liver cells: methodology and some practical applications in neurotoxicology. In: Zbinden G (ed) The brain in bits and pieces. In vitro techniques in neurobiology, neuropharmacology and neurotoxicology. MTC Verlag, Zollikon, Switzerland, pp 51-79

Honegger P, Schilter B (1993) The use of serum-free aggregating brain cell cultures in neurotoxicology. In: Chang LW (ed) Handbook of neurotoxicology. I. Basic principles and current concepts. Dekker Inc. - in press

Honegger P, Werffeli P (1988) Use of aggregating cell cultures for toxicological studies. Experientia 44:817-823

Honegger P, Lenoir D, Favrod P (1979) Growth and differentiation of aggregating fetal brain cells in a serum-free defined medium. Nature 282:305-308

Jelinek R (1982) Use of chick embryo in screening for embryotoxicity. Teratogenesis, Carcinogenesis, and Mutagenesis 2:255-261

Johnson EM (1980) A subvertebrate system for rapid determination of potential teratogenic hazards. J Environ Pathol Toxicol 4:153-156

Johnson EM, Gorman RM, Gabel BEG, George ME (1982) The hydra attenuata system for detection of teratogenic hazards. Teratogenesis, Carcinogenesis, and Mutagenesis 2:263-276

Keller SJ, Smith MK (1982) Animal virus screens for potential teratogens. I. poxvirus morphogenesis. Teratogenesis, Carcinogenesis, and Mutagenesis 2:361-374

Kimmel GL (1990) In vitro assays in developmental toxicology: their potential application in risk assessment. In: Kimmel GL, Kochhar DM (eds) In vitro methods in developmental toxicology: use in defining mechanisms and risk parameters. CRC Press, Boca Raton, USA, pp 163-173

Kistler A (1985) Inhibition of chondrogenesis by retinoids: limb bud cell cultures as a test system to measure the teratogenic potential of compounds? In: Homburger F (ed) Concepts in Toxicology. Karger, Basel, Switzerland, vol 3, pp 86-100

Kistler A (1987) Limb bud cell cultures for estimating the teratogenic potential of compounds - validation of the test system with retinoids. Arch Toxicol 60:403-414

Kistler A, Tsuchiya T, Tsuchiya M, Klaus M (1990) Teratogenicity of arotinoids (retinoids) in vivo and in vitro. Arch Toxicol 64:616-622

Kitchin KT, Schmid BP, Sanyai MK (1981) Teratogenicity of cyclophosphamide in a coupled microsomal activating/embryo culture system. Biochemical Pharmacology 30:59-64

Kitchin KT, Ebron MT (1984) Further development of rodent whole embryo culture: solvent toxicity and water insoluble compound delivery system. Toxicology 30:45-57

Klein NW, Plenefisch JD, Carey SW, Fredrickson WT, Sackett GP, Burbacher TM, Parker RM (1982) Serum from monkeys with histories of fetal wastage causes abnormalities in cultures of whole rat embryos. Science 215:66-69

Klug S, Lewandowski C, Neubert D (1985) Modification and standardization of the culture of early postimplantation embryos for toxicological studies. Arch Toxicol 58:84-88

Kochhar DM (1975) The use of in vitro procedures in teratology. Teratology 11:273-288

Koelman HJS, Jongeling AJ, van Erp YHM, Weterings PJM, Koopmans ME, Joosten HF, v.d. Dobbelsteen DJ, v.d. Aa EM, Yih TD (1991) International ring validation of the in vitro micromass teratogenicity test: preliminary results of 3 dutch laboratories. Teratology 44:30A/P39

Koeter HBWM (1993) Test guideline development and animal welfare: regulatory acceptance of in vitro studies. Reproductive Toxicology 7:117-123

Kucera P, Burnand MB (1987) Routine teratogenicity that uses chick embryos in vitro. Teratogenesis, Carcinogenesis and Mutagenesis 7:427-447

Kucera P, Burnand MB (1988) Teratogenicity screening in standardized chick embryo culture: effects of dexamethasone and diphenylhydantoin. Experientia 44:827-833

Kucera P, Cano E, Honegger P, Schilter B, Zijlstra JA, Schmid B (1993) Validation of whole chick embryo cultures, whole rat embryo cultures and aggregating embryonic brain cell cultures using six pairs of coded compounds. Toxic in Vitro 7:785-798

Langille RM, Paulsen DF, Solursh M (1989) Differential effects of physiological concentrations of retinoic acid in vitro on chondrogenesis and myogenesis in chick craniofacial mesenchyme. Differentiation 40:84-92

Laschinski G, Vogel R, Spielmann H (1991) Cytotoxicity test using blastocyst-derived euploid embryonal stem cells: a new approach to in vitro teratogenesis screening. Reproductive Toxicology 5:57-64

Manson J (1991) An overview and comparison of reproductive and developmental toxicity regulations. In: Lumley CE, Walker SR (eds) Current issues in reproductive and developmental toxicology - can an international guideline be achieved? Quay Publishing, Lancaster, UK, pp 21-34

Mummery CL, Slager HG, van Inzen W, Freund E, van den Eijnden-van Raaij AJM (1993) Regulation of growth and differentiation in early development: of mice and models. Reproductive Toxicology 7:145-154

Naya M, Kito Y, Eto K, Deguchi T (1991) Development of rabbit whole embryo culture during organogenesis. Congenital Anomalies 31:153-156

Neubert D (1982) The use of culture techniques in studies on prenatal toxicity. Pharmacology and Therapeutics 18:397-434

Neubert D, Merker HJ, Tapken S (1974a) Comparative studies on the prenatal development of mouse extremities in vivo and in organ culture. Naunyn-Schmiedeberg's Arch Pharmacol 286:251-270

Neubert D, Tapken S, Merker HJ (1974b) Induction of skeletal malformations in organ cultures of mammalian embryonic tissues. Naunyn-Schmiedeberg's Arch Pharmacol 286:271-282

New DAT (1978) Whole embryo culture and the study of mammalian embryos during organogenesis. Biol Rev 53:81-122

Newall DR, Beedles K (1993) The stem-cell test - a novel in vitro assay for teratogenic potential. Presented at the 21st Annual Conference of the European Teratology Society, Lyon, France, Teratology 48:30A

Ninomiya H, Kishida K, Ohno Y, Tsurumi K, Eto K (1993) Effects of trypan blue on rat and rabbit embryos cultured in vitro. Toxic in Vitro 7:707-717

Oglesby LA, Ebron-McCoy MT, Logsdon TR, Copeland F, Beyer PE, Kavlock RJ (1992) In vitro embryotoxicity of a series of para-substituted phenols: structure, activity, and correlation with in vivo data. Teratology 45:11-33

Palmer AK (1993) Introduction to (pre)screening methods. Reproductive Toxicology 7:95-98

Paulsen DF, Solursh M (1988) Microtiter micromass cultures of limb-bud mesenchymal cells. In Vitro Cellular & Developmental Biology 24:138-147

Peters PWJ, Piersma AH (1990) In vitro embryotoxicity and teratogenicity studies. Toxic in Vitro 4:570-576

Piersma AH, Attenon P, Bechter R, Govers MJAP, Krafft N, Schmid BP, Stadler J, Verhoef A, Verseil C (1994) Interlaboratory evaluation of embryotoxicity of eight xenobiotic compounds in the postimplantation rat embryo culture. Reproductive Toxicology, submitted

Pratt RM, Grove RI, Willis WD (1982) Prescreening for environmental teratogens using cultured mesenchymal cells from the human embryonic palate. Teratogenesis, Carcinogenesis, and Mutagenesis 2:313-318

Pratt RM, Willis WD (1985) In vitro screening assay for teratogens using growth inhibition of human embryonic cells. Proc. Natl. Acad. Sci. USA 82:5791-5794

Ranganathan S, Davis DG, Hood RD (1987) Developmental toxicity of ethanol in drosophila melanogaster. Teratology 36:45-49

Renault JY, Melcion C, Cordier A (1989) Limb bud cell culture for in vitro teratogen screening: validation of an improved assessment method using 51 compounds. Teratogenesis, Carcinogenesis, and Mutagenesis 9:83-96

Renault JY (1991) Les cultures de cellules de bourgeon de membre d'embryon de rat: un modèle pour la détection et l'étude des agents tératogènes. These pour l'obtention du grade de docteur de l'Universite Paris VII, France

Renault JY, Guillet V, Guittin P (1993) Thalidomide activity in a micromass rabbit embryo limb bud cell culture assay. Poster presented at the 21st Annual Conference of the European Teratology Society, Lyon, France, Teratology 48:32A

Robertson EJ (1986) Pluripotent stem lines as a route into the mouse germ line. Trends Genet 2:9-13

Sadler TW (1979) Culture of early somite mouse embryos during organogenesis. J Embryol exp Morph 49:17-25

Sadler TW, Horton WE, Warner CW (1982) Whole embryo culture: a screening technique for teratogens ? Teratogenesis, Carcinogenesis, and Mutagenesis 2:243-253

Sadler TW, Warner CW, Tulis SA, Smith MK, Doerger J (1985) Factors determining the in vitro response of rodent embryos to teratogens. In: Homburger F (ed) Concepts in Toxicology. Karger, Basel, Switzerland, vol 3 pp, 36-45

Sadler TW, Shum L, Warner CW, Smith MK 1988) The role of pharmacokinetics in determining the response of rodent embryos to teratogens in whole-embryo culture. Toxic in Vitro 2:175-180

Schmid BP, Trippmacher A, Bianchi A (1982) Teratogenicity induced in cultured rat embryos by the serum of procarbazine treated rats. Toxicology 25:53-60

Schmid BP (1985) Teratogenicity testing of new drugs with the postimplantation embryo culture system. In: Homburger F (ed) Concepts in Toxicology. Karger, Basle, Switzerland, vol 3, pp 46-57

Schmid BP, Trippmacher A, Bianchi A (1983) Validation of the whole-embryo culture method for in vitro teratogenicity testing. In: Hayes AW, Schnell RC, Miya TS (eds) Developments in the science and practice of toxicology. Elsevier Science Publishers, Amsterdam, New York, Oxford, pp 563-566

Schmid BP, Honegger P, Kucera P (1993) Embryonic and fetal development: fundamental research. Reproductive Toxicology 7:155-164

Schuler RL, Hardin BD, Niemeier RW (1982) Drosophila as a tool for the rapid assessment of chemicals for teratogenicity. Teratogenesis, Carcinogenesis, and Mutagenesis 2:293-301

Schuler RL, Radike MA, Hardin BD, Niemeier RW (1985) Pattern of response of intact drosophila to known teratogens. J Am Coll Toxicol 4:291-303

Seifert J (1989) Teratogenesis of polychlorocycloalkane insecticides in chicken embryos resulting from their interactions at the convulsant recognition sites of the GABA (pro)receptor complex. Bull Environ Contam Toxicol 42:707-715

Sleet RB (1992) Brine shrimp (artemia) - fish food with potential application as a prescreen to predict chemical hazard to human development. Lab Animal 21:26-36

Sleet RB, Brendel K (1983) Improved methods for harvesting and counting synchronous populations of artemia nauplii for use in developmental toxicology. Ecotoxicology and Environmental Safety 7:435-446

Sleet RB, Brendel K (1984) A flow-through hatching and cold storage system for continuous collection of freshly hatched artemia nauplii. Journal of Aquariculture & Aquatic Sciences 3:76-83

Sleet RB, Brendel K (1985) Homogeneous populations of artemia nauplii and their potential use for in vitro testing in developmental toxicology. Teratogenesis, Carcinogenesis, and Mutagenesis 5:41-54

Stahlmann R, Klug S, Foerster M, Neubert D (1993) Significance of embryo culture methods for studying the prenatal toxicity of virustatic agents. Reproductive Toxicology 7:129-143

Terlouw GDC, Namkung MJ, Juchau MR, Bechter R (1993) In vitro embryotoxicity of N-methyl-N-(7-propoxynaphthalene-2-ethyl)hydroxylamine (QAB): evidence for N-dehydroxylated metabolite as a proximate dysmorphogen. Teratology 48:431-439

Trosko JE, Chang CC, Netzloff M (1982) The role of inhibited cell-cell communication in teratogenesis. Teratogenesis, Carcinogenesis, and Mutagenesis 2:31-45

Tsuchiya T, Bürgin H, Tsuchiya M, Winternitz P, Kistler A (1991) Embryolethality of new herbicides is not detected by the micromass teratogen tests. Arch Toxicol 65:145-149

Uphill PF, Wilkins SR, Allen JA (1990) In vitro micromass teratogen test: results from a blind trial of 25 compounds. Toxic in Vitro 4:623-626

Van Maele-Fabry G, Picard JJ, Attenon P, Berthet P, Delhaise F, Govers MJAP, Peters PWJ, Piersma AH, Schmid BP, Stadler J, Verhoef A, Verseil C (1991) Interlaboratory evaluation of three culture media for postimplantation rodent embryos. Reproductive Toxicology 5:417-426

Van Maele-Fabry G, Therasse P, Lenoir E, Desager JP, Despontin K, Gofflot G, Jacobs MC, Lecart C, Berthet P, Lachapelle JM, Picard JJ (1993a) Embryotoxicity of human sera from patients treated with isotretinoin. Toxic in Vitro 7:809-815

Van Maele-Fabry G, Delhaise F, Gofflot F, Picard JJ (1993b) Developmental table of the early mouse post-implantation embryo. Toxic in Vitro 7:719-725

Walton BT (1983) Use of the cricket embryo (acheta domesticus) as an invertebrate teratology model. Fundamental and Applied Toxicology 3:233-236

Welsch F (1990) Short-term methods of assessing developmental toxicity hazard - status and critical evaluation. In: Kalter H (ed) Issues and reviews in teratology. Plenum Press, New York, USA, Vol. 5, pp 115-153

Whittaker SG, Faustmann EM (1994) In vitro assays for developmental toxicity. In: Gad SC (ed) In vitro toxicology. Raven Press, New York, USA, pp 97-122

Wiger R, Stottum A, Brunborg G (1988) Estimating chemical developmental hazard in a chicken embryo limb bud micromass system. Pharmacology & Toxicology 62:32-37

Wilk AL, Greenberg JH, Horigan EA, Pratt RM, Martin GR (1980) Detection of teratogenic compounds using differentiating embryonic cells in culture. In Vitro 16:269-276

Williams PD (1994) Scientific and regulatory considerations in the development of in vitro techniques in toxicology. In: Gad SC (ed) In vitro toxicology. Raven Press, New York, USA, pp 255-261

Zhao J, Krafft N, Terlouw GDC, Bechter R (1993) A model combining the whole embryo culture with human liver S-9 fraction for human teratogenic prediction. Toxic in Vitro 7:827-831

IN VITRO METHODS IN REGULATORY TOXICOLOGY

R. Bass

Federal Institute for Drugs and Medical Devices
Seestrasse 10-11, D-13353 Berlin-Wedding, Germany

Introduction

So-called "in vitro" experimentation to be used replacing experiments with whole animals has often been a matter rather of emotion than of data. This is no surprise, and the data thus missing cannot be presented here. In the meantime, however, a whole lot of data has been generated, investigating in detail many aspects of experiments involving biological matter at a level below that of the whole, intact organism. One possibility was the establishment of precise dose-response-relationship comparisons among and between large families of chemical substances. For such comparisons a variety of experimental endpoints is available, often involving acute cytotoxicity. Without further knowledge about the substances to be investigated and compared, e.g. about their physico-chemical attributes, interpretation of such data and comparisons becomes difficult (because of comparing e.g. different lipophilicity rather than different toxicity); this does not contribute to meaningful risk-benefit assessment and decision as is needed for medicines and other consumer products. Such data, therefore, have become of minor interest to us and will not be presented here.
The active and important interface between animal and non-animal experimentation resides somewhere else, and I shall try to describe areas of experimental endpoints in (pharmaco-) dynamics, (pharmaco-) kinetics, and toxicity, where sincere thought and work has been employed to detect, develop and validate in vitro systems or other alternatives superior to animal experimentation, so that they can be introduced into regulatory toxicology, i.e. accepted in our "rule books".

Aspects to be Considered

- The major players in risk assessment are usually the regulators and the regulated. Regulators are government administrative agencies and/or their advisors; where law-enforced, they are responsible for decisions on marketing authorization and licensing (for details and ample differences see the appropriate European Union and national legislations).
- The regulated are those developing or sponsoring product development and marketing - they are to fulfill the requirements imposed upon them or to

---

Dedicated to Prof. D. Grossklaus, former President of the BGA for his sincere engagement in animal protection remonstrate against unjustified demands. All legislation steering the marketing of products related to human health, contains escape clauses alike, so that claims can be made about specific requirements having become invalid, superflous, or replacable, e.g. by "in vitro" experimentation. This has become a major driving force for inactivating formal requirements for animal experimentation on a case by case basis, or in general.

- The consumer is interested firstly in product safety and efficacy. Safety expectations of the consumer vary greatly, as do those for efficacy. Sometimes a weak indication of "activity" is accepted, sometimes hard data on safety are disregarded by the consumer - patients constitute a large subgroup. Although being a toxicology-oriented and safety-seeking society, we probably did not attempt to educate the consumer to the extent and kind we should have. Can this be overcome through officially regulating "in vitro" toxicology?
- Animal protection has to be mentioned next. Persons acting from this point of view may be either consumers, regulators, or the regulated. Experience has shown that there is no fixed pattern as to which role will be taken on by what group. Often, those directly involved with respect to a specific product, will think and act differently than those consumers, regulators and regulated who are not involved directly.
- The persons in charge for performing - and determining which kinds of experiments actually need to be performed - are those that are knowledgeable about the advantages and drawbacks of the experimental systems available, be they animals, other biological tests, of physico-chemical or any other nature. All persons mentioned here are interested in and obliged to safety and efficacy, demanding or performing either screening, or clarifying unresolved issues from such routine-testing. At the same time they may behave like supporting animal or human rights. The resulting situation is almost never black or white, but may include chamaeleons.

The regulators and the regulated are governed by legislation like the Chemicals or Medicines Acts. It is through enforcement and fulfillment of the rules and guidelines that they have to ensure safety of the products before and after marketing. The huge variety of products, production processes, and compound classes lets us understand that rules and requirements cannot be the same for all (Fig.1). Safety is the highest ranking priority. Animal protection issues are given in a separate law, asking for action both on European and national levels, and have been cross-referenced to all legal obligations which may require animal experimentation. Animal experiments will have to be replaced, reduced or refined whenever the necessary information can be obtained "at least as well" by other means, or has become superflous. It is exactly this controversy between "protection" and "protection" issues that any regulator will find himself confronted with: the demand for safety and at the same time being denied the possibility to perform the animal experiments believed to be necessary.

For reference and further details, see European Council Directive (C.D.) 91/507/EEC and - as one national example - the German "Arzneimittelprüfrichtlinie" fulfilling §26 of the German Drug Law for testing requirements; and C.D. 86/609/EEC, attempting to harmonize national rules and guidelines on animal protection, and - as one national example - the German Animal Protection Law , amended in 1990 to redefine "animal experiment" so as to include manipulation of their genetic material (fertilized eggs or embryos), if they are used for experimental purposes,and if the animal used in conjunction with such experimentation may suffer or become damaged (§7;1 German Animal Protection Law).

The Regulators and the Regulated

- Develop and market products
- Guarantee their safety

* Finished products	* Chemicals
* Intermediate products	* Biologicals
* Raw materials	* Biotechnologicals
* Production processes	* Gene therapy
* Production facilities	* Devices
* Environmental safety	

Chemicals Act, Medicines Act, etc.

Fig. 1

Some Problem Areas of Animal Experimentation

Certain areas are more severely criticized than others. Argumentation can be seen as derived from wanting to protect any animal from suffering and pain, from reducing animal numbers in those experiments where they seem to be exceptionally high, from shielding certain animal species, from pointing out areas of animal experimentation which seem to be outdated, to scientific arguments of the usefulness of a method, irrespective of the material employed, and to the debate of extrapolation from lower to higher ranking systems and species.

- A product class, where animal experiments are actually expected to be forbidden in the near future, is that of decorative cosmetics. Regulators, including regulatory toxicologists have come to the conclusion that animal experiments can no longer be deemed necessary; animal rights have been weighed heavier than the potential increase in safety when performing studies in whole animals in addition to those "in vitro". We, the toxicologists - the regulators and the regulated - may have been surprised, but have to accept the challenge to develop non-animal experimental systems with high relevance for humans.

- Two test systems have been pointed out as having the potential to be exceptionally cruel and, therefore, to be eliminated - irrespective of their true replacability. The determination of acute toxicity (to be measured following a single administration of the test compound) as "LD 50" (lethal dose 50 - usually involving statistical methods to derive that dose, where 50% of the animals are calculated to die within 14 days of substance administration - at a 95% confidence limit). Such an endpoint does not exist in any other animal experimental set-up. The second test system is the Draize test. This test can be extremely painful through substance administration to the eyes mucous membranes, even though the experimental set-up should prevent such extreme situations. Have you ever tried to defend the requirement for this in vivo test for one class of products (medicines) when it is being replaced by "in vitro" methods in another (cosmetics)? Is this correct regulatory toxicology or not?
- Other goals of animal experimentation having been attacked recently, are those to control the pharmaceutical quality of biological medicines and solutions for parenteral administration, and as a special case, the quality control of vaccines as well as the support of their development.
- The development and safety testing of medicines has been less challenged than that of other product areas. Is this correct? Aren´t we using the same animals for all these purposes? Where should the regulatory toxicologist put his vote?
- Finally, can we, the toxicologists, regulated and regulators alike, claim to always have made the best and most appropriate use of analytical and biological data and results available, be they "in vitro" or in vivo, be they in animals or in humans, and often a combination of all of them? And if not, does it really matter which data are used inappropriately?

Examples for a Successful Switch from In Vivo to In Vitro Safety Assessment

Successful examples can be found and presented for almost any product class. Since my professional involvement is in medicines, I shall describe examples mostly from this area (Fig. 2).

- The LD50 for medicinal compounds has been replaced by a true single dose toxicity experiment. Let me remind you that it was the European regulatory toxicologists who brought about this success world-wide through the International Conference on Harmonization (ICH) activities.
- The LD50 for chemicals has not been replaced (yet), but the position assuming a high precision has been abandoned, and pragmatism brought about the same information from much less animals, and only a fraction of them bound to die from only a small portion of the substances tested; be it through the "Fixed Dose Procedure", or even more so through the "Acute Toxic Class Method". Such methods, after having been validated internationally, have to be introduced by OECD. Let me point out there are only

very few internationally acceptable cooperatives able to bring about the necessary international harmonization leading to changes in animal experimentation requirements: ICH and OECD. It is through these cooperatives that single scientists, national organisations, as well as regional (e.g. European) ones can introduce their ideas for reduction, replacement and refinement of animal experiments.

- We know that currently 80 to 85 % of animal experiments in medicine development and testing are performed on determining biological activities (pharmacodynamics). A vast number and variety of in vitro systems has been developed and is used, thus tremendously reducing the number of animal experiments. I have to mention here that the regulators are currently confronted with the task to regulate safety and general pharmacodynamic investigations by guidelines. This would make them accessible to the demands of animal protection groups which claim that this area of medicines development needs more thorough control.
- One has to keep in mind that "in vitro" studies have been developed and introduced as alternatives to be preferred over in vivo investigations not only for animals, but also for humans. The spectrum of in vitro and ex vivo investigations refining and enabling the interpretation of clinical investigations, and thereby, reducing time and number of patients to be studied prior to marketing medicines, has become large.
- Animal experiments in pharmaceutical quality control are numerous. Their replacement by physico-chemical measurements has been introduced e.g. for insuline preparations. This issue will be dealt with in more detail later.

Examples for a Successful Switch from IN VIVO to IN VITRO Safety Assessment

*	Toxicity	LD_{50} (lethality)	→	Single dose toxicity (morbidity)
		LD_{50}	→	Forfeiting assumed precision
*	Activity	"RR"	→	Receptors
*	Clinical safety	Clinical trial	→	Metabolism, Kinetics
*	Quality control (insuline)	Animal experiment	→	Physico-chemical measurement

Fig. 2

Some (Dis-) Advantages of Biological Systems

Advantages and disadvantages of "in vitro" over in vivo studies and vice versa have been described many times. In the context of this presentation (Fig. 3) the usual scaling from "whole body" to "receptors" is used to depict a subjective choice of what seems to be good or bad from case to case. For regulatory toxicology it seems important to deduce that the experience gained with one compound may not be repeatable with another. This makes it very hard or impossible to create guidance for every case successfully. From this juxtaposition the regulated and the regulator will agree that, e.g. in medicines development, testing and control methods avoiding experimental animals should be used as often as possible, whereas animal experiments should be performed as seldomly as possible and only as many as absolutely necessary. In Europe this is government-regulated on two equal levels. i.e. by the regulators of product safety, and by the regulators in charge of animal protection, both developing their needs independently. You have heard from Professor Balls what is going on in Ispra at the ECVAM Centre, and you hear my viewpoint as a product safety regulator. There is a "healthy fight" going on, we expect it to bring and keep animal experimentation at the lowest defendable level; similar activities have to be mentioned for some of the Member States of the current and future European Union as well as for Switzerland. Organized counterparts do not exist among the regulated, with the exception of personal activities. Industry companies and contract laboratories in the past have combined the two activities, usually with the toxicologist responsible for product safety. This procedure may have to be reconsidered.

Some (Dis-) Advantages of Biological Systems

System	Advantage	Disadvantage
* Whole body	homeostasis	false system
* Selected organ(s)	right choice	wrong selection
* Cells and sub-cellular organelles	quantification; dose-response	correct "environment" missing
* Enzymes, receptors, etc.	mechanistic	missing relevance

Fig. 3

Impact on Risk Assessment

Information used in product regulation is developed in two directions. Horizontally, we proceed from estimation of toxicity to assessment of safety, and, finally, to efficacy and risk-benefit management. Vertically, three issues are represented by pre-clinical safety, clinical safety, and pharmacovigilance safety (Fig. 4). Under all conditions described, the term and impact of "safety" regards humans.

Coming from different areas of product development, testing, and control, we tend to forget that different technical and scientific limitations apply. For medicines safety and efficacy can be assessed and judged from clinical data and pharmacovigilance observations, chemicals and cosmetics will have to derive their risk assessment and management without efficacy, but at least to some degree, will have available information from clinical safety and pharmacovigilance safety. Pesticides take quite another position, clinical information is not studied and pharmacovigilance safety does not play an important role for risk assessment; for risk assessment pre-clinical safety data are extrapolated from signs of toxicity observed in animals to safety assumed to exist in humans. Such differences which allow to stride for controlled findings on human safety, will inevitably put data from animal toxicity studies - with or without some additional clinical safety information - which will allow adjustment of management on both sides. This in turn will influence the actual and expected importance of information gained without animals in vitro (Fig. 5). This means that "in vitro" experimentation does influence e.g. the development of medicines greatly during the pre-clinical phase, but much less during clinical (safety) development, which is usually supported by animal toxicity studies. As pointed out earlier, "in vitro" experimentation does influence later stages of medicine development and marketing on a case by case basis, possibly even through data generated from human-derived material.

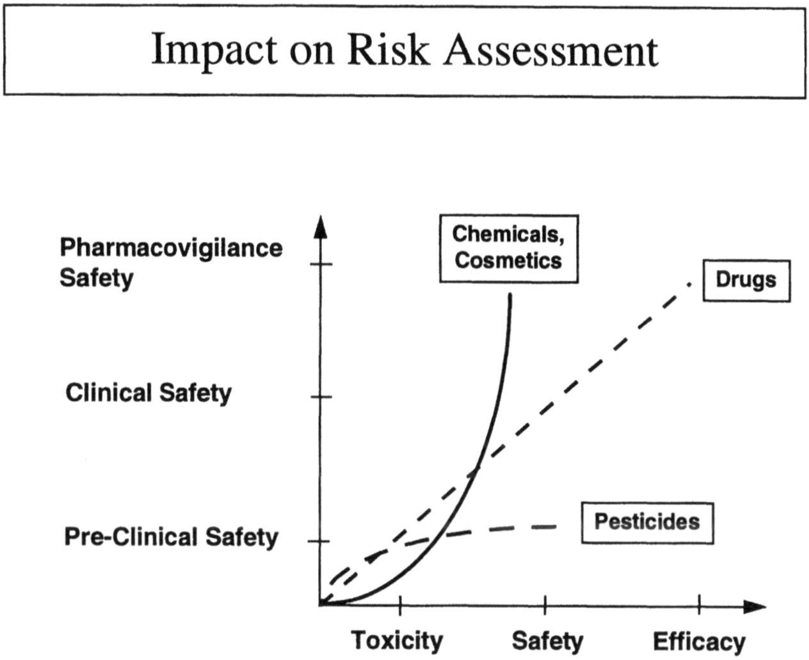

Fig. 4

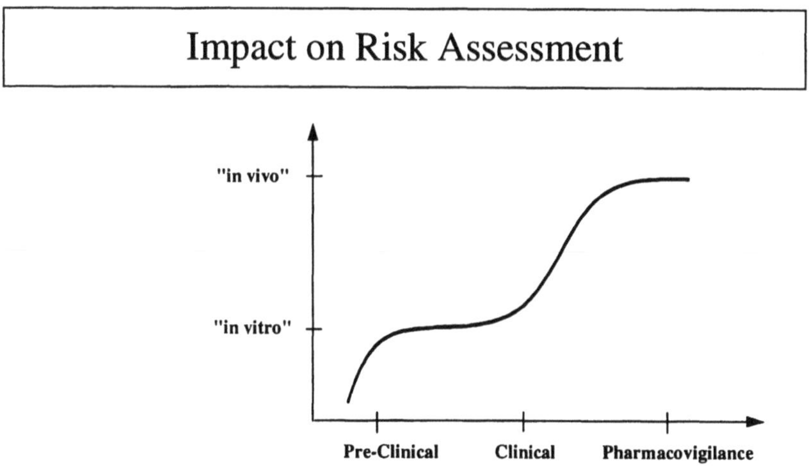

Fig. 5

To put Safety into Perspective: Drugs

Regulatory toxicology does not only influence the safety of the product. Similar to the well known development from (animal) toxicity to (human) safety, medicines are developed in "non-human" systems (in vitro and in animals) to depict all toward and untoward activities, so that this information can be used for the investigation of efficacy and safety (humans). All information generated in "non-human" systems, which are not outdated by later findings in humans, and are of significant importance for the use of the drug by patients, will influence the "safety package" available for the safe management of medicines described in Figure 6.

Drugs
- Activity / Efficacy
- Toxicity / Safety
- Quality

Safety of the Product
- Manufacture
- Reproducibility
- Approval / Licensing criteria

Safety of the Information
- Patient package leaflet
- Physician information
- Summary of product characteristics (SPC)
- Summary basis for approval

Safety of Dispensing
- OTC
- Prescription
- Limitations (patients/physicians)
- Pharmacovigilance

Fig. 6

Drugs and Risk Assessment: Role and Place of "In Vitro" Studies

As mentioned earlier, for risk and risk-benefit assessment of drugs information on pharmaceutical quality, toxicity, safety, and pharmacodynamics and efficacy is required before marketing. Medicines are generated from different backgrounds - each of them merits its own follow-up concerning the impact of "in vitro" regulatory toxicology:

- Biologicals (e.g. porcine "animal" insulin)
- Chemical substances (e.g. chemically synthesized beta-blockers)
- Biotechnologicals (e.g. recombinant "human" insulin)
- Gene therapy (e.g. diagnostics or treatment of tumors)

Two of these areas will be described to answer the question, if the same arsenal of in vivo and "in vitro" studies - involving animal or human endpoints - can be expected to be of equal importance and usefulness for their development and testing.

Biologicals

Questions concerning "in vitro" versus in vivo experimentation with biological medicines focus on the establishment and control of their pharmaceutical quality. Hormones, blood products, sera and vaccines from biological sources, and also antimicrobials are subject to quantitative determination, i.e. control of their activity ("potency tests") by biological methods. In addition qualitative control testing has to be performed to establish the identity of the substance by biological methods. Finally, qualitative control testing has to be performed to establish biologically the purity of active ingredient by proving the absence of adverse reactions ("safety tests"). The latter testing also applies for parenteralia. Animal experimentation is involved at all three levels.

In the past many pharmaceutical manufacturers waited for the regulators to change their requirements for testing, and declared the impossibility of introducing any changes themselves. The regulated are learning only now that they are in a position to actually introduce changes for their products themselves. Whoever initiates changes must ascertain that the quality of the drug remains. Changes can be achieved by deciding that a certain bioassay is no longer necessary for his product, or by developing alternative test methods. Alternative test methods may either be biological "in vitro" testing, or non-biological physicochemical methods. Changes can also be brought about by the regulators either by accepting changes proposed by the regulated or by introducing e.g. physicochemical methods instead of biological - animal experiments - into the European Pharmacopoeia Monographs. The European Pharmacopoeia Commission over the last 5 years has changed a large number of monographs, thus fulfilling the requirements of the European Union Directives on animal protection (Fig. 7).

- Concerning activity tests, a change will always require replacement of the current testing by another, validated method. Bioassays have been replaced in Europe e.g. for insulins, somatotropin and desmopressin. Bioassays plus alternative methods are specified for use of choice, e.g. for corticotrophin and oxytocins. For others, e.g. calcitonin, digitalis preparations and a variety of hormones alternative methods have not yet been accepted, validated or developed.
- Concerning safety tests, a pharmaceutical manufacturer may decide to discontinue testing or to lower the frequency of testing from each batch, based on experience validating the quality criterion to be proven. Whichever approach is chosen by the regulated, the requirements, e.g. of the "absence of abnormal toxicity" persist. Abnormal toxicity must not be found.
- Concerning safety testing monographs of the European Pharmacopoeia are changed, too. This is done either by discontinuing tests required in the past,

or by replacing tests with others. Whichever approach is chosen, the quality standard itself remains unchanged, i.e. "abnormal toxicity" must not be present. Changing monographs means also to look for simple methods, so as to fulfill the needs and requirements both of the first and the third world (both are producers and consumers at the same time - but under quite different conditions); the stride for simplification, however, will remain of limited success. All of these possibilities have been employed successfully. Testing for pyrogens, e,g, has been partially eliminated (heparins), partially replaced by the LAL-test (e.g. some penicillins and other antibiotics), and partially the LAL-test was introduced together with the Pyrogen (animal) test.

As difficult as this area of biologicals testing may look for the regulated and regulatory toxicologists alike, it is quite clear that the production of "pure" substances instead of impure "Gemische" does greatly facilitate the replacement of animal bioassays in the establishment and control of the quality of biological medicines. It was only the biotechnological production of insulin which allowed to reduce the "non-insulin" portion to such an extent as to enable the application of a validated physico-chemical quality test. This is now the European Pharmacopoeia standard requirement.

Chemical substances in medicines

Questions and problems concerning "in vitro" versus in vivo experimentation with medicines containing chemical substances as active ingredients especially concern two areas:

- Toxicity versus safety, and
- How to establish experimental models which are suitable to be used successfully on a "case by case approach", be it for diagnostics, therapeutics or antimicrobials, so that a "smallest common denominator approach" for enacting requirements can be avoided.

There is a growing tendency to investigate the "toxicity" of such chemical substances not under all circumstances up into highly toxic doses. I would like to quote Professor Robert Hess from the opening session of this congress: *"The science of "poison" is losing out."*

It has been accepted on a case by case basis (and in correlation with increasing knowledge about human exposure for the substance investigated) that studying those doses which e.g. show pharmacokinetic saturation, leads to severe toward or untoward pharmacodynamic effects incriminating homeostasis, or would require loading the organism with more than 1% of the diet, the model animal system becomes increasingly unsuitable to make valid predictions about the situation expected to exist in humans. This drastic change away from having to test animals under all possible conditions with "toxic" doses is currently being developed for carcinogenicity testing by the International Conference on Harmonization. Undoubtedly and understandably, to avoid toxic doses in long-term bioassays in vivo will greatly reduce the potential suffering of the animals. Carcinogenicity studies performed at dose levels above "meaningful" high doses often make it very difficult to interpret the results, leading to a vari-

ety of follow-up investigations otherwise unnecessary. Such situations and experience have also led the regulated and regulatory toxicologists and the regulators involved in animal rights to sincerely inquire the usefulness of the long-term carcinogenicity bioassay per se. It is my personal opinion that to require these studies routinely can be eliminated.
It is also quite clear that by establishing flexible guideline requirements, which can be adapted to the specific needs created by the test substance, the overall animal use will be rather reduced than increased. However, this may make it more difficult for the regulator controlling the application of requirements for their appropriateness to understand the rational behind the differences expected to result from such procedures. Apparingly similar substances and situations will lead to having to perform certeain testing in one case, but not in another. An increase in experimental testing may have to be expected - case by case - especially for reproductive toxicity testing and the areas of pharmaco- / toxicokinetics.

Biologicals

* Quantitative Assays

  Animal Experiment → Physico-Chemical Method?

  REPLACEMENT

* Qualitative (Safety) Tests → No. Batches to be Tested?

  REDUCTION / ELIMINATION

Fig. 7

Drugs and Experimentation: Expectations for the Future

It is of utmost importance to harmonize all areas of experimental testing. The beginning success of ICH has shown that such harmonization greatly contributes to the redcution of animal experiments without any sacrifice on behalf of drug safety for humans.
The introduction of a tripartite guideline for Reproductive Toxicity Testing to be employed in the European Union, Japan and the United States as well as by the affiliates of ICH has stopped the need for local repetition of such studies. Carcinogenicity testing, kinetics investigations, and genotoxicity testing requirements are well under way to become successful as well. Single dose

toxicity testing was harmonized so as to exclude non-rodent animals from testing and by deleting the LD50 test without even requiring a harmonized guideline.
Guidance is needed on the performance of experiments, on the fulfillment of the requirements, on assessing the results obtained, and on decision making, rather than inflexible rules.
Escape clauses must allow for scientific judgment, and to introduce scientific arguments on the replacebility of animal experiments in general and case by case. Regulated and regulatory toxicologists must accept that the areas related and under control of animal rights have moved beyond "toxicity", and now must be seen to include the whole variety of pharmacodynamic testing and pharmaceutical quality.
Moving from testing for toxicity to understanding safety has become important, also with respect to successful animal protection.

In Vitro Tests in Regulatory Toxicology: Symposium Chairman's Summing-up

M. Balls
European Centre for the Validation of Alternative Methods (ECVAM), JRC Environment Institute, 21020 Ispra (VA), Italy

This has been an excellent symposium, and the presence of such a large audience is very encouraging. The presentations of the four speakers complemented and reinforced each other, as I hoped and expected that they would. However, before drawing attention to what I consider to be some of the most important points put to us, I want to make two further scene-setting comments.

Firstly, I want to remind you or inform you that it is the European Union's short-term objective to reduce animal experimentation within the Member States of the European Union by 50% by the year 2000 (see Anon, 1994). In my opinion, this does not require the development, validation and acceptance of replacement alternative procedures and strategies. Reduction on this scale could be achieved by the more rigorous enforcement of legal requirements that strong scientific justification for a proposed programme of animal experimentation be provided, before the work is allowed to proceed. Also, as I have pointed out in my introductory remarks at this symposium (Balls, 1995), Festing (1994) has convincingly shown that a significant reduction in animal use, and an enhancement of the scientific value of any procedures necessarily performed, could be achieved, if experiments were better designed and if the data obtained were properly analysed. I was surprised and shocked to hear a senior scientist in another symposium at EUROTOX '94 admit that the data he was presenting had been analysed by the wrong statistical method and then proceed to put to the audience what were patently erroneous conclusions!

Our efforts toward producing relevant and reliable *in vitro* tests for use in toxicology and regulatory toxicity testing should be aimed at the 50% of animal procedures which would remain after that first, unnecessary, 50% had ceased.

The second point I want to make is that the Member States of the European Union are in the process of implementing a Sixth Amendment to *Directive 76/768/EC*, the Cosmetics Directive (Anon, 1993), which requires that the testing of ingredients or combinations of ingredients used in cosmetic products manufactured or marketed in the European Union should be banned, as scientifically-validated alternative methods which offer an equivalent level of protection for the consumer become available. I hope that this new law will stimulate the production of such alternative methods and that this sound and laudable principle will be extended to other forms of testing, e.g. of medicines, pesticides and industrial chemicals.

The first of our speakers, Erik Walum, representing academic *in vitro* toxicology, emphasised the need for high *quality*, and that reminded me of the theme of EUROTOX '94 - *Toxicology in Transition*. But, the question is, transition from what to what? I fear that it might be transition from bad to worse, e.g. if we were to see genetically-modified animals as the answer to all our problems. I think that a mouse will still be a mouse until we substitute all its genes with human genes, and then it will have become a human-being! No, it is non-animal alternatives which offer the escape route to transition from bad to better, but we must ourselves be more rigorous in ensuring that the quality of our own research, and of the training which makes it possible, is greatly improved.

Dr Walum went on to describe validation as the rate-limiting step between the development and acceptance of *in vitro* tests, and demonstrated what a great deal is being achieved in the MEIC scheme of the Scandinavian Society for Cell Toxicology, with its emphasis on using *human*, rather than *animal*, *in vivo* data as the basis for *in vivo/in vitro* comparisons. The skilful design of validation studies and their proper conduct and reporting are crucial to establishing the quality, i.e. the relevance and reliability, of the new methods.

Our second speaker, Leon Bruner, gave us a valuable insight into the world of the toxicologist in the cosmetic and household product industry, who is responsible for making safety assessments on hundreds, or even thousands, of products. He described the different requirements of internal safety assessment and regulatory classification, and emphasised that *in vitro* methods are already being used for making decisions in the former category. This is very important, since, although in-house and regulatory testing may indeed be different, experience gained in the first arena will undoubtedly be of immense value as we tackle the second, greater, challenge.

Dr Bruner went on to discuss relevance, and pointed out that various companies are developing their own, often quite different, tier-testing or decision-tree schemes for internal safety assessment. This must be recognised as a vital pointer concerning the regulatory acceeptance of *in vitro* tests in the future - we must be prepared to fight for flexibility and insist that it is scientifically-validated *in vitro* approaches or strategies that are accepted, not single protocols. Rigid guidelines or rules must be avoided. We already have one source of encouragement here - the US Department of Transport has now accepted two alternative methods, CORROSITEX™ and SKIN2™, as replacements for animal tests for classifying and labelling certain types of corrosive materials.

Dr Rudi Bechter then gave us the view of an industrial pharmacologist involved in the selection of candidate compounds at a relatively early stage in their development, with a particular emphasis on the need to weed out potential teratogens. He reminded us of the contrast between topical exposure, leading to effects such us eye irritation, and teratogenesis, which involves many cell types, cell differentation, cell migration, programmed cell death, time-windows of susceptibility, and maternal-fetal interactions.

Dr Bechter emphasised the importance of selecting suitable compounds for use in validation studies, and the problem of finding adequate *in vivo* data, as well as the need to accept that humans must be the recognised as the only truly valid models for other humans. He also discussed two other particularly interesting points. Firstly, he said, we must distinguish between the use of compounds to validate a test system and the use of a test system to validate compounds. That point needs further thought and discussion, but it is certainly very true that the use of test materials backed by shaky *in vivo* data in a validation study could lead to disaster - the tests being validated could be made to seem less relevant and reliable than they really were. Secondly, he said that the introduction of testing regulations has overwhelmingly been reactive, rather than proactive - we should decide what is needed and provide it, rather than waiting for a need to appear and then trying to satisfy it.

As I said in introducing our last speaker, Rolf Bass, he is one of "us" *and* one of "them", being both a scientist involved in the development and introduction of new methods and ways of thinking, and also a regulator!

Professor Bass echoed what the other three speakers had said - the challenge before us is to produce alternatives which have a high degree of relevance for humans. He emphasised that the animals used in regulatory toxicity tests are themselves replacements, so we are really trying to find new replacements for existing replacements. This is important, since there are those in some quarters who seem to think that the present animal tests have some inherent, timeless and unchangeable value, and that we who question them threaten the stability of an established system which actually works. I quoted a senior human toxicologist's views on this in my opening remarks (Balls, 1995)!

Professor Bass mentioned the great progress which has been made via the International Conferences on Harmonisation (ICH) toward unifying test guidelines for medicines and their application. He went on to refer to the danger of testing at meaningless high doses and said that rationalising dose selection in carcinogenicity testing is on the ICH agenda. This is to be welcomed, and reinforced the point made earlier in the symposium - that the rationalisation and harmonisation of guidelines and classification schemes is an essential prerequisite to the satisfactory introduction of new methods into regulatory toxicity testing.

Finally, while at one point Professor Bass mentioned rules, he made it clear that he favours flexible guidelines and recognises the need to allow for scientific judgement on a case-by-case basis. If only this wisdom could be made universal among regulators, the seeds of the revolution I seek would not only have been sown, but would be germinating!

The final question from the audience revived my fears about *Toxicology in Transition*, for Professor Bass was asked whether he saw his primary duty as a regulator as being to protect animals or to protect human beings. That is a very old-fashioned question, the Chairman intervened to say - the more we recognise the insuperable limitations of animal models and the great potential value of scientifically-based alternative methods, the more we will be able to improve the quality of protection of human-beings, while at the same time reducing both the numbers of laboratory animals needed and the suffering caused to them.

References

Anon (1993) Council Directive 93/95/EEC of 14 June 1993 amending for the sixth time Directive 76/768/EEC on the approximation of the laws of the Member States relating to cosmetic products. Off J Eur Comm L151:32

Anon (1994) Report of the Committee on Energy, Research and Technology on the role of ECVAM, a European centre for the validation and perfection of alternatives to animal testing methods. Doc EN/RR/246/24681. European Parliament, Brussels, 15 pp.

Balls M (1995) *In vitro* methods in regulatory toxicology: the crucial significance of validation. This volume.

Festing MFW (1994) Reduction of animal use: experimental design and quality of experiments. Lab Animals 28:212

Footnote: These comments represent the personal views of the author, and should not be taken to convey either the views of the other speakers in Symposium 3 at EUROTOX '94 or the position or policies of the Commission of the European Union.

Ex Vivo and *in Vitro* Models in Acetaminophen Hepatotoxicity Studies. Relationship between Glutathione Depletion, Oxidative Stress and Disturbances in Calcium Homeostasis and Energy Metabolism

S. Ellouk-Achard, V. Levresse, C. Martin, C. Pham-Huy, H. Dutertre-Catella, M. Thevenin, J.-M. Warnet and J.R. Claude

Université René-Descartes, Paris V, Faculté de Pharmacie, Laboratoire de Toxicologie, EA 207, 4 Avenue de l'Observatoire, 75270 Paris Cedex 06 (France).

The hepatotoxicity of acetaminophen (APAP) seems to be linked to several biochemical perturbations such as glutathione depletion, oxidative stress, Ca^{2+} homeostasis disturbances and/or changes in energy metabolism.

To our knowledge, no study has compared APAP toxicity simultaneously in liver subcellular fractions (*ex vivo*) and in hepatocyte primary cultures (*in vitro*). The aim of this study was to compare these two hepatic models to explore the relationship between : glutathione and oxidative stress, glutathione and Ca^{2+} homeostasis, glutathione and energy metabolism.

In order to increase the sensitivity of rat liver to APAP toxicity, animals were induced *in vivo* with 3-methylcholanthrene (3-MC) to increase the oxidation of APAP to NAPQI *via* the cytochrome P-450 pathway (Itinose *et al.*, 1989). Then a pretreatment with 1,3-bis(2-chloroethyl)-1-nitrosourea (BCNU), an inhibitor of glutathione reductase, *in vivo* in the case of the *ex vivo* study and *in vitro* in the case of the *in vitro* experiment was performed. The inhibition of glutathione reductase activity by BCNU modifies the glutathione redox system in the liver, causing an extensive GSH depletion (Nakae *et al.*, 1988). The association of these two pretreatments before APAP treatment mimics an *in vivo* APAP overdosage.

Materials and Methods

Male Wistar rats (150 - 200 g) fasted 24 hours before sacrifice were pretreated intraperitoneally with 3-methylcholanthrene (3-MC, 75 mg/kg of body weight).

In the *ex vivo* model : after 20 hours these 3-MC-pretreated rats were divided in two groups; *GSH-Depleted group* : rats receiving BCNU (80 mg/kg of body weight) one hour before APAP treatment, *Undepleted group* : rats receiving 0.9% NaCl i.p. in place of BCNU.

In both pretreated groups, rats were given 500 mg/kg i.p. of APAP (n=5). One group of rats (n = 5) without any pretreatment and receiving 0.9% NaCl i.p. in place of 3-MC, BCNU or APAP was used as *Control group*.

Three hours later rats were sacrificed after pentobarbital anesthesia (60 mg/kg of body weight). The liver was perfused with 0.9% NaCl, removed and homogenized in 9 volumes (w/v) of 1.17% KCl buffer pH 7.4. Cytosol and microsomes were then isolated.

In the *in vitro* model : after 20 hours hepatocytes from 3-MC-pretreated rats were obtained by an *in situ* collagenase perfusion. Cells were suspended in culture medium (MEM/199) supplemented with 10% FCS, plated in 6-well dishes at the density of 10^5 cells/ml then divided in two groups; *GSH-Depleted culture* : hepatocytes in primary culture treated with BCNU (100 mM) one hour before APAP treatment, *Undepleted culture* : hepatocytes in primary culture treated with phosphate buffer (PBS) in place of BCNU. In both pretreated cultures, hepatocytes were treated with APAP 5 mM during 3 hours. Hepatocytes in primary culture without any pretreatment were used as *Control culture*.

The following cytosolic parameters were evaluated in the *ex vivo* model : reduced (GSH) and oxidized (GSSG) glutathione as described by Griffith (1980), glutathione reductase activity (GSSG-Rd) by the method of Carlberg and Mannervick (1984), lipid peroxidation by the MDA procedure according to Yagi (1976), calcium content by a fluorimetric method using fluo-3-AM as described by Grynkiewicz *et al.* (1985), adenine nucleotides content using an isocratic HPLC method, Pham-Huy *et al.* (1993). Cytochrome P-450 was evaluated in microsomes by the method of Omura and Sato (1964). In the *in vitro* model the same parameters were determined in the whole hepatocytes. All data are expressed relative to protein content (Bradford, 1976). Results are expressed as means ± s.e.m. of triplicate assays. Statistical significance of the data was determined by Student's *t* -test.

Results

Validation of the pretreatments (Table I) : A significant induction of cytochrome P-450 was obtained in all cases (up to 2-fold, $p < 0.01$). No inhibition of GSSG-Rd activity was observed in the two undepleted models whereas a significant inhibition of this enzyme activity was observed in the two GSH-depleted models (76%, $p < 0.01$ and 65%, $p < 0.01$ respectively for the *ex vivo* and the *in vitro* models). This inhibition leads to a significant decrease in GSH content (up to 1.6-fold, $p < 0.05$) and a parallel significant increase in GSSG content (up to 1.5-fold, $p < 0.05$). These variations result in a significant decrease in the GSH / GSSG ratio : 0.79 ($p < 0.05$) and 0.83 ($p < 0.05$) *versus* 2.21 and 1.9 for the control group respectively for the *ex vivo* and the *in vitro* models. A good correlation between the *ex vivo* and the *in vitro* pretreatments was observed with the different parameters, specially for the GSH-depleted models : cytochrome P-450 : $r = 0.825$, GSSG-Rd : $r = 0.817$, GSH : $r = 0.834$, GSSG : $r = 0.826$, GSH / GSSG ratio : $r = 0.888$.

Table I : Validation of the pretreatments

		Control	Undepleted [3-MC]	GSH-depleted [3-MC - BCNU]
CYP-450 (nmole/mg protein)	*ex vivo*	0.34 ± 0.01	0.75 ± 0.08 **	0.69 ± 0.07 **
	in vitro	0.29 ± 0.03	0.63 ± 0.09 **	0.72 ± 0.05 **
GSSG-Rd (UI/mg protein)	*ex vivo*	528 ± 56	473 ± 34 NS	129 ± 14 **
	in vitro	382 ± 34	401 ± 41 NS	135 ± 26 **
GSH (nmole/mg protein)	*ex vivo*	128 ± 2	113 ± 3 NS	78 ± 4 *
	in vitro	112 ± 3	108 ± 2 NS	68 ± 2 *
GSSG (nmole/mg protein)	*ex vivo*	58 ± 3	63 ± 2 NS	99 ± 8 *
	in vitro	59 ± 2	53 ± 2 NS	82 ± 2 *
GSH / GSSG	*ex vivo*	2.21 ± 0.03	1.79 ± 0.02 NS	0.79 ± 0.03 *
	in vitro	1.9 ± 0.03	2.04 ± 0.07 NS	0.83 ± 0.02 *

NS Non Significant ; Significantly different from control group at : * $p < 0.05$, ** $p < 0.01$

Relationship between Glutathione and Oxidative stress, Calcium homeostasis, Energy metabolism :

In the undepleted models (Table II) : In the *ex vivo* model, APAP treatment induced in the cytosol a significant GSH decrease (21%, $p < 0.05$) without significant variation in GSSG and a slight increase in MDA content (18%, $p < 0.05$). No significant increase in calcium content and no variation in adenine nucleotides metabolism were observed. In the *in vitro* model : APAP treatment produced in the whole hepatocytes a significant GSH decrease (24%, $p < 0.05$) without significant variation in GSSG and an increase in MDA content (35%, $p < 0.01$). These variations were associated with a significant increase in calcium content (14%, $p < 0.05$). Adenine nucleotides metabolism was not modified.

Table II : Relationship between glutathione and oxidative stress, calcium homeostasis, energy metabolism in the APAP-treated undepleted models

EX VIVO		*IN VITRO*
APAP 500 mg/kg	**Parameters (%)**	APAP 5mM
79 ± 6 *	**GSH**	76 ± 5.2 *
107 ± 7.3 NS	**GSSG**	88 ± 3.1 NS
118 ± 4 *	**MDA**	135 ± 4 **
105 ± 3 NS	**$[Ca^{2+}]$**	114 ± 2 *
85 ± 1.5 NS	**ATP**	91 ± 1.7 NS
81 ± 3.3 NS	**Xanthine**	89 ± 2.1 NS
100 ± 3.1 NS	**Hypoxanthine**	99 ± 1.4 NS

Results are expressed in percent of control values ; NS Non Significant,
Significantly different from Control at : * $p < 0.05$, ** $p < 0.01$
Control values :
Ex Vivo : GSH : 113 ± 3.4 nmol/mg protein ; GSSG : 63 ± 2.4 nmol/mg protein ;
MDA : 1.87 ± 0.33 nM/mg protein ; $[Ca^{2+}]$: 101 ± 2.7 nM/mg protein ;
ATP : 21.1 ± 0.62 nM; Xanthine : 22 ± 1.31 nM; Hypoxanthine : 0.09 ± 0.01 nM
In Vitro : GSH : 108 ± 2.3 nmol/mg protein; GSSG : 53 ± 1.7 nmol/mg protein ;
MDA : 2.04 ± 0.21 nM/mg protein ; $[Ca^{2+}]$: 98 ± 1.9 nM/mg protein ;
ATP : 22.4 ± 0.59 nM ; Xanthine : 26 ± 1.36 nM ; Hypoxanthine : 0.05 ± 0.01 nM

In the GSH-depleted models (Table III) : In the *ex vivo* model, a massive depletion in GSH and in GSSG (97%, $p < 0.001$) content was observed. These variations were associated with a massive increase in MDA content (205%, $p < 0.001$) and a significant increase in Ca^{2+} content (55%, $p < 0.01$). In the *in vitro* model, the same variations as in the *ex vivo* GSH-depleted model were observed in the whole hepatocytes : a massive decrease in GSH and GSSG content (85%, $p < 0.001$ and 82%, $p < 0.001$ respectively), associated with a massive increase in MDA level (283%, $p < 0.001$) and a significant increase in calcium content (87%, $p < 0.01$). These results were accompanied with plasma membrane blebbing of the hepatocytes.

Table III : Relationship between glutathione and oxidative stress, calcium homeostasis, energy metabolism in the APAP-treated GSH-depleted models

EX VIVO APAP 500 mg/kg	Parameters (%)	*IN VITRO* APAP 5mM
3 ± 0.95 ***	GSH	15 ± 1.3 ***
3.4 ± 0.3 ***	GSSG	18 ± 2.5 ***
305 ± 12 ***	MDA	383 ± 9 ***
155 ± 3 **	$[Ca^{2+}]$	187 ± 2.8 **
58 ± 1.8 **	ATP	56 ± 3.3 **
41 ± 2.1 **	Xanthine	45 ± 2.4 **
214 ± 3.2 **	Hypoxanthine	183 ± 3.5 **

Results are expressed in percent of control values ; NS Non Significant,
Significantly different from Control at : * $p < 0.05$; ** $p < 0.01$; *** $p < 0.001$
Control values :
Ex Vivo : GSH : 78 ± 4 nmol/mg protein ; GSSG : 99 ± 5.8 nmol/mg protein ;
MDA : 3.1 ± 0.67 nM/mg protein ; $[Ca^{2+}]$: 112 ± 3.6 nM/mg protein ;
ATP : 19.4 ± 0.59nM; Xanthine : 24.2 ± 2.06 nM; Hypoxanthine : 0.10 ± 0.01 nM
In Vitro : GSH : 68 ± 1.5 nmol/mg protein ; GSSG : 82 ± 1.9 nmol/mg protein ;
MDA : 2.8 ± 0.38 nM/mg protein ; $[Ca^{2+}]$: 105 ± 3 nM/mg protein ;
ATP : 25.07 ± 0.55 nM ; Xanthine : 25.3 ± 2.3 nM ; Hypoxanthine : 0.08 ± 0.01 nM

In the two GSH-depleted models a significant decrease in ATP content was observed (42%, $p < 0.01$ and 44%, $p < 0.01$ respectively for the *ex vivo* and the *in vitro* models). In addition, in both groups, a relationship between the decrease in xanthine content (59%, $p < 0.01$ and 55%, $p < 0.01$ respectively for the *ex vivo* and the *in vitro* models) and the increase in hypoxanthine level (114%, $p < 0.001$ and 83%, $p < 0.001$ respectively for the *ex vivo* and the *in vitro* models) was observed ($r = -0.993$ and $r = -0.987$ respectively for the *ex vivo* and the *in vitro* models).

Discussion

Comparable results were obtained with the *ex vivo* and the *in vitro* models. BCNU-induced GSH depletion impairs cell defences against the toxic action of APAP by preventing NAPQI inactivation. This phenomenon results in an acute oxidative stress initiating the peroxidative decomposition of cellular membranes. The loss of membrane permeability, objectified in the *in vitro* model by the blebbing of hepatocytes, leads to an increase in the intracellular Ca^{2+} content. This Ca^{2+} homeostasis perturbation contributes to ATP degradation and, by promoting the conversion of the enzyme xanthine dehydrogenase to its oxidative form, leads to the degradation of xanthine to hypoxanthine. This conversion probably contributes to the massive production of activated oxygen species leading to the observed lipid peroxidation. This mechanism has many similarities with chemical hypoxia.

In conclusion, BCNU, as an inhibitor of glutathione reductase, greatly enhances hepatocyte sensitivity to APAP both in the *ex vivo* and in the *in vitro* models. A good correlation observed between the two models authorizes the future utilization of this *in vitro* model for further mechanistic studies.

References

Bradford MM (1976) A rapid and sensitive method for the quantification of microgram quantities of protein utilizing the principle of protein-dye binding. Anal Biochem 72: 248-254

Carlberg I and Mannervick B (1984) Glutathione reductase. Meth Enzymol 113: 484-490

Griffith OW (1980) Determination of glutathione and glutathione disulfide using glutathione reductase and 2-vinylpyridine. Anal Biochem 106: 207-212

Grynkiewicz G, Poenie M and Tsien RY (1985) A new generation of Ca^{2+} indicators with greatly improved fluorescence properties. J Biol Chem 260:: 3440-3450

Itinose AM, Leiko Doi Sakundo M and Bracht A (1989) Metabolic effects of acetaminophen. Studies in isolated perfused rat liver. Cell Biochem Funct 7: 263-273

Nakae D, Oakes JW and Farber JL (1988) Potentiation in the intact rat of the hepatotoxicity of acetaminophen by 1,3-bis(2-chloroethyl)-1-nitrosourea. Arch Biochem Biophys 267: 651-659

Omura T and Sato R (1964) The carbon monoxide binding pigment of liver microsomes. Evidence for its hemoprotein nature. J Biol Chem 239: 393-399

Pham-Huy C, Martin C, Thevenin M, Warnet JM, Ellouk-Achard S, Massicot F, Dutertre-Catella H and Claude JR (1993) Determination of ATP, ADP, AMP and other nucleotides in tissues and cell cultures by HPLC. Pharmacol Toxicol 73: 120

Yagi K (1976) A simple fluorimetric assay for lipoperoxide in blood plasma. Biochem Med 15: 212-216

Acknowledgements

The authors would like to thank Mrs M. Radionoff (Laboratoire de Toxicologie, Faculté de Pharmacie, Paris V) for this manuscript, and the "Association pour le Développement de la Recherche en Toxicologie Expérimentale" (ADRTE), Paris, France, for its financial support. Acetaminophen and BCNU were generous gifts from respectively Rhône-Poulenc Rorer (France) and Laboratoire Bristol (France).

Rabbit Renal Proximal Tubule Suspensions as a Model for Nephrotoxicity Evaluation of Native or *in situ* Metabolized β-Lactam Antibiotics

H. Dutertre-Catella[1], C. Martin[1], M. Debray[2], C. Pham-Huy[1], M. Thevenin[1], J.-M. Warnet[1], R.A. Podevin[3], J.R. Claude[1]

[1]Laboratoire de Toxicologie (EA 207) & [2]Laboratoire de Mathématiques, Faculté de Pharmacie, 75270 Paris Cedex 06; [3] INSERM U 356, F-75006 Paris (France)

Rabbit renal proximal tubule suspensions have proved a convenient model for nephrotoxicity studies, particularly for direct nephrotoxins such as Cephaloridin (Olivier *et al.*, 1992). In order to evaluate the ability of this model to metabolize certain drugs to nephrotoxic compounds, we used another ß-lactam antibiotic, Imipenem, known to be nephrotoxic through renal metabolism *via* hydrolysis by dehydropeptidase I (DHP) (Birnbaum *et al.*, 1985). The use of Cilastatin, an inhibitor of this enzyme, should confirm this metabolic pathway.

Materials and methods

Female New Zealand rabbits (2.0-2.2 kg) fed on a commercial diet and water *ad libitum* were used. Proximal tubules of renal cortex were obtained according to the method of Burg and Orloff (1962) with modifications as detailed by Podevin and Parini (1989).

Drugs : Cephaloridine (CPH) (I) was purchased as the commercialized pharmaceutical form (Ceporin, Glaxo); Imipenem (IMIP) (II) was a generous gift from MSD Laboratories (Rahway, NJ); the equimolar association of Imipenem with Cilastatin (IMIP+CILA) was purchased as the commercialized pharmaceutical form (Tienam, MSD-Chibret) (III). These drugs were incubated for three hours with rabbit renal cortex proximal tubule suspensions at the concentrations of 10 and 20 mM for I and 5, 10 and 20 mM for II and III. Tubule viability was assessed using the trypan blue exclusion test and LDH leakage evaluation. Total glutathione was determined by the method of Griffith (1980). ATP, ADP and AMP were measured using an HPLC isocratic method (Pham-Huy *et al.*, 1993) in order to assess the energy charge of the adenylate pool (ATP+1/2 ADP / ATP+ADP+AMP ratio) (Atkinson, 1968). Total, Na^+,K^+- and Mg^{++}-ATPase activities were determined with a sensitive radiometric assay (Podevin and Parini, 1989). All values are expressed relative to protein concentration (Bradford, 1976). The results are expressed as mean ± SD. Statistical significance was assessed by analysis of variance and Bonferroni's procedure (Olivier *et al.*, 1992) with a significance level of $p=0.05$.

Results

For both native (CPH) and *in situ*-activated (Imipenem) nephrotoxins, a similar dose-dependent decrease in total glutathione and energy charge (figure 1), and ATPase activities (figure 2) is observed relative to controls (in % - * $p < 0.05$; ** $p < 0.01$; *** $p < 0.001$). The equimolar association of Imipenem with a

Figure 1. Total glutathione and energy charge variations in rabbit renal proximal tubule suspensions treated with Cephaloridin (CPH), Imipenem (IMIP) and Imipenem + Cilastatin (IMIP+CILA) (in mM)

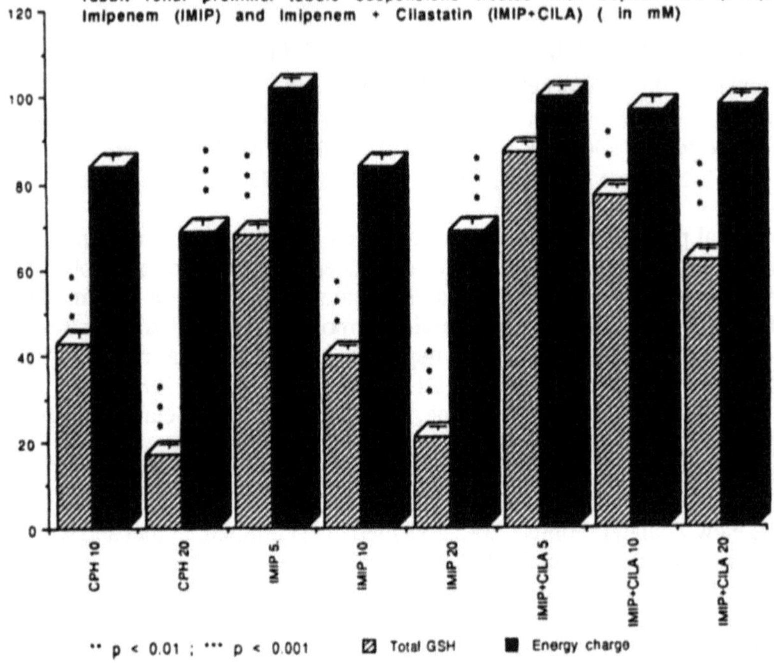

Figure 2. Total, Mg++_, Na+/K+_ ATPase activities in rabbit renal proximal tubule suspensions treated with Cephaloridin (CPH), Imipenem (IMIP) and Imipenem+Cilastatin (IMIP+CILA) (in mM)

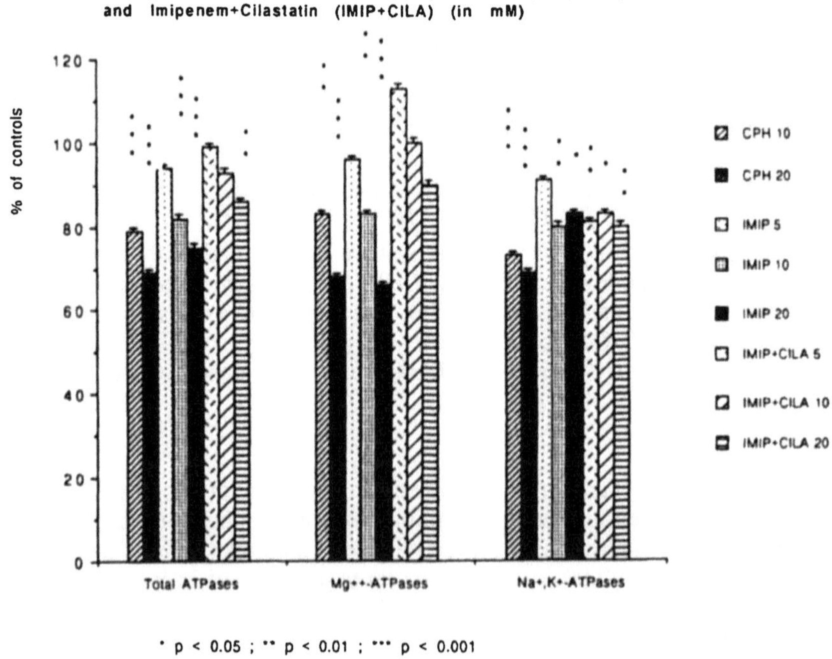

DHP inhibitor, Cilastatin, as for therapeutic use, gradually attenuates Imipenem nephrotoxicity. The recovery is complete for energy charge, whereas it is only partial for ATPase activities and GSH content.

In conclusion, this *in vitro* model using rabbit renal proximal tubule suspensions in order to detect ß-lactam nephrotoxicity is relevant for native nephrotoxins (*i.e. per se*) like Cephaloridin, as well as for *in situ*-activated nephrotoxins like Imipenem.

ATPase activities and GSH content determinations in such a model appear as suitable candidates for the evaluation of the nephrotoxic properties of drugs.

References

Atkinson DE (1968) The energy charge of the adenylate pool as a regulatory parameter. Interaction with feedback modifiers. Biochemistry 7: 4030-4034

Birnbaum J, Kahan FM & Kropp H (1985) Carbapenems, a new class of beta-lactam antibiotics. Discovery and development of Imipenem/Cilastatin. Amer J Med 78 (suppl 6A): 3-21

Bradford MM (1976) A rapid and sensitive method for the quantitation of microgram quantities of protein utilizing the principle of protein-dye binding. Anal Biochem 72: 248-254

Burg M & Orloff J (1962) Oxygen consumption and active transport in separated renal tubules. Amer J Physiol 203: 327-330

Griffith OW (1980) Determination of glutathione and glutathione disulfide using glutathione reductase and 2-vinylpyridine. Anal Biochem 106: 207-212

Olivier MF, Martin C, Dutertre-Catella H, Thevenin M, Warnet JM, Vannier B, Claude JR, Podevin RA (1992) Effects of a new Cephalosporin, Cefpirome (HR 810), on ATPase activities of rabbit renal proximal tubule suspensions : comparison with Cephaloridine and Cefotaxime. Toxicol *in vitro* 6: 175-176

Pham-Huy C, Martin C, Thevenin M, Warnet JM, Ellouk S, Massicot F, Dutertre-Catella H, Claude JR (1993) Determination of ATP, ADP, AMP and other nucleotides in tissues and cell cultures by HPLC. Pharmacol Toxicol 73 (suppl II): 120

Podevin RA & Parini A (1989) Adrenergic agonists and the Na+,K+-adenosine triphosphatase from rabbit proximal tubules and the basolateral membranes. J Pharmacol Exp Ther 250: 672-677

Acknowledgements

The authors would like to thank the "Association pour le Développement de la Recherche en Toxicologie Expérimentale" (ADRTE), Paris, France, for its financial support.

Neurobehavioral abnormalities induced by substance abuse

(Chair: Dr. J. Elsner, Switzerland)

Neurobehavioral Abnormalities Induced by Prenatal Exposure to Substances of Abuse – Stating the Problems

J. Elsner

Institute of Toxicology, ETH and University of Zurich, Schorenstrasse 16, CH-8606 Schwerzenbach

There are probably few subjects in science which induce so many controversies as the ones concerning neurobehavioral toxicology and substance abuse. The use and interpretation of neurobehavioral techniques in toxicology is matter of many debates going on at the scientific and regulatory level (Elsner 1992, Weiss & O'Donoghue 1994). Substance abuse is an issue of sociolgy, politics and criminology. Scientists, as citizens with an own view on the world and the society, are not immune against personal oppinions concerning how to deal with addiction problems and their consequences. By these, bias may be created with respect to the subject and angle of view of his or her own research or with respect to the interpretation of published research results.

It may seem ironical that there is a serious discussion going on in several countries about decriminalizing heroin and cocain, when at the same time nicotine is beeing attacked and some people would like to retract into prohibition of alcohol as well (Warburton, 1990). These contrary movements reflect a deep confusion with respect to how society should handle the human, social, economical and medical problems encountered as a consequence of drug use and abuse. Each party which represents one of the two contradictory currents, of course, uses (and maybe abuses) research results concerning health effects resulting from this practice. The ones tolerant to drug use select the good news, and the others the bad ones, and generally both kinds are readily available.

Substance abuse is a behavior which, like any other behavior, is determined by its consequences (law of effect [Dunham 1977]). In behavioristic terms, one would say that the consumption of such a substance is reinforced. Humans consume drugs in order to receive pleasure or remove discomfort. Immediate reinforcers such as relaxation, increase in productivity, neglect of unpleasant life situations, removal of pain etc. may be sufficient to develop a very strong maintenace of the consumatory behavior ("psychological dependence"). What is called "physical dependence" is governed by the same law of effect but involving additional reinforcements. Some substances induce tolerance resulting in the demand of higher or more frequent doses for the production of the same effect. Rebound phenomena and/or withdrawal symptoms produce discomfort of various degrees of severity which are viciously removed by the consumption of the same drug which produced it. These contingencies between behavior and reinforcement result in the addictive behavior observed in a high percentage of the population.

Drug consumption for pleasure is often regarded as the responsibility of each individual, its consequences, however, carry over to society. The consequence of substance abuse, involving the chronic consumption of high doses, is toxicity on the addict which finally is at public charge. Serious social problems in his or her environment also result. An even more consequential aspect ist the fact that many women in child-bearing age do not manage to free themselves from addiction during pregnancy. Even if the consumption is reduced after pregnancy recognition, harm may allready be done to the offspring before this time. During organogenesis the vulnerability of the organism to the influence of xenobiotics is increased presenting an additional risk for exposure to substances of abuse. Most substances of abuse being psychoactive, passing readily the placenta and the blood-brain barrier, risks of interference with the development of the nervous system leading to permanent neurobehavioral deficiencies are evident.

The study of neurobehavioral consequences of prenatal exposure to substances of abuse is for all these reasons an essential duty of society. While for certain substances mechanistic studies at the molecular level is scientifically appealing (see the contribution of M. Schlumpf in this issue), behavioral assessments in epidemiological studies on humans and experimental studies on animals present many advantages. Only these may inform the public and the educators about the real consequences of prenatal exposures and about the effective risks future generations undergo. Behavior reflects the integrated output of the nervous system as well as of all of the body's functions controlled by it. Behavioral testing is also nondestructive essential for human studies and favorable for animal experimentation.

Epidemiologic studies in humans

Alcohol is one of the most prevalent and probably best studied substances of abuse with respect to prenatal exposure. It is estimated that about 50% of women in childbearing age use alcohol at various levels (Wheeler 1993). In spite of a history of thousands of years of alcohol use and abuse, its teratogenicity was not recognized formally until 1967 when P. Lemoine published a first systematic observation about offspring of alcoholic mothers (Lemoine et al. 1968). The term "fetal alcohol syndrome" (FAS) was first introduced by Jones & Smith (Jones and Smith 1973). FAS was characterized by growth retardation, special facial characteristics (Clarren et al. 1987), mental retardation, hyperactivity, irritability, short attention span (Clarren and Smith 1978; Streissguth et al. 1980; Spohr and Steinhausen 1987), altered auditory function (Church and Gerkin 1988), impaired cognition and motor performance (Coles et al. 1987). Autopsies revealed microcephaly and cerebral anomalies (Clarren et al. 1978; Majewski et al. 1978). About 30% of the offspring of alcoholic mothers show the complete syndrome (Jones et al. 1974; Olegard et al. 1979; Majewski et al. 1976). About 75% of these offspring show at least partial signs (Olegard et al. 1979). Whole population studies in France (Dehaene et al. 1981), Sweden (Olegard et al. 1979), and USA (Hanson et al. 1978) indicate that about one in 750 births gives rise to a complete FAS showing the importance of the incidence.

The effects of FAS being thus relatively well investigated and described, the question concerning the consequences of moderate alcohol consumption inducing possibly fetal alcohol effects (FAE) (Hoyseth & Jones, 1989), and the determination of a no adverse effect level (NOAEL) remained to be answered. An exemplary longitudinal prospective study on a large unselected group of pregnant women resulting in a cohort of approximately 500 children has been performed by Streissguth and coworkers (Streissguth et al. 1989a; Streissguth et al. 1989b; Sampson wt al. 1989; Streissguth et al. 1989c; Streissguth et al.1994) concerning neurobehavioral effects due to prenatal alcohol exposure. It also illustrates the major methodological aspects of such an undertaking.

For the evaluation of exposure, 25 different alcohol scores were extracted from an interview with the mothers. Not only mean consumption levels (e.g. ounces of absolute alcohol per day) were investigated, but also drinking patterns were quantified by several variables (e.g. a dichotomous score of whether or not the woman reported ever drinking 5 or more drinks at a time, average drinks per occasion, maximum drinks on any occasion, monthly occasions of drinking). These scores were evaluated for two time periods: prior to pregnancy reflecting exposure during the very earliest part of pregnancy and midpregnancy exposure.

A list of 150 covariates, including use of cigarettes, caffeine and other drugs including antibiotics, maternal race, age, marital status, education, socioeconomic status, nutrition, birth order of the child, education of the father, sex of the child, postnatal events, mother-child interaction, family history of learning disabilities were considered in the statistical analysis. Appropriate neurobehavioral testing was performed on days 1 and 2, at 8 and 18 months, 4, 7 and 14 years of age. The most important result of this study is that at each age relevant neurobehavioral effects due to alcohol were discovered.

In the neonate, habituation, arrousal, body tremor and high activity, and some other observational ratings, sucking pressure and latency to suck were related to alcohol exposure. At 8 months the mental and psychomotor developmental index of the Bayley scales and at 4 years IQ scores, attentional variables and motor performance were affected by prenatal exposure. At 7 years again IQ and a pattern corresponding to memory and attentional deficits across verbal and visual modalities as well as variables reflecting poor integration and quality of responses were related to prenatal alcohol. At 14 years of age an analogous pattern was observed which was highly correlated with the results of the first 7 years of life. This suggests that the deficits had a strong tendency to be observed in the same individuals from birth up to puberty. The authors interprete their data as being consistent with the signs and symptoms of the clinical syndrome of "minimal brain damage" or "attention deficit disorder".

Another important result of this study is the conclusion that the number of drinks per occasion was a more important predictor for neurobehavioral deficits as compared to frequency of drinking occasions, and that the period prior to pregnancy recognition was generally more predicting for performance and behavior at early and late school-age. This result leads to the practical advise for potential mothers to avoid single high doses of alcohol and that prevention of alcohol consumption should be extended to the whole child-bearing age.

Caffeine is a pleasant stimulant exhibiting some dependence potential. Prenatal exposure to caffeine is predominant reaching over 80% in most surveys. An additional matter of concern is the fact that the half-life of caffeine is lengthened during pregnancy and in infants during breast-feeding (Nehlig and Debry 1994). For large doses (> 7 cups per day of coffee) caffeine ingested during pregnancy induces body-weight decreases. In a 7 year prospective study, no effects induced by prenatal caffeine exposure on the neurobehavioral development of children were found (Barr and Streissguth 1991). However, in experiments with monkeys, Gilbert and Rice (1991a, 1991b, 1994) demonstrated that prenatal caffeine affects body weight, somatic measurements, and blood and urine clinical chemistry, and also the behavior of the offspring, among which feeding patterns and some response patterns in an operant conditioning paradigm. In view of the high prevalence level of caffeine, these conflicting results between animal and human studies leave a certain uneasiness.

Opiate abuse results in the US in an estimated annual number of up to 10'000 heroin exposed babys (Hans 1989). The immediate consequence of prenatal opiate exposure is the neonatal abstinence syndrome (NAS), characterized by irritability, tremulousness, sweating, stuffy nose, uncoordinated suck with feeding difficulties, diarrhea, and vomiting (Wheeler 1993). It is observed a few hours after birth and may last weeks up to months. Behavioral difficulties in this critical period of life may induce developmental deficits which may add to long-term developmental neurobehavioral deficits. Unfavorable stimulation by heroine addicted mothers may also be the reason why school performance of heroine exposed children fares less well with teachers reporting behavior dysfunction and inattentiveness (Wilson 1989). Methadone exposed children which also show pronounced NAS appear to be less at risk for cognitive effects at 2 years with the exception of those living at the lowest socioeconomic levels (Hans 1989). This difference may be due to the fact that mothers who undergo a methadone treatement may be able to give to her child a better structured environment than untreated heroin addicts.

Cocaine abuse has probably shown the most spectacular increase in the last years estimated to be up to tenfold since1983. In the United States an estimate of annually 100'000 babys are born from cocaine consuming mothers (Singer et al. 1991). An estimated 50% or more of pregnancies at some of the inner-city hospitals in New York are reported to be cocaine exposed (Dow-Edwards 1991), a figure which is probably not ready to drop.

Reports concerning effects due to prenatal cocaine exposure are somewhat inconclusive mainly due to differences in proper sample selection and confounder control, and obscured by polydrug exposure. A recent review, however (Singer et al 1991), concludes that prenatal cocaine has been linked to growth retardation and several neurobehavioral abnormalities in the neonate such as impaired habituation skills. Another review (Dow-Edwards 1991) describes transient neurological abnormalities including tremors, rigidity, hyperactivity, abnormal EEG, seizures abnormal state orientation, altered sensory function and motor ability. By 6 months of age, most of the neurological problems appear to be resolved. Concerning long-term effects, a recently published three-year study by Griffith et al. (1994) reports effects on verbal reasoning, but also this study is obscured by polydrug abuse.

Organic solvents and gasoline are abused by a behavior for which the term "glue sniffing" is used, but which involves many other application modes. Prenatal exposure to high levels of abused organic solvents such as toluene are found in recent studies to induce a similar picture as the one induced by alcohol, including the facial abnormalities observed in the fetal alcohol syndrome (Pearson et al. 1994). One may expect similar neurobehavioral effects as the ones induced by prenatal alcohol.

One important fact in substance abuse is that generally more than one substance is abused which complicates the picture in humans. While statistical procedures can partially seperate the effects due to the contributions of the individual substances, some uncertainties will always remain. Interaction effects are possible and should receive more attention in future research. In addition, the determination of the exposure level is often problematic in humans done in most studies by interviews or questionaires. Recent developments in analysis of maternal or fetal hair may remedy this problem (e.g. for cocaine [Grant et al 1994] or for heroin or nicotine [Kintz and Mangin 1993]).

Another difficulty of epidemiological investigations are the heterogenicity of human subjects with respect to genetical, socioeconomical, educational factors and to the quality of mother-child interactions. Specially in situations of drug abuse of the mother these aspects are crucial. Special characteristics of the personality and the life situation of a substance abusing mother, the pharmacological changes induced by the substance on the behavior of the mother and the newborn may induce direct or indirect effects on the psychological development of a child.

The information which may be obtained from the statistical analyses of the multifactorial modulation of neurobehavioral effects in well designed studies in humans is of particular interest by itself, shedding light on the complex relations between the parameters involved. For the determination of the strict causalities between exposures and developmental effects, however, experimental animal studies are an important complement.

Methodological issues in animal studies

In animal studies, most of abovementioned extraneous and indirect effects can be avoided (Spear 1994). For laboratory animals, the genetical, environmental, nutritional and exposure history can be closely controlled. Indirect effects on the offspring induced by behavioral changes of the dam can be eliminated by cross-fostering procedures. However, methodological issues of behavior testing are an important aspect of animal experimentation determining by far the ability to extrapolate results to humans.

In comparison to preclinical research of medical drugs for which often little is known about possible neurobehavioral effects in the developing organism, research in substances of abuse have a methodological advantage. Often some information is already available from human studies concerning the type of behavior or the molecular and physiological structures which may be affected in the animal model leading to experimental setups which are specifically desinged for this purpose.

In addition, behavioral measures in animals may be complemented by neurochemical investigations and by studies of other systems related to neurobehavioral phenomena. One example is the study of effects in the central and peripheral nervous system, as well as of immunological alterations induced by prenatal exposure to benzodiazepines as demonstrated by the contribution of Schlumpf et al. (this issue). This work also shows that neurobehavioral developmental toxicology does not need strongly increased doses in order to observe any effects in animals if the appropriate assessment methodology is used. At high doses there is a high chance that other toxic mechanisms may overshadow the analysis of subtle deficits induced in the nervous system and expressed by behavioral changes. Using appropriate experimental techniques, effects can be described and quantified which occur at levels at or very close to the ones humans are normally exposed to.

There are mainly two fundamental approaches in the investigation of neurobehavioral effects in laboratory animals. In the first approach, tests are designed according to the natural behavioral repertoire of the animal under study. Most early age measures such as surface righting, negative geotaxis, olfactory discrimination, swimming behavior etc. belong to this category. Some tests in adolescents and adults such as open field and different other motor activity tests, as well as rotarod, and sexual behavior may also be included. Several protocols group such tests into batteries for screening neurobehavioral toxicity by observational methods of scoring simple measures of the integrity of sensory functions, reflexes, autonomic signs, motor responses and muscular strength (Alder and Zbinden 1983; Moser 1989; Foss 1994). Developmental neurobehavioral toxicity was also formalized in study protocols for fulfilling safety assessment guidelines of new pharmaceuticals (Barlow 1985; Tanimura 1985; Suter and Schön 1986). In these, simple learning tests were also included such as active and passive avoidance and startle habituation (Wier 1994).

In the second approach, advanced tests investigate cognitive performances specifically designed for modelling human performances of learning, memory and attention and often encompass operant paradigms. Understanding the underlying mechanisms of behavior common to organisms including humans is the key feature of this approach, fulfilling requirements of face and construct validity (Willner 1984). Modern computer technologies allow microanalyses of schedule controlled behavior by which neurobehavioral deficiencies in many different behavioral features can be assessed during the same session (Elsner 1986; Elsner 1988). Such an approach in animals probably comes closest to what is measured in a neurologic examination or in cognitive and mental testing in humans (Laties 1994). In addition, specific findings from human studies may also be modelled in animals if the appropriate technology is applied (Elsner 1991).

In a collaborative study on rats involving 10 different European laboratories, a broad spectrum of tests assessing a total of 167 independent variables were applied to investigate physical and neurobehavioral effects induced by prenatal alcohol exposure (Elsner et al submitted). A meta-analysis evaluated the sensitivity of the different test categories. Operant conditioning techniques including visual discrimination, delayed spatial alternation and DRL and DRH procedures measuring learning performance, short-term memory, attention and response speed were clearly more sensitive with respect to the exposure level at which effects were observed than the other measures including physical development and physiological and somatic variables. This result can be compared to the human study reported above in which attentional and memory components as well as the IQ were the most sensitive test outcomes after moderate levels of prenatal alcohol exposure.

Conclusion

Science is the art of knowledge. Considering all the uncertainties in the studies of neurobehavioral effects on the offspring induced by prenatal exposure to substances of abuse, one may ask the question concerning whether we posess sufficient knowledge about the causalities involved in order to decide about the necessary actions to be taken. It is certain that the children of strongly drug-addicted mothers are clearly at a disadvantage. Alcohol in large amounts produces the fetal alcohol syndrome involving severe neurobehavioral deficits including attention deficit disorder and mental retardation. Strong abuse of organic solvents such as toluene leads to similar abnormalities. Opiates induce a severe and cocaine a milder neonatal abstinence syndrome.

However, with the exception for alcohol and for cigarette and marihuana smoking (see the contribution of P. Fried in this issue) for which children were followed in well designed prospective studies to the age of up to14 years, there are many questions remaining open concerning the longterm development of neurobehavioral characteristics and the strict causalities involved for children exposed to moderate levels of substances of abuse. Investigations in animals at the behavioral and molecular level add scientific information, still including extrapolation problems.

The situation is similar to the one of developmental effects due to environmental pollutants such as lead for which certain decisions about costly measures have to be reached without having a full proof of the causalities involved (CDC 1991). If the intellectual and behavioral competence of future generations is at stake, it is better to start prevention and treatment programs for the reduction of risks to the neurobehavioral development of our children even if only partial information is available. Meanwhile, research programs taking advantage from progress made in any involved scientific discipline, should be financed for a continuous search for the truth.

References

Alder S, Zbinden G (1983) Neurobehavioral tests in single- and repeated-dose toxicity studies in small rodents. Arch Toxicol 54: 1-23

Barlow SM (1985) United Kingdom: regulatory attitudes toward behavioural teratology testing. Neurobehav Toxicol Teratol 7: 643-646

CDC (1991) Preventing lead poisoning in young children. Centers for Disease Control and Prevention, U.S. Department of Health and Human Services, Atlanta, USA

Church MW, Gerkin KP (1988) Hearing disorders in children with fetal alcohol syndrome: Findings from case reports. Pediatrics 82: 147-154

Clarren SK, Alvord SC, Sumi SM, Streissguth AP, Smith DW (1978) Brain malformations related to prenatal exposure to ethanol. J Pediatr 92: 64-67

Clarren SK, Sampson PD, Larsen J, Donnell DJ, Barr HM, Bookstein FL, Nartin DC, Streissguth AP (1987) Facial effects of fetal alcohol exposure: Assessment by photographs and morphometric analysis. Am J Med Gen 26: 651-666

Clarren SK, Smith DW (1978) The fetal alcohol syndrome. N Engl J Med 298: 1063-1067

Coles CD, Smith IE, Falek A (1987) Prenatal alcohol exposure and infant behavior: Immediate effects and implications for later development. Adv Alc Subst Abuse 6: 87-104

Dehaene P, Crepin G, Delahousse G, Querleu D, Walbaum R, Titran M, Samaille-Villette C (1981) Epidemiological aspects of the foetal alcoholism syndrome. 45 cases. Nouv Presse Med 10: 2639-2643

Dow-Edwards DL (1991) Cocaine effects on fetal development: A comparison of clinical and animal research findings. Neurotoxicol Teratol 13:347-352

Dunham P (1977) The nature of reinforcing stimuli. In: Handbook of operant behavior, Prentice-Hall, Englewood Cliffs, NJ, pp 98-124

Elsner J (1986) Testing strategies in behavioral teratology. III. Microanalysis of behavior. Neurobehav Toxicol Teratol 8: 573-584

Elsner J (1991) The tactile-kinesthetic system of rats as an animal model for minimal brain dysfunction. Arch toxicol 65: 465-473

Elsner J (1992) Animal models for human behavioral deficiencies during development. NeuroToxicol, 13: 65-76

Elsner J, Alder S, Zbinden G (1988) Interaction between ethanol and caffeine in operant behavior of rats. Psychopharmacology 96: 194-205

Elsner J, Hodel B, Bruinink A, Suter KE, Ziegler C, Ulbrich B, Pritzl M, Bornhausen M, Strauss A, Lämmer S, Schäppi U, Fitzgerald RE, Youmbi M, Schlumpf M, Lichtensteiger W, Cagiano R, Cuomo V, Lacomba C, de Salvia MA, Govoni S, Haglid K, Wang S, Rosengren LE. Evaluation of neurobehavioral methods in rats: results of an european interlaboratory study on effects after prenatal ethanol exposure. Submitted to Neurotoxicol teratol

Foss JA (1994) The application of a functional observational battery and motor activity test in safety assessment studies. In: Weiss B, O'Donoghue JL (eds) Neurobehavioral Toxicity, Analysis and Interpretation, Raven Press, New York, pp 137-143

Gilbert SG, Rice DC (1991a) Somatic development of the infant monkey following in utero exposure to caffeine. Fundam Appl Toxicol 17:454-65

Gilbert SG, Rice DC (1991b) Effects of chronic caffeine consumption in pregnant monkeys (Macaca fascicularis) on blood and urine clinical chemistry parameters. Fundam Appl Toxicol 16:299-308

Grant T, Brown Z, Callahan C, Barr H, Streissguth AP (1994) Cocaine exposure during pregnancy: improving assessment with radioimmunoassay of maternal hair. Obstet Gynecol 83: 524-531

Griffith DR, Azuma SD, Chasnoff IJ (1994) Three-year outcome of children exposed prenatally to drugs. J Am Acad Child Adolesc Psychiatry 33: 20-27.

Hans SL (1989) Developmental consequences of prenatal exposure to methadone. In: Hutchings DE (ed.) Prenatal abuse of licit and illicit drugs. Ann N Y Acad Sci 562: 195-207

Hanson JW, Streissguth AP, Smith DW (1978) The effects of moderate alcohol consumption during pregnancy on fetal growth and morphogenesis. J Pediatr 92: 457-460

Hoyseth KS, Jones PJH (1989) Minireview: Ethanol induced teratogenesis - characterization, mechanisms, and diagnostic approaches. Life Sci 44: 643-649

Jones KL, Smith DW (1973) Recognition of the fetal alcohol syndrome in early infancy. Lancet 2: 999-1001

Jones KL, Smith DW, Streissguth AP, Myrianthopoulos NC (1974) Outcome in offspring of chronic alcoholic women. Lancet 1: 1076-1078

Kintz P, Mangin P (1993) Evidence of gestational heroin or nicotine exposure by analysis of fetal hair. Forensic Sci Int 63:99-104

Laties VG (1994) Schedule-controlled operant behavior during the hazard identification stage of neurotoxicity risk assessment. In: Weiss B, O'Donoghue JL (eds) Neurobehavioral Toxicity, Analysis and Interpretation, Raven Press, New York, pp 253-262

Lemoine P, Harousseau H, Boteyru JP, Menuet JC (1968) Les enfants de parents alcooliques. Anomalies observées à propos de 127 cas. Quest medical 25: 477-482

Majewski F, Bierich JR, Loser H, Michaelis R, Leiber B (1976) Clinical aspects of pathogenesis of alcohol embryopathy. Muench Med Wochenschr 118: 1635-1642

Majewski F, Fischbach H, Peiffer J, Bierich JR (1978) Interruption of pregnancy in alcoholic women. Dtsch Med Wochenschr 103: 895-898

Moser VC (1989) Screening approaches to neurotoxicity: a functional observational battery. J Am Coll Toxicol 8:85-93

Nehlig A, Debry G (1994) Consequences on the newborn of chronic maternal consumption of coffee during gestation and lactation: a review. J Am Coll Nutr 13: 6-21

Olegard R, Sabel KG, Aronsson M, Sandin B, Johansson PR, Carlsson C, Kyllermann M, Iversen K., Hrbek A (1979) Effects on the child of alcohol abuse during pregnancy. Retrospective and prospective studies. Acta Paediatr. Scand. Suppl. 275: 112-121

Pearson MA, Hoyme HE, Seaver LH, Rimsza ME (1994) Toluene embryopathy: delineation of the phenotype and comparison with fetal alcohol syndrome. Pediatrics 93:211-215

Sampson PD, Streissguth AP, Barr HM, Bookstein FL (1989) Neurobehavioral effects of prenatal alcohol: Part II. Partial least squares analysis. Neurotoxicol Teratol 11: 477-492

Singer LT, Garber R, Kliegman R (1991) Neurobehavioral sequelae of fetal cocaine exposure. J Pediatr 119: 667-672

Spear LP (1994) Issues in developmental toxicology. In: Weiss B, O'Donoghue JL (eds) Neurobehavioral Toxicity, Analysis and Interpretation, Raven Press, New York, pp 59-69

Spohr HL, Steinhausen HC (1987) Follow-up studies of children with fetal alcohol syndrome. Neuroped 18: 13-17

Streissguth AP, Barr HM, Sampson PD, Bookstein FL, Darby BL (1989b) Neurobehavioral effects of prenatal alcohol: Part I. Research strategy. Neurotoxicol Teratol 11: 461-476

Streissguth AP, Bookstein FL, Sampson PD, Barr HM (1989c) Neurobehavioral effects of prenatal alcohol: Part III. PLS analyses of neuropsychological tests. Neurotoxicol Teratol 11: 461-476

Streissguth AP, Landesman-Dwyer S, Martin JC, Smith DW (1980) Teratogenic effects of alcohol in humans and laboratory animals. Science 209: 353-361

Streissguth AP, Sampson PD, Barr HM (1989a) Neurobehavioral dose-response effects of prenatal alcohol. In: Hutchings DE (ed.) Prenatal abuse of licit and illicit drugs. Ann N Y Acad Sci 562: 145-158

Streissguth AP, Sampson PD, Olson HC, Bookstein FL, Barr HM, Scott M, Feldman J, Mirski AF (1994) Maternal drinking during pregnancy: attention and short-term memory in 14-year-old offspring - a longitudinal prospective study. Alcohol Clin Exp Res 18: 202-218

Suter KE, Schön H. Testing strategies in behavioral teratology: I. Testing battery approach. Neurobehav Toxicol Teratol 8: 561-566

Tanimura T (1985) Guidelines for developmental toxicity testing of chemicals in Japan. Neurobehav Toxicol Teratol 7: 647-652

Warburton DM (1990) Heroin, cocaine and now nicotine. In: Warburton DM (ed) Addiction controversies, Harwood, London, pp 21-35

Weiss B, O'Donoghue JL (eds) (1994) Neurobehavioral Toxicity, Analysis and Interpretation, Raven Press, New York

Wheeler SF (1993) Substance abuse during pregnancy. Prim Care 20:191-207

Wier PJ (1994) Neurobehavioral evaluations: integral components of developmental toxicity testing. In: Weiss B, O'Donoghue JL (eds) Neurobehavioral Toxicity, Analysis and Interpretation, Raven Press, New York, pp 71-83

Willner P (1984) The validity of animal models of depression. Psychopharmacology 83: 1-16

Wilson SL (1989) Clinical studies of infants and children exposed prenatally to heroin. In: Hutchings DE (ed.) Prenatal abuse of licit and illicit drugs. Ann N Y Acad Sci 562: 183-194

Prenatal Exposure to Marihuana and Tobacco During Infancy, Early and Middle Childhood: Effects and an Attempt at Synthesis

P.A. Fried

Department of Psychology, Carleton University, Ottawa, Ontario, Canada K1S 5B6

In the past two decades a notable increase has occurred in the number of investigations devoted to examining the effects on the behavioural development of offspring who have been prenatally exposed to a range of substances that women may ingest or absorb during pregnancy. Among the most common of the non-medicinal drugs used by pregnant women are cigarettes and marihuana. The former is the subject of a relatively extensive amount of research, whereas the latter is striking for the paucity of objective information available. In this report I will briefly review, in a selective fashion, the body of literature that examines the consequences for the infant and young child of *in utero* exposure to these two substances and focus upon the results arising from a long-term, ongoing prospective study underway in Ottawa, Canada. Much of the work and interpretations presented in this chapter have appeared in publications arising from the Ottawa study and have been the subject of two recent reviews (Fried, 1993; Fried, in press).

The Ottawa Prenatal Prospective Study (OPPS)

Aside from one or two studies, all of the information pertaining to the behavioural effect of prenatal exposure to marihuana in children beyond the toddler stage is limited to the reports coming from the Ottawa Prenatal Prospective Study (Fried et al. 1980). Because of this, the protocol and the limitations of this Canadian work is described in some detail. Additional information can be found elsewhere (Fried et al. 1980).

As recently as 1980, the only information describing the effect marihuana may have upon the pregnant user and her offspring was limited to two, polydrug case reports. This paucity of information, the results of animal work (reviewed in Fried, 1984; Dalterio and Fried, 1992) the extent of usage among women of reproductive age and the cooperation of the teaching hospitals in the Ottawa area combined to establish the need and set the climate and opportunity for the inception of the OPPS in 1978.

Birth data have been collected in a prospective fashion from approximately 700 women residing in the Ottawa, Canada region. For a number of pragmatic reasons it has been feasible to have followed-up only approximately 200 offspring of this sample. Pregnant women volunteered after being informed of the study by a variety of means including via their physicians, by notices in the waiting rooms of obstetricians, or by notices in the reception rooms of prenatal clinics in the major hospitals in Ottawa. The information that was disseminated by the various means did not, at this juncture, mention marihuana but rather discussed, in general terms, how life-style habits during pregnancy may influence the developing fetus. Upon contacting our research facility, the potential subject was given further details about the particular habits we were interested in – marihuana, alcohol, and cigarettes. It was emphasized that, for purposes of comparison, we also wished to recruit women who used any of these substances to a very small degree or not at all. After volunteering and signing an informed consent, the mother-to-be was interviewed once during each of the trimesters remaining in her pregnancy by a trained female interviewer.

This procedure of recruiting volunteers in the fashion described above has both strengths and weaknesses that pervade the entire OPPS. The self-selection procedure limits the extent to which generalizations can be made in terms of epidemiological information collected with the possibility of selection bias being the most obvious interpretative problem. However, as noted elsewhere (Fried et al. 1980, Fried et al. 1984) on several key demographic variables including parity, age, and family income, the volunteer sample participating in the OPPS is quite similar to non-participating women living in the Ottawa area giving birth in the hospitals taking part in our study.

The recruitment procedure used has the advantage of increasing the likelihood of the reliability of self-report (elaborated below) and of increasing the probability of a long-term commitment to the study. Aside from subjects who have moved from the Ottawa area (about a third), a retention rate of over 95% has been maintained over the past decade.

During each of the prenatal interviews information was collected on such variables as socioeconomic status, mother's health (both current and before pregnancy), the health history of the father, obstetrical history of previous pregnancies, a 24-hour dietary recall (including an assessment of caffeine intake), as well as past and present drug use patterns with particular detailed information being gathered with respect to marihuana, cigarettes and, alcohol. For the establishment of the use of these three drugs, information was gathered both for the year preceding the pregnancy and for each trimester of the pregnancy. Further details of the interview and the categorization of the various drugs have been described previously (Fried et al. 1980).

There was an extensive range of marihuana use in the sample and the drug was not used by a similar number of subjects across the range of usage. As a result of these factors, for descriptive and some statistical purposes, the marihuana data were treated categorically. There were the non-users, irregular users (one joint or less per week), moderate users (two to five joints per week) and heavy users (more than five joints per week).

The women who smoked marihuana regularly during their pregnancy differed from the remainder of the sample on a number of factors that potentially influence the development of the offspring and were dealt with by various statistical procedures. These possible confounding factors included lower socioeconomic level, less formal education and increased cigarette smoking. Although no difference in parity was noted, the heavy users were 3.2 years younger than the non-users. In terms of nutritional adequacy and weight gain during pregnancy there were no differences among the four groups.

The self-report procedure used in the OPPS to assess drug habits raises the critical issues of validity and reliability. Despite the obvious shortcomings of this mode of assessing drug use, at the time of the collection of data (primarily between 1979 and 1983) no practical alternative was available. Today, laboratory tests can measure the presence of metabolites of marihuana up to one to two weeks from the time of use. The use of both the interview and biological assessment approaches are critically discussed in a well reasoned paper by Day and Richardson (1991). In the OPPS, procedures were undertaken to enhance the likelihood of accurate data collection. A congenial relationship in a non-stressful, familiar environment (typically the mother's home) between the interviewer and the individual being interviewed has been part of the protocol of the OPPS and the same female interviewer "followed" the mother-to-be during her entire pregnancy. A second procedure designed to enhance the accuracy of the self-reports was the number of times the same drug related questions are asked. The questionnaire was administered once during each trimester and during each of the interviews, the questions pertaining to drug use for each three month period of pregnancy that had passed and for the 12 months prior to pregnancy were repeated, permitting a test-retest reliability measure.

Marihuana

Prevalence & Pattern of Use.

In exploring the impact of maternal use of marihuana upon the young offspring, one is immediately struck by the paucity of objective information. This is surprising when one considers that marihuana is far from a recent substance in terms of usage by members of society. References exist to its use in civilizations thousands of years ago (Abel, 1980) and, in fact, the drug was part of Eastern European medical folklore dealing with pregnancy for many centuries.

The very limited number of contemporary scientific studies that focus upon marihuana's potential long term behavioural teratological effects upon the developing fetus becomes of major concern when one considers the number of women of reproductive age that use this drug. In some cases, marihuana may be the only potentially teratogenic substance used, in other cases it may be used with other legal potentially teratogenic agents (e.g. alcohol and tobacco) while in further instances marihuana may be combined with other illegal substances that are under extensive investigation for their possible role in affecting the unborn child. For example, the majority of the studies using samples in which the long term consequences of *in utero* exposure to cocaine are being determined, report the use of that substance being highly correlated with marihuana (e.g. Chasnoff et al. 1992; Frank et al. 1988). Although it is possible (although not always) to control, to a certain extent, marihuana's impact by statistical means, clearly, knowing the singular role of cannabis upon the dependent variable in question is of great importance in interpreting the nature of the contribution of other substances.

A recent survey conducted by the National Institutes of Health indicated, that among women, marihuana usage reported during the previous year was 8.5% and during the previous month, 5% (National Institute on Drug Abuse, 1991). Among those of reproductive age the usage figures were considerably higher: 17.5% during the previous year and 8.5% during the previous month.

Several reports have examined the extent of use of marihuana during pregnancy but frequently the prevalence rates cited may not be representative of that drug's use in the general population as sampling procedures involve populations that are selectively biased towards drug use. On the basis of either interviews or urine screens conducted prenatally or post-partum, a rate of 27% was reported among a high-risk, predominantly non-white, Boston inner-city sample (Zuckerman et al. 1989). In another high risk sample in Pittsburgh (Day et al. 1991) a random sampling of women from an outpatient prenatal clinic found a 30% rate. At the Yale New Haven Hospital with a relatively low risk sample the rate at any time during pregnancy was found to be 10% (Hatch et al. 1986). In another low-risk population in the Seattle area, the rate was 17% (Streissguth et al. 1989). In a comparison between public health clinics and private obstetrical offices located in Florida the rate, based on urine screens, was quite similar with 12.4% in the former and 11.3% in the latter group (Chasnoff et al. 1990). In contrast, in Chicago (MacGregor 1990), based on urine screens at the time of admission into the labor-and-delivery unit a marked difference or marihuana rates was noted between clinic patients (32%) and private patients (7.5%)

In our own work in Ottawa, Canada, among predominantly middle-class volunteers (Fried et al. 1984, Fried et al. 1985) in the year before pregnancy 80% did not use any marihuana, 12% used the drug irregularly, 3% smoked two to five joints per week, and 5% smoked more than that amount. After the recognition of pregnancy, usage declined significantly, although during each of the three trimesters, the percentages remained relatively constant. Approximately 6% reported irregular use, 1% reported smoking two to five joints per week, and 3% continued to smoke a greater amount. In a trend that is similar to that noted among cigarette users (described later in this review) the heaviest users were the most likely to re-establish prepregnancy levels of consumption in the year following the birth of the baby.

In spite of the fact that marihuana is the most widely used illicit drug by pregnant women (Day and Richardson, 1991), as mentioned at the outset of this article, there is a notable lack of information about its long term consequences. The major reasons for this state of affairs lie in the ethical and practical difficulties surrounding the quasi-experimental research paradigm (Kilbey and Asghar, 1992) that is necessitated by research with human subjects in this area. Clearly, drugs cannot be administered to gravid women and so exact doses or amounts utilized and the precise timing of such use are impossible to quantify. Further, potentially confounding factors (including other drug use and socioeconomic factors) cannot be controlled by random assignment to groups. Human studies, particularly those investigating the long term effects of *in utero* exposure, have to be based on volunteer samples with reports of drug use being gathered either before (i.e. prospectively) or after (i.e. retrospectively) birth. These limitations are severe ones and, although a degree of control can be attained with statistical procedures, the interpretation and conclusions drawn from the research must be undertaken with the appropriate caveates in mind.

Neurobehavioural Observations

The literature pertaining to the behavioural effects of prenatal marihuana exposure is relatively sparse and, although suggestive is far from definitive. In 1980, in the first published report, four day old babies born to 12 regular users in the OPPS were examined (Fried 1980) and the findings were replicated in a subsequent, considerably larger study using the Ottawa sample (Fried and Makin 1987). Prenatal exposure to the drug was associated with decreased rates of visual habituation, increased tremors, frequently accompanied by exaggerated startles which were both spontaneous and in response to minimal, external stimulation. Similar observations were noted at 9 and 30 days using the Prechtl neurologic assessment (Fried et al. 1987). Further, at 9 days, increased hand-to-mouth behaviour was found among the babies born to the marihuana users. These possible indicants of impairments in nervous system state regulation and/or mild withdrawal were noted by some others (Chasnoff 1990) but not by all (Tennes et al. 1985, Richardson et al. 1989). Other signs of alterations in nervous system integrity have also been associated with *in utero* marihuana exposure. Sleep cycling and motility in newborns differed between marihuana and non-marihuana babies (Scher et al. 1988) and disturbed sleep patterns were still associated with prenatal exposure when the offspring were 3 years of age (Dahl et al. 1989).

The observations with the OPPS sample in the newborn period were, as will be described below, the only significant associations noted with prenatal marihuana exposure for a number of years as the children continued to be tested.

The children in the OPPS were examined at one year of age (Fried and Watkinson 1988) using the Bayley Scales (Bayley 1969). No adverse effects of prenatal marihuana exposure were noted. The Bayley test consists of three components. The Mental Developmental Index (MDI) assesses sensory perceptual abilities, early acquisition of object constancy, memory, problem solving, vocalization, and the onset of words. The Psychomotor Developmental Index (PDI) assesses gross and fine motor movement. The Infant Behaviour Record (IBR) evaluates the infant's attitudes, interests and temperament. The failure to find a relationship between the one year old infant's behaviour and maternal marihuana use is consistent with other reports assessing the children at the same age (Tennes et al. 1985, Astley and Little 1990).

Studies with their principal objective being the neurobehavioural assessment of *in utero* marihuana exposure in children older than a year are, with one exception, limited to those based on the data derived from the

OPPS. At 24 months, prenatal marihuana exposure was not negatively correlated with overall scores on the Bayley test (Fried and Watkinson 1988). At this age, using the Reynell Developmental Language Scale (Reynell 1977) a negative association with a measure of language comprehension but not language expression (Fried and Watkinson 1988) was observed but this association did not persist after statistically adjusting for other variables, especially ratings of the home environment.

At three years of age, children in the Ottawa sample (Fried and Watkinson 1990) were administered the Reynell test of language expression and comprehension as well as the McCarthy Scales of Children's Abilities (McCarthy 1972). This latter instrument is based upon six scales: verbal, perceptual, quantitative, general cognitive (a composite of the three previous scales) memory, and motor. As found when the children were a year younger, after controlling for potentially confounding variables, prenatal marihuana exposure was not significantly associated with any of the outcome variables. In a recent report from Pittsburgh based on three year old children born to women of generally lower social status (Day et al. 1994) prenatal marihuana exposure was not found to impact upon the overall composite score of the Standford-Binet Intelligence Scale. However, there was a significant effect on the short-term memory subscale. When the sample was broken down into white and African-American children, marihuana negative effects on verbal reasoning and short-term memory were noted in the latter group of children. Preschool/day-care attendance offset the marihuana associated deficit in the white children but not in the African-American children suggesting the importance of postnatal environmental conditions.

At four years of age the OPPS sample were given the test battery that was administered a year earlier plus the Peabody Test (Dunn and Dunn 1981) of receptive vocabulary and a series of motor tests (Fried and Watkinson 1990). General, global intellectual measures were not related to prenatal cannabis exposure, congruent with the findings of another study in which alcohol rather than marihuana was the primary drug of interest (Streissguth et al. 1989). However, in the OPPS sample on tests of verbal ability (both the McCarthy subscale and the Peabody) and memory, the children of regular marihuana users were significantly inferior to other children. This relationship persisted after statistically controlling for a host of potentially confounding factors including the home environment. In terms of the neurobehavioural dimensions that appear vulnerable to prenatal marihuna exposure the results are strikingly similar to those noted very recently in three year olds as described above (Day et al. 1994). Within the OPPS, this negative relationship was the first reported association beyond the neonatal stage. The observation of a neurobehavioural significant effect at this age (and not earlier) may indicate that the degree and type of deficits noted can be identified only when normal neurological development has proceeded to a certain level of maturity and when complex behaviour can be examined at a more specific, rather than global level. This maturation hypothesis reflects the notion that the effects of prenatal exposure to marihuana are subtle and that their consequences on complex behaviour are not manifested and/or cannot be tested before 4 years. This line of thinking will be elaborated later in this chapter.

The difficulty in unraveling the long-term consequences of *in utero* marihuana exposure becomes very apparent when one examines the data gleaned from the cognitive and language assessment of the 5 and 6 year old OPPS participants (Fried et al. 1992). These children were given the same battery as when they were 4 but, unlike the findings at 48 months, upon statistical analysis, no relationship was observed at either 5 or 6 years of age between any of the subscales of the McCarthy or the Peabody tests and maternal marihuana use.

The reason for the disparity of observations is not at all clear. One possibility may be the increasing effect of environmental variables. As the children get older, they are exposed to an increasing similarity of postnatal influences that bear on cognitive development. For example, by 5 years of age 89 % of the non exposed children and 87% of the exposed children had a year of formal schooling. Could it be that this common feature would tend to overwhelm some of the quite subtle differences on memory and verbal abilities noted at an earlier age?

Indirect evidence of the possible influence of ubiquitous, relevant environmental factors may be seen by the age corrected 'catching up' scores of the marihuana children. The McCarthy verbal and memory scores at 4 and 5 years of age were essentially unchanged for the non-exposed children. At both ages these children scored 1 to 1.5 standard deviation (sd) above age norms. On the same subscales, the marihuana children improved their scores by approximately half a sd between the ages of 4 and 5 to 1 sd above the age norm at 60 months. Thus, the post-natal influence of school may have served to overcome the marihuana associated observations noted at 4 years. This would certainly be entirely consistent with the findings from Pittsburgh described earlier (Day et al. 1994) in which the negative association with prenatal marihuana exposure in the verbal and memory domains were attenuated by attendance of preschool or day-care in the white children of the sample.

Instruments that provide a general description of cognitive abilities may not be capable of identifying nuances in neurobehaviour that may discriminate between the marihuana and non-marihuana exposed children.

However, tests that examine specific characteristics that may underlie cognitive performance may be more appropriate and successful. This approach to assessing the consequences of prenatal marihuana exposure was examined in a recent study (Fried et al. 1992) in which impulse control and sustained attention was examined in 6 year olds and in a study assessing specific aspects of cognitive performance (O'Connell and Fried 1991).

The children were assessed using two forms of a computerized vigilance task using a one-button solid-state console (McClure and Gordon 1983). In order to examine the child's ability to withhold responding, a 6-second DRL (Differential Reinforcement of Low Rate Responding) schedule was employed. Under this regimen a reinforcement (points displayed on a screen) would be obtained when a button press occurred six seconds after the emission of a previous response. Responses that occurred prior to the termination of this six seconds were not reinforced and served to reset the timer so that six seconds of no button pressing would have to elapse before the next button press would result in a reinforcement. Thus, on this DRL 6-second schedule, a child would receive a reinforcement for every button press emitted after an interval of six seconds.

Three sets of data were obtained: the absolute number of responses, the total number of rewarded responses and an efficiency ratio which was obtained by dividing the number of rewarded responses by the total number of responses.

The same apparatus was used to examine sustained attention. A series of single digit numbers was shown on the screen at a rate of one per second. They were displayed for 200 milliseconds with an 800 millisecond interval between each signal. Each subject was asked to press a button whenever the target stimulus appeared on the display screen among a series of randomly presented numbers. The scores were the number of correct responses, the number of omissions (missed target stimuli) and the number of commissions (button press to non-target stimuli or false alarms). The scores were computed for each of three three-minute blocks and then totaled for the overall 9-minute trial.

As an additional facet in this work, parents assessed the child's impulsivity/inattention at home by using portions of the Conners' Parent Rating Scale-48 (CPRS-49) (Conners 1989). This 48 item behavioural symptom checklist was completed by the child's mother at the time of testing using a four point rating system. The scale yields six behavioural clusters one of which – the Impulsive-Hyperactive Scale – was used for this assessment. The four items which enter into this scale include excitable, impulsive; restless or "squirmy"; wants to run things; and restless, always on the go.

The results suggested that prenatal marihuana exposure was not associated with poorer impulse control as the children of the heavy marihuana users were not deficit in the delay task in either the number of rewards or the efficiency ratio. In the vigilance task the commission errors were very similar among all three marihuana groups (again suggesting no impairment in impulse control) but the omission errors and the number correct were differentiated, in a dose related fashion, among the children of the various marihuana groups. Further, across temporal epochs within the vigilance task, only the children in the heavy marihuana category increased their omission errors. The overall increase in omission errors and the greater number towards the end of the vigilance task may reflect a deficit in sustained attention.

There was a significant tendency for the women who used marihuana heavily during pregnancy to rate their children as being more impulsive/hyperactive. The nature of the scale emphasizes overall activity rather than attention behaviour. Although consistent with the more objective measurements there is a difficulty in interpreting these results. The fact that women in the heavy marihuana group tended to identify their children as more problematic in this domain may be an accurate reflection of the child's behaviour or it may represent the mother's perception and attitude toward this behaviour. Do the present observations indicate a true behavioural difference in the attention related domain or is there a lowered parental tolerance? Ratings by other observers such as teachers and additional assessments of maternal parenting attitudes and expectations might help to clarify this issue.

In a preliminary report (O'Connell and Fried 1991) examined the school aged (6-9 years of age) offspring of regular marihuana users and matched (in terms of alcohol and cigarette use during pregnancy) controls participating in the OPPS school age on a battery of neurobehavioural tests. These included assessment of intellectual abilities, visual perceptual skills, distractibility, memory, language comprehension, academic achievement, visual motor skills and parental rating of behaviour.

Measures which discriminated between the study groups and on which the children of the marihuana users scored more poorly included parental behaviour ratings (particularly conduct problems), visual perceptual and visual memory tasks, language comprehension and distractibility. It is striking that these behaviours are ones that have cropped up in our work with these children at earlier ages and in the report on three year olds from the Pittsburgh sample (Day et al. 1994). On the other hand, the data from the OPPS work are not without interpretative complications. For the measure of visual memory and language comprehension, the mother's age at the child's birth potentiated the effect of cannabis use to produce lowered scores for children of young, cannabis-using mothers relative to children of young, non-using mothers. Further, when controlling for the influence of the mother's age at delivery, mother's self-rated personality – the marihuana cohort being higher on neuroticism and lower on agreeableness and conscientiousness – and the home environment in which greater aggression and less supervision was present in the marihuana homes, the discriminating variables were no longer statistically significant.

Whether the inclusion of the personality and home environment variables as statistical controls is appropriate is a difficult issue and has been discussed elsewhere (Fried and Watkinson 1988; O'Connell and Fried 1991) and is considered later in this report. Briefly, the important question is whether this inclusion results in a too conservative approach to the data analysis. The finding of differing personality and home environment ratings between the users and non-users of marihuana may well be viewed in a transactional framework (Sameroff and Chandler 1975). This model states that the developmental outcomes are the product of both maternal and child characteristics and the relationship between the mother and child characteristics is a reciprocal one. Thus home environment measures and personality characteristics may be an outcome in themselves arising from interactions with a behaviourally altered child.

Two additional points have to be kept in mind in interpreting the findings with respect to prenatal marihuana exposure described in this article. The women in the Ottawa work represent a very low-risk sample. There is a considerable body of literature (animal and human) to suggest that the drug's effect is potentiated in a higher risk environment (reviewed in Fried 1993) and thus one must be very cautious in extrapolating the present observations to other marihuana using populations. There is also the concern that the potency of marihuana preparations, in terms of tetrahydrocannabinol content, has increased several fold (Elsohly and Elsohly 1989) since the entrance of pregnant women into the Ottawa study in the late seventies and early eighties. This increase in potency adds to the importance of interpreting the present results as representing conservative observations.

Attempt at a Synthesis.

What can one conclude from the material described up to this point? On the surface it appears that the only conclusive statement would be that, if there are long term consequences of prenatal exposure to marihuana, such effects are very subtle. However, the data may allow conclusions that go somewhat beyond this level.

The marihuana findings may be summarized in the following manner. In the newborn and neonate, although far from definitive, there appears to be an association between nervous system state regulation and prenatal exposure to marihuana. However, between 6 months and three years of age no neurobehavioural consequences of marihuana have been reported although, at two years of age, language comprehension was lower among the children of cannabis users prior to statistical control for the home environment. At 4 years, tests of verbal ability and memory statistically discriminated between the offspring of regular marihuana users and the remainder of the children in the OPPS sample. The same domains of vulnerability were noted in three year olds in the high risk Pittsburgh sample (Day et al. 1994). At 5 and 6 years of age, prenatal marihuana exposure was not associated with global tests of cognition and language after statistically controlling for potentially confounding data. However, at approximately these ages and slightly older, tests that examined more specific aspects of behaviour did appear to suggest a relationship between performance and *in utero* exposure to marihuana. In school-aged children, a deficit in sustained attention was noted on a task that differentiated between impulsivity and vigilance. Further, parental ratings of behaviour indicated greater problems (particularly in the area of inattention and conduct) among the children of cannabis users. Finally, visual perceptual, visual memory, language comprehension and distractibility discriminated between the 6-9 year old offspring of marihuana and non-marihuana users. The latter findings did not remain statistically significant upon the inclusion of maternal personality and home environment conditions as potential confounds although this statistical control (as discussed above) may be inappropriate.

Two issues that arise from this data are the seeming absence of prenatal marihuana cognitive effects prior to four years of age and the question of whether there is any common theme among the effects and trends noted at fours years and beyond.

Dealing with the latter issue first, the areas of vulnerability that have emerged over the course of the OPPS (as well as the finding at three years of age in the Pittsburgh sample [Day et al. 1994]) are ones that are quite consistent with the cognitive construct that several authors have termed executive function (Luria 1966; Duncan 1986; Welsh and Pennington 1988). This function is defined as the cognitive ability to maintain an appropriate problem-solving set for attainment of a future goal and involves the integration of cognitive processes. The executive function behaviours noted to be negatively associated with prenatal marihuana exposure include those that involve self-regulatory abilities (the dysfunction possibly manifesting itself in the form of behavioural problems), the ability to maintain attention (noted as impairments in vigilance and distractibility) and the ability to act on accumulated knowledge (poorer performance on facets of language and memory).

Executive function is thought to serve as a marker of prefrontal lobe function and thus it may be that this part of the central nervous system may be particularly vulnerable to prenatal marihuana exposure. Frontal lobe development is not an all or none phenomenon but appears to be a multistage process as is executive functioning (Walsh & Pennington 1988). Although aspects of executive functioning are present in infants and toddlers (e.g. object permanence behaviour) many aspects of prefrontal functioning are not apparent or are difficult to test until children approach or reach school-age. This would certainly be very congruent with the OPPS results.

An important further property of executive functioning is that it is disassociated from measures of global intelligence. This is consistent with the observation of the sparing of IQ after frontal lobe damage (Damasio 1979) and may reflect the fact that traditional, global intelligence tests evaluate overlearned information and established cognitive sets. One of the consistent findings noted among the children in the OPPS (and in the Pittsburgh sample [Day et al. 1994]) was that prenatal marihuana exposure was not associated with a lowering of general IQ.

Recent observations from diverse fields within the general marihuana literature also implicate the frontal lobe in that drug's effects. The discovery of receptors for cannabinoid substances in the mammalian brain (including humans) provides very convincing evidence for the direct action of marihuana on certain mental processes (e.g. Herkenham et al. 1990; Matsuda et al. 1990, Herkenham et al. 1991) in long-term, chronic adult users that includes fragmentation of thought, difficulty in short term memory tasks, disturbances in attention, concentration and judgment – tasks that are associated with frontal lobe functioning. Recently, Australian workers (Solowij 1993) have reported that among chronic users the ability to evaluate stimuli and reject irrelevant stimuli is impaired and these alterations in cognitive ability do not reverse themselves even after 5 years of abstinence (Solowij, personal communication, October, 1993).

In the rat, within different regions of the cortex, the frontal area has been reported to contain the highest density of binding sites (Herkenham et al. 1991). In 9 chronic adult users, Tunving (1986) reported reduced blood flow throughout the cerebral cortex. Intriguingly, only 2 of the users were not polydrug users and in these cases in which marihuana was the drug uniquely used, the prefrontal area was the most affected (Lundqvist, personal communication, July 1993). Finally, Struve and his co-workers (Struve et al. 1989, Struve et al. 1993) have recently reported that chronic, daily use of marihuana results in a marked alteration in alpha activity primarily in the frontal region even after prolonged cessation of use.

Together then, a suggestive, although, at this stage, highly speculative picture is beginning to emerge. The behavioural evidence gathered primarily from the children participating in the OPPS over the past years, the delayed temporal appearance of the observed effects and, the recent findings linking altered frontal lobe functioning with chronic marihuana exposure is certainly compatible with the notion that prenatal marihuana exposure may result in altered frontal lobe functioning in the offspring. One of the next steps in this picture is to examine the children in the OPPS in tasks that are thought to be particularly sensitive to frontal lobe dysfunction. These include tests of problem solving that require cognitive flexibility, route finding tasks, measures of distractibility and attention, and working memory. These assessments are presently underway.

Cigarettes

Prevalence & Patterns of Smoking

The significant number of women who smoke during pregnancy is one of the major reasons for cigarette smoking being among the most widely researched topics in the area of teratology. During the late eighties and early nineties, relatively consistent findings have been reported with respect to the proportion of the population that smoke in Europe and North America and the changing trends among different subgroups of smokers. In Canada, the results of a recent national survey by the government (Health and Welfare 1988) indicate that the general prevalence of smoking by individuals 15 years and older was declining among all age groups. Whereas 41% of the adult population smoked on a daily basis in the mid nineteen seventies, 30% were daily smokers a decade later with the largest reduction being among males. Their use diminished from 49% to 32% over this time period. Although females also followed this downward trend, the degree of change from 32% to only 29% was substantially less. Particularly pertinent to the topic at hand is that, among women of reproductive age there was essentially no change in the proportion smoking over this decade with approximately one third reporting that they smoked on a regular basis. In fact, a recent press release from Ottawa said, for the first time in any Western country, in this particular age group the number of women who smoked outnumbered similar aged males. Prevalence of use by women of reproductive age is similar in other countries such as the United States

(National Institute on Drug Abuse 1991) and Sweden (Cnattingius 1989). As an example, in the U.S., the 1990 NIDA National Household Survey on Drug Abuse (1991) 31% of women between the ages of 18-34 smoked at least a package of cigarettes a day in the past month. It is perhaps important to note that in many African and Asian countries, contrary to the trends noted in Europe and North America, the number of young male and female smokers are increasing substantially reflecting, at least in part, the vigourous advertising campaigns aimed at young adults by the marketers of cigarettes.

An important facet of smoking habits and gender has to do with the proportion of smokers consuming a package a day or more. For example in Canada, the relative increase in the proportion of these heavy smokers during the past decade was 31% for males as compared to 57% for females (Health and Welfare 1988). The relevance of this observation for the present work can be found in a Swedish study (Cnattingius, 1989) which noted that one of the predictors for continued smoking during pregnancy was the amount smoked prior to pregnancy. Certainly many smokers have switched to cigarettes with lower tar and nicotine content but often these individuals have counteracted this potential benefit by smoking more of them. This change in smoking habit has considerable implications as the relationship between maternal smoking and immediate and long term effects upon offspring may be related to carbon monoxide which increases directly with the number of cigarettes smoked (Peacock et al. 1991).

Cigarette use by pregnant women in nonghetto urban regions has generally been found to range from 22 to 28% (Fried et al. 1984; 1985; Stewart and Dunkley 1985; Streissguth et al. 1983; Williamson et al. 1989). In our own work – the Ottawa Prenatal Prospective Study (OPPS) – among volunteer, primarily middle-class participants, the percentage of heavy smokers diminished from 12% in the year before pregnancy to seven percent in the first trimester and to five percent by the third trimester (Fried et al. 1985). On the other hand, we have noted that half of the light smokers (less than a package a day) and 90% of the heavy smokers continued to smoke to some degree once pregnancy was confirmed. However, as will be discussed later, even a reduction in smoking is of benefit to the fetus since risk factors associated with a host of related outcomes are dose dependent. As such, the success or failure of programs designed to encourage pregnant women to alter their smoking habits must be evaluated not only in terms of total cessation but also by whether a reduction in smoking has taken place (Seckerwalker et al. 1992).

The relationship between smoking during pregnancy with a variety of adverse pregnancy outcomes including lowered birth weight have been the subjects of a large number of well controlled studies. These have been recently reviewed (Fried 1993) and will not be discussed here as the focus of the present article is on the neurobehaviour of the offspring.

Neonatal Behavioural Observations

Although not totally consistent (e.g. Richardson et al. 1989), there are a number of reports noting relationships between maternal smoking during pregnancy and altered behaviour among a number of dimensions in the neonate. The findings from several studies are suggestive of auditory system involvement. In the OPPS, prenatal cigarette exposure was associated with decreased auditory habituation in 3-6 day old babies (Fried and Makin 1987). A similar finding has been reported elsewhere (Saxton 1978) coupled with the observation of decreased auditory responsiveness. Poorer auditory responsiveness has been noted by other researchers in infants less than a week of age (Picone et al. 1982; Jacobson et al. 1984).

Maternal use of cigarettes has been reported to be associated with increased tremors (Fried and Makin 1987) and poorer autonomic regulation (Picone et al. 1982) – symptoms consistent with mild withdrawal.

The offspring of women in the Ottawa study were assessed at 9 and 30 days examining reflexes, autonomic regulation, limb movements and posture (Fried et al. 1987). At the younger age, the features distinguishing the babies born to smokers from those born to nonsmokers were an increased incidence of fine tremor and some heightened motoric reflexes. At 30 days of age, aspects of muscle tonus differentiated the two groups with increased tonicity and heightened motoric reflexivity noted in the smoking group. The hypertonicity remained statistically associated after controlling for birth weight indicating that the relationship was not due to reduced fetal growth. Increased startles and tremors were also associated with maternal smoking in the 30 day old infants and, together with the hypertonicity, appear suggestive of nervous system excitation (Fried et al. 1987).

Several reports have linked maternal smoking with hyperactivity in young children (Denson et al. 1970; Nichols and Chen 1981) and poor orientation and impaired attention in a vigilance task (Streissguth et al. 1984; Kristjansson et al. 1989; Fried et al. 1992b). Research with the OPPS sample (Kristjansson 1992; Fried et al. 1992b) has examined this aspect in some detail in older children and will be discussed later in this article. At this juncture it may be sufficient to point out the consistency between these longer term associations of maternal smoking and the proposal of increased nervous system excitation mentioned above. Further, research has demonstrated that habituation and recovery of attention such as appear to be affected in infants born to smokers is a better predictor of childhood cognitive competence than traditional infant developmental tests (Bornstein and Sigman 1986). However, even if one makes the reasonable assumption that development is a continuous process with common substrates underlying behavioural performances (Bornstein and Sigman 1986) such a cause and effect postulation at this stage is highly speculative and further, the impact of the child's environment on behaviour must not be relegated to a minor role (Sameroff and Chandler 1975).

The instrument used in most research centres to examine the relationship between maternal smoking and effects in the baby has been the Bayley Scales of Mental and Motor Development (Bayley 1969) which is divided into three components as described earlier in this article.

In one study, with relatively few heavy smokers and after statistically adjusting for alcohol, caffeine and gestational age, no effects of maternal nicotine were noted in eight month old babies (Streissguth et al. 1980). However, in a pilot study using babies derived from the Ottawa sample, 13 month old babies born to smokers were observed to a have a decrease in verbal comprehension (assessed by a cluster derived from the mental portion of the Bayley scales) after adjusting for confounding variables (Gusella and Fried 1984).

A more comprehensive examination of 217 one year old children in the OPPS was undertaken, carefully controlling for confounding factors that may affect the outcome measurements. Potential confounding variables that were considered statistically included family income, maternal age and education, maternal caffeine use, protein and caloric intake during pregnancy, alcohol and marihuana use, complications during pregnancy, maternal and paternal health history, birth weight, parity, and an assessment of the home environment (Fried and Watkinson 1988).

Cigarette use contributed significantly to the prediction of the Mental Developmental Index (MDI) scores. The average MDI of the babies born to the non-smokers was 109.5 whereas the average score of the toddlers born to heavy smokers was 96.5 with the average for babies born to light smokers intermediate. Further, when examining the negative relationship between maternal smoking and MDI values, 39% of the heavy smokers had infants with scores of under 85 (more than one standard deviation below the mean) in contrast to 6% of the remaining subjects. No relationship was noted between the Psychomotor Developmental Index (PDI) and maternal smoking but, importantly, there was a significant negative association between auditory related items on the Infant Behaviour Record (IBR) and *in utero* exposure to cigarette smoke. This observation is consistent with the infant observations described above in which altered auditory responsiveness was noted in neonates born to maternal smokers. This longitudinal continuity of findings supports the interpretation of a direct effect of prenatal smoking rather than postnatal, maternal lifestyle habits including postnatal second hand smoke. Clearly, however, it doesn't preclude the possibility that postnatal exposure also contributes to the observations.

At two years of age, children from the OPPS were examined (Fried and Watkinson 1988) using the Bayley test (1969) and Reynell Developmental Language Scale (Reynell 1977). At 24 months, smoking during pregnancy was associated with auditory related behaviours as measured by the IBR of the Bayley but maternal smoking no longer contributed a unique significant explanation of variance of the MDI or the comprehension and expressive language scores on the Reynell test. The children born to the heavy smokers scored more poorly on all measurements compared to the children of non-smokers (e.g on the MDI 106 versus 119) but factors other than maternal smoking statistically 'overwhelmed' the smoking effect. In part, this was due to the fact that the outcomes used at this age were relatively global in nature and did not focus upon various components of behaviour.

In this study examining the 24 month old infants, the assessment of the home environment and parental involvement was carried out using the HOME scale (Caldwell and Bradley 1979). The scores on the HOME were highly positively associated with the 24 month MDI and the Reynell tests and negatively correlated with maternal smoking. A woman who smoked during pregnancy is more likely to have a lifestyle that includes less maternal/child involvement (Mackey and Fried 1981; Goodine and Fried 1984) and this may well impact upon the child's cognitive development. This combination resulted in the loss of unique explanatory power of the mother's prenatal smoking habits and suggests that the postnatal environment acted as an intervening variable.

Although it is possible that prenatal smoking has no measurable effect on two year old cognitive measures this appears unlikely. As the children got older and were assessed using tests that could access more subtle and complex behaviour, prenatal cigarette exposure was, in fact, found to be consistently related to particular outcomes, some of which were in the same domain as observed in infancy.

As described in a recent review (Rush and Calahan 1989), the relationship between lowered cognitive skills in childhood and maternal smoking is relatively consistent with most investigators reporting the association to be either statistically significant or to approach significance. In the Ottawa study, three year old children were assessed (Fried and Watkinson 1990) using the Reynell language test plus the McCarthy Scales of Children's Abilities (McCarthy 1972). Using statistical procedures that controlled for such potential confounding variables as other drug use, family income and education, nutrition, maternal age, parity, birth weight and the home environment, the results revealed a negative dose response relationship between maternal smoking and many of the subscales as well as the general cognitive score. Among the subscales, the verbal component of the McCarthy test appeared to be the most affected by maternal smoking. Even though, as at 24 months, the HOME scores were highly related to both the outcome measures and maternal smoking, the latter remained significantly associated after control for the HOME with cognitive measures at three years of age which assessed more complex and subtle aspects of the child's performance than was possible at two years of age.

At four years of age the same sample was given a similar test battery (Fried and Watkinson 1990) plus a widely used measure of receptive vocabulary – the Peabody Picture Vocabulary Test (Dunn and Dunn 1981) and a series of motor tests. The results were strikingly similar to those obtained when the children were a year younger with the performance of the smoking group negatively associated with the cognitive measures in a dose response fashion. The three language associated tests (the McCarthy verbal subscale, the Reynell expressive scale and the Peabody) differentiated the heavy, light and non-smokers to the greatest degree.

It is noteworthy that the findings in the younger children have persisted in five and six year olds (Fried et al. 1992a). At these ages the children were given the same test battery as at forty eight months. Once again, a negative, dose-response association was observed between both the overall cognitive score and language tests with statistical significance remaining after controlling for potential confounding factors that included maternal use of other drugs, parental education, birth weight and the home environment.

In order to extend and futher examine the association between maternal cigarette smoking and the poorer performance by the children on language, language based and auditory tests, a study was undertaken examining whether exposure during fetal development is related to central auditory processing (McCartney et al. in press). In this work the SCAN (A Screening Test for Auditory Processing Disorders) (Keith 1986) was administered to 6 to 11 year old subjects in the OPPS. This instrument requires the child to listen, via heaphones, to tape recorded stimuli under three conditions: filtered words, auditory figure ground and competing words. In the Filtered Words subtest the words were low passed filtered so as to sound muffled and the listner had to fill in the missing elements in order to comprehend the word. In the Auditory Figure Ground subtest, the words were presented at eight decibels above background noise (indistinct party conversation). In the Competing Words subtest word pairs were presented simultaneously on stereo channels with one word in one ear, the other word in the opposite ear. The child was asked to repeat pairs of words leading with one ear and then the other. The total scores for the three subtests formed a global, composite SCAN score.

Maternal smoking was significantly linearly associated with poorer performance on the overall, composite SCAN score and, particularly, on the Competing Words subtest although all three subtests exhibited the same trends. The significant associations remained after adjusting for other drug use, demographic variables and passive smoke exposure both during pregnancy and posnatally. As all the SCAN tasks are imitative and make perceptual rather than linguistic demands the deficits noted in this work suggest that *in utero* exposure to smoke affects fundemental aspects of auditory processing – ones which may manifest themselves as impaired performance on auditory-based cognitive and memory tasks. Further, an impairment of auditory processing at this basic level is consistent with the observations, described earlier, of altered, non-verbal auditory responsiveness in neonates born to maternal smokers.

The underlying possible neurophysiological mechanisms underlying the association between maternal smoking and auditory functioning have been speculated upon elsewhere (McCartney et al. 1994). Briefly stated, the efferent component of the auditory system which serves to increase the sharpness of tuning and the sensitivity of the basilar membrane is a cholinergic system involving nicotinc receptor sites (Pickles 1988). The number of binding sites for nicotine are highly correlated with nicotinic acetylcholine receptor sites and these binding sites have been reported very early in fetal development, increasing with gestational age (Cairns and Winnocott 1988). The presence of these nicotine binding sites in the fetus thus has implications for the effect of maternal smoking on cholinergic mediated systems. It is tantalizing, although highly speculative, to hypothesize that the exposure of the fetus to nicotine and its putative effect upon the efferent auditory system may be a contributing physiological mechanism underlying the negative association between auditory/language performance and maternal smoking.

As noted earlier, the infant data derived from the OPPS sample and work with older children from other research centres have also suggested an association between maternal smoking during pregnancy and aspects of impulsivity in the offspring. In the OPPS this aspect has been examined in two seperate studies. In the first (Kristjansson et al. 1989), four to seven year old children were assessed on auditory and visual continuous performance tasks (CPT) in which children were told to press a game button when the critical stimulus (a picture in the visual task and a word in the auditory task) occurred and to withold responses when the non-critical stimulus occurred. Two types of errors could be analyzed. Commission errors (false alarms) are typically thought to reflect impulsivity whereas omission errors (misses) are usually interpreted as reflecting inattention. While participating in this study, the children were seated upon a movement sensor so that the activity level of the child could be objectively ascertained. Cigarette smoking during pregnancy was related to an increased activity level and increased errors of commission suggesting that deficits in attention among the offspring of smokers may reflect impulsive responding and increased overall activity. Of considerable interest in this task was the observation that the auditory task revealed greater impairments among the children born to smokers than did the visual task.

In a second study assessing attentional behavior in six year old children the OPPA subjects were examined using the response inhibition task and the sustained vigilance task described in the marihuana portion of this article (Fried et al. 1992). There was a marked association between prenatal smoking and impulsivity. In the impulse control situation in which the child had to refrain from emitting nonreinforced responses, the efficiency ratio, a measure of the ratio of the number of rewards to the number of responses, was significantly lower among the children born to heavy smokers and the number of responses emitted by these children was also greater. In the sustained attention (vigilance) task, the children of heavy smokers had a slightly higher number of omission errors but a significantly greater number of commission errors compared to the nonsmokers. In fact, the average commission scores of the children born to the heavy smokers were within the range that are classified as abnormal (Gordon and McClure 1983) i.e. the fifth percentile of the normal population illustrating that the commission error rate among these children may be more than mere statistical significance.

General Conclusions

Although the evidence is clearer with respect to maternal use of cigarettes, both of the drugs considered in this chapter are implicated as affecting the developing fetus. Further, it would appear that marihuana and cigarettes affect the offspring in a differential manner. *In utero* cigarette exposure is negatively associated with cognitive performance, various facets of auditory dependent behaviour and impulsivity. Marihuana may affect a number of neonatal behaviours and facets of cognitive behaviour in the older child under conditions in which complex demands are made which may reflect prefrontal lobe functioning.

Separating *in utero* effects from postnatal effects becomes more and more problematic as the child gets older. This is a very complicated issue and underlies much of the difficulty in interpreting the behavioural teratological findings (Kilbey and Asghar, 1992) in which the effects are subtle.

One of the dilemmas in making conclusions about the consequences of drug exposure during fetal development is that the effects noted in the offspring are, in fact, not due solely to the drug in question but also to the lifestyle and parent-child interaction that are often concomitant of a particular drug habit. Looking for the statistically unique contribution of a drug after controlling for so called confounding factors may well obscure the reality of the drug effect(s). Furthermore, there is a reciprocal relationship between the baby's behaviour and the parent's interaction with their infant. This transactional state of affairs (Sameroff and Chandler 1975) may well serve to exacerbate particular *in utero* effects. For example, if the infant tends to be hypertonic and less cuddly, the parents may respond by less physical contact with the baby which, in turn may decrease the baby's relaxation when held.

Although statistically, a significant amount of variance may be accounted for by the drug(s) in question, the amount of the variability attributable to the drug(s) is relatively small compared to other factors and decreases as the child gets older. In our own work spanning more than a decade, non-drug life-style habits account for up to 35 percent of the cognitive outcome variability (Fried & Watkinson, 1990) but the behavioural effects uniquely associated with maternal drug use (tobacco or alcohol or marihuana) only ranges from 1.5 to 8 percent after the variance due to other potentially confounding factors is partialled out. In other laboratories the figure can be less. This low proportion of unique, explained variance should not be interpreted as indicating that maternal drug use is of little significance. Far from it. Not only does it have real, measurable effects as described above, it is also one of the few variables that can realistically be modified – more so than other life-style factors such as socioeconomic status which impinge on the mother and child. Furthermore, rarely does a drug act in isolation or in a, statistically speaking, unique fashion. It interacts with a host of other factors including other drugs, other environmental and genetic risk factors.

In other publications (e.g. Fried and Watkinson 1988) arising from the OPPS, it has been argued that it is more likely that the drug's real association with the behavioural outcomes in question may lie between the drug's unique contribution (after potential confounds are considered) and its zero-order correlation (with no potential confounds considered). In the latter approach, variance attributable to drugs may be as high as 12 percent, whereas, as stated earlier, the unique contribution is often in the region of 1 or 2 percent. The likely contribution or influence of the drug may well fall between these two figures.

However, the small proportion of unique variance attributable to maternal drug use does lead to a variety of interpretative problems and emphasizes the importance of longitudinal investigations in which suspected drug effects from maternal usage can be examined across many ages. If one notes effects in the very young infant along particular dimensions of behaviour and continues to see effects in related spheres as the offspring gets older more, confidence can exist in attributing some of the findings to the *in utero* exposure.

In interpreting the evidence presented in this chapter, this interactional state of affairs must not be forgotten and should be an integral part in drawing conclusions about both marihuana's and cigarette's effects upon the mother, fetus and child.

Finally, the domains that are affected, even if to a relatively small extent, by *in utero* exposure to marihuana and cigarettes are ones that have the potential to impact markedly upon the growing child's milieu. For example, in the school environment, the child is required to attend critical stimuli, refrain from impulsive and intrusive behaviour and be able to retain and recall material particularly in the language domain. Thus, the attention-related behaviours, the impairments in the auditory/language, memory, behavioral and cognitive areas associated with the drugs considered in this article have clear implications for performance in the classroom setting.

Acknowledgements

The OPPS has been supported in the past by grants from the National Research Council of Canada, National Health and Welfare, Canada and a Carleton University Social Science Research Grant. In recent years and presently the work is supported by a National Institute on Drug Abuse grant DA-04874. The work described could not have been carried out without the continuing and ongoing assistance of my long-time associates over the years, including B. Watkinson, H. Linttell, and R. Gray and the enduring patience of the mothers and their children during the past decade.

Abstract

Both marihuana and cigarettes appear implicated, in a differential fashion, in the neurobehaviour of infants and children born to women who used these substances during pregnancy. In a low-risk upper middle class sample, marihuana use was associated, in the newborn, with mild withdrawal symptoms and some autonomic disruption of nervous system state regulation. However, between 6 months and 3 years of age no behavioural consequences of marihuana exposure (once confounding factors were controlled) were noted. At four years of age, although global tests of intelligence did not differentiate exposed from non-marihuana exposed children, verbal ability and memory were associated with *in utero* marihuana exposure. At five and six years of age these general areas were also noted to be associated with maternal cannabis use as was sustained attention. These areas of neurobehavior that appear affected by marihuana exposure during fetal development are ones that are consistent with the cognitive construct of 'executive functioning' which is thought to be a marker of prefrontal lobe functioning. Consistent with the observations derived from these children is that prefrontal functioning may not be apparent until approximately four years of age and that executive functioning is disassociated from measures of global intelligence.

Exposure to cigarettes during pregnancy appears to be associated with neurobehavioural deficits in the auditory domain. In the newborn this is manifested by decreased responsivity to sound and altered auditory habituation. Between the ages of one and 11 years the performance on auditory related tasks (verbal memory, language, auditory processing) were consistently the domains that differentiated the cigarette exposed from the non exposed children. The possible role of the cholinergic mediated efferent auditory system is discussed. Also associated with *in utero* exposure to cigarettes were general cognitive performance and parental reports and objectively derived measures of impulsivity. The striking degree of consistency over the years lends strength to the interpretation that the observations in childhood have, at least as their partial etiology, the prenatal exposure to cigarettes.

However, in interpreting the evidence presented it must be recognized that the alterations in the child's behaviour may well affect the parenting behaviour. This potential transactional interaction must remain an integral part of drawing conclusions about both marihuana and cigarette's effects.

References

Abel E (1980) Marijuana: The First Twelve Thousand Years. New York: Plenum Press

Astley S, Little R (1990) Maternal marijuana use during lactation and infant development at one year. Neurotoxicol Teratol 12:161-168

Bayley N (1989) Bayley Scales of Infant Development. New York: Psychological Corporation

Bornstein MH, Sigman MD (1986) Continuity in mental development from infancy. Child Devel 57:251-274

Cairns NJ, Winnacott S (1988) [^{3}H](-)nicotine binding sites in fetal human brain. Brain Res 475:1-7

Caldwell BM, Bradley RH (1979) Home observation of measurement of the environment University of Arkansas, Little Rock

Chasnoff IJ (1990) Cocaine use in pregnancy: Effect on infant neurobehavioural functioning. Paper presented at the American Society for Pharmacology and Experimental Therapeutics. Washington

Chasnoff IJ, Griffith DR, Freier C, Murray J (1992) Cocaine/polydrug use in pregnancy: Two-year follow-up. Pediatrics 89: 284-289

Chasnoff IJ, Landres HJ, Barrett ME (1990) The prevalence of illicit drug or alcohol use during pregnancy and discrepancies in mandatory reporting in Pinellas county, Florida. New Engl J Med 322:1202-1206

Cnattingius S (1989) Smoking habits in early pregnancy. Addictive Behav 14:453-457.

Conners CK (1989) Manual for Conners' Rating Scales. Toronto: Multi Health Systems

Dahl RE, Scher MS, Day NL, Richardson G, Klepper T, Robles N (1988) The effects of prenatal marijuana exposure: evidence of EEG-sleep disturbances. J Dev Behav Pediatr 9: 333-339

Dalterio SL, Fried PA (1992) The effects of marijuana use on offspring. In: Sonderegger TB (ed) Perinatal substance abuse. Research findings and clinical implications. Baltimore: The John Hopkins University Press, pp. 161-183

Damasio A The frontal lobes In: Heilman KM, Valenstein E (eds) Clinical neuropsychology. New York: Oxford University Press, pp.360-412

Day N, Richardson G (1991) Prenatal marijuana use: Epidemiology, methodological issues and infant outcome. In: Chasnoff I (ed). Clinics in perinatology. Philadelphia: W.B. Saunders, pp. 77-92

Day N, Richardson GA, Goldschmidt L et al. (1994) The effect of prenatal marijuana exposure on the cognitive development of offspring at age three. Neurotoxicol Teratol 16: 169-176

Day N, Sambamoorthi U, Taylor P, et al (1991) Prenatal marihuana use and neonatal outcome. Neurotox Teratol 13:329-334.

Denson R, Nanson JL, McWatters MA (1970) Hyperkinesis and maternal smoking. Child Dev 41: 667-682

Duncan J (1986) Disorganization of behaviour after frontal lobe damage. Cog Neuropsychol 3: 271-290, 1986

Dunn LM, Dunn LM (1981) Peabody Picture Vocabulary Test-Revised. American Guidance Service, Circle Pines, MN

Elsohly MA, Elsohly HN (1989) Marijuana: Analysis and detection of use through urinalysis. In: Redda K, Walker C, Barnett G (eds) Cocaine, marihuna and designer drugs: chemistry, pharmacology, and behaviour. Boca Raton, FL: CRC Press Inc., p.145

Frank DA, Zuckerman BS, Amaro H et al. (1988) Cocaine use during pregnancy: Prevalence and correlates. Pediatrics 82: 888-895

Fried PA Marihuana use by pregnant women: Neurobehavioural effects in neonates. Drug Alcohol Depend. 6: 415-424

Fried PA (1984) Prenatal and postnatal consequences of marihuana use during pregnancy. In Yanai J (ed) Neurobehavioural teratology. New York: Elsevier, pp.275-285.

Fried PA (1993) Prenatal exposure to tobacco and marijuana: Effects during pregnancy, infancy and early childhood. Clin Obstet Gynecol 36: 319-337

Fried PA, Barnes MV, Drake ER (1985) Soft drug use after pregnancy compared to use before and during pregnancy. Amer J Obstet Gynecol 151: 787-792

Fried PA, Innes KE, Barnes MV (1984) Soft drug use prior to and during pregnancy: A comparison of samples over a four-year period. Drug Alcohol Depend 13: 161-176

Fried PA, O'Connell CM, Watkinson B (1992) 60- and 72-month follow-up of children prenatally exposed to marijuana, cigarettes, and alcohol: cognitive and language assessment. J Dev Behav Pediatr.13: 383-391

Fried P A, Makin J E (1987) Neonatal behavioural correlates of prenatal exposure to marihuana, cigarettes, and alcohol in a low risk population. Neurotoxicol Teratol 9:1-7

Fried P A, Watkinson B 1(1988) 12- and 24-month neurobehavioural follow-up of children prenatally exposed to marihuana, cigarettes and alcohol. Neurotoxicol Teratol 10:305-313

Fried P A, Watkinson B (1990) 36- and 48-month neurobehavioural follow-up of children prenatally exposed to marihuana, cigarettes and alcohol. Dev Behav Pediatr 11:49-58

Fried P A, Watkinson B, Dillon R F, Dulberg C S (1987) Neonatal neurological status in a low-risk population after prenatal exposure to cigarettes, marijuana and alcohol. Dev Behav Pediatr 8:318-326

Fried P A, Watkinson B, Grant A, Knights R M (1980) Changing patterns of soft drug use prior to and during pregnancy: a prospective study. Drug Alcohol Depend 6:323-343

Fried PA, Watkinson B, Gray R (1992) A follow-up study of attentional behaviour in 6-year-old children exposed prenatally to marihuana, cigarettes, and alcohol. Neurotoxicol Teratol14: 299-311

Goodine LA, Fried PA (1984) Infant feeding practices: pre- and postnatal factors affecting choice of method and duration of breastfeeding. Can J Public Hlth 75: 439-444.

Gordon M, McClure D F (1984) Gordon diagnostic system: interpretive supplement. Goldon, Co, Clinical Diagnostics

Gusella JL, Fried PA (1984) Effects of maternal social drinking and smoking on offspring at 13 months. Neurobehav Toxicol Teratol 6:13-17

Hatch EE, Bracken MR (1986) The association of marijuana use in pregnancy on fetal growth. Am J Epidemiol 124: 986-993

Health and Welfare Canada's Health Promotion Survey: Technical Report (1988) Statistics Canada, Ottawa.

Herkenham M, Lynn AB, Johnson MR, Melvin LS, De Costa BR, K.C. Rice (1991) Characterization and localization of cannabinoid receptors in rat brain: A quantitative *in vitro* autoradiographic study. J Neuroscience 11: 563-583

Herkenham M, Lynn AB, Little MD et al. (1990) Cannabinoid receptor localization in brain. Proc Natl Acad Sci USA 87: 1932-1936

Jacobson SW, Fein GG, Jacobson JL, Schwartz PM, Dowler JK (1984) Neonatal correlates of prenatal exposure to smoking, caffeine, and alcohol. Infant Behav Dev 7:253-265

Keith RW (1986) SCAN-A screening test for auditory processing disorders. San Antonio, TX: The Psychological Corporation, Harcourt Brace Jovanovitch, Inc.

Kilbey M.M, Asghar, K. (eds)(1992) Methodological Issues in Epidemiological, Prevention, and Treatment Research on Drug-Exposed Women and Their Children. National Institute on Drug Abuse Research Monograph 117. DHHS Pub. No. ADM 92-1881. Washington, DC: Supt. of Docs., U.S. Govt. Print. Off.

Kristjansson EA, Fried PA, Watkinson B (1989) Maternal smoking during pregnancy affects children's vigilance performance. Drug Alcohol Depen 24:11-19

Luria A R (1966)Higher Cortical Functions in Man. New York: Basic

MacGregor SN, Sciarra JC, Keith L, Sciarra JJ (1990) Prevalence of marijuana use during pregnancy: a pilot study. J Reprod Med 23: 1147-1149

Mackey S, Fried PA (1981) Infant breast and bottle feeding practices: some related factors and attitudes. Can J Public Hlth 72:312-318

Matsuda LA, Lolait SJ, Brownstein MJ, Young A.C, Bonner TI. (1990) Structure of a cannabinoid receptor: functional expression of the cloned cDNA. Nature 346: 561-563

McCarthy P (1972) McCarthy Scales of Children's Abilities. New York: The Psychological Corporation

McCartney JS, Fried PA, Watkinson B (1994) Central auditory processing in school-age children prenatally exposed to cigarette smoke. Neurotoxicol Teratol 16:269-276

McClure FD, Gordon M (1983) Performance of disturbed hyperactive and non-hyperactive children on an objective measure of hyperactivity. J Abnorm Child Psychol 168: 26-33

National Institute on Drug Abuse (1991) National household survey on drug abuse: Main findings 1990. Rockvill,MD: Department of Health and Human Services; DHSS publication no. 91-1788.

Nichols PL, Chen TC (1981) Minimal Brain Dysfunction: A Prospective Study. Lawrence Erlbaum Associates, Hillsdale, NJ.

O'Connell CM, Fried PA (1991) Prenatal exposure to cannabis: A preliminary report of postnatal consequences in school-age children. Neurotoxicol Teratol 13: 631-639

Peacock JL, Bland JM, Anderson HR, Brooke OG (1991) Cigarette smoking and birth weight – type of cigarette smoked and a possible threshold effect. International Journal of Epidemiology 20: 405-412.

Pickles JO (1988) An Introduction to the Physiology of Hearning (2nd ed.) San Diego, CA: Academic Press

Picone TA, Allen LH, Olsen PN, Ferris ME (1982) Pregnancy outcome in North American women. II Effects of diet, cigarettes smoking, stress and weight gain on placentas, and on neonatal physical and behavioural characteristics. Amer J Clin Nutr 36: 1214-1224

Reynell J (1977) Reynell Developmental Language Scales- Revised. Windsor, England: NFER Publishing

Richardon GA, Day NL, Taylor PM (1989) The effect of prenatal alcohol, marijuana and tobacco exposure on neonatal behaviour. Inf Behav Dev 12: 199-209

Rush D, Calahan KR (1989) Exposure to passive cigarette smoking and child development. A critical review. In: Hutchings DE (ed) Prenatal abuse of licit and illicit drugs Ann N Y Acad Sci 562:74-100

Sameroff A, Chandler M (1975) Reproductive risk and the continuum of caretaking causality. In: Horwitz, F.; Hetherington, E.; Scarr-Salaptek, S; and Siegal, G., eds. Review of Child Development Research Vol. 4. Chicago; University of Chicago Press, pp. 187-243.

Saxton D (1978) The behaviour of infants whose mothers smoke in pregnancy. Early Human Devel 2/4:363-369.

Scher MS, Richardson GA, Coble PA, Day NL, Stoffer D (1988) The effects of prenatal alcohol and marijuana exposure: disturbances in neonatal sleep cycling and arousal. Pediatr Res 24: 101-105

Seckerwalker RH, Solomon LJ, Flynn BS et al (1992) Training obstetric and family practice residents to give smoking cessation advice during prenatal care. Amer J Obstet Gynecol 166:1356-1363

Solowij N, Michie PT, Fox AM (1993) Differential impairments of selective attention due to frequency and duration of cannabis use. Paper presented at the International Cannbis Research Society, Toronto, Ontario, Canada

Stewart PJ, Dunkley GC (1985) Smoking and health care patterns among pregnant women. Can Med Ass J 133: 989-994

Streissguth AP, Barr HM, Martin DC, Herman CS (1980) Effects of maternal alcohol, nicotine and caffeine use during pregnancy on infant mental and motor development at eight months. Alcohol: Clin Exp Res 4:152-164

Streissguth AP, Barr HM, Sampson PD, Darby BL, Martin DC (1989) IQ at age 4 in relation to maternal alcohol use and smoking during pregnancy. Dev Psychol 25:3-11

Streissguth AP, Darby BL, Barr HM, Smith JR, Martin DC (1983) Comparison of drinking and smoking patterns during pregnancy over a six year interval. Amer J Obstet Gynecol 145:716-724.

Streissguth AP, Martin DC, Barr HM, Sandman BM, Kirchner GL, Darby BL (1984) Intrauterine alcohol and nicotine exposure: attention and reaction time in 4-year-old children. Dev Psychol 20: 533-541

Struve FA, Straumanis JJ, Patrick G, Price L (1989) Topographic mapping of quantitative EEG variables in chronic heavy marihuana users: Empirical findings with psychiatric patients. Clin EEG 20: 6-23

Struve FA, Straumanis JJ, Patrick G et al. (1993) Topographic quantitative EEG sequelae of chronic cumulative THC exposure: Recent and continuing studies. Paper presented at International Cannabis Research Society, Toronto, Canada

Tennes K, Avitable N, Blackard C et al. (1985) Marijuana: prenatal and postnatal exposure in the human. In: Pinkert TM (ed) Current Research on the Consequences of Maternal Drug Abuse. National Institute on Drug Abuse Research Monograph 59. DHHS Pub. No. (ADM) 85-1400. Washington, DC: Supt. of Docs., U.S. Govt. Print. Off., pp. 48-60.
Tunving K, Thulin SO, Risberg J, Warkentin S (1986) Regional cerebral blood flow in long-term heavy cannabis use. Psychiat Res 17: 15-26
Welsh MC, Pennington BF (1988) Assessing frontal lobe functioning in children: Views from developmental psychology. Develop Neuropsychol 4: 199-230
Williamson DF, Serdula MK, Kendrick JS, Binkin NJ (1989) Comparing the prevalence of smoking in pregnant and nonpregnant women, 1985 to 1986. JAMA 261;70-74.
Zuckerman B, Frank D, Hingson R, et al (1989) Effects of maternal marijuana and cocaine use on fetal growth. New Engl J Med 320:762-768

DELAYED DEVELOPMENTAL NEURO- AND IMMUNOTOXICITY OF BENZODIAZEPINES

M. Schlumpf, R. Parmar, E.E. Bütikofer, S. Inderbitzin, A.R. Salili, A.A. Schreiber, H.R. Ramseier[1], *H. van Loveren*[2] *and W. Lichtensteiger*
Institute of Pharmacology and Institute for Immunology and Virology[1], University of Zürich, Zürich, Switzerland and Institute of Public Health and Environmental Protection[2], Bilthoven, Holland

Benzodiazepines are the most frequently prescribed class of psychotropic drugs in the United States, in Europe and estimated worldwide (Bergman and Dahl-Puustinen, 1989, Ruiz et al., 1989, 1993, Miller and Mahler, 1991). Benzodiazepines have also been known for their low toxicity since their introduction into medical practice. However, the developing organism differs in many respect from the adult and drug actions during early life might differ from actions on the adult organism in a qualitative and quantitative way. In addition, developing organs or systems that are not a designed target for this class of drugs might unexpectedly be sensitive to the chemical during early developmental periods. It therefore seemed to be appropriate to reconsider the potential action of benzodiazepines during ontogeny in the sense of increased knowledge of developmental processes and with refined methodology.

Although non medical use of benzodiazepines is more often seen in a young population, potential drug problems connected to pregnancy and childbirth are rarely mentioned in discussions on the matter. Benzodiazepines in pregnancy are discussed in the context of obstetric use solely. Recently, a benzodiazepine embryofetopathy (Laegreid et al., 1989) has been reported in infants of mothers with high benzodiazepine intake during pregnancy. The syndrome displays several grades of severity from dysmorphic signs to behavioral dysfunctions, delayed mental and motor development and perceptual disorders. The human neonate, in addition, suffers from the "floppy infant syndrome" with thermoregulatory difficulties, low Apgar scores, apneic spells, poor sucking and cardiorespiratory difficulties (Bavoux et al., 1982, Speight, 1977, 1978). Withdrawal symptoms including high irritability, tremors, hyperactivity and hypertonicity threaten the newborn barely recovered from the perinatal"floppy" symptomatology (Rementeria et al., 1977, Marquez and Alonso, 1993).

The perinatal effects occur under direct influence of the drug; they are distinct from *delayed developmental* effects. Delayed toxicity is disclosed at any time in the postnatal period including adult life; it occurs in complete absence of the drug and is thought to be derived from a primary as yet undefined defect in early life based on drug contact.

This chapter will review delayed developmental toxic effects of benzodiazepines on the nervous and immune system of the rat. It is known for some time, that the nervous system of animals exposed to these drugs during development is functionally impaired (Kellogg et al., 1980, 1983, Pankaj and Brain, 1991), and drug sensitivity is altered (Schlumpf et al., 1989). Possible neurochemical correlates of behavioral dysfunction have been analyzed. Data on changes in number or affinity of benzodiazepine binding sites are inconclusive (see below). Disturbances in the functional qualities of the $GABA_A$/benzodiazepine receptor complex appear to represent a better correlate to neurobehavioral toxicity. In addition, other neurotransmitter systems may be involved as well.

More recently, the fetal immune system emerges as a most sensitive target for developmental actions of benzodiazepines (Schlumpf et al., 1989a, 1990, 1993, 1993a). Immunotoxicity has becone a recognized side - effect of drugs and environmental chemicals (Descotes, 1988). Delayed immune dysfunction in offspring has been observed after perinatal exposure to alcohol (Norman et al., 1989, 1991) and to environmental chemicals like the 2,3,7,8- tetrachloro-dibenzo-p-dioxine (Fine et al., 1990). Our group reported on severe and prolonged suppression of immune responses in prenatally diazepam exposed rats. Pregnant rats are treated with 1.25 mg/kg body weight diazepam dissolved in olive oil or with the vehicle from gestational day (GD) 14 to GD 20 (Schlumpf et al., 1989, 1992, 1993, 1993a, Schreiber, 1993, Schreiber et al., 1993, Schreiber et al., 1993a). Drug concentrations reached in adult and fetal tissues under this repeated drug treatment are in the nanomolar range and comparable to levels reached in humans after oral or parenteral dose. On GD 23, the day of birth, no drug was detectable in mother or pups (Schlumpf et al, 1992). This review summarizes our data on the delayed developmental neuro- and immunotoxic effects of diazepam in Long Evans rats.

DEVELOPMENTAL NEUROTOXICITY

Neurobehavioral deficits

Nest odor behavior

Sucking difficulties is one of the main symptoms in the "floppy-infant syndrome", that occurs in neonates, whose mothers had been exposed to benzodiazepines during pregnancy (Bavoux, et al., 1982, Gillberg, 1977, Cree et al., 1973, Speight, 1977, 1978, Guillois et al., 1992). Olfactory stimuli are important for the attachment of the rat pup to the nipple. They also help the young pup during the early phase of postnatal development, when his eyes are still closed, to orient towards the nest (Bauer, 1983, Blass et al., 1979).

The development of olfactory guided behavior can be tested by giving the rat pup a choice between two compartments, one containing bedding of its own nest and the other one containing clean bedding. The animal will preferentially turn towards the familiar olfactory cue at a very early age. In the test, the pup is placed in the middle of a grid over a plexiglas box divided into two compartments with either fresh or home bedding, and the percentage of time spent over the familiar cue within a given time period (3 min.) is measured. Zivic Miller Sprague Dawley rat pups whose mothers had been exposed to diazepam (2.5mg(kg s.c. from gestational day 14 to 20) (fig 1.) were tested on different postnatal days (PN). The performance of prenatally drug exposed rats became comparable to controls only at PN 13, the time of the eye-opening, when the behavioral repertoire of the animal has changed markedly (Schlumpf et al., 1989). The behavioral delay of about one week characterizes the animal as less competent in recognizing olfactory cues. Nest odor behavior requires a high degree of sensorimotor integration. The behavioral deficit is not due to persistent drug or metabolites in the brain of the animal and hence reflects a toxic effect on developmental processes. High densities of benzodiazepine bindindg sites are present in the fetal olfactory bulb (Fig. 3) (Schlumpf et al., 1983). Furthermore, mRNA's of $\alpha 1/\beta 2/\gamma 2$ $GABA_A$ receptor subunits have been identified in the olfactory bulb at a time in development that coincides with the treatment period.

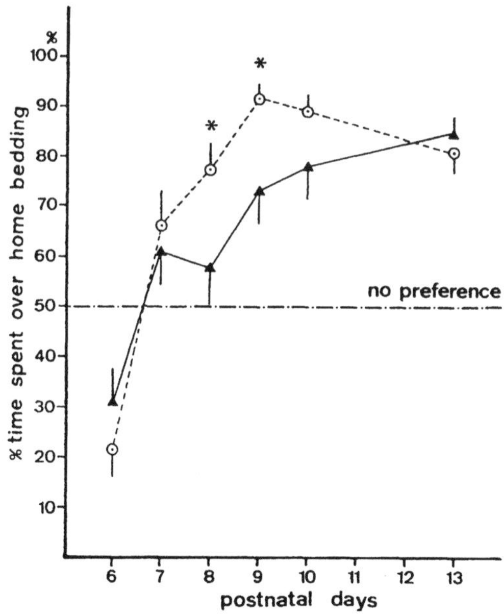

Fig. 1. Nest odor preference. The percentage of a 3 minutes observation time spent by the rat pups over home bedding vs. clean bedding at different postnatal days. Solid triangles: offspring of dam treated with diazepam; open circles: control offspring. Performance at postnatal days 8 and 9 is significantly better in controls ($p < 0.01$), Schlumpf et al., 1989).

Defects in social behavior, learning impairment

Deficits in social behavior like the highly disturbed pattern of infant ultrasonic calling was observed in prenatally diazepam exposed rat pups removed from the nest (Cagiano et al., 1990). The defective pattern of infant rat ultrasonic calling

could, as suggested by the authors, play an important role in the production of enduring behavioral abnormalities in adult rats exposed prenatally to benzodiazepines. Early developmental milestones, like the righting reflex, eye opening, appearance of hair and others were found to be retarded in Swiss mice progeny exposed to high doses of chlordiazepoxide (20mg/kg) or midazolam (2.5 or 5mg/kg) (Pankaj and Brain, 1991).
In more complex learning situations like six choice simultaneous discrimination task or extinction of active avoidance behavior, prenatally diazepam exposed animals needed more time and produced more errors in a dose dependent way. However, learning simple tasks, e.g. acquisition of successive discrimination tasks or passive avoidance task remained unaffected (Gai and Grimm, 1982, Frieder et al., 1984, Lauer et al., 1987).
Lower doses of diazepam, i.e., 1.0 or 2.5 mg/kg, given to the pregnant dam during the last gestational third, also led to altered behavior in adult male progeny in anxiety tests (Kellogg,et al., 1991). Most of these studies identify the last gestational third as the most sensitive period for the development of delayed learning disabilities in the *rat*. The vulnerable period coincides with the early development of benzodiazepine binding sites (Schlumpf et al., 1983).

Drug sensitivity

Diazepam

In principle, prenatal exposure to a drug might alter drug sensitivity to the same class of drugs and to other drugs later in life. Some observations indicate that this may also be the case with benzodiazepines. Taylor and coworkers (Taylor et al. 1983) previously reported on delayed effects of ethanol on responses to diazepam. Their studies show that *in utero* exposure to ethanol produces offspring which are hyper-responsive to hypothermic effects of two psychotropic drugs: diazepam and pentobarbital (Taylor et al., 1983).

We tested the hypothermic response to a single dose of diazepam (5mg/kg, i.p.) and to pentobarbital (40mg/kg, i.p.) in adult (2 to 3 months old) male and female offspring of diazepam treated dams (Schlumpf et al., 1989). Barbiturates interact with the $GABA_A$ receptor complex via an allosteric binding site, in analogy to benzodiazepines (Richards et al., 1991). Females were tested on the day of estrus as checked by vaginal smears. After drug injection, rectal temperature was recorded every 15 minutes. The hypothermic reaction to diazepam was significantly enhanced in the treatment group (Fig. 2), while after pentobarbital injection, only female offspring showed a (non-significant) tendency towards lower temperature values. *In utero* exposure to diazepam appears to produce offspring which will be more sensitive to the drug as adults (Schlumpf et al., 1989).

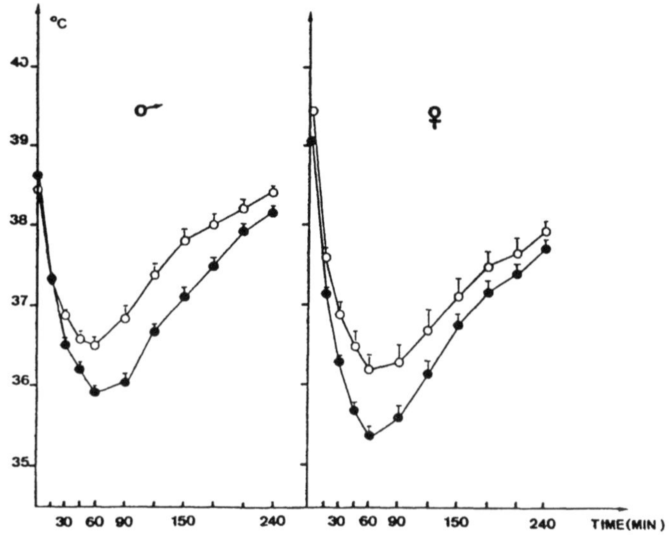

Fig. 2. Hypothermia induced by diazepam (5 mg/kg s.c.) in adult male and female offspring of rats treated with diazepam (solid symbols) or vehicle (open symbols) from GD 14 to GD 20. In males, significant differences are seen at all time-points after diazepam injection (ANOVA, $p < 0.01$), except for the first two (15 and 30 min) and the last two time points (210 and 240 min). In females, the differences are significant at 15, 30, 45, 60 and 90 min ($p < 0.01$) and at 120, 150, 180 min ($p < 0.05$) (Schlumpf et al., 1989).

Neurochemical correlates to behavioral disturbances

$GABA_A$ / benzodiazepine receptor

The benzodiazepine-induced modulation of brain function has a cellular and molecular correlate i.e., the benzodiazepine receptor. The identification of glycoproteins in neuronal cell membranes as specific high affinity binding sites for benzodiazepines, using tritiated diazepam as a ligand, led to the notion that these sites were allosteric control elements of the $GABA_A$ receptor (Squires and Braestrup, 1977, Möhler and Okada, 1977). Through this receptor γ-aminobutyric acid (GABA), the major inhibitory neurotransmitter of the central nervous system, produces most of its effects on neuronal excitability (Stephenson, 1987). The structural and functional coupling of benzodiazepine receptors to the $GABA_A$ receptor, an integral part of a chloride channel, explains the modulatory effects of benzodiazepines on GABA-gated chloride influx leading to an increase in negative charge and hyperpolarization. By potentiating GABA action, benzodiazepines can reduce the level of anxiety, vigilance, muscle tension and convulsions (Haefely, 1990, Möhler et al., 1990). The $GABA_A$ receptor is the site for modulatory action by benzodiazepines, barbiturates, anesthetics, steroids and picrotoxin (Richards et al., 1991).

For GABAergic signalling the mammalian brain recruits various settings of modulatory properties. Molecular cloning of the $GABA_A$ receptor subunits revealed 16 different genes that are divided into distinct classes: α, β, γ and ro (Haefely, 1994) and regulated independently. Immunoprecipitation experiments of the major $GABA_A$ receptor subtypes in situ demonstrated a high abundance of receptors containing the α1, the β2/β3 subunits and the γ2 subunit (Benke, 1991). Changes in the regulation of $GABA_A$ subunit mRNA expression had been discussed as one possible mechanism underlying alterations of receptor function after both prolomged ethanol and chronic benzodiazepine administration (Montpied et al., 1991, Mhatre and Ticku, 1992, Kang and Miller, 1991).

Development of $GABA_A$ / benzodiazepine receptors in the fetal brain

With biochemical techniques, benzodiazepine binding sites were detected 8 days before birth in rat brain homogenates (Braestrup and Nielsen, 1978, Mallorga et al., 1980, Regan et al., 1980). Our autoradiographic investigation revealed marked differences between individual brain regions at different developmental stages (Schlumpf et al., 1983).

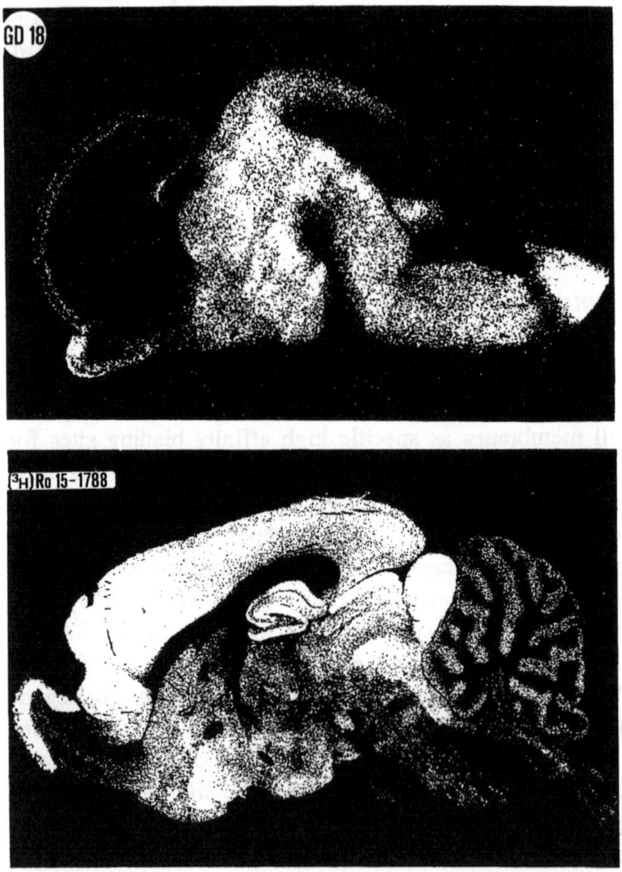

Fig.3. Sagittal section through GD 18 fetal (above) and adult rat brain. Autoradiographic localization of benzodiazepine binding sites using the agonist 3H flunitrazepam or the antagonist 3H Ro 15-1788. In the fetal brain binding sites are present in lower brain stem, di- and mesencephalon, olfactory bulb and neocortex. Note the shift of benzodiazepine binding sites from brain stem (fetus) into forebrain and neocortex (adult).

The developmental pattern of benzodiazepine receptors includes an early presence of these high affinity binding sites in fetal brainstem with a gradual switch off in these areas and a concomitant spreading into forebrain regions and neocortex (Fig. 3) A strong parallelism between the development of benzodiazepine sites, the $GABA_A$ receptor subunit mRNA's (Schlumpf et al., 1992) and the GABAergic system (Lauder et al., 1986) in several fetal brain regions suggests developmental interactions between GABAergic system, GABAergic drugs and their targets. In view of the functional state of the system during fetal life, the ratio between different subunit mRNA's in individual brain regions is of interest, because it provides some (indirect) information on the possible availability of subunits for the receptor assembly. The nature of such interactions is difficult to predict because of the continuosly changing regional localizations and dynamic changes in functional qualities of the $GABA_A$ receptor complex as examplified by the high variability in $GABA_A$ subunit mRNA expression in fetal brain.

Changes in the $GABA_A$ / benzodiazepine receptor complex

The neurochemical basis of behavioral dysfunction caused by prenatal benzodiazepine exposure has not been clarified as yet. It seems logical to suppose alterations in GABAergic systems and in the $GABA_A$ receptor complex. This idea is supported i.a. by the data indicating a change in sensitivity to benzodiazepines.

One possible basis for functional deficits might be sought in long-term changes of benzodiazepine receptor number or else, in a defective response of the functionally coupled $GABA_A$/benzodiazepine receptor/chloride channel protein.

In utero exposure to psychoactive drugs had been associated with altered binding characteristics of relevant receptor sites (Rosengarten and Friedhoff, 1979, Jason et al., 1981). Results on *in utero* exposure to benzodiazepines are not consistent. From the several studies it has become clear, that prenatal diazepam does not induce dramatic changes in ligand binding affinity (Kd) or maximal receptor density (Bmax) in offspring, inspite of the relatively high doses used in most studies (Kellogg, 1988, Lauer, et al., 1987, Rothe and Bigl, 1989, Miller et al., 1989). Exposure to 2.5mg/kg or 10 mg/kg of diazepam from GD 13 to 20 (Kellogg, 1988) induced a 25% increase in KD and a 20% decrease of Bmax in adult offspring. A decrease in both Kd and Bmax for ^{3}H-diazepam binding in 1 year old offspring of prenatally diazepam exposed rats (5.0 - 7.5 mg/kg from GD 15 to GD 20) was reported by Livezey (Livezey et al., 1985).

The flux of Cl^- ions in the $GABA_A$ receptor complex can be altered as a consequence of prenatal benzodiazepine action. Exposure to 2.5mg/kg diazepam during fetal life enhanced the sensitivity to GABA of GABA-mediated Cl^- uptake in synaptosomal preparations in male progeny (Kellogg et al., 1991). Analysis of muscimol stimulated Cl^- uptake in prenatally lorazepam exposed adult mice (2mg/kg from GD 13 -20 (Chesley et al., 1991) revealed a decrease in uptake, a fact that is interpreted as a consequence of decreased $GABA_A$ receptor function. The opposite results seen in $GABA_A$ receptor function of prenatally benzodiazepine exposed mice and rats might be drug and/or species related.

$GABA_A$ receptor subunit expression

Recent investigations on expression of mRNAs encoding for different $GABA_A$ receptor subunits revealed that chronic treatment of pregnant rats with diazepam during a comparatively short period of 7 days causes changes of α1 subunit mRNA expression in the offspring: At 8 days postnatally, a transient depression of α1 mRNA in a number of brain regions including neocortex was observed (Salili et al., 1994). In adult offspring, expression of this mRNA was enhanced in selected regions such as anterior cingulate and frontal cortex, and normalized in other areas, e.g., parietal cortex (Tab. 1). It is interesting that the postnatal reduction of α1 mRNA expression was seen in a rather wide range of regions, whereas the increased expression in adulthood appears to be limited to a small number of cortical and subcortical areas. For an interpretation of the functional significance of these changes, the entire subunit compositon of the receptor should be known. In principle, the enhancement of α1 subunit mRNA expression would fit with the observed hypersensitivity of the developmentally diazepam exposed animal to this drug in adult life (Schlumpf et al., 1989).

Table 1. Expression of α1 subunit mRNA of $GABA_A$ receptor in adult male rat cerebral cortex (CX) after prenatal exposure to diazepam (1.25 mg/kg/d, GD 14-GD 20).

PRENATAL TREATMENT	ANTERIOR CINGULATE CX % of control	FRONTAL CX % of control	PARIETAL CX % of control
VEHICLE	100 ± 4.1	100 ± 2.2	100 ± 11.2
DIAZEPAM	145.4 ± 16.3 **	146.0 ± 27.0 *	97.2 ± 11.4

In situ hybridization with S^{35} -labeled 60mer oligonucleotide. Mean ± S.D. as percentage of relative optical density of vehicle-exposed controls, n = 4 rats out of 4 different litters per group. * $p < 0.05$, ** $p < 0.02$ (Student's t test).

Functional state of opioid circuitry in basal ganglia

It seems conceivable that the interaction of benzodiazepines with developing GABAergic systems may in turn influence the ontogeny of other transmitter systems. We were particularly interested in the functional state of the circuitry of the basal ganglia, which play a central role in sensory-motor integration and reward mechanisms. In basal ganglia, dynorphin is contained in a subset of GABAergic neurons that project to the midbrain (Vincent et al., 1982, Gerfen and Young, 1988, Gerfen et al., 1990). The functional state of these neurons was assessed by analysis of prodynorphin mRNA expression. In situ hybridization for prodynorphin mRNA was performed in brain sections from adult (2 months) male offspring of time-pregnant LE rats that had been treated with diazepam (1.25 mg/kg/day) or vehicle from GD 14 to GD 20. The acute responsiveness of dynorphin-containing neurons of adult offspring was checked by injection of d-amphetamine (8 mg/kg) which typically elicits an increase in prodynorphin mRNA expression (Smith and McGinty, 1994). Brain sections of corresponding antero-posterior levels of all 4 experimental groups were arranged on the same autoradiogram; relative optical density was determined by image analysis (MCID, IMAGING RESEARCH) and expressed as percentage of control (= prenatal vehicle, adult saline). Baseline prodynorphin mRNA expression was not affected by prenatal diazepam exposure. However, the acute increase in prodynorphin mRNA expression elicited by d-amphetamine, was abolished in olfactory tubercle, ventromedial and dorsolateral caudate-putamen, and significantly reduced in dorsomedial caudate-putamen (Fig. 4. Analogous changes in nucleus accumbens did not reach statistical significance). These data point to an alteration in the functional responsiveness of striatal output systems of adult rats following prenatal diazepam exposure (S. Inderbitzin, M. Schlumpf and W. Lichtensteiger, in prep.).

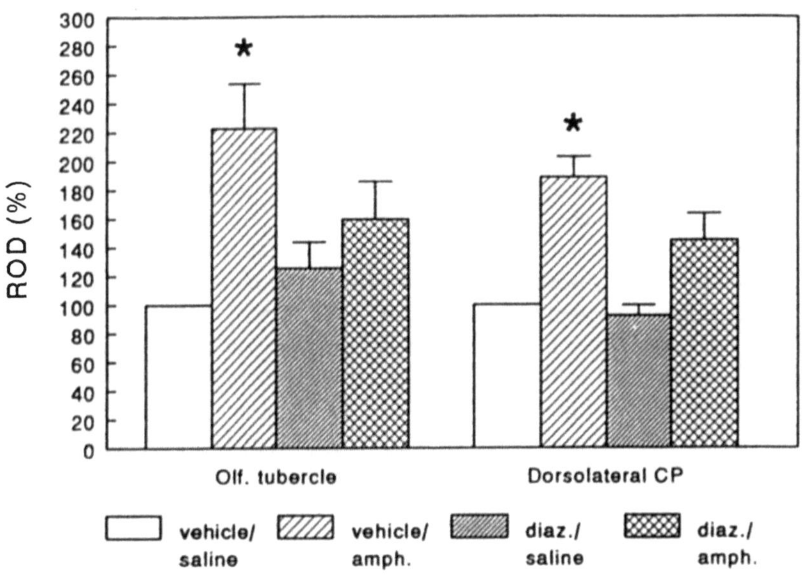

Fig. 4. Prodynorphin mRNA expression as percentage of control (vehicle) values in relative optical densities in two brain regions of adult male rats exposed prenatally to diazepam (diaz) or vehicle, studied 6 hr after injection of amphetamine (amph., 8mg/kg) or saline

DEVELOPMENTAL IMMUNOTOXICITY

Immunotoxicology, a young subdiscipline of toxicology, is defined as the study of adverse effects of chemicals on the immune system (Descotes, 1988, 1990). There is some concern today that exposure to pharmaceutical drugs and environmental chemicals might be potentially hazardous to the immune system and induce clinically relevant immune-mediated diseases. While there is increasing knowledge on toxicity to the adult immune system, the evaluation of *developmental immunotoxicity* of drugs and/or chemicals has so far been limited to a few selected substances. Work done in the field calls for attention, as developmental immunotoxic effects might differ in quality and quantity from immunotoxic effects on the adult organism.

Benzodiazepines

Delayed developmental immunotoxic effects of diazepam were first noted as unexpected side effects during investigation of neurobehavioral and neurotoxic effects of the drug in Long Evans rats exposed in our laboratory to low doses of the drug during prenatal life. In the context of developmental immunotoxicity, benzodiazepines represent a special case insofar as specific high affinity peripheral type benzodiazepine receptors (PBR) are present on immune cells.

We chose a treatment schedule where time-pregnant Long Evans rats were subcutaneously injected with a comparatively low chronic dose of diazepam, i.e., 1.25 mg /kg diazepam in olive oil from gestational day 14 to 20. Plasma and tissue values were higher in pregnant vs. non pregnant female rats (Schlumpf et al., 1990, 1992). Fetal diazepam concentrations were close to maternal in brain and liver. Concentrations reached after seven daily 1.25 mg/kg doses are in the high nanomolar range for the dam (plasma 100 nM, brain 314 nM) and the fetus (fetal brain 144 nM). This concentration range is met after dosages of 10 to 40 mg diazepam in humans (Greenblatt and Shader, 1974, Costa, 1983). With the dosage used in our animal model, we found no indication for an accumulation of the drug or its desmethylated metabolite. Drug levels declined rapidely during the first 24 hours after the last injection and at the time of birth the drug and its major metabolites became undetectable (Schlumpf et al., 1992).This fact excludes any direct effect of the drug or of its metabolites during postnatal development.

Effects on cellular and humoral immune responses

The proliferative response of lymphocytes to allogeneic and polyclonal (mitogenic) stimulation was greatly depressed in offspring of diazepam treated rats during the entire postnatal development and was normalized only in adulthood (Schlumpf et al., 1989a). The proliferative response was measured in mixed lymphocyte cultures (Ramseier, 1987): Responder spleen cells (Long Evans rats, major histocompatibility complex (MHC Rt-2) were stimulated with irradiated allogeneic DA rat (MHC Rt-4) spleen cells. Polyclonal stimulation was done with concanavalin A (ConA). The failure to respond to allogeneic or polyclonal stimulation was demonstrated as early as PN (postnatal day) day 6. Proliferation rates remained reduced by 50% and more until about 2 months of life, i.e., into young adulthood. Thereafter, they gradually returned to control values at around PN 90.

The capacity for antibody formation was tested in the plaque forming assay (antibody formation against sheep red blood cells (SRCB)). Antibody formation following SRCB stimulation was significantly reduced in prenatally diazepam-exposed animals at four weeks of age. Animals were immunized by injecting SRCB intraperitoneally and plaque-forming cells were determined according to the Jerne-Nordin Technique (Jerne and Nordin, 1963) using splenic lymphocytes taken 4 days after SRCB injection. The reduction by about 50% of B-lymphocyte capacity to produce antibodies against SRCB is present in the young animal at weaning age and coincides with the maximum inhibition of T-lymphocyte proliferative responses (Tabl. 2, Bütikofer et al., 1993).

Table 2. Effect of prenatal diazepam exposure on antibody formation against sheep red blood cells by splenocytes of 4 week-old rat offspring

Prenatal treatment	Males	Females
Vehicle	22.4 ± 9.1 (9)	33.8 ± 9.7 (9)
Diazepam *	10.8 ± 5.8 (8)	12.9 ± 5.6 (9)

Number of plaque forming cells (PFC) x 1000 per spleen, 4 days after SRCB injection (mean ± S.E.M. and number of animals). * different from vehicle in factorial ANOVA for $p < 0.05$ (Bütikofer et al., 1993).

One possible explanation for impaired immune responses during the postnatal period might be sought in a general delay of maturation. However, an analysis of blood cells (leucocytes, erythrocytes, thrombocytes) and of surface markers in splenocytes (Schreiber et al., 1993) revealed no difference between treated and control animals. These findings pointed to a more specific type of dysfunction and prompted us to analyze different cytokines. Several prominent cytokines of the cytokine cascade initiated by allogeneic or mitogenic immune stimulation, were selected for analysis in prenatally diazepam-exposed rats.

The deficient cytokine cascade

Successful immune responses are the result of an optimal interplay between cells and soluble mediators or factors, the cytokines. Cytokines are produced by many different cells, but the main sources are immune cells. These soluble factors participate in the development and maturation of the various immune cells, but they also play a decisive part in triggering, regulating and terminating the immune response.

TNF-α. At a very early stage of inflammatory or immune reaction TNF-α is released predominantly by stimulated macrophages (Beutler et al., 1985) and then controls the production of other cytokines like, e.g., the inflammatory cytokines (interleukine-1) IL-1 and (interleukine-6) IL-6. Unstimulated macrophages express mRNA for TNF-α, but synthesis of the protein dose not occur until triggered, e.g., by components of bacteria (lipopolysaccharid (LPS)) leading to transcription of TNF-α and to liberation of the protein (Old, 1985, Beutler et al., 1985, Beutler and Cerami, 1988).

Determination of most rat cytokines is based on bioassays; only recently, antibodies against some rat cytokines are becoming available. TNF-α liberation by LPS -stimulated mixed splenocytes in prentally diazepam exposed rats was determined by measuring the lytic capacity of the incubation medium for lysis of LM cells (Schreiber et al., 1993, Schlumpf et al., 1992) At the age of two weeks TNF-α was found to be greatly suppressed. This effect was still present at 4 weeks of age. In contrast, in the young adult animal at 8 weeks, an overshoot of TNF-α release was observed (Fig. 6). Interestingly, this reduction in TNF-α release by mixed splenocytes was not seen with preparations of isolated spleen macrophages.

Interleukin-1 (IL-1). This cytokine again is predominantly produced by monocytes and macrophages, but various other cells such as fibroblasts, endothelial cells and keratinocytes also release IL-1 (Dinarello, 1991). IL-1 also enfoldes a broad spectrum of activities. It regulates cell development in bone marrow and thymus and induces fibroblasts and endothelial cells to release growth factors and IL-6. During an immune response, IL-1 induces mature T-cells to divide and to produce a diversity of cytokines like IL-2, IL-4, IL-5 and IL-6. Together with TNF-α, IL-1 also regulates antibody production by B-cells (Di Giovine and Duff, 1990, Balkwil and Burke, 1989).

IL-1 was analyzed in supernatants of LPS activated spleen cells with a bioassay measuring the proliferative activity for D10 G4 cells. Production of this cytokine was significantly reduced in male offspring of diazepam treated dams throughout the whole observation period from PN 6 to PN 83. In female offspring, the reduction was only statistically significant in the young adult animal. In contrast to TNF-α, release of IL-1 remained impaired at the young adult stage (Fig. 5, Schlumpf et al., 1993a).

Interleukin-2 (IL-2). IL-2 is well recognized for its central position in the evolvement of the cytokine cascade. This product of stimulated T-cells is autostimulatory by stimulating T cell activation. IL-2 further enhances T lymphocyte action on macrophages by increasing the production of interferon-γ (INFγ). IL-2 in addition promotes growth of natural killer (NK) cells and B cells and increases production of immunoglobulins (Haas et al., 1990, Balkwil and Burke, 1989).

Prenatal diazepam treatment also affected IL-2 liberation (Schlumpf et al.,

1993a), as measured by the proliferative action of splenocyte supernatants on the IL-2 dependent CTLL-2 cell line (Ramseier et al., 1993, 1993a). IL-2 was significantly lower in prenatally diazepam exposed male and female rats during early postnatal (PN6 - 28) and peripubertal periods (PN 34 - 61) (Fig 5). Values normalized in adulthood.

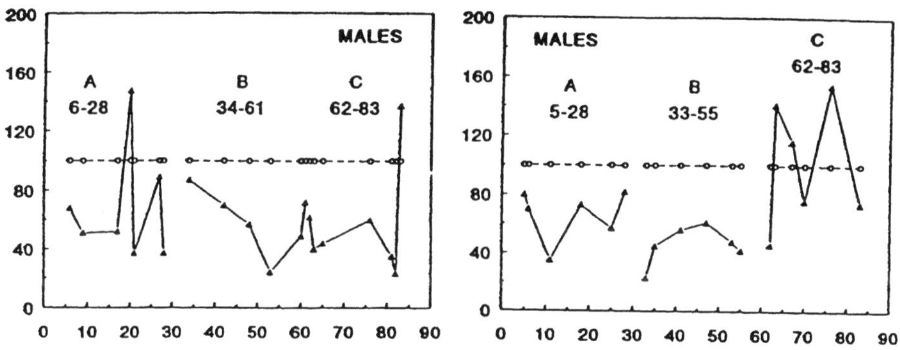

Fig. 5. Left: LPS-stimulated release of interleukin-1 (IL-1), right: Con A-stimulated release of IL-2 by splenocytes from male offspring of LE rats exposed to diazepam or to vehicle (values of diazepam exposed offspring are given in % of control). Analysis of developmental periods for IL-1: preweaning period (A, PN6 to PN 28): no significant difference (n.s.). Postweaning period: (B, PN 34 - 61): males $p < 0.05$, females n.s. (not shown). Adult period (C, PN 62 - 83): males as well as females: $p < 0.05$. Analysis of developmental periods for IL-2: Preweaning period (A): males (PN 5 - 28) $p < 0.05$, females (PN 3 - 27) $p < 0.01$ (not shown). Postweaning period (B): males (PN 35 - 55) $p < 0.05$, females (PN 35 - 54) n.s. (not shown). Adult period (C):males (PN 62 - 83) as well as females (PN 60 - 84) n.s. For female values see Schlumpf et al., 1993a.

Interleukin-6 (IL-6). IL-6 is a later stage cytokine in the immune response. IL-6 is a multifunctional cytokine important for B cell stimulation and was known for a long time under the name "B-cell stimulatory factor 2". As we found a greatly diminished antibody production in the progeny of diazepam treated rats, it appeared important to investigate the secretory pattern of this cytokine in our animals. IL-6 is produced by activated lymphoid and non-lymphoid cells, i.e., T cells and monocytes/ macrophages, but also by fibroblasts and endothelial cells. Monocytes and T cells differ with regard to mRNA expression in response to stimuli: IL-6 mRNA expression in monocytes is present 5 hours after stimulation, T cells express maximum mRNA only after 24 to 48 hours (Wong and Clark, 1988).

Release of IL-6 by mixed splenocytes, determined in supernatants of LPS-stimulated spleen cells by measuring the proliferative action on 71D1 cells (Schreiber et al., 1993a), was likewise severely reduced in drug exposed animals. As in the case of IL-1 the reduction persisted into adulthood (Fig. 6). An analogous reduction of LPS-stimulated IL-6 release was observed in a purified macrophage preparation from prenatally drug-exposed rats. In contrast, spleen lymphocytes exhibited a more complex picture, with a slight but significant inhibition at 12 hr of incubation and an enhancement at 24 hr of incubation.

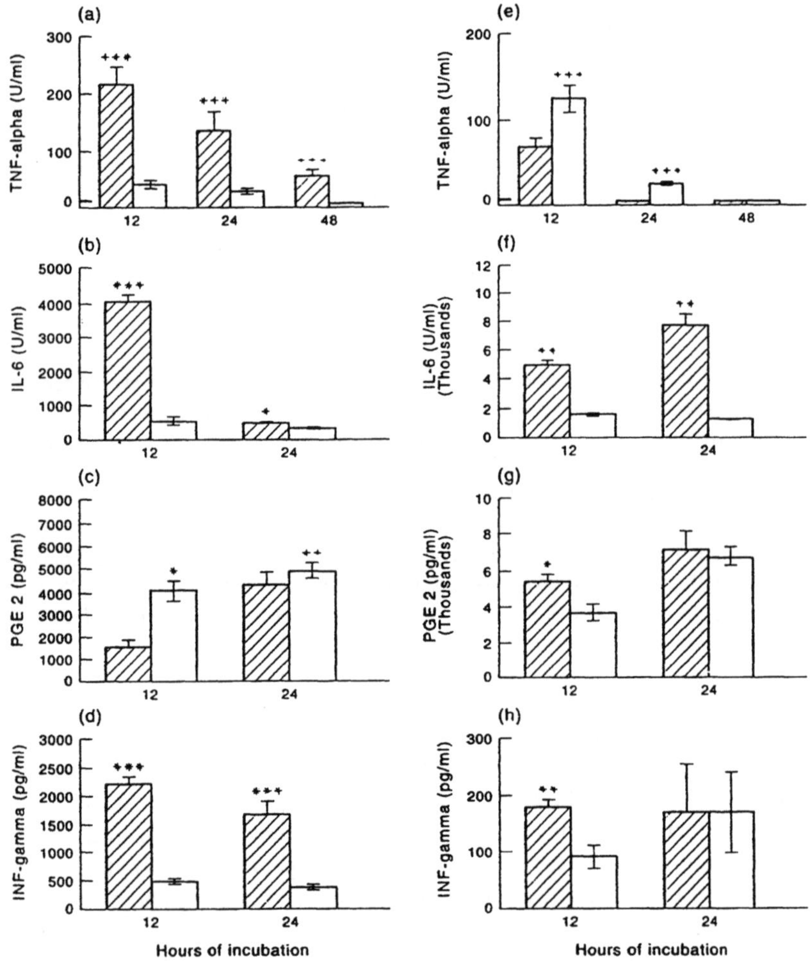

Fig. 6. Effect of diazepam treatment of pregnant rats on LPS-stimulated release of TNF-α, IL-6, PGE2 and IFNγ in male offspring (0.1 ug/ml LPS, 12, 24 and (TNF-α) 48 hours of incubation) (open bars); values from offspring of vehicle treated dams (striped bars). Left, offspring at 2 weeks of age, right, offspring at 8 weeks. Significant differences vs. prenatally vehicle-exposed controls: * $p< 0.05$, ** $p< 0.005$, *** $p< 0.0005$ (Schreiber et al., 1993, 1993a, Schlumpf et al., 1994).

Interferon-γ (IFNγ) and Prostaglandin E2 (PGE2). It becomes evident from this set of data that the whole cytokine cascade is affected by prenatal diazepam exposure, and it is not easy to locate the primary defect. The changes in TNF-α, IL-1 and IL-6 point to a primary defect in macrophages. However, the absence of a decrease of TNF-α in the purified spleen macrophage preparation remained intriguing and suggested that the regulatory interactions at the level of macrophages and between different cell types might play a role. This prompted an analysis of interferon-γ (IFNγ) and prostaglandin E2 (PGE2).

Interferons are produced by T cells, NK cells and a variety of other cells. INFγ induces TNF-α secretion in macrophages and hence reinforces the immune response (Vidovic et al., 1990). PGE2, a major product of arachidonic acid metabolism in macrophages (Frey et al., 1986) can be induced by a variety of stimuli including LPS (Spatafora, 1991), IL-1 (Dinarello, 1987) and TNF-α (Spatafora, 1991). PGE2 plays an important role in the regulation of TNF-α production. When it is released under the influence of TNF-α, it inhibits LPS-induced TNF-α production (Endres et al., 1991).

IFNγ and PGE2 were determined in supernatants of LPS stimulated spleen cells from 2 and 8 week old male offspring using the rat INFγ ELISA test kit (Holland Biotechnology, Leiden) and the PGE2 ELISA test kit (Advanced Magnetics, Cambridge, MA, USA) (Schreiber, 1993, Schlumpf et al., 1994). IFNγ was decreased in the immature prenatally drug exposed animals at 2 weeks of age, and remained below control levels until 8 weeks of life (Fig. 6). (Schreiber 1993, Schlumpf et al., 1994 and 1994 in press). PGE2 relase by splenocytes of diazepam treated animals presented a mirror image of TNF-α release, being higher than control levels at 2 weeks and reduced at 8 weeks (Fig.6 Schreiber, 1993, Schlumpf et al.,1994). This indicates that the regulatory interaction between TNF-α and PGE2 is functional in these animals.

Deficient host resistance

In order to assess the significance of the cellular dysfunctions for host defense in vivo, offspring of diazepam exposed dams were tested for their resistance to infection by *Trichinella spiralis* (Schlumpf et al., 1994, 1994a). This infection model requires the participation of immune cell populations (T-cells, B-cells and macrophages) we had identified as functionally defective (Schlumpf et al., 1992, Schlumpf et al., 1993, 1993a, Bütikofer et al., 1993, Schreiber et al.,1993 and 1993a).

Ingested Trichinella larvae enter the small intestine and mature sexually. After copulation, the adult females penetrate the intestinal mucosa, producing larvae that migrate via lymphatics and blood vessels to host striated muscles. Here, they are walled off by a host derived capsule and remain viable for extended periods. The expulsion of adult worms from the gut is mediated by an immunologically specific inflammation of the bowel. The resistance to various life cycle stages of *Trichinella spiralis* depends on T cells for the expulsion of worms during the first phase of infection, and on B- and T cells for limitation of female parasite fecundity and killing of migrating larvae (Van Loveren et al., 1994). Macrophages bind to and damage newborn larvae. Several classes of antibodies have been implicated in this process: IgG participates in rapid expulsion of ingested larvae and in systemic resistance to migrating larvae, while the role of IgA is not clear as yet (Van Loveren et al., 1994).

Male offspring of rats treated with diazepam (1.25mg/kg s.c.) from GD 14 to GD 20 were tested in young adulthood. 42 days after the infection, the diazepam

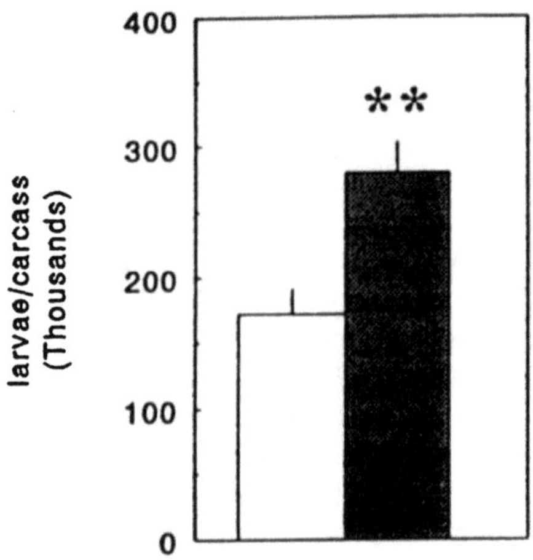

Fig. 7. Decreased host resistance. Total number of larvae per carcass in adult (3.5 month-old) male offspring of rats treated with diazepam (1.25mg/kg from GD 14 to GD 20 (filled bars) or with vehicle (open bars).Mean ± SEM, n = 8 (3 litters) for controls and prenatally drug-exposed animals. Offspring of diazepam treated dams exhibited a significant increase in number of larvae (** $p< 0.005$, Schlumpf et al., 1994).

exposed group demonstrated a significantly impaired host resistance to *T. spiralis*. The number of larvae, as indicated by analysis of digested carcasses and by morphometry, was significantly increased in progeny of diazepam treated dams versus control animals (Fig. 7). Alterations in host defense were also reflected by changes in antibody titers with a decrease in IgG and a rise in IgA (Schlumpf et al., 1994a).

Downregulation of peripheral type benzodiazepine receptor (PBR)

The peripheral benzodiazepine receptor (PBR), also referred to as mitochondrial benzodiazepine receptor (MBR) or mitochondrial DBI (diazepam binding inhibitor) receptor complex (mDRC), was first described as a diazepam binding site in rat peripheral tissues with a pharmacological profile distinct from the site associated with the central $GABA_A$ receptor (Braestrup and Squires, 1977, Regan et al.,1981 Schoemaker et al., 1981, 1983, Kozikowski et al., 1993). Although PBR is found in the central nervous system, it is more abundant in peripheral tissues (Shoemaker et al., 1983). Unlike the membrane associated central benzodiazepine receptor (CBR), the PBR is primarily localized on mitochondrial outer membranes in steroidogenic tissues (Krueger and Papadopoulos, 1993, Papadopoulos, 1993).

Schematically, the PBR can be represented with two ligand gated binding domains: the isoquinoline site (IQ) that is identified by high affinity binding of the prototype ligand PK 11195 (an isoquinoline carboxamide), and the benzodiazepine site identified by the binding of Ro5-4864.

The high affinity site for PK 11195 is diagnostic for the PBR complex. The affinities of benzodiazepines for PBR vary over several orders of magnitude in different species (Cymerman et al., 1986, Awad and Gavish, 1987, Eshleman and Murray, 1989). In rats and mice, benzodiazepines potently inhibit PK 11195 binding and the dissociation constant of Ro5-4864 is similar to that of PK 11195, i.e., in the low nanomolar range (Herz, 1993). The rank order of benzodiazepine potency for competitive displacement of ^{3}H PK 11195 is: Ro5-4864 (Ki: 5nM) > diazepam (Ki: 30 nM) > flunitrazepam and clonazepam (Ki: 2 μM). In most non-rodent species PBR exhibit low affinity for benzodiazepines. Bovine or bovine like receptors (PBR_b) have a comparatively low affinity for 1,4-benzodiazepines including Ro5-4864, diazepam and clonazepam. Human PBR is a PBR_b -like receptor (Broaddus and Bennett, 1990). The highest affinity for both types of receptors, the central $GABA_A$ and the peripheral type, is shown by the imidazopyridine alpidem (Langer et al., 1990). Species differences in benzodiazepine affinities might be due to structural differences in common binding site elements or to differences in the subunits of the PBR (Parola and Yamamura, 1993)

When possible pathogenetic mechanisms are discussed in developmental studies, the biochemical changes induced by the direct action of the drug on fetal tissues should be distinguished from biochemical and structural alterations underlying the delayed dysfunctions observed during postnatal development. In the case of the prenatal action of benzodiazepines on the developing immune system, these two types of mechanisms have not been clarified as yet.

Since immune cells, i.a., macrophages, express PBR (Schlumpf et al., 1993), it seemed possible that the developmental action of benzodiazepines was mediated by PBR. This idea is supported by the observation that prenatal treatment with the specific PBR agonist Ro 5-4864 was also followed by postnatal depression of cellular immune responses (Schlumpf et al., 1990). Similarly, it seemed conceivable that changes in PBR expression and/or in the regulation of endogenous PBR ligands might account in part for the immune dysfunctions occurring during postnatal development.

As a first approach to the question, the number and affinity of PBR on immune cells was studied in male offspring after prenatal diazepam exposure (Schlumpf et al., 1993). ^{3}H PK 11195 was used as a ligand. For saturation assays ^{3}H PK 11195 was added in increasing amounts (concentration range 0.1 - 10 nM) to membrane fractions of splenic macrophage preparations. In competition experiments, 0.6 nM ^{3}H PK 11195 was incubated with cold ligands in appropriate concentration ranges (Schlumpf et al., 1993).

Prenatal exposure to diazepam was followed at 8 weeks of age by a marked decrease in maximal binding capacity (Bmax) in spleen macrophage membranes

Table 3. Specific binding of ^3H PK 11195 to splenic macrophage membranes of prenatally diazepam-exposed 8 week-old rat offspring

	Vehicle		Prenatal diazepam	
	Kd nM	Bmax pmol/mg prot.	Kd nM	Bmax pmol/mg prot.
MALES	0.64 ± 0.25	36.4 ± 11.3 (6)	1.14 ± 0.55	15.8 ± 8.7 (6)
FEMALES	1.62 ± 0.50	33.0 ± 6.0 (5)	1.54 ± 0.79	12.6 ± 3.6 (5)*
BOTH SEXES	1.08 ± 0.29	34.9 ± 6.5 (11)	1.32 ± 0.45	14.3 ± 4.8 (11)**

Mean ± S.E.M., number of assays. Bmax different from control for * $p< 0.05$, ** $p< 0.02$. (Schlumpf et al., 1993).

from offspring of both sexes (Tab. 3). Earlier postnatal stages could not be studied because of the small size of the spleen and hence the small number of macrophages available. Membranes of a mixed splenic cell preparation exhibited an increase of the dissociation constant at 2 and 8 weeks of life (Schlumpf et al., 1993). These data point to a deregulation of the system of endogenous ligands and PBR in prenatally diazepam exposed rats.

DISCUSSION AND CONCLUSIONS

Delayed developmental neurotoxicity

Most crucial for the induction of benzodiazepine-induced short- or longterm behavioral abnormalities in rats is the timing of the treatment. Pre- and perinatal treatment produces behavioral changes that significantly differ from those seen after drug administration to the adult. More recently, the last gestational third has been identified as the most sensitive period for the development of behavioral disabilities in locomotor, social and learning behaviors, in sensory motor integration, sensory system functions and in hormone dependent functions.

The human neonate suffers from the "floppy infant syndrome". In the rat pup, righting reflex and other developmental landmarks are altered or delayed, pup locomotion is retarded and sensory motor integration is delayed and impaired as a consequence of prenatal or perinatal benzodiazepine exposure (Bavoux et al., 1982, Pankaj and Brain, 1991, Kellogg et al., 1980, 1988, Schlumpf et al., 1989). Although the prenatal drug schemes, used by different laboratories, varied considerably with respect to drug concentrations, treatment windows and drug application, there is ample agreement on the disadvantages that accrue for the animal from prenatal benzodiazepine exposure. The absence of drug or active metabolites in the neonate rat (Schlumpf et al., 1992) clearly demonstrates that the behavioral dysfunctions seen during postnatal life are due to developmental defects rather than the persistence of the drug.

The adult prenatally benzodiazepine exposed animal is no longer competitive in complex behavior and learning tasks and fails in social behavior paradigms like anxiety tests. (Laviola et al., 1992, Barbier et al., 1991). The supersensitivity observed in the adult progeny of benzodiazepine treated dams is suggestive of changes in neural circuits involving GABAergic systems (Taylor et al., 1983, Schlumpf et al., 1989). Changes in benzodiazepine receptor numbers and characteristics or defective function of the $GABA_A$/benzodiazepine receptor complex are discussed as a neurochemical basis for behavioral disturbances.

Several studies made it clear, that prenatal benzodiazepines do not induce dramatic changes in ligand affinity nor benzodiazepine receptor densities (Rothe and Bigl, 1989, Kellogg, 1988). However, an alteration of the $GABA_A$ receptor complex is indicated by the evidence for changes in GABA mediated chloride (^{36}Cl) uptake by synaptosomes of rats and mice exposed to benzodiazepines in utero (Kellogg et al., 1991, Chesley et al., 1991).

Changes in the functional properties of the $GABA_A$ receptor complex are thought to derive from altered receptor subunit combinations, which are known to occur in the course of chronic drug treatment. A decrease of $\alpha 1$ and $\gamma 2$ subunit expression is noted after chronic lorazepam treatment in adult rats (Kang and Miller, 1991). According to our preliminary findings, in utero diazepam treatment causes a transient decrease of $\alpha 1$ subunit mRNA expression in the neonate and an enhanced expression of this mRNA in selected brain regions of adult offspring (Salili et al., 1994). As mentioned above, the increase of $\alpha 1$ subunit expression, a subunit exhibiting the structural requirements for benzodiazepine binding, would be in keeping with the findings on benzodiazepine supersensitivity in adult, prenatally benzodiazepine exposed rat offspring.

$GABA_A$ receptor function might also be affected indirectly through neurosteroids whose synthesis is influenced by actions on the PBR (Papadopoulos, 1993). Depending on the affinities of the drug towards the $GABA_A$ receptor complex and towards the PBR, the part played by indirect actions through PBR and neurosteroids may be more or less important. For the high affinity ligands of PBR, like the benzodiazepines diazepam and flunitrazepam or the imidazopyridines alpidem and zolpidem, indirect ways of targeting $GABA_A$ receptor via PBR and steroid sythesis, might be expected.

In development, the action of benzodiazepines is complicated by the regional variations in densities of the central benzodiazepine receptors, the functional immaturity of the $GABA_A$ receptor complex exemplified by its changing subunit pattern and the early presence of the PBR in fetal cns and peripheral organs, whereby the role of this receptor in developing cells is unknown. Interactions with PBR sites on peripheral immature steroidogenic tissues, such as testes, ovaries or adrenals may result in hormonal unbalance and hence shape central nervous system development from the periphery.

Delayed developmental immunotoxicity

Our more recent observations in developmental toxicology further classify diazepam as developmental immunotoxicant. Benzodiazepines have earlier been found to exert acute inhibitory actions on immune functions of adult animals and humans at comparatively high doses (Descotes, 1982, Zavala et al., 1990, Taupin et al., 1991). These effects are seen during the presence of the drug and should be distinguished from delayed developmental effects that become manifest in the absence of the drug. The mechanism of action of the compound in fetal tissue that is responsible for the delayed dysfunction, remains unknown. A thorough analysis of the phenomenology of defects in the postnatal animal seemed to be necessary before an analysis of the primary mechanisms in the fetus could be undertaken.

Cellular and humoral immune responses were severly suppressed in prenatally diazepam exposed rats over a postnatal period of more than two months, i.e., until the animal is young adult (Schlumpf et al., 1989 , Bütikofer et al., 1993). Both types of immune responses are reduced to about 50% of their normal values. Impaired intercellular signalling between immune cells represents a possible basis for the long-lasting deficits in immune response.

The cytokine network was found to be disrupted by *in utero* exposure to benzodiazepines. Both, macrophage-derived and T lymphocyte-derived cytokines were affected. Different cytokines exhibited different types of deviant patterns: 1., an initial phase of depression of mitogen-stimulated release followed by normalization (IL-2) or even an overshoot of (TNF-α) in young adulthood, or 2., a suppression of cytokines (IL-1, IL-6, IFNγ) that extended into adulthood (Schreiber et al., 1993, 1993a, Schlumpf et al., 1992, 1993a, 1994). Depending upon the type of cytokine, the manifestation of the dysfunction appeared to be relatively independent (IL-6) or highly dependent (TNF-α) on interactions between different immune cell types (Figs. 5 and 6).

TNF-α release represents a good example for the role of intracellular and intercellular regulatory mechanisms: In isolated splenic macrophages, where the interplay between macrophages and T lymphocytes is missing, down-regulation of mitogen-stimulated TNF-α liberation by prenatal diazepam exposure was not observed.

The impaired release of IFNγ observed in these offspring points to a disturbed interaction between T lymphocytes and macrophages. In addition, alterations of autoregulatory processes are indicated by the changes in PGE2 production. PGE2 is known to down-regulate LPS-induced TNF-α release (Renz et al., 1988). Hence, the reduction of LPS-stimulated TNF-α release by splenocytes of immature, prenatally diazepam-exposed offspring might possibly be a consequence of increased PGE2 production.

In contrast to TNF-α, liberation of IL-6 was depressed in mixed splenocyte preparations as well as in isolated macrophage preparations, indicating that this effect is less dependent on regulatory cues from other cell types (Schreiber et al., 1993a). This cytokine, which is also produced by lymphocytes and by a variety of other cells and is secreted at a later stage in the cytokine cascade, is interesting with respect to our observation on reduced antibody formation by B cells (Bütikofer et al., 1993) because it induces the maturation of plasma cells and production of immunoglobulins (Kirchner et al., 1993).

The time-course of changes in the release of the T lymphocyte-derived cytokine IL-2 corresponds closely to that of T cell proliferation (Schlumpf et al., 1989). Alterations in IL-2 release could represent either a consequence of disturbed macrophage function or a result of a pathological change in T cells due to a drug action during fetal life.

In summary, prenatal exposure to a low dose of diazepam results in prolonged disturbance of the cytokine cascade. There appear to exist two groups of functionally interrelated mediators with different types of long-term reactions to early benzodiazepine exposure, i.e., IL-1/IFNγ/IL-6 that remain affected into adulthood, and IL-2/TNF-α/PGE2 which recover.

Impaired host defense. The *Trichinella spiralis* model has proven valuable in demonstrating consequences of adult and developmental exposure to immunotoxic chemicals (van Loveren et al., 1994).The reduced host resistance towards *T. spiralis* observed in offspring of diazepam-treated rats demonstrates that the multiple deficiences we detected in the cell biology of macrophages, T and B lymphocytes result in an impairment of immune function in the situation of an acute infection (Schlumpf et al. 1994 and 1994a). To note: host defence had been tested in young adult male offspring, i.e., at a time when part of the cytokines (TNF-α, IL-2 and PGE2) had already equalized control values.

Interaction with the peripheral type benzodiazepine receptor (PBR) The initial mechanism of drug action on the fetal immune system that leads to the delayed effects, remains unknown. However, an action on the PBR and interaction with endogenous PBR ligands in the fetus seems possible, in particular because prenatal treatment with the selective PBR agonist Ro 5-4864 also causes a postnatal depression of cellular immune responses (Schlumpf et al., 1990), and because of the fact that specific binding sites exist on macrophages (Zavala et al. 1984, Schlumpf et al., 1990). The delayed effects of the prenatal treatment on PBR, i.e., the down-regulation of ^{3}H-PK 11195 binding to spleen macrophage membranes (Schlumpf et al., 1993), and the decrease in affinity of membranes of mixed splenocytes (Schlumpf et al., 1990, 1993) suggest a disturbed functional state of the system of endogenous ligands and PBR at the time of manifest immune dysfunction.

The possibility of an interaction with endogenous ligands such as DBI (diazepam binding inhibitor), which are known to interact with benzodiazepines at the PBR, exists, since PBR and DBI (or closely related immunoreactive material) have been found to coexist in all major human lymphocyte sets and in monocytes, with a significant correlation between PBR densitiy and abundance of DBI immunoreactivity (Rocca et al., 1993).

In the case of benzodiazepine effects on the developing immune and nervous system, a common primary defect might consist in interactions with the peripheral type benzodiazepine receptor (PBR) that is present in early fetal life and to which the benzodiazepine diazepam binds with high affinity. If this were the case, species differences with regard to vulnerability of the target systems might be expected, as benzodiazepines bind with different affinities to the benzodiazepine binding subunit of the PBR receptor complex of different species (bovine (human) PBR_b and rodent type PBR). Various other drugs like the imidazopyridines (alpidem, zolpidem) or the arylindolacetamides are high affinity ligands to the PBR and hence, do have the potential to interfere with developing neuronal, endocrine or immune systems *via* this receptor. Our data indicate that an interaction with the fetal PBR leads to a severe disturbance of receptor regulation which persists for a long time in the absence of an exogenous agonist.

The recent analyses of developmental toxicity, with emphasis on the functional capacity of several organ systems, clearly designate benzodiazepines as potentially harmful to the developing organism. The data discussed in this review indicate that the developing nervous and immune system are particularly sensitive. Additional developmental alterations have been observed in neuroendocrine systems (Segovia et al., 1991, Perez-Laso et al., 1994). The effects of prenatal benzodiazepine exposure occur in the absence of classical reprotoxic effects such as changes in litter size, weight gain, or structural effects such as malformations, that are most generally accepted, routinely measured and evaluated.

The delayed toxic effects on nervous, (neuro)endocrine and immune systems are not life-threatening. But they curtail life quality. And their effects on future generations, e.g., in the context of reproduction or immune function, cannot be judged on the basis of present knowledge.

REFERENCES

Awad, M. and Gavish, M. Binding of ^{3}H Ro5 - 4864 and ^{3}H PK 11195 to cerebral cortex and peripheral tissues of various species: species differences and heterogeneity in peripheral benzodiaze pine binding sites. J. Neurochem. 49: 1407 - 1414, 1987

Barbier, P., Breteaudeau, J., Autret, E., Bertrand, P., Foussard-Blampin, O. and Breteau, M. Effects of prenatal exposure to diazepam on exploration behavior and learning retention in mice. Dev. Pharma col. Ther. 17: 35 - 43, 1991

Benke, D., Mertens, S., Trzeciak, A., Gillessen. D. and Möhler, H. $GABA_A$ receptor display associa tion of $\gamma 2$ subunit with $\alpha 1$- and $\beta 2/\beta 3$ subunits. J. Biolog. Chem. 266: 4478 -4483, 1991

Blass, E.M., Hall, W.G. and Teicher, M.H. The ontogeny of suckling and ingestive behaviors. Progr. in Psychbiology and Physiology 8: 243 - 299, 1979

Bauer, J.H. Effects of maternal state on the responsiveness to nest odors of hooded rats. Physiology and Behav. 30: 229 - 232, 1983

Bavoux, F., Olivier, C. and Pons, G. Effects chez le nouveau-né des benzodiazepines dministrées a la mere. Journées parisiennes de pédiatrie 123 - 134, 1982

Bergman, U. and Dahl-Puustinen, M.L. Use of prescription forgeries in a drug abuse surveillance net work. Eur.J. Clin. Pharmacol. 36: 621 - 623, 1989

Beutler, B., Greenwald, J.D., Hulmes, M., Chang, Y.C.E., Pan, J., Mathison, R., Ulevitch, R. and Cerami, A. Identity of tumour necrosis factor and the macrophage secreted factor cachetin. Nature 316: 552 - 554, 1985

Beutler, B. and Cerami, A. Tumour necrosis, cachexia, shock and inflammation: A common mediator. Am. Rev. Biochem. 57: 505 - 518, 198

Braestrup, C. and Nielsen, M. Ontogenetic development of benzodiazepine receptors in the rat brain. Brain Research 147: 170 - 173, 1978

Braestrup, C. and Squires, R.F. Specific benzodiazepine receptors in rat brain characterized by high-affinity ^{3}H diazepam binding. Proc. Natl. Acad. Sci. USA 74:3805 - 3809, 1977

Broaddus, W.C. and Bennett, J.P. Peripheral benzodiazepine receptors in human glioblastomas:phar macological characterization and photoaffinity labeling of ligand recognition site.Brai Res. 518: 199 - 208, 1990

Bütikofer, E.E., Lichtensteiger, W. and Schlumpf, M. Prenatal exposure to diazepam causes sex-dependent changes of the sympathetic control of rat spleen. Neurotoxicology and Teratology 15: 377 - 382, 1993

Chesley, S., Lumpkin, M., Schatzki, A., Galpern, W.R., Geenblatt, D.J., Shader, R.I. and Miller, L.G. Prenatal exposure to benzodiazepines. 1. Prenatal exposure to lorazepam in mice alters open-field activity and $GABA_A$ receptor function. Neuropharmacology 30: 53 - 58, 1991

Cree, J.E., Meyer, J. and Hailey, D.M. Diazepam in labour: Its metabolism and effect on the clinical condition and thermogenesis of the newborn. Brit. Med. J. 4: 251 - 255, 1973

Cagiano, R., De Salvia, M.A., Persichella, M., Renna, G., Tattoli, M. and Cuomo, V. Behavioral chan ges in the offspring of rats exposed to diazepam during gestation. Europ. J. Pharmacology 177: 67 - 74, 1990

Costa, E. The Benzodiazepines: From molecular biology to clinical practice. Raven Press, New York, 1983

Cymerman, U., Pazos, A. and Palacios, J.M. Evidence for species differences in "peripheral" benzodia zepine receptors: an autoradiographic study. Neurosci. Lett. 66: 153 - 158, 1986

Descotes, J., Tedone, R. and Evreux, J.C. Suppression of humoral and cellular immunity in normal mice by diazepam. Immunology Lett.5: 41 - 43, 1982

Descotes, J., Immunotoxiclogy of drugs and chemicals, Elsevier, Amsterdam, New York, Oxford, 1988

Descotes, J. Drug-induced immune diseases. Drug-induced disorders Volume 4, Elsevier, Amsterdam, New York, Oxford, 1990

Di Giovine, F.S. and Duff, G.W. Interleukin-1: The first interleukin. Immunology Today 11: 13 - 20, 1990

Dinarello, C.A. Inflammatory cytokines: Interleukin-1 and tumour necrosis factor as effector molecules in autoimmune diseases. Curr. Opinion in Immunol. 3: 941 -948, 1991

Endres, S., Fülle, H.J., Sinha, B., Stoll, D., Dinarello, C.A., Gerzer, R. and Weber, P.C. Cyclic nucleotides differentially regulate the synthesis of tumour necrosis factor-α and interleukin-1β by human mononuclear cells. Immunology 72: 56 - 60, 1991

Eshleman, A.J., and Murray, T.F. Differential binding properties of the peripheral type benzodiazepine ligands ^{3}H PK 11195 and ^{3}H Ro5-4864 in trout and mouse brain membranes. J. Neurochem. 53: 494 - 502, 1989

Frieder, B., Epstein, S. and Grimm, V. The effects of exposure to diazepam during various stages of gestation or during lactation on the development and behavior of rat pups. Psychopharmacology 83: 51 - 55, 1984

Gai, N. and Grimm, V. The effect of prenatal exposure to diazepam on aspects of postnatal develop ment and behavior in rats. Psychopharmacology 78: 225 - 229, 1982

Greenblatt, D.J., Shader, R.I. Benzodiazepines in clinical practice. p 23 Raven Press, New York, 1974

Fine, J.S., Gasiewicz, T.A. and Silverstone, A.E. Lymphocyte stem cell alterations following perinatal exposure to 2,3,7,8-tetrachlorodibenzo-p-dioxin. Mol Pharmacol. 35 : 18 - 25, 1989

Fine, J.S., Silverstone, A.E. and Gasiewicz, T.A. Impairment of prothymocyte activity by 2,3,7,8-tetrachlorodibenzo-p-dioxin. J. Immunol. 144: 1169 - 1176, 1990

Frey, J., Janes, M., Engelhardt, W., Afting, E.G., Geerds, C. and Möller, B. Fc gamma(γ) -receptor - mediated changes in the plasma membrane potential induce prostaglandin release from human

fibroblasts. Eur. J. Biochem. 158: 85 - 89, 1986
Gillberg, C. "Floppy infant syndrome" and maternal diazepam. Lancet II: 244, 1977
Gerfen, C.R. and Young, W.S.I. Distribution of stratonigral and striytopallidal peptidergic neurons in both patch and matrix compartments: an in situ hybridization histochemistry and fluorescent retrograde tracing study. Brain Res. 460: 161 - 167, 1988
Gerfen, C.R., Engber, T.M., Mahan, L.C., Susel, Z., Chase, T.N., Monsma, J.M. and Sibley, D.R. D1 and D2 dopamine receptor-regulated gene expression of sriatonigral and striatopallidal neurons. Science 250: 1429 - 1432, 1990
Guillos, B., Sizun, J., Simon, I. and Moulin, M. Benzodiazepines dans le dernier mois de grossesse: attention au nouveau-né. J. Gynecol. Obstet. Biol. Reprod. 21: 261 - 262, 1992
Haas, W., Kaufmann, S. and Martinez, A.C. The development and function of T cells. Immunology Today 11: 340 - 343, 1990
Haefely,W. Benzodiazepine receptor ligands: Structural and Functional Differences in : Benzodiazepi nes: Current Concepts. (Hindmarch, I., Beaumont, G., Brandon, S. and Leonard, B.E., eds.) pp 1 - 18, J. Wiley and Sons LTd., 1990
Haefely, W. Allosteric modulation of $GABA_A$ receptor channel: a mechanism for interaction with a multitude of central nervous sytem functions. in: The challenge of neuropharmacology (Möhler, H. and Da PRada, M., eds.) p 15 - 41, Editiones Roche, Basel, 1994
Hertz, L. Binding characteristics of the receptor and coupling to transport proteins. In: Peripheral Benzodiazepine Receptors (Giesen-Crouse, E., ed.) p 27 - 57, Academic Press, 1993
Jason, K.M., Cooper, T.B. and Friedman, Z. Prenatal exposure to imipramine alters early behavioral development and beta-adrenergic receptors in rats. J. Pharmacol. Exp. Ther. 217: 461 - 466, 1981
Jerne, N.K. and Nordin, A.A. Plaque formation in agar by single antibody- producing cells. Science 140: 405 - 407, 1963
Kang, I.,and Miller, L.G. Decreased $GABA_A$ receptor subunit mRNA concentrations following chronic lorazepam administration. Brit. J. Pharmacol. 103: 1285 - 1287, 1991
Kellogg, C., Tervo, D., Ison, J., Parisi, T. and Miller, R.K. Prenatal exposure to diazepam alters behavioral development in rats. Science 207: 205- 207, 1980
Kellogg, C.K., Chisholm, J., Simmons, R.D., Ison, I.R. and Miller, R.K. Neural and behavioral consequences of prenatal exposure to diazepam. in : Drugs and hormones in brain development (Schlumpf, M. and Lichtensteiger, W., eds.) Monograph in Neural Sci 9: 119 - 129, Karger, Ba sel,1983
Kellogg, C.K. Benzodiazepines: Influence on the developing brain. Progr. in Brain Res 73:207 - 228, 1988
Kellogg, C.K., Primus, R.J. and Bitran, D. Sexually dimorphic influence of prenatal exposure to diazepam on behavioral responses to environmental challenge and on γ-aminobutyric acid (GABA)-stimulated chloride uptake in the brain J. Pharmacol. Exp. Ther. 256: 259 - 265, 1991
Kirchner, H., Kruse, A., Neustock, P. and Rink, L. Cytokine und Interferone: Botenstoffe des Immunsystems. Spektrum Akademischer Verlag, Heidelberg, Berlin, Oxford, 1993
Kozikowski, A.P., Ma, D., Brewer, J., Sun, S., Costa, E., Romeo, E. and Guidotti, A. Chemistry, bind ing affinities and behavioral properties of a new class of antineophobic mitochondrial DBI receptor complex (mDRC) ligands. J. Med. Chem. 36: 2908 - 2920, 1993
Krueger, K.E. and Papadopoulos, V. The role of mitochondrial benzodiazepine receptors in steroidoge nesis. In : Peripheral benzodiazepine receptors (Giesen-Crouse, E., ed.) Acad. Press, p 89 - 108, 1993
Laegreid, L., Olegard, R., Walström, J. and Conradi, N. Teratogenic effects of benzodiazepine use during pregnancy. J. Pediatr. 114: 126 - 131, 1989
Langer, S.Z., Arbilla, S., Tan, K., Lloyd, K.G., George, P., Allan, J. and Wick A.E. Selectivity for omega-receptor subtypes as a strategy for the development of anxiolytic drugs. Pharmacopsychiatry 23: 103 - 107, 1990
Laviola,G. Pick, C.G., Yanai, J. and Alleva, E. Eight-arm maze performance, neophobia, and hippo campal cholinergic alterations after prenatal oxazepam in mice. Brain Res. Bull. 29: 609 - 616, 1992
Lauder, J.M., Han, V.K.M., Henderson, P., Verdoorn, T. and Towle, A.C. Prenatal ontogeny of the GABAergic system in the rat brain: an immunocytochemical study. Neuroscience 19: 465 - 493, 1986

Lauer, J.A., Adams, P.M. and Johnson, K. Perinatal diazepam exposure: behavioral and neurochemical consequences. Neurotoxicology and Teratology 9: 213 - 219, 1987

Livezey, G.T., Radulowacki, M., Isaac, L. and Marzynski, T.J. Prenatal exposure to diazepam results in enduring reactions in brain receptors and deep slow wave sleep. Brain Res. 334: 361 - 365, 1985

Mallorga,P., Hamburg, M., Tallman, J. and Gallager, D. Ontogenetic changes in GABA modulation of brain benzodiazepine binding. Neuropharmacology 19: 405 - 406, 1980

Marquez, P.G. and Alonso, M.J.B. Sindrome de abstinencia a benzodiacepinas an el recien nacido. Casos clinicos. An. Exp.Pediatr. 38: 555 - 556, 1993

Mhatre, M.C. and Ticku, M.K. Chronic ethanol administration alters γ-aminobutyric $acid_A$ receptor gene expression. Molec. Pharmacol. 42: 415 - 422, 1992

Miller, G.M., Weill, C.L., Roy, R.B. and Gaver, A. Lorazepam administration during embryonic deve lopment alters $GABA_A$ receptor binding and function. Developmental Brain Res. 44: 241 - 246, 1988a

Miller, N.S. and Mahler, J.C. Addiction to and dependence on benzodiazepines: Diagnostic confusion in clinical practice and research studies. J. Substance Abuse Treatment: 61 - 67, 1991

Möhler, H. and Okada T. Benzodiazepine receptor demonstration in the central nervous system. Science 198: 849 - 851 1977

Möhler, H., Malherbe, P., Draguhn, A. and Richards, J.G. $GABA_A$-Receptors: Structural Requirements and Sites of Gene Expression in Mammalian Brain. Neurochemical Research 15: 199 - 207, 1990

Montpied, P. Morrow, L.A., Karanian, J.W., Ginns, E.I., Martini, B.M. and Paul, S.M. Prolonged ethanol inhalation decreases γ-aminobutyric $acid_A$ receptor subunit mRNAs in the rat cerebral cortex. Molec. Pharmacology 39: 157 - 163, 1991

Norman, D.C., Chang, M.P., Wong, C.M.K., Branch, B.J., Castle, S. and Taylor, A.N. Changes with age in the proliferative response of splenic T cells from rats exposed to ethanol in utero. Alcoholism: Clinical and experimental Res. 15: 428 - 432, 1991

Norman, D.C., Chang, M.P., Castle, S.C., van Zuylen, J.E. and Taylor, A.N. Diminished proliferative response of con-A blast cells to interleukin-2 in adult rats exposed to ethanol in utero. Alcohol. Clin. Exp. Res. 13: 69 - 72, 1989

Old, L.J. Tumor necrosis factor: Polypeptide mediator network. Nature 326: 330 - 331, 1985

Olsen, J.M.M., Ciliax,B.J., Mancini, W.R. and Young, A.B. Presence of peripheral-type benzodiazepi ne binding sites on human erythrocyte membranes. Eur. J. Pharmacol. 152: 47 - 53, 1988

Papadopoulos, V. Peripheral-type benzodiazepine/diazepam binding inhibitor receptor: biological role in steroidogenic cell function. Endocrine Rev. 14: 222 - 240, 1993

Pankaj, V. and Brain, P.F. Effects of prenatal exposure to benzodiazepine- related drugs on early development and adult social behaviour in swiss mice. 1. Agonists. Gen. Pharmac. 22:33 - 41, 1991

Parola, A.L. and Yamamura, H.I. Molecular properties of mitochondrial benzodiazepine receptors. in: Peripheral benzodiazepine receptors (Giesen-Crouse, E., ed) p 3-26, Academic Press, 1993

Perez-Laso, C., Valencia, A., Rodriguez-Zafra, M., Cales, J.M., Guillamon, A. and Segovia, S. Perinatal administration of diazepam alters sexual dimorphism in the rat acessory olfactory bulb. Brain. Res. 634: 1 - 6, 1994

Ramseier, H.R. Reconstitution of primary mixed-lymphocyte culture reactions inhibited with class 1 antibodies by a product of histoincompatible spleen cells. Cellular Immunol. 110: 95 - 106, 1987

Ramseier, H.R., Lichtensteiger, W. and Schlumpf, M. In vitro inhibition of cellular immune responses by benzodiazepines and PK 11195. Immunopharm Immunotox. 15 (5): 557 - 562, 1993

Ramseier, H.R., Lichtensteiger, W., Schweidler, R. and Schlumpf, M. Release of IL-1 and IL-2 and expression of IL-2 receptor as a function of postnatal age and sex. Europ. Cytokine Network 4: 453 - 459, 1993a

Regan, J.W., Roeske, W.R. and Yamamura, H.I. The benzodiazepine receptor: Its development and its modulation by gamma aminobutyric acid. J. Pharmacol. Exp. Ther. 212: 137 - 143, 1980

Rementeria, J.L. and Bhatt, K. Withdrawal symptoms in neonates from intrauterine exposure to diazepam. J. of Pediatrics 90: 123 - 126, 1977

Renz, H., Gong, J.H., Schmidt, A., Nain, M. and Gemsa, D. Release of tumor necrosis factor-α from macrophages. J. of Immunol. 141: 2388 - 2394, 1988

Richards, J.G., Schoch, P.and Haefely, W. Benzodiazepine receptors: new vistas. Seminar in Neu roscience 3: 191- 203, 1991

Rocca, P., Bellone, G., Benna, P., Bergamasco, B., Ravizza, L. and Ferrero, P. Peripheral-type benzodiazepine receptors and diazepam binding inhibitor-like immunoreactivity distribution in human peripheral blood mononuclear cells. Immunopharmacology 25: 163 - 178, 1993

Rosengarten, H., and Friedhoff, A. J. Enduring changes in dopamin receptor cells of pups from drug administration to pregnant and nursing rats. Science 203: 1133 - 1135, 1979

Rothe, T. and Bigl, V. The ontogeny of benzodiazepine receptors in selected regions of the rat brain: Effect of perinatal exposure to diazepam. Neuropharmacology 28: 503 - 508, 1989

Ruiz, I., Offermanns, J., Fuentes, P. and Castillo, M. Utilization of benzodiazepines in Chile during 1982 - 1986. Eur. J. Clin. Pharmacol. 37: 139 - 143, 1989

Ruiz, I., Offermanns, J., Lanctot, K.L. and Busto, U. Comparative study on benzodiazepine use in Canada and Chile. J. Clin. Pharmacol. 33 : 124 - 129, 1993

Salili, R., Schlumpf, M. and Lichtensteiger, W. Pre- and postnatal expression of mRNA encoding for $\alpha 1$, $\alpha 5$, $\beta 2$ and $\gamma 2$ - subunits ofGABA$_A$ receptors in the rat cns after prenatal exposure to benzodia zepines. Experientia 50 S12 - 26, 1994

Schindler, R., Ghezzi, P. and Dinarello, C. A. IL-1 induces IL-1. IV. IFNγ suppresses IL-1, but not lipopolysaccharide induced transscription of IL-1. J. Immunol. 144: 2216 - 2222, 1990

Schlumpf, M., Richards, J.G., Lichtensteiger, W. and Möhler, H. An autoradiographic study of the development of benzodiazepine binding sites in rat brain. J. Neuroscience 3: 1478 -1487, 1983

Schlumpf, M., Ramseier, H.R., Abriel, H., Youmbi, M., Baumann, J.and Lichtensteiger, W. Diazepam effects on the fetus. Neuro-Toxicology 10: 501 - 516, 1989

Schlumpf, M., Ramseier, H.R. and Lichtensteiger, W. Prenatal diazepam induced persisting depression of cellular immune responses. Life Sciences 44: 493 - 501, 1989a

Schlumpf, M., Parmar, R., Ramseier, H.R. and Lichtensteiger, W. Prenatal diazepam induced immuno suppression: Possible involvement of peripheral benzodiazepine site. Dev. Pharmacol. Ther. 15: 178 -185, 1990

Schlumpf, M. Parmar, R., Schreiber, A., Ramseier, H.R., Bütikofer, E., Abriel, H., Barth, M., Rhyner, T. and Lichtensteiger, W. Nervous and immune system as targets for developmental effects of benzodiazepines. Dev. Pharm. Ther. 18: 145 - 158, 1992

Schlumpf, M., Parmar, R. and Lichtensteiger, W. Prenatal diazepam induced persisting downregulation of peripheral ($\omega 3$) benzodiazepine receptors on rat splenic macrophages. Life Sciences 52: 927 - 934, 1993

Schlumpf, M. Lichtensteiger, W. and Ramseier, H.R. Diazepam treatment of pregnant rats differential ly affects Interleukin-1 and Interleukin-2 secretion in their offspring during different phases of postnatal development. Pharmacology and Toxicology 73: 335 - 340, 1993a

Schlumpf, M., Bütikofer, e.E., Schreiber, A.A., Parmar, R., Ramseier, H.R. and Lichtensteiger, W. Delayed developmental immunotoxicity of prenatal benzodiazepines. Toxicol. in vitro, 1994, in press

Schlumpf, M., Lichtensteiger W. and van Loveren, H. Impaired host resistence to *Trichinella spiralis* as a consequence of prenatal treatment of rats with diazepam. Toxicology, in press 1994a

Schreiber, A.A. The effect of prenatal diazepam exposure on cytokine production by rat splenocytes. Thesis, Philosophical faculty II, University of Zürich, Zürich, 1993

Schreiber, A.A., Frei, K., Lichtensteiger, W. and Schlumpf, M. The effect of prenatal diazepam expos ure on TNF-α production by rat splenocytes. Agents Actions 38: 265 - 272, 1993

Schreiber, A.A., Frei, K., Lichtensteiger, W. and Schlumpf, M. Alterations in interleukin-6 production by LPS-and Con A-stimulated mixed lymphocytes, spleen macrophages and lymphocytes in prenatally diazepam-exposed rats. Agents Actions 39: 166 - 173, 1993a

Segovia, S., Perez-Laso, C., Rodriguez-Zafra, M., Cales, J.M., Del Abril, A., De Blas,M.R., Collado, P., Valencia, P. and Guillamon, A. Early postnatal diazepam exposure alters sex differences in the rat brain. Brain Res. Bull. 26: 899 - 907, 1991

Shoemaker, H., Bliss, M. and Yamamura, H.I. Specific high affinity saturable binding of ^{3}H Ro5-4864 to benzodiazepine binding sites in rat cerebral cortex. Eur. J. Pharmacol. 71: 473 - 475, 1981

Shoemaker, H., Boles, R.G., Horst, W.D. and Yamamura, H.I. Specific high affinity binding sites for ^{3}H Ro 5-4864 in rat brain and kidney. J. Pharmacol. Exp.Ther. 225: 61- 69, 1983

Smith, A.J.W. and Mc Ginty, J.F. Acute amphetamine or metamphetamine alters opiod peptide mRNA expression in rat striatum. Mol. Brain. Res. 13: 165 - 170, 1992

Spatafora, M., Chiappara, G., D'Amico, D., Volpes, D., Melis, M., Pace, E. and Merendino, A. Effect of indomethacin on the kinetics of tumour necrosis factor alpha gene expression by human blood monocytes. Pharmacol. Res. 23: 247 - 257, 1991

Speight, A.N.P. Floppy-infant syndrome and maternal diazepam and/or nitrazepam. Lancet II: 878, 1977

Speight, A:N:P. Diazepam in pre-eclamptic toxaemia. Brit. Med. J. 1: 1420 - 1421, 1978

Squires, R.F. and Braestrup,C. Benzodiazepine receptors in rat brain. Nature 266: 732 - 734 1977

Stephenson, F.A. Understanding the $GABA_A$ receptor: a chemically gated ion channel. Biochem. J. 249: 21 - 32, 1987

Taupin, V., Jayais, P., Descamps-Latscha, B., Cazalaa, J.B., Barrier, G., Bach, J.F. and Zavala, F. Benzodiazepine anesthesia in humans modulate the interleukin-1β, tumor necrosis factor-α and interleukin-6 responses of blood monocytes. J Neuroimmunology 35: 13 - 19, 1991

Taylor, A.N., Branch, B.J„ Kokka, N. and Poland, E.P. Neonatal and long-term neuroendocrine effects of fetal alcohol exposure. Monographs in Neural Sciences (Schlumpf, M. and Lichtensteiger, W., eds.) 9: 140 - 152, Karger, Basel, 1983

Van Loveren, H., Luebke, R.W. and Vos, J.G. Assessment of immunotoxicology with the parasitic infection model *Trichinella spiralis* Methods in Immunology (Burleson, G. ed.) Wiley, 1994, in press Vidovic, M., Sparacio, S.M., Elovitz, M. and Beneviste, E.N. Induction and regulation of class II major histocompatibility complex mRNA expression in astrocytes by interferon-gamma and tumor necrosis factor α. J. Neuroimmunol. 30: 189 - 200, 1990

Vincent, S.R., Hökfelt, T., Christensson, I. and Terenius, L. Dynorphin - immunoreactive neurons in the central nervous system of the rat. Neurosci. Lett. 33: 185 - 190, 1982

Wong, G.G. and Clark, S.C. Multiple actions of Interleukin-6 within a cytokine network. Immunology Today 9: 137 - 139, 1988

Zavala, F., Haumont, J. and Lenfant, M. Interaction of benzodiazepines with mouse macrophages. Eur. J. Pharmacol. 106: 561 - 566. 1984

Zavala, F., Taupin, V. and Descamps-Latscha, B. In vivo treatment with benzodiazepines inhibits murine phagocyte oxidative metabolism and production of interleukin-1, tumour necrosis factor and interleukin-6. J. Pharmacol. Exp. Ther. 256; 442 - 450, 1990

Changes in EEG of freely-moving rats caused by three-generation organophosphate treatment

Nagymajtényi L., Schulz H., Dési I.
Department of Public Health, Albert Szent-Györgyi Medical University, 6720 Szeged, Dóm tér 10, Hungary

The organophosphorous compounds, including dimethoate, dichlorvos and parathion-methyl, have been widely used since the middle of the fifties and the beginning of the sixties. Their way of action has been investigated by several authors thus it is known that the reason of the symptoms in both acute and chronic intoxications (i.e. salivation, muscle fibrillation, etc.) is the acetylcholine overweight caused by the inhibition of the enzyme cholinesterase. Based on the mechanism of action of these compounds it was also stated that the clinical symptoms are to be treated with atropine and the enzyme reactivator PAM (WHO, 1986, 1988, 1989, 1993).

The organophosphorous compounds can also alter the electrophysiological activity of the central nervous system. Burchfiel and his coworkers (1976) registered EEG disturbances of monkeys lasting even for a year after a single large dose of sarin. Changes of the complex EEG and its frequency bands of rats treated subchronically with bromophos was seen in the experiments done by Dési (1983). In Metcalf and Holmes's study (1969) and in Brown's investigation (1971) different EEG changes of workers exposed to organophosphates were also seen.

In our previous acute experiments it was found that the changes of some neurophysiological parameters - EEG, evoked potentials - were more expressed than the biochemical ones (e.g. inhibition of cholinesterase), and even a large dose of the antidote- atropine - could not prevent that (Dési et al. 1991) The same correlation between physiological and biochemical changes was seen in the subchronic experiments when the rats were treated with low doses (1/50 and 1/100 LD50) of organophosphates (Nagymajtényi et al.1988) The mean frequency and the EEG index were significantly increased, while the mean amplitude was significantly decreased, showing an increased bioelectrical function of brain.

In the present study investigated the functional-neurotoxic effect of low doses of three widely used organophosphorous insecticides, dimethoate, dichlorvos and parathion-methyl in a three-generation experiment.

The Research was supported by the Hungarian OTKA grant No 26/92

Methods

Wistar rats - 6 animals/groups - were used. They were kept in plastic cages under standard circumstances, fed with standards rat food and water ad libitum. The doses of the compounds were the following:

	1/100	1/75	1/50	1/25	LD50
Dimethoate	7.00	10.50	14.00	28.00	mg/kg
Dichlorvos	0.97	1.46	1.94	3.88	"
Parathion-methyl	0.22	0.33	0.44	0.88	"

The male and female rats were treated by gavage with the above doses continuously 5 times a week, during the pregnancy and lactation the female rats 7 times a week. Control animals were orally given only saline.

The rats were observed continuously and some general toxicological parameters (e.g. body weight, relative weight of the internal organs, etc.) were registered.

At the age of 10 weeks chronic silver electrodes were placed on the dura of the primary somatosensory, visual and auditory cortex of the rats anaesthetized with urethane (1000 mg/kg ip.). The EEG of the freely moving animals was recorded at the age of 12 weeks.

Having put the rats into isolated boxes, they were given an adaptation period of 15-20 minutes. After this period the spontaneous EEG from the somatosensory cortex was recorded for 5 minutes. Then, the spontaneous EEG of the visual cortex was recorded for 5 minutes and after 5 stimulations (80 lux, 1 Hz) the EEG of this region was taken for another 5 minutes. After that the EEG of the auditory cortex was recorded for 5 minutes which was repeated after 5 acoustical stimulations (60 dB, 1 Hz).

The EEG taken was simultaneously analyzed using a numerical EEG program. The measured numerical parameters were mean amplitude, mean frequency, the EEG index (a quotient of the two fast and the two slow frequency bands).

For statistical evaluation analysis of variance (ANOVA) was performed. For comparison of changes in the measured parameters of the animals in various ways during the three generations, we used relative numbers which were calculated on the basis of the original values.

The brain cholinesterase activity was measured with the Ellen's method.

Results

The general toxicological parameters (body weight, etc.) of the treated groups did not differ from those of the controls, and no symptoms of the chronic organophosphate intoxication were seen even in the groups treated with the largest doses. The inhibition of the brain cholinesterase was significant only in the groups treated with the largest doses (1/25 LD50) of the compounds.

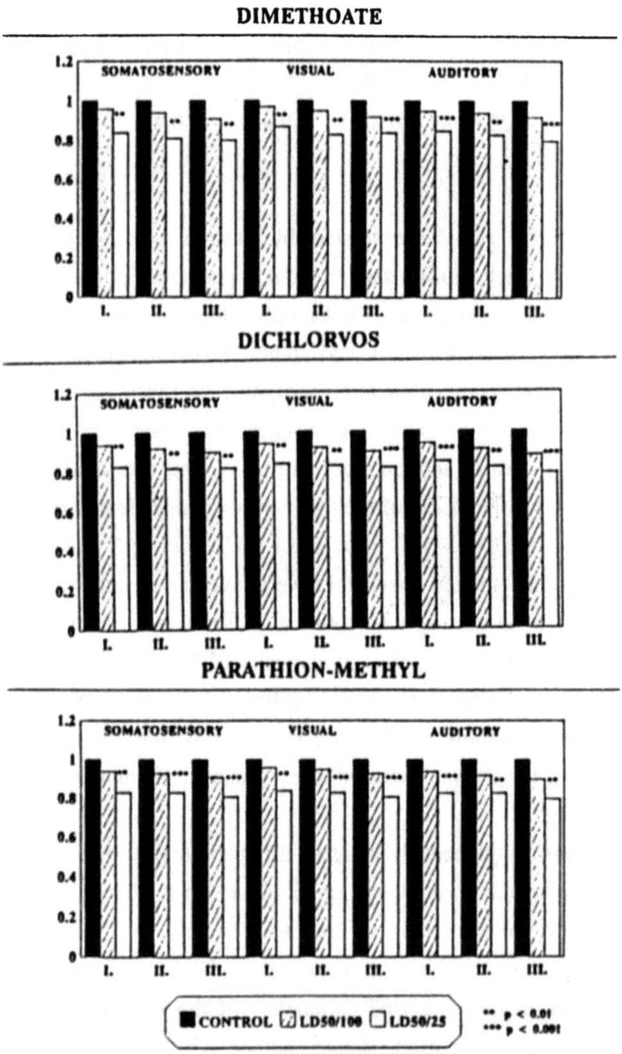

Fig. 1 The changes of EEG mean amplitudes caused by the organophosphate treatment

Comparing the mean amplitude of the spontaneous EEGs of the treated rats to that of the controls, a dose-dependent decrease was seen which was significant in the groups treated with the larger doses (Fig. 1). Although the decrease of the mean amplitudes in the second and third generations was not significant compared to the data of the identically treated groups of the first generation, the changes of the mean amplitudes were more expressed. There were only small differences between the data from the three cortical regions and the trends of decrease of the mean amplitudes caused by identical doses of all the three compounds in the generations was more or less equivalent.

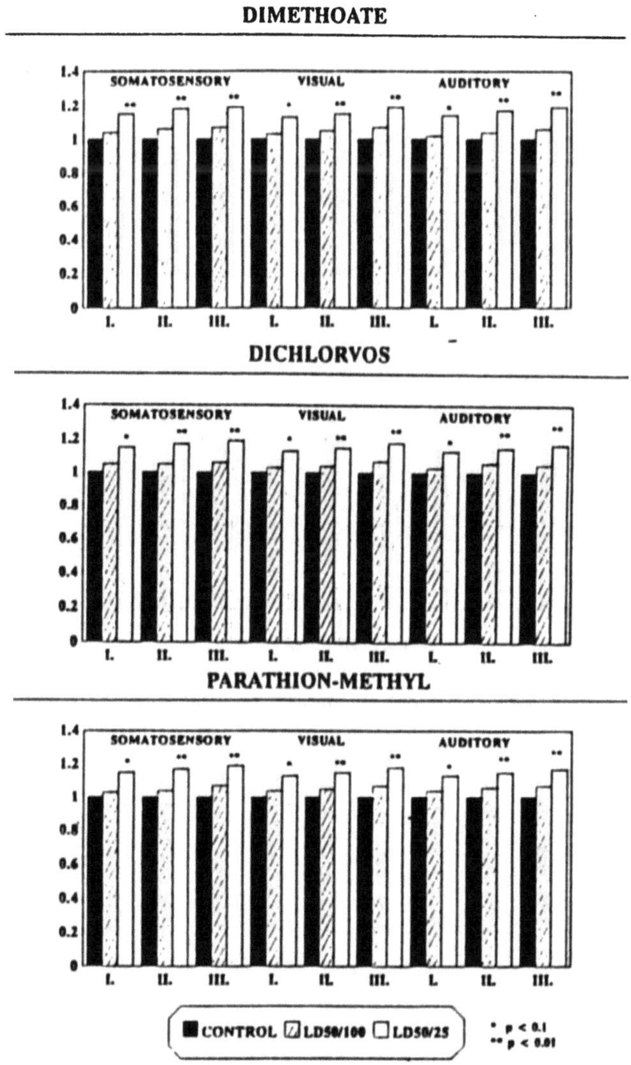

Fig.2 The changes of the EEG mean frequencies caused by the organophosphate-treatment

The mean EEG frequencies of the treated rats showed a dose-dependent increase compared to those of the control animals which, for the larger doses, was also significant (Fig. 2). More but not significantly expressed increase of the mean frequencies was found in the second and third generations compared to the first one. The alteration of the mean frequencies of the three cortical regions was practically the same in all the investigated groups given the same doses of the substances.

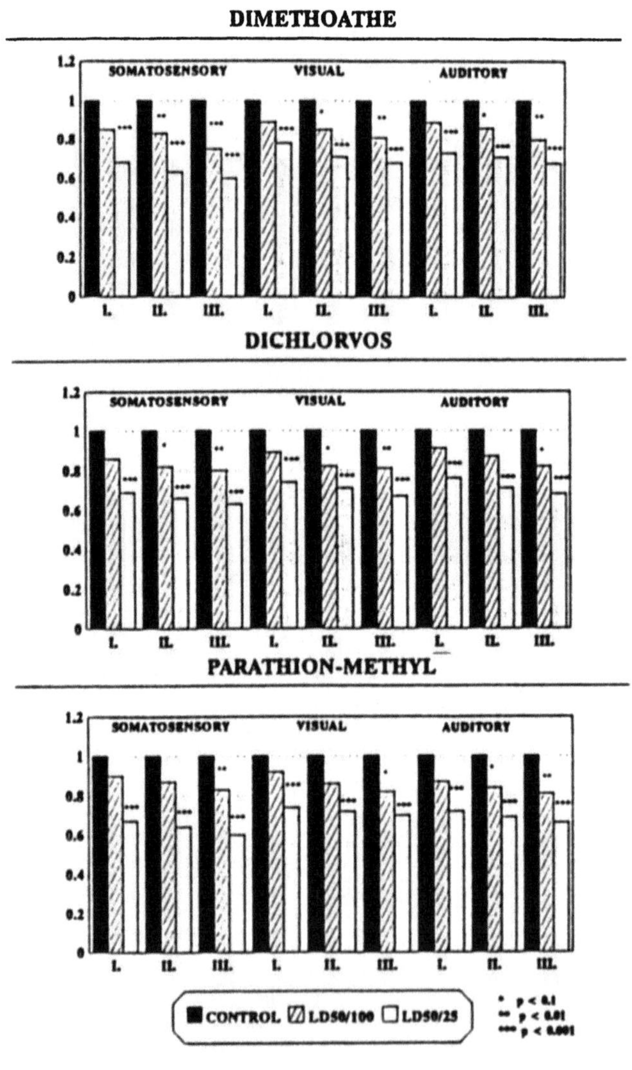

Fig. 3 The changes of the EEG indexes caused by the organophosphate treatment

From the investigated parameters, the changes of the EEG index were the most prominent (Fig. 3). Increases of this index in all the groups treated with all the doses of the compounds were significant, and became even more expressed in the second and third generation. There were practically no differences between the effects of identical doses of the compounds on the three investigated areas

Discussion

Based on our data it can be stated that the organophosphates alter the electrophysiological function in all the investigated cortical areas. The changes were practically dose dependent and were significant in case of the higher doses, and all changes - even the non-significant ones - became more expressed in the second and third generations. The effect of identical doses of the three compounds were more or less the same. It became also clear that relatively low doses of certain widely used organophosphorous compounds (the lowest dose, 1/100 LD50, was only 2-3 times higher than the NOEL for these chemicals) can affect the bioelectrical activity in different parts of the brain, and this effect can be more serious with the successive exposed generations. It is also important to mention that all these bioelectrical-functional changes were detected in absence of any clinical symptom of the chronic organophosphate exposure.

We have also found that the alteration of the electrophysiological function of the brain, especially in the groups treated with the smaller doses of the compounds, was more expressed than the inhibition of the brain cholinesterase. Based on these data, the biochemical effects of the substances and those on the cortical bioelectrical activity, although both dose-dependent, seem to be independent of each other.

Emphasizing the fact that, since the beginning of the widespread use of organophosphates, already the second or third human generation has been exposed to residues in food, sprays applied in the household, etc., we think that the importance of the chronic low level - multigeneration - exposure of the human population to these compounds must be taken into more serious consideration.

References

Brown H.R. (1971) Electroencephalographic change and disturbance of brain function following human organophosphate exposure. North-West Med. 70, 845-849.

Burchfiel, J.L., Duffy, F.H., Sim, V,M. (1976) Persistent effects of sarin and dieldrin upon the primate electroencephalogram. Toxicol. Appl. Pharmacol., 35, 365-379.

Dési, I.(1983) Neurotoxicological investigation of pesticides in animal experiments. Neurobehav. Toxicol. Teratol. 5, 503-515.

Dési, I., Nagymajtényi L., Lorencz R. Molnár Zs. (1991) The effects of organophosphorous compounds on the central nervous system of rats. Arch. Toxicol. 48, 33-37

Metcalf, D.R., Holmes, J.H. (1969) EEG, psychological and neurological alterations in humans with organophosphorous exposure. Ann. N.Y. Acas. Sci. 160, 357-365.

Nagymajtényi L., Dési I., Lorencz R. (1988) Neurophysiological markers as early signs of organophosphate neurotoxicology. 10, 429-434.

WHO. Organophosphorous Compounds - A general Introduction, Geneva, 1986, pp 181

WHO. Dichlorvos, Geneva, 1988, pp 86

WHO. Dimethoate, Geneva, 1989, pp 56

WHO. Parathion-methyl, Geneva, 1993, pp 87

DEVELOPMENTAL NEUROTOXICITY OF CARBON MONOXIDE

M.R. Carratu, R. Cagiano, M.A. De Salvia, L. Trabace, V. Cuomo

Institute of Pharmacology, School of Medicine, University of Bari, Policlinico, Piazza G. Cesare, I-70124 Bari, Italy

ABSTRACT

The results of recent studies dealing with behavioral and electrophysiological dysfunctions induced in rat offspring by prenatal exposure to relatively mild CO concentrations (75 and 150 ppm) have been reviewed in the present paper. In particular, gestational exposure to CO produced the following behavioral changes: (i) reduction in the minimum frequency of ultrasonic calls emitted by rat pups removed from their nest; (ii) decrease in ultrasonic responsiveness to a challenge dose of diazepam; (iii) alterations in the acquisition of an active avoidance task. The electrophysiological changes induced by prenatal CO in peripheral nervous system are as follows: (i) increase in time constants of sodium current inactivation; (ii) increase in the percentage of the maximum number of activatable sodium channels at the normal resting potential; (iii) negative shift of sodium current reversal potential. Moreover, results of ongoing researches on peripheral myelinogenesis in rats exposed to CO during gestation will be also summarized.

Key words: carbon monoxide, development, neurotoxicity, rat

INTRODUCTION

Carbon monoxide (CO), a toxic gas produced by incomplete combustion of carbonaceous materials, represents one of the hundreds of constituents of cigarette smoke (*Longo, 1982*). Experimental data suggest that the fetus and developing organism are especially vulnerable to CO (*Mactutus, 1989*). In particular, developmental exposure to CO has been reported to induce delays in behavioral development and central electrophysiological changes at concentrations below those producing acute effects in adult animals (*Dyer et al., 1979; Mactutus and Fechter, 1984; Mactutus, 1989*).

Our recent studies have been designed in order to further investigate the behavioral (*Di Giovanni et al., 1993*) and electrophysiological (*Carratù et al., 1993*) outcomes to prenatal CO exposure. The results of these findings are summarized in the present paper.

With regard to the behavioral approach, the ultrasonic emission of rat pups removed from the nest was investigated. This behavioral response represents a valuable test for detecting subtle emotional alterations produced by adverse treatment during development (*Cuomo et al., 1987*). Moreover, the effects of in utero exposure to CO on ultrasonic vocalizations emitted by rats during active avoidance learning were evaluated. This response seems to be a sensitive indicator of the emotional state of the animal during the acquisition of an active avoidance task (*Cuomo et al., 1992*).

As far as the electrophysiological approach is concerned, our interest in evaluating the possible changes in membrane excitability at PNS level was based on experimental data showing that electrical excitability occurs at very early stages during ontogeny (*Salkoff and Tanouye, 1986; Salkoff et al., 1987*) and that the responses of excitable cells to electrical stimuli change during the course of further development (*Spitzer, 1979*). Since exposure of immature organisms to toxic compounds may have consequences for maturation of voltage operated ion channels which control membrane excitability, the voltage clamp analysis on myelinated nerve fibers of the rat sciatic nerve was used as an accurate tool to explore subtle changes in the selective conductance pathways involved in the action potential generation.

Finally, the influence of gestational exposure to CO on peripheral myelinogenesis has been investigated. During development myelin undergoes morphologic changes to achieve the best packing arrangement. Differences in morphology reflect differences in chemical composition until a mature form is attained; these include synthesis of myelin-protein components which are targeted to the appropriate sites of myelin assembly (*Morell et al., 1994*). Among the major protein components, myelin basic protein (MBP) is a highly charged soluble intracellular protein whose expression is limited to the major dense line, thus playing a structural role in maintaining the integrity and compactness of the myelin sheath. Therefore, experimental designs were devoted to evaluation of myelin sheath thickness growth and MBP expression in rat pups exposed to CO during gestation.

CHANGES IN BEHAVIORAL PARAMETERS

Dam and male pup weight gain, pregnancy length, and litter size at birth are not significantly affected by prenatal exposure to CO (75 and 150 ppm). Moreover, CO exposure does not influence the number of dams giving birth and postnatal mortality of male pups.

Table 1 summarizes the effects produced by CO exposure on ultrasonic calling of rat pups removed from their nest and on ultrasonic vocalizations emitted by adult rats during the acquisition of an active avoidance task. Ultrasonic vocalizations of rat pups (5- and 10-day-old animals) were recorded according to the technique previously described by *Cagiano et al. (1986)*. The following parameters of the ultrasonic emission were evaluated: rate of calling, maximum and minimum frequency, duration. The active avoidance acquisition and ultrasonic vocalization were investigated in 90-day-old rats according to the technique described by *Cuomo et al. (1992)*.

TABLE 1

BEHAVIORAL CHANGES PRODUCED IN RATS BY PRENATAL EXPOSURE TO CO

Behavioral parameters	**CO concentrations** 75 ppm	150 ppm
Rat pup ultrasonic emission		
Minimum frequency	-	⇓
Maximum frequency	-	-
Rate of calling	-	-
Diazepam-induced decrease in the rate of calling	-	⇓
Duration	-	-
Active avoidance acquisition and ultrasonic emission		
Conditioned avoidance responses	-	⇓
Rate of calling	-	-

⇓ : decrease; -: no change. Derived from *Di Giovanni et al., 1993.*

Rat pups exposed to CO (150 ppm) during gestation exhibited a significant decrease in the minimum frequency of ultrasonic calls; other parameters of the ultrasonic emission such as the rate of calling, maximum frequency and duration were not affected by prenatal exposure to CO. The decrease in ultrasonic emission elicited by the administration of diazepam was significantly attenuated in male pups prenatally exposed to 150 ppm of CO (*Di Giovanni et al., 1993*). Since benzodiazepine-GABA-receptor-chloride channel complex is involved in the mediation of rat pup ultrasonic vocalization (*Insel et al., 1986*), changes in this behavioral parameter induced by prenatal exposure to CO could be due to an impaired ontogeny of benzodiazepine-GABA-receptor-chloride channel complex.

In this regard, previous findings suggested that developmental exposure to CO can affect the ontogeny of central GABAergic neuron function in rats (*Fechter et al., 1986*).

The changes in rat pup ultrasonic calling induced by prenatal CO could have a role in the production of long lasting behavioral abnormalities of adult rats exposed to this chemical during gestation. In fact, changes in ultrasonic emission of rat pups could alter maternal behavior, which in turn might affect the behavior of the offspring.

Moreover, rats exposed to CO (150 ppm) exhibited a significant impairment in the acquisition of an active avoidance schedule, whereas ultrasonic calls emitted by rats during the intertrial interval- or conditioned stimulus-periods of the active avoidance task were not significantly affected. Since the present data indicate that ultrasonic calling during an active avoidance is not influenced by prenatal CO, learning disruption produced by gestational exposure to this chemical does not seem to be linked to changes in the emotional state of animals.

CHANGES IN ELECTROPHYSIOLOGICAL PARAMETERS

The voltage clamp analysis on myelinated nerve fibers showed that prenatal exposure to CO did not affect the amplitude and the time to peak of the inward sodium current, whereas it induced significant changes in the time course of inactivation. In fact, in 40-day-old rats prenatally exposed to CO the relative contribution of slow and late components to the overall inward current was notably increased. Moreover, the percentage of the maximum number of activatable sodium channels at the normal resting potential was increased. In both 40- and 270-day-old rats exposed to CO during gestation, the equilibrium potential of sodium current was lower, thus reflecting changes in the ratio of extra- and intracellular sodium ion concentrations (*Carratù et al., 1993*). These findings, summarized in Table 2, suggest that gestational CO exposure may produce changes in the rate of ion channel development. The slow

TABLE 2

CHANGES IN SODIUM CURRENT ACTIVATION AND INACTIVATION IN RATS PRENATALLY EXPOSED TO CO.

Parameters	**CO concentrations** 75 ppm	150 ppm
Time to peak	-	-
Reversal potential	⇓	⇓
Time constant of fast inactivation	⇑	⇑
Time constant of slow inactivation	⇑	⇑
Percentage of activatable Na channels at resting potential	⇑	⇑

⇑ : increase; ⇓ : decrease; - : no change. Derived from *Carratù et al., 1993*.

kinetic of sodium inactivation observed in 40-day-old rats exposed to CO seems to be reminiscent of immature channels present in undifferentiated neurons. The functional specialization of the different channels is probably a consequence of the high degree of biochemical and morphological differentiation of mature neuronal cells. The functional specialization of the different channels may involve synthesis and insertion of new channel proteins or modification of pre-existing channels (*Spitzer, 1979*). Therefore, CO-induced changes could reflect an effect on isolated parts of the Na channel macromolecule or perturbation in the microenvironment of the integral channel proteins leading to conformational changes.

CHANGES IN PERIPHERAL MYELINOGENESIS

Our recent findings (unpublished observations) have shown that in rat pups born to dam exposed to CO (75 and 150 ppm) throughout gestation a linear relationship was found between growth in myelin sheath thickness and age as in control rats, thus indicating that in both nonexposed and CO-exposed rats myelin development occurs *pari passu* with age. However, in 40-day-old rats exposed to 150 ppm of carbon monoxide throughout gestation myelin sheath thickness was found to be significantly decreased in comparison with control rats. A linear relationship was also found between myelin sheath thickness and axon diameter in both nonexposed and CO-exposed rats; however, the slope of the line relating these two physiological variables in CO-exposed rats was found to be significantly different in comparison with control rats, thus indicating that growth in axon diameter and myelin sheath thickness is asynchronous in the offspring of dam exposed to this gas throughout gestation. Immunohistochemistry with polyclonal antibody to MBP has shown that in 150 ppm CO-exposed rats developmental expression of this protein was notably reduced in comparison with control rats; in 75 ppm CO-exposed rats an impairment in the rate of MBP expression rather occurred with a progressive recovery to control values. Collectively, the results provide the first straightforward evidence that prenatal exposure to relatively mild levels of carbon monoxide impairs myelinogenesis in the rat and hint at CO influence on the broad array of signals between Schwann cells and axons which promote mutual growth and differentiation. These regulatory mechanisms (reviewed in *De Vries, 1993*) involve synthesis of neurite growth-promoting factors by Schwann cells and mytogenic signal for Schwann cell division by the growing axons; therefore, prenatal CO exposure could disrupt any of these regulatory mechanisms leading to an impairment of myelinogenesis. To date, the regulatory mechanisms involved in myelin gene expression have not been so far elucidated. A few experimental evidences suggest that a cAMP-mediated signal from the axon to the Schwann cell can play an important role in myelin gene expression (*Poduslo, 1993*); a CO interference with a signal transduction pathway involving second messengers in myelination cascade could be well envisaged on the basis of recent studies highlighting a role for this gas as a physiological regulator of cyclic GMP (*Verma et al., 1993*) which, in turn, may regulate cAMP breakdown (*Goy, 1991*).

CONCLUSIONS

The results of the present paper show that prenatal exposure to relatively mild concentrations of CO markedly affects both central and peripheral nervous system functions. As far as the mechanism involved in CO-mediated neurotoxicity, cellular hypoxia could be only partially involved since it does not account for the delayed appearance of outcomes, i.e. after fetal COHb has been cleared from the circulation. Recent data indicate that oxidative stress in vulnerable brain regions and at the mitochondria after reoxygenation could explain the delayed neurotoxicity of CO poisoning (*Thom, 1992; Zhang and Piantadosi, 1992*); the mechanism would involve generation of partially reduced oxygen species (PROS) from mitochondria with consequent late energy failure. Whether both central and peripheral nervous system dysfunctions observed in rat pups born to dam exposed to CO throughout gestation may be caused by impaired recovery of energy metabolism and/or by the deleterious effects of PROS on mitochondrial function can be only speculative; future studies designed to clarify molecular targets and specific mechanisms of CO-mediated neurotoxicity could provide insight into the poorly understood late neurological sequelae of exposure to this gas early in development.

ACKNOWLEDGEMENTS

This work has been supported by the CNR-ENEL Project-Interactions of Energy Systems with Human Health and Environment- and by MURST.

REFERENCES

Cagiano, R., Sales, G.D., Renna, G., Racagni, G. and Cuomo, V. (1986) Ultrasonic vocalization in rat pups: effects of early postnatal exposure to haloperidol. *Life Sci.* **38**: 1417-1423.

Carratù, M.R., Renna, G., Giustino, A., De Salvia, M.A. and Cuomo, V. (1993). Changes in peripheral nervous system activity produced in rats by prenatal exposure to carbon monoxide. *Arch. Toxicol.* **67**, 297-301.

Cuomo, V., De Salvia, M.A., Maselli, M.A., Santo, L. and Cagiano, R. (1987) Ultrasonic calling in rodents: a new experimental approach in behavioral toxicology. *Neurotoxicol. Teratol.* **9**: 157-160.

Cuomo, V., Cagiano, R., De Salvia, M.A., Mazzoccoli, M., Persichella, M. and Renna, G. (1992) Ultrasonic vocalization as an indicator of emotional state during active avoidance learning in rats. *Life Sci.* **50**: 1049-1055.

De Vries, G.H. (1993) Schwann cell proliferation. ***In***: *Peripheral Neuropathy*. P.J. Dyck, P.K. Thomas (Eds.), W.B. Saunders Company, Philadelphia, Pennsylvania, pp 290-298.

Di Giovanni, V., Cagiano, R., De Salvia, M.A., Giustino, A., Lacomba, C., Renna, G. and Cuomo, V. (1993) Neurobehavioral changes produced in rats by prenatal exposure to carbon monoxide. *Brain Res.* **616**, 126-131.

Dyer, R.S., Eccles, C.U., Scott Swartzwelder, H., Fechter, L.D. and Annau, Z. (1979) Prenatal carbon monoxide and adult evoked potentials in rats. *J. Environ. Sci. Health* **13**: 107-120.

Fechter, L.D., Mactutus, C.F. and Storm, J.E. (1986) Carbon monoxide and brain development. *Neurotoxicology* 7(2): 463- 474.

Goy, M.F. (1991) cGMP: the wayward child of the cyclic nucleotide family. *TINS* **14**, 293-299.

Insel, T.R., Hill, J.L. and Mayor, R.B. (1986) Rat pup ultrasonic isolation calls: possible mediation by the benzodiazepine receptor complex. *Pharmacol. Biochem. Behav.* **24**: 1263-1267.

Longo, L.D. (1982) Some health consequences of maternal smoking: issues without answers. *Birth Defects* **18**: 13-21.

Mactutus, C.F. and Fechter, L.D. (1984) Prenatal exposure to carbon monoxide: learning and memory deficits. *Science* **223**: 409-411.

Mactutus, C.F. (1989) Developmental Neurotoxicity of nicotine, carbon monoxide, and other tobacco smoke constituents. *In: Prenatal abuse of licit and illicit drugs.* D.E. Hutchings (Ed.), The New York Academy of Sciences, New York, pp 105-122.

Morell, P., Quarles, R.H. and Norton, W.T. (1994) Myelin formation, Structure, and Biochemistry. *In*: *Basic Neurochemistry*. G.J. Siegel, B.W. Agranoff, R.W. Albers, P.B. Molinoff (Eds.), Raven Press, New York, pp 117-143.

Poduslo, J.F. (1993) Regulation of myelin gene expression in the peripheral nervous system. *In*: *Peripheral Neuropathy*. P.J. Dyck, P.K. Thomas (Eds.), W.B. Saunders Company, Philadelphia, Pennsylvania, pp 282-289.

Salkoff, L.B. and Tanouye, M.A. (1986). Genetics of ion channels. *Physiol. Rev.* **66**: 301-329.

Salkoff, L.B., Butler, A., Wei, A., Scavarda, N., Baker, K., Pauron, D. and Smith, C. (1987). Molecular biology of the voltage-gated sodium channel. *Trends in Neuroscience* **10**: 522-526.

Spitzer, N.C. (1979). Ion channels in development. *Annu. Rev. Neurosci.* **2**: 363-397.

Thom, S.R. (1992) Dehydrogenase conversion to oxidase and lipid peroxidation in brain after carbon monoxide poisoning. *J. Appl. Physiol.* **68**, 997-

Verma, A., Hirsch, D.J., Glatt, C.E., Ronnet, G.V. and Snyder, S.H. (1993) Carbon monoxide: a putative neural messenger. *Science* **259**, 381-384.

Zhang, J. and Piantadosi, C.A. (1992) Mitochondrial oxidative stress after carbon monoxide hypoxia in the rat brain. *J. Clin. Invest.* **90**, 1193-1199.

Safety assessment of over the counter (OTC) products

(Chair: Dr. B. Schmid, Switzerland)

The Safety Assessment of Over-The-Counter (OTC) Products

B. Schmid
Zyma SA, Preclinical Development & Drug Safety, CH-1260 Nyon, Switzerland

1. Summary

In recent years many compounds have become available as Over-The-Counter (OTC) medicine, for self-medication, which were originally marketed as prescription medicines only. This has been coupled with changes in the consumers' behaviour to self treatment and the governments commitment to restrict rising health costs. The safety and safe use of OTC products are important considerations for their wide distribution which is paralleled by diminishing medical supervision.

OTC products were generally registered when toxicological requirements differed from the stringent guidelines of today, and their safety assessment thus often involves ongoing evaluation as new data become available. For this, human data must also be integrated for these assessments. However, in some areas, such as mutagenicity and carcinogenicity testing, human data are hardly obtainable and testing in animals or cells, is ultimately the only reliable source for extrapolation to man.

Routine toxicological methods have mainly been developed for the assessment of new pharmaceutical entities and often provide 'yes / no' answers only. Data of such tests are, therefore, not directly applicable to the safety assessment of OTC products. Additional information on the kinetics in animal and man are indispensable for such extrapolation. On the other hand, modifications of original test protocols for the assessment of new formulations are indispensable.

It was the aim of this workshop to discuss general approaches of safety assessments within the important new area of OTC products, as well as to present specific cases of recently arisen safety issues.

2. The Self-Medication Market

Over the past three decades the role of the consumer has changed from a "passive patient" to a "health manager", and today the consumers are increasingly aware of the treatment options available to them. Self-medication products, or so-called Over-The-Counter (OTC) products, play a vital role in this new health care environment so that there is a growing demand for safe, more effective OTC products with new indications. This increasing commitment to self-medication is also shared by governments as they try to restrict rising health care costs. Nevertheless, one has to be aware that the European self-medication market is only in its faltering beginnings, as this is underlined by the fact that the first European OTC-forum was held in November of last year.

OTC products can either undergo new development or can be switched from the prescription-only status to the OTC status. Legal as well as registration requirements for both procedures differ from continent to continent, and, within Europe, within different member states. Also, it is important to remember that a switched product can continue to retain prescription business, and that a switch can, for example, be limited to one particular indication, form, strength and/or pack size. Examples for major compound switches in the United Kingdom during the last years are presented in Table 1.

TABLE 1 : Prescription-Only-Medicines to OTC status changes

° Acrivastine	° Ibuprofen (oral)
° Acyclovir	° Ibuprofen (topical)
° Astemizole	° Ketoprofen
° Cetirizine	° Loperamide
° Crotamiton	° Loratadine
° Dextromethorphan	° Mebenzadole
° Dihydrocodeine+paracetamol	° Nicotine chewing gum and patches
° Hydrocortisone (topical)	° Terfenadine
° Hyoscine butylbromide	° Vaginal imidazoles

Source : OTC Today, 1992

Generally the leading markets for non-prescription medicines situate in the areas of cough & cold / respiratory, digestive / intestinal, pain relieves, skin treatments, vitamins, minerals & nutritional supplements.

3. Safety Evaluation Programs

In contrast to the development of a new chemical entity, basic information on the safety of a potential new OTC product is, in general, already available. Actually, for an OTC product the safety aspects are very early centered around aspects of "hazard" and risk, rather than potential and potency (for definitions see Table 2), except if an entirely new ingredient would be developed.

TABLE 2
POTENTIAL refers to the inherent toxicological properties of a chemical, i.e. the possibility that toxicity can occur, with no concern for its likelihood or severity.
POTENCY is a measure of the relative toxicity of a chemical and can be used for the ranking of chemicals and/or their classification.
HAZARD is a quantitative expression of the adverse effects elicited by a chemical under defined conditions of exposure.
RISK is the probability that an event will occur given a particular condition of exposure. It is expressed as the product of the hazard and the likelihood of exposure, where exposure is estimated for a specific population.

The quantitative and qualitative aspects of this information varies, however, considerably. The most important part of an OTC product safety evaluation program is, in fact, the initial overall process of defining and developing an adequate data package as to the potential hazards associated with the manufacture, distribution and sale of a given product.

For a proper safety assessment data and investigative methods expand beyond the pre-clinical section and the following domains should be considered :

- Animal investigations
- Pharmacokinetics, pharmacodynamics
- Single case studies
- Prospective, controlled studies
- Epidemiological studies

It is important to note that the safety assessment of OTC drug products is a continues process and frequently involves ongoing evaluation. If in any of the above indicated areas new data are generated and become available an update and reassessment of the new situations is required.

3.1. The Role of Pharmacokinetics

It is apparent that of all these investigations the kinetic part plays a central role within the frame of safety assessment. Not only allows it the measurement of given drug concentrations of a specific target site where, for example, a potential safety problem has been identified, but also represents a unique mean in extrapolating safety data from animal to man. Also kinetic parameters from man can be retrospectively used for the initialisation of more sophisticated animal experiments, or experiments which are ethically not conceivable in man.

Further, pharmacokinetic investigations play an ever more important role in explaining unexpected side-effects due to interactions with already marketed drugs. For example, the histamine H_1-receptor antagonist terfenadine is a widely used drug and is available without prescription in many countries. It is a prodrug and undergoes almost complete first pass metabolism to the active acid metabolite and the inactive dealkylated metabolite (Mathews et al., 1991). There are now reports indicating an association between the use of the drug and altered cardiac repolarisation (Peck et al., 1993). Coadministration of ketoconazole, an inhibitor of cytochrome P450-mediated drug metabolism, was identified as an important risk factor (for review, Kivistö et al., 1994). Another example is the H_2-antagonist cimetidine which was rejected for OTC approval by FDA due to concerns about drug-drug interactions (Script, 1994). This with particular attention to theophyllin, phenytoin and warfarin interactions, which are all metabolised by the cytochrome P450 system. The drug interactions with cimetidine are caused by inhibition of the hepatic microsomal system and plasma concentrations of compounds like theophyllin can increase several fold by this, and lead to relevant side-effects (for review see Penston and Wormsley, 1986).

3.2. The Special Case of Safety Assessments for Mutagenicity / Carcinogenicity and Reproductive Toxicity

Since clinical investigations often do not allow the assessments for the measurements of mutagenic and/or carcinogenic effects, pre-clinical data in these areas are of particular importance. Only recently, some well-known active ingredients of OTC products, such as the H_1-antagonist doxylamine succinate, paracetamol, anthraquinones, have undergone reassessment for their safety concerning mutagenicity and/or carcinogenicity, as now data have arisen concern for genotoxicity or carcinogenicity. The topic of these three cases is fully discussed in this volume by Müller and Kasper, and Skare et al. It should be stressed again that positive findings in one of the genotoxicity tests do not automatically preclude the use of a given product, and that additional toxicological information is required.

Paracetamol, for example, shows that indirect mechanisms via depletion of the liver's glutathion-pool could be responsible for the observed genotoxic effects thus allowing further use if the threshold does not exceeded therapeutic levels (see this volume : Müller and Kasper).

A somewhat similar situations holds true for the overall assessments of reproductive toxins as they are mainly to be carried out based on animal data. One of the difficulties is that drug induced effects in the reproductive cycle strongly depend on exposure to an agent during a specific 'sensitive' period in development. On the other hand, toxicological data basis (e.g. Catalogue of teratogenic agents) are being developed providing direct access to information on potential risks of agents.

4. The Testing of New Formulations

The pharmaceutical form of an active ingredient is very important in the area of OTC products. The cosmetic aspects can play a dominant role, or, on the other hand, also highly sophisticated delivery forms such as patches or chewing gums are being developed and marketed. The choice of these pharmaceutical forms will, of course, heavily influence the bioavailability of the active ingredient, and, as a consequence, its toxicity and efficacy. Also, the composition of the final formulation may have different toxicity than the individual components, as synergistic effects cannot be excluded.

The testing of new formulations may require modifications on already existing routine methods, or apply new, alternative, methods, (i.e. human cell cultures). Further validation of the modified or alternative tests against human data is then indispensable. A good example for modifications has recently been reported on the contact dermatitis maximisation test, making the testing of transdermal delivery systems possible (Maurer and Hess, 1989, Maurer, 1991).

An area where the use of alternative method has strong scientific ground, is the area of local tolerance. A variety of *in vitro* methods have been developed for establishing the safety of ingredients where ocular dermal irritancy, sensitisation and percutaneous absorption are of major concern (for review see Balls). Each of those methods has its own advantages and disadvantages must be selected in accordance with the physico-chemical characteristics of a new formulation and the intended use in man (e.g. long/short term exposure, site of application).

In addition to the toxicological testing of new formulations, pharmacological activity (if adequate animal or cell models are available), and testing of microbial preservative activity and stability testing must also be considered. These elements are not only essential in the selection of a new appropriate formulation but also contribute effectively to the improvement of already existing formulations (see Figure 1).

Further substantial efforts will be needed to meet the consumer requirements for safe and efficacious OTC products.

FIGURE 1 : Schematic Outline for the Testing of New Formulations

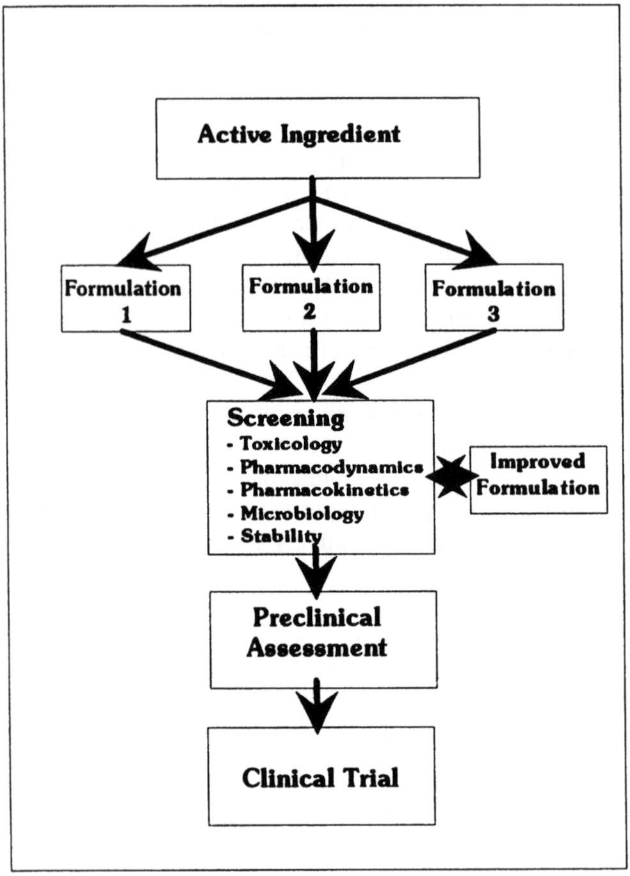

7. References

-Balls M, Blaauboer B, Brusick D, Frazier J, Lamb D, Pemberton M, Reinhardt Ch, Roberfroid M, Rosenkranz H, Schmid B, Spielmann H, Stammati AL, Walum E (1990) Report and Recommendations of the CAAT/ERGATT Workshop on the Validation of Toxicity Test Procedures. ATLA 18: 313-337.

-Balls M (1992) For beauty lives with kindness - some thoughts on cosmetics testing and safety assessment. International Journal of Cosmetic Science 14: 199-277.

-Grygiel JJ, Miners JO, Drew R, Birkett DJ (1984) Differential effects of cimetidine on theophylline metabolic pathways. European Journal of Clinical Pharmacology 26: 335-340.

-Kivistö KT, Neuvonen PJ, Klotz U (1994) Inhibition of Terfenadine Metabolism - Pharmacokinetic and Pharmacodynamic Consequences. Clin. Pharmacokinet. 27 (1): 1-5.

-Madden V (1992) Safety first as medicines. OTC Today Summer Ed. 26, 12-13.

-Mathews DR, Mc Nutt B, Okerholm R (1991) Torsades de pointes occurring in association with terfenadine use. JAMA 266: 2375-2376.

-Maurer Th, Hess R (1989) The Maximization Test - Updating of the Standard Protocol and Validation of a Modified Protocol. Fd. Chem. Toxicol. 27: 807-811.

-Maurer Th (1991) Kontaktallergenitätsprüfungen. In Arzneimitteltoxikologie von Robert Hess, 139-145.

-Müller L, Kasper P (1994) OTC Pharmaceuticals and Genotoxicity Testing : The Paracetamol, Anthraquinone, and Griseofulvin Cases (this volume).

-Peck CC, Temple R, Collins JM (1993) Understanding consequences of concurrent therapies. JAMA 269: 1550-1552.

-Penston J, Wormsley KG (1986) Adverse Reactions and Interactions with H_2-Receptor Antagonists. Medical Toxicology 1: 192-216.

-SCRIP (1994) Tagamet OTC rejected again in US, Ed. 1946, 20-21.

-Shepard Th. (1983) Catalogue of Teratogenic Agents. The Johns Hopkins University Press, Baltimore, Maryland 21218.

-Skare JA, Murphy VA, Bookstaff RC, Thompson GA, Heise MA, Horowitz ZD, Powell JH, Parkinson A, St. Peter JV (1994) Safety Assessment of OTC Drugs : Doxylamine Succinate (this volume).

OTC Pharmaceuticals and Genotoxicity Testing: The Paracetamol, Anthraquinone, and Griseofulvin Cases

L. Müller, P. Kasper
Federal Institute for Drugs and Medical Devices, Seestr. 10, D-13353 Berlin, Germany

Summary

Genotoxic effects are hardly assessable in an exposed population but are generally considered to be serious due to their unpredictable effects on subsequent generations and to the link between genotoxicity and cancer. Lack of knowledge about a genotoxic / carcinogenic potential has to be stated for numerous compounds which are often in pharmaceutical use known for a long time. A thorough testing programme like it is done for new compounds is essential for such compounds that are not completely unsuspicious with respect to being reactive with macromolecules or that have the potential to generate reactive metabolites in the body.

Paracetamol, *anthraquinone-containing preparations*, and *griseofulvin* are examples for pharmaceuticals that have been in use for a long time but for which genotoxicity testing revealed a possible deleterious potential only recently. The Federal Health Office / Federal Institute for Drugs and Medical Devices therefore imposed new studies upon companies marketing these compounds in the last years. These studies in part led to a more thorough description of possible adverse effects or even restrictions for use. Paracetamol exhibits a genotoxic potential in vitro and in vivo probably via indirect, cytotoxicity or enzyme inhibition-mediated effects. Further studies will have to clarify whether a threshold could be established and whether effects do not occur at therapeutic dose levels. Genotoxicity data on the mixed group of anthraquinones reveal positive and negative findings. Compounds such as lucidin, danthron, emodin supposedly have a genotoxic and carcinogenic potential. Further studies with anthraquinone-containing plant preparations will have to clarify the content and genotoxic activity of the preparations and the active ingredients. Lucidin- and danthron-contaning preparations are currently no longer in use now whereas restrictions apply for other anthraquinone-containing laxatives. Griseofulvin is acknowledged in the meantime as an aneugen for somatic and germ cells. It is in vitro effective in concentrations that correspond to therapeutic plasma levels.

Necessity for mutagenicity testing with special emphasis on OTC-(over-the-counter) products

Mutagenicity, or in a broader sense genotoxicity is a field in toxicology which arose only 20 years ago and which has attracted an enormous amount of research since. In the beginning it was hoped that simple mutagenicity testing would allow an accurate predicition of carcinogenicity. The prediction of heritable germ cell effects was the second goal. For these two purposes, mutagenicity testing has been incorporated in regulatory safety assessment in many countries and many systems have evolved in which genotoxicity could be measured.

OTC pharmaceuticals are products which are generally not taken under supervision of a physician and must therefore have a very high safety standard for use. Many of these pharmaceuticals have been on the market for decades and are therefore presumably safe for use. However, safety is not reliably assessable for areas in which data on side effects in humans are hardly if ever obtainable. This is especially relevant for the areas of mutagenicity and carcinogenicity. In these areas, animal tests and even in vitro tests are generally carried out for detection and description of effects. Results of these tests are also used in the absence of any reliable data from humans for risk assessment and risk management procedures.

In the following, some examples of pharmaceuticals are given which have now been on the market for a long time, and with the exception of griseofulvin are available without prescription, but for which genotoxicity testing revealed only recently that there may be a risk that has not been considered before. The examples selected are *paracetamol, anthraquinone preparations* (mainly laxatives), and *griseofulvin*. For pharmaceutical products on the market, which contain these compounds the Federal Health Office / Federal Institute for Drugs and Medical Devices has urged the pharmaceutical industry in the last years to carry out new studies in the field of genetic toxicology for a more precise description of the genotoxic potential of these compounds. Risk-benefit analyses subsequently led to new warnings in the patient information and restrictions in the mode and duration of use for most indications for which the above named chemicals are used in medicine. The following paragraphs reflect only the most important parts of the genetic toxicology of the compounds discussed. It would go beyond the limits of this essay to cover all the published literature on genotoxicity of paracetamol, anthraquinones, and griseofulvin.

Paracetamol

General toxic and cytotoxic effects of paracetamol in animals have been known for a long time. It has also been known that such effects may lead to sometimes irreversible liver damage and can occur at doses which exceed the therapeutic dose only slightly.

During the last years a number of genotoxicity test results added a new quality of toxic effects to the already known spectrum of side effects. The final part of the story started with the publication of elevated chromosomal aberration frequencies observed in the peripheral lymhocytes of a number of volunteers who had taken 4g paracetamol within 8 hours (Kocisova et al., 1988). These findings were confirmed by a second group (Hongslo et al., 1991). A very thorough reinvestigation employing a placebo group as well and the determination of plasma levels revealed no effects (Kirkland et al., 1992).

Table 1: A selection of reports on genotoxic effects of paracetamol

endpoint	*cell type / species*	*results*	*reference*
gene mutations	mouse lymhoma cells, Tk-gene	increase with concentration, LOEC: 500mcg/ml	Hazleton Microtest, internal report, 1992
chromosomal aberrations	in vitro, various cell lines	increase with concentration, LOEC: 150mcg/ml	Müller et al., 1991 and others
chromosomal aberrations	lymphocytes from volunteers, different group sizes	small increase after oral gavage of 4 g within 8 hours small increase, same regimen as above no increase found, same regimen as above	Kocisova et al, 1988 Hongslo et al., 1992 Kirkland et al., 1992
DNA-adducts	mouse: liver, kindney	Adducts detected at 300mg/kg orally; pretreatment with diethylmaleate for glutathion depletion	Hongslo et al., 1994
micronuclei	rat	small but consistent increase at 3x900mg/kg orally, no lower dose tested	Hazleton Microtest, internal report 1993

In vivo, DNA-damage and DNA-adducts were found in mouse liver and kidney, but only after depletion of the glutathion pool (Hongslo et al., 1994). In an unpublished study, paracetamol caused an increase of micronucleus rates in the bone marrow of rats following oral treatment with a high dose of 3 x 900 mg/kg body weight within 8 hours (report Hazleton/Microtest, 1993). The effects were relatively small (within the range of a threefold increase of the spontaneous rates), however, they were observed in almost all animals. In the same study, paracetamol exerted toxic effects to the hematopoietic system. The administered dose led to plasma concentrations of approx. 250 mcg/ml (1.7 mM) in rats. This corresponds to approx. the 10- to 20-fold of the therapeutic concentration. In vitro effects (e.g. replication inhibition, chromosomal aberrations) also occur in this range and below (starting at 0.5 mM). It remains unclear whether paracetamol can also induce such effects within the therapeutic concentration range. Basically no threshold limit can be expected for genotoxic effects. However, a number of studies for paracetamol would make a threshold limit plausible:

(I) It is known that the mechanisms of detoxification of paracetamol can be overloaded (depletion of the glutathion-pool). Given this situation the electrophilic reactive metabolite of paracetamol, N-acetyl-p-benzoquinone (NAPQI) is no longer efficiently inactivated. NAPQI is known to induce genotoxic effects in high concentrations in different test systems. Sufficient levels of glutathion guarantee an efficient and fast inactivation of NAPQI .

(II) An inhibition of ribonucleotide reductases by paracetamol with the consequence of a significant reduction of the nucleotide pool has been described (Hongslo et al., 1990). As a result, via disturbances in the replication system and an increased chance of errors in the replication process, mutations could evolve. Such a mechanism is thought to depend on the concentration and will most probably exhibit a threshold. Consequently, cells with a high level of the free tyrosil-radical of ribonucleotide reductase were resistant to the genotoxic action of paracetamol (Hongslo et al., 1990). In vitro studies showed that inhibition of replicative DNA-synthesis by paracetamol could be neutralized with ascorbic acid in the therapeutic dose range (Honglso et al., 1989). Lower concentrations of paracetamol than necessary for depletion of the glutathion-pool efficiently inhibit this enzyme in vitro. These results make it plausible that the inhibition of the ribonucleotide reductase might be a major mechanism of the observed genotoxic action of paracetamol.

(III) The cytotoxic and genotoxic effects of paracetamol are also discussed on the basis of influences on the intracellular Ca^{2+}-content (Ray et al., 1990; Shen et al., 1992). Increased intranuclear Ca^{2+}-levels activate endonucleases and lead to DNA-fragmentation, LDH-leakage and cell death.

The findings on the genotoxicity of paracetamol reported to date need further clarification regarding their significance for therapeutic administration. The in vivo genotoxicity studies with paracetamol lack the investigation of comprehensive dose-effect relationships as well as the simultaneous investigation of relevant toxicity parameters. It is of special interest to clarify to what extent the various threshold-based mechanisms of the genotoxic effects can be seen in combination with the reported effects in animal experiments on bone marrow. Therefore, the induction of micronuclei in the bone marrow of rats by paracetamol and a dose/effect ratio down to the therapeutic range (a maximum of approx. 4 g/day) will now be investigated. The cytotoxicity of paracetamol to the bone marrow, hepatic functions (glutathione, transaminases etc.), pathomorphologic and pathohistologic changes in the liver, and the urine status will be assessed concomitantly. This experiment being conducted at present will hopefully allow a final evaluation of the question whether or not the genotoxic effects are considered to be concomitant effects of a cytotoxic action in dosages without therapeutic relevance.

Does paracetamol exhibit a carcinogenic potential?

Indications of genotoxicity are always discussed in connection with a possible carcinogenicity of the respective substance. For paracetamol the problem of carcinogenic effects is currently not considered of high priority, as the most recently performed carcinogenicity feeding studies in rats and mice (NTP-Report 394, 1991) with a high daily intake of paracetamol revealed no substantial indications of a carcinogenic effect (the data can be considered to be a further indication of the fact that genotoxic effects of paracetamol are connected with a threshold value and only occur as a consequence of other toxic effects).

Current risk management for paracetamol

The summary of product characteristics (SPC) currently reflects the situation and serves as an information for the physician. Here, the genotoxic potential of paracetamol is described in the following way: *Although there is at present no justified indication for a genotoxic potential of paracetamol within the therapeutic dose range, there are results published needing clarification in this respect. There are plausible explanations for the known genotoxic effects of paracetamol which imply a threshold mechanism. Further studies will be done which are designed to elucidate whether the threshold for genotoxic effects of paracetamol will not be exceeded when administered therapeutically.*

Anthraquinones

Anthraquinones, i.e. hydroxyanthracene derivatives, are constituents of many plant preparations with laxative and other properties. There is no doubt that at least the laxative anthraquinone-phytopharmaceuticals are efficient and therefore constitute a valuable alternative to synthetic compounds. But they also constitute a very reactive group of compounds some of which have been clearly identified as DNA-damaging agents.

Figure 1 gives some of the most frequently found anthraquinones in phytopharmaceuticals. Most of these compounds were mutagenic to bacteria (Ames-test), i.e. positive results were obtained in at least one strain (e.g. Brown and Brown, 1979; Tikkanen, 1983; Krivobok et al., 1992; Westendorf et al., 1990). Some compounds were genotoxic and / or carcinogenic to mammalian cells (table

2). In vitro, the effects were often dependent on, or enhanced with, competent metabolization (S9-mix, hepatocytes). Anthraquinone-glycosides as present in many preparations could be converted into the corresponding aglycon in the intestine. This requires an appropriate consideration of the contribution of glycosidic anthraquinones to the specific genotoxic potential of the aglycon.

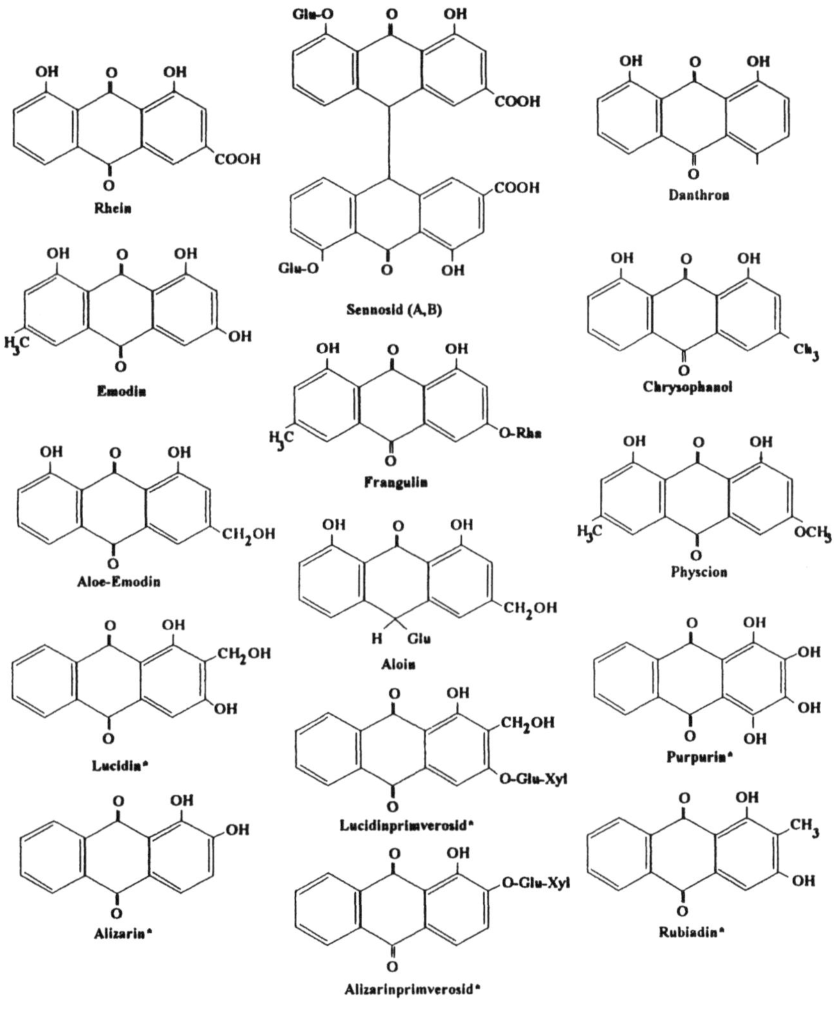

Fig. 1: Some important anthraquinones (hydroxyanthracene-derivatives) present in various phytopharmaceuticals (*, non-laxative anthraquinones present in Rubia tinctorum)

Anthraquinones present in Rubia tinctorum L.

Clear evidence for genotoxic and carcinogenic activity is available for anthraquinones present in Rubia tinctorum (figure 1). These anthraquinones, namely lucidin, alizarin, purpurin are genotoxic in vitro (Kawasaki et al., 1992; Westendorf et al., 1989). A metabolite of one of the main anthraquinones, alizarinprimverosid is the carcinogenic 1-hydroxyanthraquinone (Blömeke et al., 1992; Mori et al., 1990). Rubia tinctorum anthraquinones have no laxative properties and the efficiency of Rubia tinctorum in the treatment of bladder stones was seriously questionned. For these reasons, Rubia-preparations could no longer be marketed in Germany from 1993 onwards.

Table 2: Important genotoxicity/carcinogenicity test results with anthraquinones and mammalian cells

test compound	*cell type/species*	*endpoint*	*effect*	*reference*
lucidin	V79-cells rat hepatocytes rat	gene mutations DNA-repair DNA-adducts	small increase increase adducts in liver	Westendorf et al., 1988 Westendorf et al., 1990 Poginsky et al., 1991
1-hydro-xyanthra-quinone	rat	tumorigenicity	large bowel, liver and stomach	Mori et al., 1990
danthron	rat hepatocytes rat / mouse	DNA-repair tumorigenicity	positive rat: intestine mouse: liver and intestine	Westendorf et al., 1990 Mori et al., 1985; 1986
emodin	V79-cells V79-cells, rat hepatocytes rat liver	gene mutations gene mutations, DNA-repair DNA-binding	negative both endpoints positive no binding detected	Bruggeman and Hoeven, 1984 Westendorf et al., 1990 Bösch et al., 1987
aloe-emodin	in vitro and in vivo, various cells and species	gene mutations and chromosomal aberrations, DNA-repair	positve in vitro, negative in vivo	Westendorf et al., 1990 several unpublished reports (Madaus AG)
sennosides (A,B)	in vitro and in vivo, various cells and species rat	gene mutations and chromosomal aberrations tumorigenicity	completely negative no increased tumor incidence	several unpublished reports (Madaus AG) Sokolowski et al., 1992

Other anthraquinones

The synthetic laxative danthron (chrysazin) was the first compound of the group for which carcinogenic effects in mice and rats were published (Mori et al., 1985; 1986). Soon thereafter it was taken off the market. A number of publications concerning the genotoxic and carcinogenic potential of the anthraquinone group followed. A culminating point was reached when a positive correlation between anthranoid laxative abuse (diagnosed as pseudomelanosis coli) and colorectal cancer incidence was published (Siegers et al., 1993).

The structural requirements for a genotoxic activity of anthraquinones and anthraquinone-glycosides have not been clearly outlined up to now. There is some evidence that a methyl-group at the 6-position plays a part in the genotoxic activity though this group is neither modified nor eliminated by hepatic enzymatic systems (Krivobok et al., 1992). Westendorf et al. (1990) found that in mammalian cells only anthraquinones with 2 hydroxy groups in the 1,3 positions or with a hydroxymethyl sidechain were mutagenic.

It is a major shortcoming that most of the plant preparations on the market have not been characterized up to now with respect to the content and chemical identity of the anthraquinones present. Moreover valid in vitro and in vivo tests for genotoxicity and carcinogenicity are lacking for almost all preparations. At present it is therefore not possible to give a thorough evaluation of the possible risk associated with the intake of such products. This statement is valid with one exeption: comprehensive in vivo examinations of a defined extract of senna pod were negative for mutagenicity and carcinogenicity. This extract contained mainly the sennosides A and B but also 0.15% aloe-emodin and 0.006% emodin (Sokolowski et al., 1992).

Current risk management for anthraquinone-laxatives

Monographs and Summary of Products Characteristics (SPC) for anthraquinone-containing products are in the final stage within Germany and Europe. The announcement of the Federal Health Office / Federal Institute for Drugs and Medical Devices communicated on June 20, 1994 contains the following measures to avoid drug-associated risks:

The only indication for use is obstipation. As a safety precaution laxative anthraquinones should not be taken for more than two weeks, they should not be given to children below age 12, and they should not be given during pregnancy and lactation.

Further clarification on the genotoxic / carcinogenic potential of the laxative anthraquinones emodin, aloe-emodin, 2-hydroxyemodin, and physcion will probably be reached in the course of the studies that have to be carried out following this announcement. Companies marketing anthraqinone-containing laxatives have to submit data on the content of anthraquinones in their preparations and have to submit the results of a step-wise genotoxicity testing programme for these pharmaceuticals beginning with in vitro tests and continuing with valid in vivo tests if in vitro positive results are observed.

Griseofulvin

Griseofulvin is an efficient fungistatic agent which is distributed systemically after oral administration. It is preferentially deposited in the keratin precursor cells and since it is tightly bound to keratin, the cells become resistant to fungal invasions. It is used for ringworm infections of the skin, nails, and hair. Treatment often has to be carried on for several months.

Early toxicology studies revealed hepatocellular necrosis and liver tumors by the compound in mice and some thyroid tumors in rats (Rustia and Shubik, 1978). The liver effects were attributed to the disturbances in porphyrine metabolism followed by the appearance of protoporphyrine pigment in hepatocytes and bile canaliculi with the consequence of a proliferation of bile duct epithelia (Hurst and Paget, 1963; Yokoo et al., 1979). Thyroid tumors apparently were a consequence of the depletion of thyroxin with a subsequent stimulation of the thyroid stimulating hormone. Both effects are considered to be of no relevance for the therapeutic administration. Early studies also report a suppression of spermatogenesis, an embryotoxic and a teratogenic action of griseofulvin.

A number of more recent investigations were carried out concerning the genotoxicity of griseofulvin. It is now evident that griseofulvin interacts with the formation of microtubuli and therefore has the potential of disturbing the correct distribution of chromosomes into daughter cells during cell division (colchicine-like effect).

Table 3: Griseofulvin (GF)-induced aneuploid metaphases in human lymphocytes treated in vitro (Hoechst, unpublished report; average of two experiments)

treatment	*S9 mix*	*number of cells analyzed*	*aneuploid cells (n)*	*aneuploid cells (%)*
untreated control	-	2000	37	1.85
solvent control (DMSO)	-	2000	49	2.45
GF 1mcg/ml	-	2000	88	4.40
GF 10 mcg/ml	-	1500	122	6.10
untreated control	+	2000	35	1.75
solvent control (DMSO)	+	2000	41	2.05
GF 100 mcg/ml	+	2000	57	2.85
GF 120 mcg/ml	+	2000	131	6.55

Following a review of the genetic toxicology of griseofulvin by DeCarli and Larizza (1988), new studies were initiated. In the course of studies requested by the German Federal Health Office in 1989/1990 the induction of aneuploid cells was observed in vitro. Human lymphocytes were treated in vitro with griseofulvin for 48 hours. A concentration of 10 mcg/ml severely inhibited the cell replication and a concentration as low as 1 mcg/ml significantly increased the number of aneuploid metaphases (table 3).

More recently carried out studies have shown similar results. Griseofulvin was a potent inducer of micronuclei in vitro in Chinese Hamster V79-cells (Seelbach et al., 1993). Although no effects were found in an early micronucleus test with bone marrow of mice (Bruce and Heddle, 1979) recent studies are positive for aneuploidy in mouse oocytes (Mailhes et al., 1993; Marchetti and Mailhes, 1994). After superovulation with human chorionic gonadotropin, griseofulvin causes severe meiotic delay and aneuploid oocytes. Doses of more than 1 g/kg body weight were clearly effective. From these studies it is evident that in mice griseofulvin or its metabolites can reach the female germ cells and are able to damage them.

Table 4: Lowest observed effect concentrations / doses (LOEC/LOED) in vitro and in vivo for griseofulvin in aneuploidy experiments and pharmacokinetic data

study type	*species/cells*	*concentrations/doses/effects*
aneuploidy induction in human lymphocytes in vitro (Hoechst, internal report)	human primary cells in vitro	1 mcg/ml doubles the incidence of aneuploid metaphases 10 mcg/ml cause severe mitotic delay
micronucleus induction in Chinese hamster V79 cells (Seelbach et al., 1993)	permanent cell line in vitro	10 mcg/ml induces six-times higher micronuclei frequencies than in control cells; 3.1 mcg/ml is negative
meiotic delay and aneuploidy in mouse oocytes (Mailhes et al., 1993)	mouse oocytes	meiotic delay: effects found at the lowest dose of 750 mg/kg aneuploid oocytes: disputable effects at 1 g/kg, clear effects at 2.5 g/kg
rat pharmacokinetics (Venkataram and Rogers, 1988)	rat plasma	peak plasma level: ca. 1.5 mcg/ml 5 hours after a single oral dose of 100 mg/kg
human pharmacokinetics (Schäfer-Korting et al., 1985)	human plasma, skin blister fluid	plasma: 1.34 mcg/ml 2-4 hours after 500 mg griseofulvin orally skin blister fluid: 0.84 mcg/ml

Interaction with the spindle apparatus as supposed for griseofulvin is a dose-dependent mechanism which usually allows the postulation or even detection of a no-effect dose. Safety margins much lower than for the usual genotoxic carcinogen where the DNA serves as target may apply. Based on the effective concentrations in vitro and pharmacokinetics in animals and man (table 4) it is evident that an appropriate safety margin could not be calculated for griseofulvin.

Human plasma levels peak after a single oral dose of 500mg at about 1.4 mcg/ml. This is a concentration which efficiently induces effects in human lymphocytes in vitro. It seems that much higher doses are needed for rats in order to reach the same plasma levels, i.e. 100 mg/kg. No data on plasma levels are available for the mouse, the species in which effects on germ cells were found.

Risk management for Griseofulvin

No efficient alternative for griseofulvin in the treatment of certain fungal infections is available on the market up to now. In view of the possible risk, a therapeutic benefit of the drug is still seen. Therefore, the appropriate approach is a communication of the possible risk associated with griseofulvin to the physicians and patients. A possible risk reduction by shortening of treatment is not appropriate here due to the fact that in dependence on the rate of growth of nails (4-6 months) a long-term treatment is necessary for a therapeutic success. Here, the physician has the responsibility to weigh possible risks against the benfit of the drug.

General Implications for OTC products

Three cases, namely *griseofulvin*, *paracetamol* and *anthraquinones* illustrate that genotoxicity may be a newly discovered property of pharmaceuticals which have been on the market for a long time and have been generally considered as safe if used appropriately. Since genotoxic and / or carcinogenic effects are hardly assessible in an exposed population of users, testing as performed for new compounds is basically indicated also for those compounds on the OTC market which have not been subjected to such a testing programme. Table 5 gives the usual programme required in the US, Japan, and the EC.

If a compound is negative in such a programme, which is not very costly, a significant step forward has been taken for exclusion of a genotoxic and carcinogenic potential. Positive findings in genotoxicity tests do not totally preclude the use of the incriminated compound as a pharmaceutical. The classical concept which implies that there is no safe dose for a DNA-damaging compound is not valid for all genotoxins and does not preclude the existence of mechanisms with an underlying threshold.

Table 5: Core Test Batteries for Genotoxicity Testing of Pharmaceuticals in Japan, the US, and the EC

Japan *3 tests*	EC *4 tests*	USA *3 or 4 tests*
bacterial assay	bacterial assay	bacterial assay
	eucaryotic cell mutation assay (preference: mammalian cells)	mammalian cell mutation assay (preference: mouse lymphoma assay)
chromosomal aberrations in vitro	chromosomal aberrations in vitro	assay for chromosomal damage only needed if a test other than the mouse lymphoma assay is chosen above
micronucleus test with bone marrow	in vivo test (preference: bone marrow)	in vivo test (preference: chromosomal aberrations in bone marrow)

The data for the aneugen griseofulvin show that a threshold may exist, but may be reached therapeutically. The paracetamol example shows that indirect mechanisms, which may involve the establishment of a threshold, could be responsible for the observed genotoxic effects thus allowing further use if the threshold is reliably not exceeded therapeutically. The data on anthraquinone-containing products show that a market regulation towards presumably less dangerous preparations is possible without a total condemnation of this important and effective group of phytotherapeutics. Especially for phytotherapeutics this requires an identification of the active ingredient, an appropriate determination of the content of the incriminated compound in the preparations, and the determination of exposure levels during therapy. This clearly leads towards a more reliable standardization and characterization of these compound mixtures and will thus improve the safety aspect of the pharmaceutical.

References

Ames BN, Durston WE, Yamasaki E, Lee FD (1973) Carcinogens are mutagens: a simple test system combining liver homogenates for activation and bacteria for detection. Proc Natl Acad Sci USA 70: 2281-2285

Blömeke B, Poginsky B, Schmutte C, Marquardt H, Westendorf J (1992) Formation of genotoxic metabolites from anthraquinone glycosides, present in Rubia tinctorum L.. Mutation Res 265: 263-272

Bösch R, Friedrich U, Lutz WK, Brocker E, Bachmann M, Schlatter Ch (1987) Investigations on DNA binding in rat liver and in Salmonella and on mutage-

nicity in the Ames test by emodin, a natural anthraquinone. Mutation Res 188: 161-168
Brown J, Brown R (1976) Mutagenesis by 9-10-anthraquinones and related compounds in salmonella typhimurium. Mutation Res 40: 203-224
Bruce WB, Heddle JA (1979) The mutagenic activity of 61 agents as determined by the micronucleus, salmonella, and sperm abnormality assay. Can J Genet Cytol 21: 319-334
DeCarli L, Larizza L (1988) Griseofulvin. Mutation Res 195: 91-126
Hongslo JK, Bjoernstad C, Schwarze P, Holme JA (1989) Inhibition of replicative DNA synthesis by paracetamol in V79 Chinese hamster cells. Toxic In Vitro 3: 13-20
Hongslo JK, Bjoerge C, Schwarze PE, Broegger A, Mann G, Thelander L, Holme JA (1990) Paracetamol inhibits replicative DNA synthesis and induces sister chromatid exchange and chromosomal aberrations by inhibition of ribonucleotid reductase. Mutagenesis 5: 475-480
Hongslo JK, Broegger A, Bjoerge C, Holme JA (1991) Increased frequency of sister chromatid exchange and chromatid breaks in lymphocytes after treatment of human volunteers with therapeutic doses of paracetamol. Mutation Res 261: 1-8
Hongslo JK, Smith CV, Brunborg G, Soederlund EJ, Holme JA (1994) Genotoxicity of paracetamol in mice and rats. Mutagenesis 9: 93-100
Hurst EW, Paget GE (1963) Protoporphyrin, cirrhosis and hepatoma in the livers of mice given griseofulvin. Brit J Derm 75: 105-112
Kawasaki Y, Goda Y, Yoshihira K (1992) The mutagenic constituents of Rubia tinctorum. Chem Pharm Bull 40: 1504-1509
Kirkland DJ, Dresp JH, Marshall RR, Baumeister M, Gerloff C, Gocke E (1992) Normal chromosome aberration frequencies in peripheral lymphocytes of healthy human volunteers exposed to a maximum daily dose of paracetamol in a double blind trial. Mutation Res 279: 181-194
Kocisová J, Rossner P, Binková B, Bavorová H, Srám RJ (1988) Mutagenicity studies on paracetamol in human volunteers. I. Cytogenetic analysis of peripheral lymphocytes and lipid peroxidation in plasma. Mutation Res 209: 161-165
Krivobok S, Seigle-Murandi F, Steiman R, Marzin DR, Betina V (1992) Mutagenicity of substituted anthraquinones in the Ames/Salmonella microsome system. Mutation Res 279: 1-8
Mailhes JB, Marchetti F, Aardema MJ (1993) Griseofulvin-induced aneuploidy and meiotic delay in mouse oocytes: effect of dose and harvest time. Mutation Res 300: 155-163
Marchetti F, Mailhes JB (1994) Variation of mouse oocyte sensitivity to griseofulvin-induced aneuploidy and meiotic delay during the first meiotic division. Env Molec Mutagen 23: 179-185
Mori H, Sugie S, Niwa K, Takahashi M, Kawai K (1985) Induction of intestinal tumors in rats by chrysazin. Brit J Cancer 52: 781-783

Mori H, Sugie S, Niwa K, Yoshimi N, Tanaka T, Hirono I (1986) Carcinogenicity of chrysazin in large intestine and liver of mice. Jpn J Cancer Res 77: 871-876

Mori H, Yoshimi N, Iwata H, Mori Y, Hara A, Tanaka T, Kawai K (1990) Carcinogenicity of natural occuring 1-hydroxyanthraquinone in rats: induction of large bowel, liver and stomach neoplasms. Carcinogenesis 11: 799-802

Müller L, Kasper P, Madle S (1991) Further investigations on the clastogenicity of paracetamol and acetylsalicylic acid in vitro. Mutation Res 263: 83-92

National Toxicology Program Report No. 394, 1991

Poginsky B (1991) Evaluation of DNA-binding activity of hydroxyanthaquinones occuring in Rubia tinctorum L.. Carcinogenesis 12: 1265-1271

Ray SD, Kamendulis LM, Gurule MW, Yorkin RD, Corcoran GB (1990) Ca2+ antagonists inhibit DNA fragmentation and toxic cell death induced by acetaminophen. In: Biological Reactive Intermediates IV, Widmer CM, et al. (Eds.), Plenum Press, New York, pp. 699-705

Rustia M, Shubik P (1978) Thyroid tumours in rats and hepatomas in mice after griseofulvin treatment. Brit J Cancer 38: 237-249

Schäfer-Korting M, Korting HC, Mutschler E (1985) Human plasma and skin blister fluid levels of griseofulvin following a single oral dose. Eur J Clin Pharmacol 29: 109-113

Siegers CP, von Hertzberg-Lottin E, Otte M, Schneider B (1993) Anthranoid laxative abuse - a risk for colorectal cancer? Gut 34: 1099-1101

Shen W, Kamendulis LM, Ray SD, Corcoran GB (1992) Acetaminophen-induced cytotoxicity in cultured mouse hepatocytes: Effects of Ca2+-endonuclease, DNA-repair, and glutathion depletion inhibitors on DNA fragmentation and cell death. Toxicol Appl Pharmacol 112: 32-40

Seelbach A, Fissler B, Strohbusch A, Madle S (1993) Development of a modified micronucleus assay in vitro for detection of aneugenic effects. Toxic in vitro 7: 185-193

Sokolowski AL, Montin G, Nilsson A, Olofsson I-M, Sjöberg P (1992) Two-year carcinogenicity study with senna in the rat. Abstract, Sixth Int Congress of Tox, Rome, 28.6.-3.7.92

Tikkanen L, Matsushima T, Natori S (1983) Mutagenicity of anthraquinones in the Salmonella preincubation test. Mutation Res 116: 297-304

Venkataram S, Rogers JA (1988) Characteristics of drug-phospholipid coprecipitates II: bioavailability studies of griseofulvin in rats. J Pharm Sci 77: 933-936

Westendorf J, Poginsky B, Marquardt H, Groth G, Marquardt H (1988) The genotoxicity of lucidin, a natural component of Rubia tinctorum L., and lucidinyl-ether, a component of ethanolic rubia extracts. Cell Biol Toxicol 4: 225-239

Westendorf J, Marquardt H, Poginsky B, Dominiak M, Schmidt J, Marquardt H (1990) Genotoxicity of natural occurring hydroxyanthraquinones. Mutation Res 240: 1-12

Yokoo H, Craig RM, Harwood TR, Cochrane C (1979) Griseofulvin-induced cholestasis in swiss albino mice. Gastroenterology 77: 1082-1087

Safety Assessment of OTC Drugs: Doxylamine Succinate

J.A. Skare,[1] V.A. Murphy,[1] R.C. Bookstaff,[1] G.A. Thompson,[1] M.A. Heise,[1] Z.D. Horowitz,[1] J.H. Powell,[1] A. Parkinson,[2] and J.V. St. Peter.[3]
[1]The Procter & Gamble Company, Cincinnati, OH 45242, USA
[2]University of Kansas Medical Center, Kansas City, KS 66160, USA
[3]Drug Evaluation Unit, Minneapolis MN 55415, USA

Safety assessment of over-the-counter (OTC) drug products frequently involves ongoing evaluation as new data become available on the active ingredients used in these products. This is particularly true for drugs that have been in use for many years and were developed prior to the routine use of certain types of toxicological studies such as the two-year rodent bioassay for evaluation of carcinogenicity. Presented here is a case study involving doxylamine succinate (DOX). DOX is an antihistamine of the ethanolamine class which has a long history of safe and effective use in OTC allergy products, multi-symptom cold medications, and sleep aids and which has recently been the subject of two-year rodent bioassays.

In 1980 it was reported that methapyrilene, an antihistamine structurally unrelated to DOX, produced malignant liver tumors in rats (Lijinsky et al. 1980). Before it became known that the activity of methapyrilene appeared to be specific to the rat and was not shared by structurally similar antihistamines (Lijinsky et al. 1983, Brennan and Creasia 1982, Lijinsky and Kovatch 1986, Habs et al. 1986), the U.S. National Toxicology Program (NTP) initiated rodent carcinogenicity studies with several other antihistamines that are structurally unrelated to methapyrilene. DOX was one of the antihistamines chosen for study, and two-year bioassays in both rats and mice were conducted at the U.S. National Center for Toxicological Research (NCTR).

The relevance of the DOX rodent bioassay results for human risk assessment was deliberated by the U.S. Food and Drug Administration (FDA), which sought the input of two separate advisory committees of independent experts, the Pulmonary and Allergy Drugs Advisory Committee and the Nonprescription Drugs Advisory Committee, in reaching their conclusions. As a result of these deliberations, the FDA concluded that doxylamine is safe and effective for OTC use as an antihistamine, and doxylamine was included in the final monograph for OTC antihistamine drug products (U.S. Fed. Regist., Jan. 28, 1994). This case study will discuss the rodent bioassays conducted with DOX as well as additional studies that have been conducted to address the relevance of the rodent bioassay data for human risk assessment. These additional studies support the conclusion of the FDA's advisory committees that DOX is not likely to have human carcinogenic potential (U.S. Fed. Regist., Jan. 28, 1994).

Genotoxicity Test Results

DOX has been tested in several genotoxicity screening assays (Budroe et al. 1984, Casciano et al. 1986, Zieger et al. 1987, NTP 1987, Muller et al. 1989). These studies utilized a variety of endpoints of genotoxic (DNA reactive) activity both *in vitro* and *in vivo*. DOX showed no evidence of mutagenic activity or clastogenic activity in *in vitro* assays (Ames test, HGPRT locus mutation assay in Chinese hamster ovary cells, chromosomal aberration and sister chromatid exchange (SCE) assays in Chinese hamster ovary cells, SCE assay in cultured human lymphocytes). Additionally, no evidence of genotoxic activity was found when DOX was tested *in vivo* (Chinese hamster micronucleus, mouse embryo SCE, mouse fetal micronucleus).

Inconclusive data in an *in vivo* chromosome aberration assay in mouse embryos is of questionable significance because the response was marginal and seen in embryos only at doses producing clear signs of toxicity to the dams (Muller et al. 1989). A marginal response was obtained in an *in vitro* unscheduled DNA synthesis (UDS) assay in rat hepatocytes (Budroe et al. 1984). The UDS assay has been repeated, and the results did not show any evidence of unscheduled DNA synthesis at doses comparable to and higher than those used by Budroe et al. (1984) (The Procter & Gamble Company, unpublished results). Thus, the results of genotoxicity assays suggest that DOX does not possess genotoxic activity.

NCTR Rodent Bioassay Results

Rat

In the rat bioassay conducted by the NCTR, DOX was administered in the feed to groups of male and female Fischer 344 (F344) rats at dose levels of 0, 500, 1000, and 2000 parts per million (ppm) (based on the free amine) for two years with an interim sacrifice conducted at 65 weeks (Jackson and Blackwell 1993). There were no increases in neoplastic lesions in female rats. Tumors in male rats that showed a statistically significant increase versus the control group in a trend test or in pairwise comparisons are shown in Table 1. These included liver neoplasms which were found only in the high-dose group. A trend test was significant ($p \leq 0.05$) for increased incidence of hepatocellular adenoma and carcinoma with increasing doses of DOX, but the increased incidence of either lesion alone in the high-dose group was not significant compared with controls. When adenomas and carcinomas were combined, the incidence in the high-dose group was significant at the 5% but not 1% level of significance.

Table 1. Incidence of hepatocellular tumors and mononuclear cell leukemia in male rats administered doxylamine[a]

	DOXYLAMINE (CONCENTRATION IN DIET)			
	None (Control)	500 ppm	1000 ppm	2000 ppm
- Hepatocellular adenoma	0/57+	0/57	0/57	3/57
- Hepatocellular carcinoma	0/57+	0/57	0/57	2/57
- Mononuclear cell leukemia	18/57	21/57	30/57+	23/57

[a]Incidence is the number of animals with the tumor/number of animals examined overall, including terminal sacrifice, dead or moribund, and 65-week sacrifice animals. The significance of the "continuity corrected" Cochran-Armitage trend test is indicated beside the control incidence. The significance of Fisher's Exact Test comparing each group with controls is indicated beside that group's incidence. Degree of significance: + = $p \leq 0.05$. These data are reproduced from Jackson (1991) with permission from the author.

Hepatocellular tumors (adenomas and carcinomas) are not uncommon tumors in F344 rats. In studies conducted by the NTP the mean incidence of hepatocellular tumors in control males was 4.1% (range, 0-12%) and 1.0% (range, 0-6%) for hepatocellular adenomas (termed neoplastic nodules in some studies) and carcinomas, respectively (Haseman et al. 1990). Although there are no rigid statistical decision rules that are employed by the NTP in interpretation of rodent carcinogenicity data, for common tumors (occurring at an incidence of at least 1% in historical controls), a statistical significance of $p<0.01$ is generally associated with tumor findings that have been concluded to be treatment-related (Haseman 1983, Haseman 1984). This criterion was not met for the liver tumors in male rats and thus we conclude that the liver tumors in male rats represent equivocal and not conclusive evidence of a treatment-related effect.

The only other tumor type that showed a statistically significant increase in incidence in male rats was mononuclear cell leukemia. However, these tumors are extremely common in F344 rats, only the mid-dose showed a statistically significant increase, and there was no evidence of a dose response. Furthermore, the incidence at the mid-dose was comparable to the incidence in male controls from a concomitant NTCR study (Littlefield et al. 1989) and within the historical control range reported by the NTP (Haseman et al. 1990). These factors suggest that the mononuclear cell leukemias were not treatment-related.

Mouse

In the mouse bioassay conducted by the NCTR, DOX was administered in diet to groups of male and female B6C3F1 mice at dose levels of 0, 190, 375, and 750 ppm (based on free amine) for two years with an interim sacrifice conducted at 65 weeks (Jackson and Sheldon 1993). Tumors that showed a statistically significant increase versus the control group in a trend test and in pairwise comparisons included hepatocellular adenomas and thyroid follicular cell adenomas (Table 2). DOX produced a significant increase in hepatocellular adenomas in the mid- and high-dose groups in males and in the high-dose group in females. There was no increase in the incidence of hepatocellular carcinomas in male mice and no hepatocellular carcinomas observed in any female mice. Thyroid follicular cell adenomas also were increased in treated mice of both sexes. There were no thyroid follicular cell carcinomas observed.

Table 2. Incidence of hepatocellular tumors and thyroid follicular cell tumors in mice administered doxylamine[a]

	DOXYLAMINE (CONCENTRATION IN DIET)			
Males	None (Control)	190 ppm	375 ppm	750 ppm
- Hepatocellular adenoma	6/60+++	12/60	17/59++	31/60+++
- Hepatocellular carcinoma	4/60	4/60	6/59	3/60
- Thyroid follicular cell adenoma	1/58+	0/57	11/57++	5/58

Females	None (Control)	190 ppm	375 ppm	750 ppm
- Hepatocellular adenoma[b]	0/58+++	3/60	0/60	9/60++
- Thyroid follicular cell adenoma	0/56+++	0/60	2/57	8/59++

[a]Incidence is the number of animals with the tumor/number of animals examined overall, including terminal sacrifice, dead or moribund, and 65-week sacrifice animals. The significance of the "continuity corrected" Cochran-Armitage trend test is indicated beside the control incidence. The significance of Fisher's Exact Test comparing each group with controls is indicated beside that group's incidence. Degree of significance: + = $p \leq 0.05$; ++ = $p \leq 0.01$; +++ = $p \leq 0.001$.

[b] No hepatocellular carcinomas were found in female mice

These data are reproduced from Jackson (1991) with permission from the author.

Studies on the Mode of Action of Doxylamine

Treatment of rodents with compounds that induce microsomal enzymes has been associated with the development of both thyroid and hepatocellular tumors in rodent bioassays. The following study was conducted to evaluate DOX as an inducer of liver microsomal enzymes and to determine if DOX affects thyroid hormone balance in B6C3F1 mice. Groups of male and female mice were administered DOX in the diet at dose levels of 0, 40, 375, 750, and 1500 ppm (based on the free amine). In addition, a group of male and female mice received sodium phenobarbital at a dietary level of 375 ppm (expressed as free acid). Half of the animals in each group were treated for 7 days and sacrificed on day 9 and half were treated for 15 days and sacrificed on day 17. Livers were collected from 16 mice/sex/group/time point and 4 livers/sex/group/time point were pooled for the preparation of microsomes. In addition, blood was collected from the mice used for liver microsomal preparation and from an additional 20 mice/sex/group/time point (total of 36 mice) and 4 blood samples/sex/group/time point were pooled for analysis of serum thyroid stimulating hormone (TSH), thyroxine (T4) and triiodothyronine (T3) by radioimmunoassay.

Liver microsomes were analyzed for changes in the activity of P450 enzymes belonging to the CYP1A, CYP2B, CYP2E, CYP3A and CYP4A subfamilies. These particular enzymes were examined to categorize DOX into one of five classes of liver microsomal enzyme inducers based on the pattern of P450 enzymes induced. DOX was also evaluated as an inducer of T4-glucuronosyltransferase activity. Quantitation of enzymatic activity involved either fluorimetric assays (7-pentoxyresorufin O-dealkylation, 7-ethoxyresorufin O-dealkylation) or HPLC procedures with either UV detection (testosterone oxidation and chloroxazone 6-hydroxylation) or radiometric detection (lauric acid hydroxylation).

The results of the study indicate that DOX is a phenobarbital-type inducer of liver microsomal enzymes and that DOX is as effective and as potent as phenobarbital as an inducer in B6C3F1 mice. DOX caused a marked induction of CYP2B enzymes as demonstrated by large increases in O-dealkylation of 7-pentoxyresorufin (up to 38-fold) and 16β-hydroxylation of testosterone (up to 6.9-fold) which are indicative of CYP2B induction The response seen after 15 days of treatment is shown in Figure 1. The response after 7 days of treatment was similar to that seen in Figure 1 (data not shown).

Like phenobarbital, DOX caused a modest (up to 3.3-fold) induction of CYP3A and CYP2A enzymes catalyzing certain hydroxylations of testosterone other than 16β-hydroxylation. At the doses tested, DOX did not appear to induce P450 enzymes in the CYP1A, CYP2E or CYP4A enzyme subfamilies (data not shown). None of the enzyme-inducing effects of DOX could be distinguished from those of phenobarbital.

In male mice DOX and phenobarbital caused a ~50% increase in T4-glucuronosyltransferase activity (Figure 1). This conjugating activity was higher and less inducible in female mice. Although the increases in T4-glucuronosyltransferase activity were small when expressed on a pmol/mg microsomal protein basis (Figure 1), DOX also produced an increase in mg microsomal protein per liver (approximately 80-90% at the 1500 ppm dose). Thus, T4-glucuronosyltransferase activity was increased 2.8-fold and 2.1-fold in male and female mice given 1500 ppm DOX, respectively, when expressed as total hepatic activity. As has been reported by McClain (1989), phenobarbital also showed greater increases when expressed as total hepatic activity (data not shown).

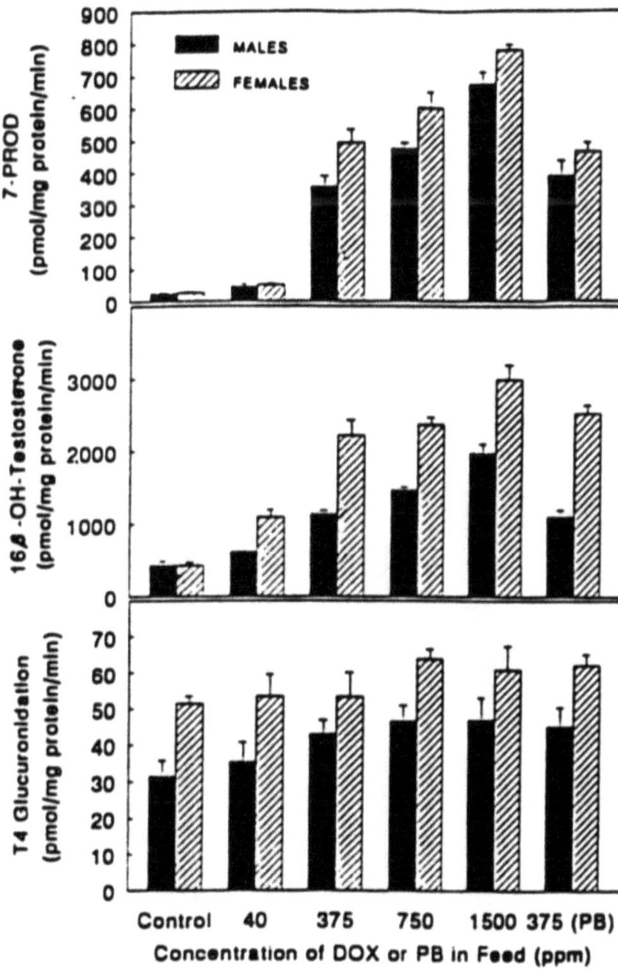

Fig. 1. Effect of 15 days of doxylamine (DOX) or phenobarbital (PB) treatment on 7-pentoxyresorufin O-dealkylation (7-PROD), 16β-hydroxylation of testosterone, and thyroxine (T4) glucuronidation activity per mg liver microsomal protein in B6C3F1 mice. Values are the mean ± standard deviation.

DOX caused a dose-dependent increase in serum TSH levels in B6C3F1 mice. TSH levels in controls were higher in males than in females, and increased 3.0 fold and 4.8 fold in males and females, respectively, after 15 days of exposure to 1500 ppm DOX in diet (Figure 2). A similar pattern was seen after 7 days of treatment, with 3.8-fold and 4.2-fold increases observed in males and females in the 1500 ppm DOX group, respectively (data not shown). Phenobarbital at 375 ppm also increased serum TSH levels, with increases of 1.3-fold and 1.8-fold observed in males and females, respectively, after 15 days of treatment (Figure 2). Serum T4 levels were decreased approximately 20-30% in both males and females given DOX, although the decreases were not dose-related (Figure 2). Similar decreases of approximately 20% were seen after 7 days of DOX treatment (data not shown). Phenobarbital also decreased serum T4 approximately 20% (Figure 2). There were no treatment-related changes in serum T3 levels observed for either DOX or phenobarbital (Figure 2).

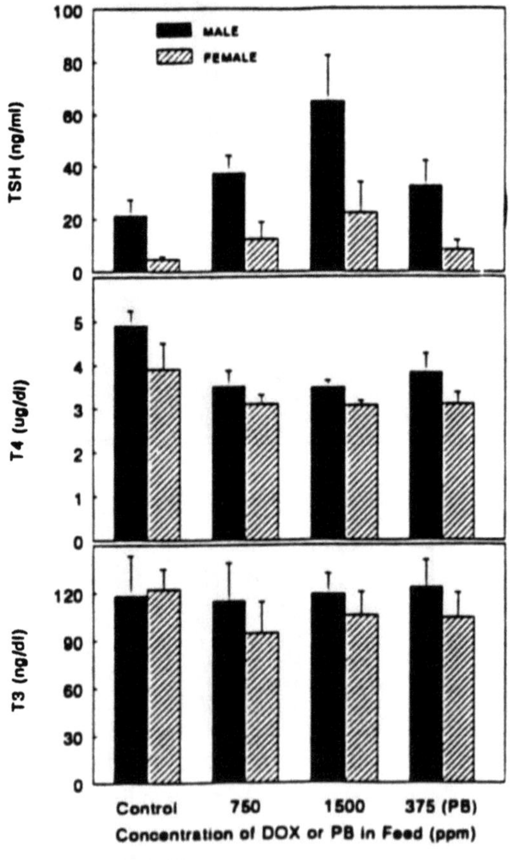

Fig. 2. Effect of 15 days of doxylamine (DOX) or phenobarbital (PB) treatment on serum thyroid stimulating hormone (TSH), thyroxine (T4), and triiodothyronine (T3) concentrations in B6C3F1 mice. Values are the mean ± standard deviation.

Evaluation of Enzyme Induction and Thyroid Function in Humans

In order to characterize the human response to therapeutic doses of DOX with respect to effects on enzyme induction and thyroid hormone balance, a study was conducted in healthy adult male volunteers. Sixteen subjects were evaluated per treatment group using a parallel study design. Subjects in each of three treatment groups received oral doses of 12.5 mg DOX, placebo, or 30 mg phenobarbital every 6 hours for 17 days. Induction of cytochrome P450 activity was assessed based on the pharmacokinetics of antipyrine and its metabolites, and on the urinary recovery of 6β-hydroxycortisol and cortisol. Thyroid function was assessed from measurements of serum concentrations of T4, T3, reverse T3, T3 resin uptake, TSH, and the derived Free Thyroxine Index.

Antipyrine has been widely used as a marker of P450 enzyme activity (St. Peter and Awni 1991, Poulsen and Loft 1988, Danhof and Teunissen 1984, Hartleb 1991, Loft 1990). It is predominantly eliminated by oxidative metabolism to three major metabolites (4-hydroxyantipyrine, norantipyrine, and 3-hydroxymethylantipyrine) which undergo glucuronide conjugation. Although the cytochrome P450 enzyme(s) responsible for the metabolism of antipyrine has not been definitively established, it appears that all drugs that induce oxidative metabolism in humans also significantly increase antipyrine clearance (Park and Kitteringham 1989, Park and Kitteringham 1990). 6β-Hydroxycortisol, a metabolite of cortisol, is another commonly used marker of enzyme induction (Park and Kitteringham 1990, Park and Breckenridge 1981, Ohnhaus et al. 1989, Saenger, 1983). This endogenous substance has been shown to be a specific marker for cytochrome CYP3A induction (Ged et al. 1989, Waxman et al. 1988). The expression of 6β-hydroxycortisol urinary excretion as the ratio of 6β-hydroxycortisol/cortisol adjusts for variations in endogenous production of cortisol, the substrate for the enzyme.

Measurements of the parameters discussed above were made 7 days prior to initiation of drug treatment (baseline) and on days 1, 8, and 15 of treatment. On the days when antipyrine pharmacokinetics were measured, a 500 mg dose of antipyrine was administered i.v. over 5 minutes and plasma samples were obtained at 0 (prior to dosing), 0.5, 1, 2, 4, 6, 12, 24, 28, and 72 hours. Pooled urine samples were collected at 1-6, 6-24, 24-48, and 48-72 hours for HPLC assay of antipyrine and antipyrine metabolites. Urinary excretion of 6β-hydroxycortisol and cortisol was measured on days 1, 8, and 15 of treatment by HPLC assay of pooled urine samples collected at 0-24 hours.

Mean antipyrine total clearance for each of the three treatment groups is shown in Figure 3. Antipyrine total clearance was significantly increased by phenobarbital treatment. This finding is consistent with previously published findings with this drug (Park and Kitteringham 1989, Loft et al. 1987). Analysis of the formation clearances for antipyrine metabolites in the phenobarbital group indicated that only norantipyrine formation clearance was significantly increased over time. The increase in antipyrine total clearance in the phenobarbital-treated subjects was accompanied by a corresponding decrease in terminal exponential half life for antipyrine (data not shown). In contrast, for the DOX-treated subjects, no change in antipyrine total clearance compared to placebo was observed, indicating no change in P450 enzyme activity (i.e., no enzyme induction).

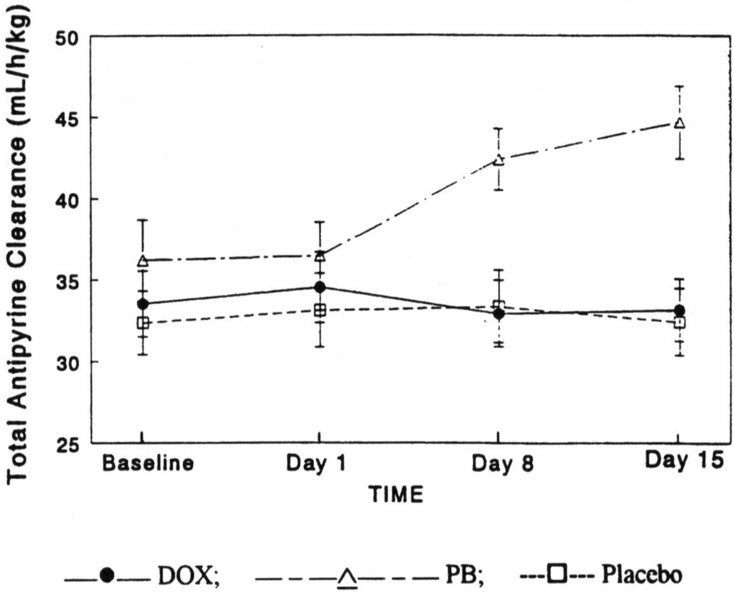

Fig. 3. Effect of doxylamine (DOX), phenobarbital (PB), and placebo on the total antipyrine clearance as a function of treatment time in human male subjects. Values are the mean ± standard error.

Changes from baseline in the excretion of 6β-hydroxycortisol adjusted for cortisol production showed a significant increase relative to placebo only in the subjects treated with phenobarbital for 15 days (Table 3). Previous studies have also indicated that phenobarbital administration results in a significant increase in 6β-hydroxycortisol excretion (Ohnhaus et al. 1989). No difference between the DOX and placebo groups was observed for urinary recovery of 6β-hydroxycortisol adjusted for cortisol production. These findings are consistent with an enzyme-inducing effect of phenobarbital in the study subjects and no measurable effect with DOX.

The results of serum TSH, T4, and T3 analyses are shown in Table 4. There were no significant differences relative to baseline among the treatment groups for any of these parameters. Other thyroid function parameters that were measured (reverse T3, T3 resin uptake, and the derived Free Thyroxine Index) also showed no clinically significant changes (data not shown).

Table 3. Effect of doxylamine (DOX), phenobarbital (PB), and placebo on the ratio of 6β-hydroxycortisol to cortisol in urine as a function of treatment time in human male subjects

Time	DOX	PB	Placebo	Statistical Analyses[a]
Baseline	5.56±2.23[b]	5.10±1.49	6.07±2.31	
Day 1	6.17±2.00	4.04±1.57	5.71±3.02	PLAC DOX PB
Day 8	5.94±2.14	7.16±3.63	6.75±3.64	PLAC DOX PB
Day 15	7.37±3.56	7.92±3.79	6.05±2.33	PLAC DOX PB

[a] Analysis of change from Baseline; DOX is doxylamine; PB is phenobarbital; and PLAC is placebo. Common underline indicates no significant difference. There were no significant differences between treatment groups on Day 1 or Day 8. On Day 15 PB was significantly increased relative to PLAC.
[b] Values are the mean ± standard deviation for 16 subjects.

Table 4. Effect of doxylamine (DOX), phenobarbital (PB), and placebo on serum TSH, T4, and T3 concentrations as a function of treatment time in human male subjects.

	DOX	PB	Placebo
TSH (mIU/L)			
- Screening[b]	2.63±0.92[a]	2.46±0.93	2.50±1.06
- Baseline	1.76±0.65	1.35±0.61	1.49±0.56
- Day 1	3.03±1.12	2.53±0.93	2.58±0.77
- Day 8	3.44±1.63	2.78±1.24	2.54±0.89
- Day 15	3.29±1.37	2.58±1.15	2.56±0.86
T_4 (μg/dl)			
- Screening[b]	7.68±0.96	7.49±1.24	7.19±1.20
- Baseline	7.94±1.26	7.92±0.93	7.58±1.00
- Day 1	7.93±1.38	7.96±1.11	7.72±1.01
- Day 8	7.41±1.01	7.73±1.08	7.63±0.94
- Day 15	7.69±1.23	7.80±0.92	7.76±0.97
T_3 (ng/dl)[c]			
- Baseline	130.1±19.3	126.3±19.1	125.9±21.5
- Day 1	123.2±15.3	121.8±19.4	119.9±15.2
- Day 8	119.8±14.4	124.0±18.2	112.2±11.7
- Day 15	125.9±12.1	126.4±18.6	114.6±17.7

[a] Values are the mean ± standard deviation for 16 subjects. There were no statistically significant differences in change from baseline between treatment groups.
[b] Screening TSH and T4 values were obtained within 28 days of study initiation. Only subjects with values within the normal range were included in the study. Normal ranges: TSH 0.4-4.2 mIU/L; T4 5.3-10.5 ug/dl.
[c]Screening values were not obtained for T3. Normal range: 70-204 ng/dl.

Human Relevance of the Rodent Data

The results of the rodent study presented here indicate that DOX is a phenobarbital-type inducer of liver microsomal enzymes and produces changes in thyroid hormone balance in B6C3F1 mice. The clinical data indicate that there were no measurable differences in parameters indicative of enzyme induction and no effects on thyroid hormone function in humans administered therapeutic doses of DOX. These findings are important in the interpretation of the rodent bioassay results for the purposes of human risk assessment.

There is substantial evidence that microsomal enzyme-inducing compounds can alter thyroid function in rodents and that this is related to the development of thyroid tumors (Hill et al. 1989, McClain 1989, Capen 1992., McClain 1992). Certain microsomal enzyme inducers have been shown to increase the biliary excretion of thyroid hormones in rats, and this appears to be due to an increase in T4-UDP-glucuronosyltransferase activity and an increase in bile flow. This results in a decrease in circulating thyroid hormone levels and increased secretion of pituitary TSH. TSH stimulation of the thyroid gland results in compensatory changes of the follicular cells that include hypertrophy, hyperplasia, and ultimately, thyroid follicular cell tumors in rodents. Phenobarbital is a well studied example of a microsomal enzyme inducer that produces changes in thyroid function in rodents (McClain 1989, McClain et al. 1989) and has been shown to promote the development of rat thyroid tumors in an initiation-promotion model (Hiasa et al 1982, Tsuda et al. 1983, Diwan et al. 1985). A number of liver microsomal enzyme inducing agents have also been shown to increase the incidence of thyroid follicular cell tumors in rat carcinogenicity bioassays (McClain 1989). Given the changes in thyroid hormone function and the increase in T4-UDP-glucuronosyltransferase activity observed with DOX in the mouse, the mode of action of DOX in producing thyroid adenomas is likely similar to that described for other microsomal enzyme inducers.

In contrast to the rodent, there is no evidence linking human use of drugs that are microsomal enzyme inducers with an increased incidence of thyroid neoplasia (Curran and DeGroot 1991). Clinical studies with phenobarbital suggest that therapeutic doses of this drug have only a marginal effect on thyroid hormone function despite the fact that therapeutic doses of phenobarbital induce microsomal enzyme activity in humans. Epidemiologic studies of patients receiving long-term phenobarbital therapy have shown no increased risk of thyroid neoplasia (Clemmesen and Hjalgrim-Jensen 1978, White et al. 1979, Clemmesen and Hjalgrim-Jensen 1981, Shirts et al 1986, Olsen et al. 1989). Furthermore, even in situations where humans have marked changes in thyroid function and elevated TSH levels, as is the case in areas with a high incidence of endemic goiter due to iodine deficiency, there is little, if any, increase in the incidence of thyroid cancer (Doniach 1970, Curran and DeGroot 1991).

Many non-genotoxic compounds that induce hepatic microsomal enzyme activity in rodents also increase the incidence of liver tumors in rodents with chronic treatment (Williams 1990). Although induction of hepatic cytochrome P450 enzymes and liver tumor promotion are closely associated (Lubet et al. 1989), there is insufficient evidence to support microsomal enzyme induction *per se* as a cause for the development of liver tumors. Nevertheless, P450 induction serves as an indicator that adaptive changes in cellular function have occurred in the liver. In the case of doxylamine, the mouse bioassay results also indicate the presence of other toxicity in the liver (hepatocellular necrosis) (Jackson and Sheldon 1993). A review of the literature suggests that under conditions where such adaptive and/or toxic changes do not occur, the development of tumors will also not occur (Grasso et al. 1991).

Despite the demonstrated susceptibility of rodents to the development of liver tumors when treated with certain microsomal enzyme inducers, evidence is lacking to suggest that humans treated with microsomal enzyme inducing drugs are at increased risk for the development of liver cancer. This is best exemplified by the epidemiologic data on phenobarbital. The weight of the epidemiologic evidence indicates that phenobarbital is not associated with an increased risk of liver cancer when patients receiving Thorotrast, a known human liver carcinogen, are excluded (Clemmeson and Hjalgrim-Jensen 1978, White et al. 1979, Clemmesen and Hjalgrim-Jensen 1981, Shirts et al. 1986, Olsen et al. 1989). An International Agency for Research on Cancer (IARC) review concluded that phenobarbital "should be classified among the compounds for which, at present, the epidemiological evidence weighs so strongly against the possibility that it is carcinogenic to man that one can disregard, for practical purposes, laboratory evidence that is of doubtful generality" (IARC 1985). We conclude that the data indicating a DOX-related increase in liver and thyroid tumors in rodents is also not generalizable to humans. In summary, the rodent data on mode of action, together with the clinical data, support the conclusion made by FDA's advisory committees that DOX is not likely to have human carcinogenic potential.

References

Brennan LM and Creasia DA (1982) The effects of methapyrilene hydrochloride on hepatocarcinogenicity and pentobarbital-induced sleeping time in rats and mice. Toxicol. Appl. Pharmacol. 66, 252-258.

Budroe JD, Shaddock JG and Casciano DA (1984) A study of the potential genotoxicity of methapyrilene and related antihistamines using the hepatocyte/DNA repair assay. Mutat. Res. 135, 131-137.

Capen CC (1992) Pathophysiology of chemical injury of the thyroid gland. Toxicol. Lett. 64/65, 381-388.

Casciano DA, Schol HM, Shaddock JG and Feuers RJ (1986) Failure of methapyrilene and selected analogues to induce mutations in the Chinese hamster ovary mutagenesis assay system. Environ. Mutagen. 8 (Suppl. 6), 16.

Clemmesen J and Hjalgrim-Jensen S (1978) Is phenobarbital carcinogenic? A follow-up of 8078 epileptics. Ecotoxicol. Environ. Safety 1, 457-470.

Clemmesen J and Hjalgrim-Jensen S (1981) Does phenobarbital cause intracranial tumors? A follow-up through 35 years. Ecotoxicol. Environ. Safety 5, 255-260.

Curran PG and DeGroot LJ (1991) The effect of hepatic enzyme-inducing drugs on thyroid hormones and the thyroid gland. Endocrine Rev. 12, 135-150.

Danhof M and Teunissen WE (1984) Antipyrine as a model drug to assess oxidative drug metabolizing activity in man. Pharm. Int. 5, 11-15.

Diwan BA, Palmer AE, Ohshima M and Rice JM (1985) N-nitroso-N-methylurea initiation in multiple tissues for organ-specific tumor promotion in rats by phenobarbital. J. Natl. Cancer Inst. 75, 1099-1105.

Doniach I (1970) Aetiological consideration of thyroid carcinoma. In: Tumors of the thyroid gland. Vol. 6. D Smithers (ed) E & S Livingstone, London, 55-72.

Ged C, Rouillon JM, Pichard L, Combalert J, Bressot N, Bories P, Michel H, Beaune P and Maurel P (1989) The increase in urinary excretion of 6β-hydroxycortisol as a marker of human hepatic cytochrome P450IIIA induction. Br. J. Clin. Pharmacol. 28, 373-387.

Grasso P, Sharratt M and Cohen AJ (1991) Role of persistent, non-genotoxic tissue damage in rodent cancer and relevance to humans. Annu. Rev. Pharmacol.Toxicol. 31, 253-287.

Habs, M, Shubik, P and Eisenbrand, G (1986) Carcinogenicity of methapyrilene hydrochloride, mepyramine hydrochloride, thenyldiamine hydrochloride and pyribenzamine hydrochloride in Sprague-Dawley rats. J. Cancer Res. Clin.. Oncol. 111, 71-74.

Hartleb M (1991) Drugs and the liver. Part II. The role of the antipyrine test in drug metabolism studies. Biopharm. Drug Dispos. 12, 559-570.

Haseman JK (1983) A reexamination of false-positive rates for carcinogenicity studies. Fund. Appl. Toxicol. 3, 334-339.

Haseman JK (1984) Statistical issues in the design, analysis and interpretation of animal carcinogenicity studies. Environ. Health Perspect. 58, 385-392.

Haseman JK, Arnold J and Eustis SL (1990) Tumor incidences in F344 rats: NTP historical data. In: Pathology of the Fischer Rat - Reference and Altas. Boorman, GA, Eustis SL, Elwell, MR, Montgomery Jr. CA, and MacKenzie W (eds.) Academic Press, 555-564.

Hiasa Y, Kitahori Y, Ohshima M, Fujita T, Yuasa T, Konishi N and Miyashiro A (1982) Promoting effects of phenobarbital and barbital on development of thyroid tumors in rats treated with N-bis (2-hydroxypropyl) nitrosamine. Carcinogenesis 3, 1187-1190.

Hill RN, Erdreich, LS, Paynter, OE, Roberts PA, Rosenthal SL and Wilkinson CF (1989) Thyroid follicular cell carcinogenesis. Fund. Appl. Toxicol. 12, 629-697.

IARC Scientific Publications No. 65 (1985) Interpretation of negative epidemiological evidence for carcinogenicity. Proceedings of a symposium held in Oxford 4-6 July 1983. N.J. Wald and R. Doll (eds.), 151-160.

Jackson CD (1991) Chronic study of doxylamine in Fischer 344 rats and B6C3F1 mice. NCTR Technical Report for Experiment 406 and 407. April.

Jackson CD and Blackwell BN (1993) Two-year toxicity study of doxylamine succinate in the Fischer 344 rat. J. Am. Coll. Toxicol. 12, 1-11.

Jackson CD and Sheldon W (1993) Two-year toxicity study of doxylamine succinate in B6C3F1 mice. J. Am. Coll. Toxicol. 12, 311-321.

Lijinsky W, Reuber MD and Blackwell BN (1980) Liver tumors induced in rats by oral administration of the antihistaminic methapyrilene hydrochloride. Science 209, 817-819.

Lijinsky W, Knutsen G and Reuber MD (1983) Failure of methapyrilene to induce tumors in hamsters and guinea pigs. J. Toxicol. Environ. Health 12, 653-657.

Lijinsky W and Kovatch RM (1986) Carcinogenicity studies of some analogs of the carcinogen methapyrilene in F344 rats. J. Cancer Res. Clin. Oncol. 112, 57-60.

Littlefield NA, Gaylor DW, Blackwell BN and Allen RR, (1989) Chronic toxicity/carcinogenicity studies of Gentian Violet in Fischer 344 rats: two-generation exposure. Fd. Chem. Toxicol. 27, 239-247.

Loft S (1990) Metronidazole and antipyrine as probes for the study of foreign compound metabolism. Pharmacol. Toxicol. 66 (Suppl. VI), 1-31.

Loft S, Sonne J, Poulsen HE, Petersen KT, Jorgensen BG and Dossing M (1987) Inhibition and induction of metronidazole and antipyrine metabolism. Eur. J. Clin. Pharmacol. 32, 35-41.

Lubet RA, Nims RW, Ward JM, Rice JM and Diwan BA (1989) Induction of cytochrome P450b and its relationship to liver tumor promotion. J. Am. Coll. Toxicol. 8, 259-268.

McClain RM (1989) The significance of hepatic microsomal enzyme induction and altered thyroid function in rats: implications for thyroid gland neoplasia. Toxicol. Pathol. 17, 294-306.

McClain RM (1992) Thyroid gland neoplasia: non-genotoxic mechanisms. Toxicol. Lett. 64/65, 397-408.

McClain RM, Levin AA, Posch R and Downing JC (1989) The effect of phenobarbital on the metabolism and excretion of thyroxine in rats. Toxicol. Appl. Pharmacol. 99, 216-228.

Muller L, Korte A and Madle S (1989) Mutagenicity testing of doxylamine succinate, an antinauseant drug. Toxicol. Lett. 49, 79-86.

National Toxicology Program Results Report (1987) Results and status information on all NTP chemicals. July 7.

Ohnhaus EE, Breckenridge AM and Park BK (1989) Urinary excretion of 6β-hydroxycortisol and the time course measurement of enzyme induction in man. Eur. J. Clin. Pharmacol. 36, 39-46.

Olsen JH, Boice JD, Jensen JPA and Fraumeni, JF (1989) Cancer among epileptic patients exposed to anticonvulsant drugs. J. Natl. Cancer Inst. 81, 803-808.

Park BK and Breckenridge AM (1981) Clinical implications of enzyme induction and enzyme inhibition. Clin. Pharmacokinet. 6, 1-24.

Park BK and Kitteringham NR (1989) Relevance and means of assessing induction and inhibition of drug metabolism in man. In Progress in Drug Metabolism, Gibson GG, (ed.) Taylor & Francis, New York, 1-60.

Park BK and Kitteringham NR (1990) Assessment of enzyme induction and enzyme inhibition in humans: toxicological implications. Xenobiotica 20, 1171-1185.

Poulsen HE and Loft S (1988) Antipyrine as a model drug to study hepatic drug-metabolizing capacity. J. Hepatol. 6, 374-382.

Saenger P (1983) 6β-Hydroxycortisol in random urine samples as an indicator of enzyme induction. Clin. Pharmacol. Ther. 34, 818-821.

Shirts SB, Annegers JF, Hauser WA and Kurland LT (1986) Cancer incidence in a cohort of patients with seizure disorders. J. Natl. Cancer Inst. 77, 83-87.

St. Peter JV and Awni WM (1991) Quantifying hepatic function in the presence of liver disease with phenazone (antipyrine) and its metabolites. Clin. Pharmacokinet. 20, 50-65.

Tsuda H, Fukushima S, Imaida K, Kurata Y and Ito N (1983) Organ-specific promoting effect of phenobarbital and saccharin in induction of thyroid, liver, and urinary bladder tumors in rats after initiation with N-nitrosomethylurea. Cancer Res. 43, 3292-3296.

U.S. Fed. Regist. (Jan. 28, 1994) Vol. 59, No. 19, 4216-4218.

Waxman DJ, Attisano C, Guengerich FP and Lapenson DP (1988) Human liver microsomal steroid metabolism: identification of the major microsomal steroid hormone 6-beta-hydroxylase cytochrome P-450 enzyme. Arch. Biochem. Biophys. 263, 424-436.

White SJ, McLean AEM and Howland C (1979) Anticonvulsant drugs and cancer. A cohort study in patients with severe epilepsy. Lancet 2, 458-461.

Williams GM (1990) Epigenetic mechanisms of liver tumor promotion. In: Mouse Liver Carcinogenesis: Mechanisms and Species Comparisons. Stevenson, DE et al. (eds) Alan R. Liss, Inc., New York, 131-145.

Zeiger E, Anderson B, Haworth S Lawlor T, Mortelmans K and Speck W (1987) Salmonella mutagenicity tests III. Results from the testing of 255 chemicals. Environ. Mutagen. 9 (Suppl. 9), 1-110.

Safety Assessment of OTC Products - A Regulatory View

P. Sjöberg
Division of Pharmacology & Toxicology, Medical Products Agency, Box 607, 751 25 Uppsala, Sweden

When considering whether a medicinal product can be granted OTC status i.e. used as a non-prescription drug, a number of criteria need to be addressed and assessed. Conceivably, these criteria should be essentially similar world-wide. However, as judged by considerable national differences in the availability of non-prescription drugs, there are inconsistencies in how regulatory agencies assess the suitability of products for self-medication. The main criteria used by the Swedish Medical Products Agency are i) suitability for self-care indications ii) treatment should not mask serious symptoms or diseases iii) there should be adequate proof of safety at normal use, misuse and overdosing iv) absence of CNS stimulation, tolerance or dependency.

Suitability fo self-care

A medicinal product which is suitable for self medication should be used for the treatment of illness which the customer or the patient can diagnose himself or recognize the symptoms when the doctor has made the diagnosis. This should be feasible with a high degree of accuracy. Moreover, the individual should be able to assess the result of the treatment. Examples of generally accepted indications for self-care medication are eczema, fever, constipation, cough, nasal catarrh/nasal congestion, various types of pain/ache, travel sickness and pollen allergy (runny nose/sneezing/eye complaints).

Masking of serious symptoms/diseases

The criterion that treatment with an OTC product should not mask serious symptoms/diseases is illustrated by the application for OTC status of H2-antagonists for prevention and treatment of dyspepsia (epigastric pain and discomfort). In a recent study (Bytzer et al. 1994), it was found, not unexpectedly, that empirical treatment with ranitidine as opposed to endoscopy misses peptic ulcer cases and also tumours. The empirical strategy postponed the diagnosis of malignant disorders by a few weeks and no cancers were missed after a longer follow up. Thus, it needs to be assessed whether self-medication with H2-antagonists will delay the consultation with a physician to such an extent that the risks associated with peptic ulcer and tumours will increase to an unacceptable level.

Proof of safety

Obviously, any product being considered for OTC sale should have a low toxicity profile. It should be devoid of serious side effects in the normal population. Type A side effects, i.e. those related to the pharmacological effect, may be acceptable if they are associated with identified risk groups. Moreover, if there is a risk of overdosing, the overdose adverse reaction profile should be known and acceptable. Paracetamol is an example of an OTC product which is associated with considerable toxicity when overdosed or misused. When paracetamol was introduced as an OTC product in Sweden in 1988, its ability to induce severe liver toxicity when acutely overdosed was well documented. Since then, additional data have become available which question the OTC status of paracetamol-containing products. The Swedish Poison Information Centre reported an incidence of 35-50 cases/year of unintentional overdosing (repeated intake of higher than recommended dose) (Person et al. 1991). A retrospective study of 5,000 case reports of hospitalized intoxications from the same period (1988-1990), revealed 10 cases of severe or very severe intoxications with paracetamol. It was estimated that the true number of severe or very severe intoxications due to unintentional overdosing was twice as high. In contrast to overdosing with e.g. salicylates, there are no symptoms which warn the patient for continued use of high doses of paracetamol. The low margin of safety and the serious effects which may occur during unintentional overdosing, suggest the need for an in depth analysis of what can be done to minimize the risk associated with this OTC product

A low potential for toxicity also includes the absence of significant interaction problems. For products which undergo extensive metabolism, information on responsible isoenzymes are a prerequisite when assessing the potential for metabolic interactions. Terfenadine is a compound which has OTC status in a number of countries and for which there is a significant potential for metabolic interaction. Concomitant administration of compounds which are potent inhibitors of the cytochrome P-450 enzyme CYP3A4, such as ketoconazole and erythromycin, may cause QT-prolongation and cardiac arrhythmias including torsade de pointes (Honig et al. 1993; Woosley et al. 1994). The same effects are reported in overdosed patients and in patients with impaired hepatic function. It would appear that some regulatory agencies are too concerned about these properties of terfenadine to grant the drug OTC status.

Another safety requirement for OTC products is that they should not be associated with significant risks for developmental toxicity. Unfortunately, such risks can seldom be analyzed from clinical data and the assessment of risk has to be based almost entirely on data from animal studies. A similar situation is at hand when assessing the risk for carcinogenicity. Since extrapolations from animals to man have to be done, it is particularly difficult to define or determine what constitutes a significant risk. In assessing the results of developmental toxicity studies one also needs to consider the statistical power of the animal tests. The commonly used number of animals only allows the detection of a 5-12 fold increase in malformations, a 3-6 fold increase in *in utero* deaths and a 15-25% decrease in fetal weight (Kavlock 1993).

Loratadine, an H1-antagonist which is used in pollen allergy, examplifies an OTC compound which at high doses in rats induces reproductive disorders, dystocia and fetal deaths. These effects are seen at systemic exposure levels more than 10 times higher than those achieved in man (Swedish PDR 1994). Despite the low power of the animal tests, these safety margins have been considered to be sufficiently high. For a number of the nonsteroidal anti-inflammatory drugs (NSAID's) there are human data available on reproductive disturbances. These drugs, which inhibit prostaglandin synthesis, may inhibit uterine contractions and delay the delivery. They may cause pulmonary hypertension and neonatal respiratory insufficiency by contraction or closure of the ductus arteriosus *in utero*. Furthermore, the platelet function in the fetus might be inhibited (Swedish PDR 1994). When confronted with such information one would tend to consider these drugs too toxic for granting them OTC status. However, since the clinical experience with these drugs is very extensive, there is a consensus that the benefit of having relatively potent NSAID's available as non-prescription drugs outweighs the risks described above and also other risks associated with this type of drugs.

If the assessment of developmental toxicity rests to a considerable extent on pre-clinical data, the assessment of carcinogenic risk does so entirely. Products under consideration for OTC status are generally non-genotoxic. An exception to this rule is paracetamol, a compound which at high doses also causes liver tumours (Flaks and Flaks 1983; Flaks et al. 1985). A detailed analysis of these effects and their relevance to therapeutic use is discussed by others during this workshop. When discussing the potential carcinogenicity of OTC products it should be realized that almost 50% of the pharmaceuticals tested in conventional bioassays turn out positive in at least one species (Bentley et al. 1992). The mechanisms by which the majority of these non-genotoxic compounds induce tumours can be classified into three categories; i) compounds which cause an excessive secretion of trophic hormones controlling endocrine organs ii) compounds which activate specific receptors involved in growth and differentiation and iii) compounds which cause chronic cell injury resulting in cell proliferation and hyperplasia (Purchase 1992). The rationale for allowing the use of animal carcinogens as prescription only drugs or as OTC compounds is based on the fact that mechanistic information is available and that pharmacokinetics and exposure levels have been thoroughly elucidated to allow a meaningful risk assessment and the determination of acceptable safety factors. This is the case for paracetamol and loratadine.

Absence of CNS stimulation, tolerance and dependency

An additional property which needs to be addressed when assessing whether a compound should be granted OTC status or not is its capability of inducing CNS stimulation and tolerance. Nicotine possesses both attributes, but is nevertheless available prescription-free since nicotine substitutes nicotine from tobaccoo, the use of which is much more harmful. Again, this emphasizes that there is no rule without exception and that the circumstances under which a drug is used and, in particular, the available clinical experience will determine to what extent generally accepted rules can be overlooked.

Conclusions

A number of safety criteria need to be fulfilled before granting a medicinal product OTC or non-prescription drug status. The treatment should not mask serious symptoms or diseases and the compound should have an acceptable toxicity profile when used at recommended doses as well as when overdosed and used concomitantly with other drugs. OTC products should be devoid of significant developmental toxicity and the risk for tumour development should be negligible. In order for the regulator to accomplish a prompt and fair evaluation of an application for OTC status, these issues should be addressed in significant detail; the assessment being accompanied by good quality data.

References

Bentley P, Baier H, Krug E (1992) Results from carcinogenicity tests in two rodent species - do they improve or confuse decision making? - an industry viewpoint. In McAuslane JAN, Lumley CE, Walker SR (eds) The carcinogenicity debate. Quay Publishing, London, pp 111-20

Bytzer P, Hansen JM, Schaffalitzky de Muckadell OB (1994) Empirical H2-blocker therapy or prompt endoscopy in management of dyspepsia. Lancet 343 (April 2):811-816

Flaks A, Flaks B (1983) Induction of liver cell tumours in IF mice by paracetamol. Carcinogenesis 4:363-68

Flaks B, Flaks A, Shaw APW (1985) Induction by paracetamol of bladder and liver tumours in the rat. Acta Path Microbiol Immunol scanc Sect A 93:367-77

Honig PK, Wortham DC, Zamani K et al. (1993) Terfenadine-ketoconazole

interaction. Pharmacokinetic and elctrocardiographic consequences. JAMA 269:1513-18
Kavlock RJ (1993) Indirect effects on development through maternal toxicity. In Sundvall A et al (eds) Developmental toxicology - preclinical and clinical data in retrospect. Tryckgruppen, Stockholm, pp 37-43
Person H, Sjöberg G, Torell E (1991) Förgiftningsrisk vid subakut terapevtisk överdosering av paracetamol. Läkartidningen 88:3678-80
Purchase I (1992) Current knowledge of mechanisms of carcinogenicity: genotoxins versus non-genotoxins. In McAuslane JAN, Lumley CE, Walker SR (eds) The carcinogenicity debate. Quay Publishing, London, pp 3-24
Woosley RL, Chen Y, Freiman JP et al. (1993) Mechanism of the cardiotoxic actions of terfenadine. JAMA 269:1532-36

Drug toxicity findings and no action taken

F.W. Wolff
Department of Medicine, The George Washington University,
School of Medicine, Washington, D.C. U.S.A.

In 1634 Fabricius Hildanus, town physician at Bern, stated "I find it my duty to think and complain against the lying false chemists and rascals that are the destroyer of man ... the physician should have known about the property of the drug ... (Hildanus 1963).

In 1811 Andrew Mathias wrote on the nature of mercurial disease, produced by the therapeutic use of mercury (Mathias 1811). As late as 1912 there was a discussion of the frequent occurrence of death following salvarsan infusion for syphilis (Hoffman 1913).

European Directive No. 75318/EEC states that drugs intended for long-term use should be tested in two animal species, which are exposed for a period of six months. Elsewhere in the Directive, one finds emphasis on the fact that the form of exposure should be akin to that for the intended human use (Ascher 1994).

There appears to be little emphasis on the role of drug metabolism in the selection of the appropriate animal species for preclinical toxicity tests, the administration of a minute amount of drug to a normal volunteer (in the olden days the researcher him or herself), the results of study of its metabolites by modern analytical techniques, that will aid in the selection of the animal species for toxicity tests.

Finally, clinical testing must not be commenced until all the results of animal studies, including detailed histological reports, are at hand and show no problem.

The nucleoside analogue fialuridine (FIAU) underwent clinical trial for hepatitis B in the United States in 1993, in some 15 patients. After five patients had died of heptic toxicity, the clinical trial was stopped. A long-term feeding study in rats then (!) uncovered liver toxicity. A high powered panel was called to investigate this disaster. Its report absolves the investigators, and makes strange reading to a toxicologist (Marshal 1994).

Western countries used to look to the United States Food and Drug Administration for in-depth review and assessment of preclinical toxicology and pharmacology testing. The more-than-four-year delay in the warning concerning the use of Aspirin in little children and Reye´s Syndrome was, perhaps, responsible for as many as 1700 unnecessary death (Davis 1992).

Immitrex (Summatriptan Succinate), a novel and effective drug for migraine, has been associated with some forteen serious reactions, including death in a one year period in the United States, and larger numbers elsewhere (FDA 1994).

Toxicity testing is traditionally assumed to be limited to preclinical and premarketing testing, prior to the use approval of a drug in man. This situation has changed because of less-than-effective regulatory supervision of new drug introduction, and poorly supervised writing of labels for physicians and patients.

The drug adverse reaction database, including deaths, has become an important component of toxicity testing. Although there are some who belittle the significance of the adverse reaction reporting system, examples abound showing its usefulness. An outstanding example was the spontaneous reaction reporting of phocomelia after thalidomide (Weicker 1962).

Novel molecules may have predictive toxicity based on structure-function relationships, laboratory tests, and library searches. Examples abound where findings of toxicity were not taken cognizance of or followed up.

Sulfonylureas and biguanides are still used widely in maturity onset diabetes. In the only lengthy controlled study to date (Klimt 1970), these drugs have been shown to increase cardiovascular mortality, compared to diet or diet and insulin.

Phenformin-related lactic acidosis led to its withdrawal in the United States in 1977 (Avery 1980). The closely related drug metformin, with some 255 cases of lactic acidosis on record (FDC 1994), is now being considered for introduction in the United States, and like phenformin, is already on the market in many european countries.

Drugs such as benoxaprofen (Oralflex), flexin, zomepirac sodium (Zomac), midazolam hydrochloride (Versed), triazure (Azarabine), flecainide (Tambocor), are reported to have serious toxicity, but through oversight, were either permitted on the market by the United States Food and Drug Administration, or did not have sufficient warning in the label (Sigelmann 1994).

The following lessons are appropriate for toxicologists involved in drug development:

1. Animal toxicity tests remain a vital component and require being assessed fully, before a drug is tried in man.
2. Drug regulatory bodies, in spite of good intentions, are imperfect. The writing of drug labels has to take cognizance of the general poor training of practicing physicians in toxicology, pharmacology, and therapeutics, and the difficulties to keep up to date.

4. Scientists in all of the above disciplines must bear in mind the overriding rule not to hurt patients, and must insist on remaining part of it throughout the drug development process. Their primary loyalty must be to science and man, and not to commercial advantage.

References

Ascher W Sir (1994) Personal communication

Avery GS (1980) Drug treatment. Second Edition, 506

Davis LD, Buffler P (1992) Reduction of deaths after drug labelling for risk of Reye´s syndrome. Lancet 340:1042

FDC Reports (1994) 56:12-16

Food and Drug Administration (1994) FOI report

Hildanus F (1963) Von der Furtrefflichkeit und Nutz der Anatomie. Aarau, Leipzig, HR Sauerlander & Co

Hoffmann E (1913) Dermatolog. Ztschr. 19:1043

Klimt R et al (1970) Increased cardiovascular mortality with sulfonylureas. Diabetes 19, 2:744-830

Marshal E (1994) Hepatitis study: drug trial deaths deemed unavoidable. Science 264:1530

Mathias A (1811) The mercurial disease. Philadelphia: Howard Parker

Sigelmann DW (1994) Turning the table on drug companies. Exposing deficiencies in FDA regulations. Trial 3:72

Weicker H and Hungerland H (1962) Deutsch. Med. Wschr. 87:992

Species specificity of organ toxicity

(Chair: Dr. W. Butler, United Kingdom)

Species Specificity of Organ Toxicity

W.H. Butler

BIBRA Toxicology International, Woodmansterne Road,

Carshalton, Surrey SM5 4DS, Great Britain.

Introduction

The species specificity of target organ toxicity has been the basis for the understanding of toxic mechanism and development of new active chemicals in the areas of pharmaceutics and agrochemicals. The demonstration of selective toxicity is evident in comparative studies in both animals and plants. The development of selective herbicides is a clear example from the plant kingdom.

In this symposium we will be considering specific examples from mammalian toxicity which will illustrate the wide ranging nature of selective toxicity. The diversity of the effects reflect the differing anatomy, physiology, pharmacology and biochemistry between species which have been recognized for decades as well as the more recent behavioural studies.

What is becoming increasingly obvious is the requirement to demonstrate the mechanisms of species specificity of toxicity in the course of the development of a new product if a claim is made that the findings are to be extrapolated to man. In the area of developing pharmaceutics comparative pharmacology is normal but comparative studies are less common in the field of agrochemicals where long-term studies are frequently demanded. In the field of carcinogenicity some success has been achieved in demonstrating the species specificity of a $\alpha 2\mu$ globulin nephropathy and renal adenomas (Dietrich & Swenberg, 1991) and thyroid tumours following enzyme induction in rat (Alison *et al.* 1994).

While not wishing to anticipate the remarks of the speakers in this symposium I would like to briefly illustrate some areas where species specificity is important in identifying hazards and estimating the risk to man. There are many examples illustrating species specific organ toxicity and I will only select a few of these from my own experience to illustrate some general points before the presentations of the speakers.

It is self evident that there are profound anatomical differences between mammalian species. A example is that the common laboratory rat is albino and the absence of pigment will modify the response to a number of agents which may accumulate in melanin or effect ocular toxicity. Compounds will induce a light dependant retinal atrophy in the absence melanin. Examples of compounds showing this effect is clonidine (Noell *et al.* 1966; Weisse *et al.* 1971) as well as the new anti-epileptic vigabatrin (Butler *et al.* 1987). In the latter case albino rats developed a retinal atrophy while pigmented rats had normal retinas (Butler *et al.* 1987) as did pigmented mice, dogs and primates (Gibson *et al.* 1990). Therefore it was reasonable to suggest that man was not at risk from this adverse effect.

Studies with the β_2 agonist, medroxalol have shown that leiomyomas of the uterus and other β_2 agonist (Sells & Gibson, 1987) induce leiomyomas of the mesovarian ligament. The incidence of the lesions is clearly dose dependant and depends on the presence of the appropriate receptor. The neoplastic effect can be blocked by the administration of the β_2 antagonist propanolol (Jack *et al.* 1983). Therefore it is possible to demonstrate species difference in response depending upon receptor sites (Apperley *et al.* 1978) and relate this to both anatomical structure and receptor sites in man.

Differences in the biochemical profile clearly modify the response of animals. The induction of hepatic peroxisomes and the concomitant hepatocyte proliferation is species specific (Bentley *et al.* 1993; Lake *et al.* 1989). This can be shown in increasing liver weight, peroxisome specificity enzymes and labelling index (Lake *et al.* 1989; Parzefall *et al.* 1991). Such differences correlate with the ability to induce tumours. It is of interest in the extrapolation across species that there is little evidence to suggest that peroxisomes are inducible primates (Lake *et al.* 1989) or man (Lake, B.G., personal communication).

A further interesting example of species specificity is the effects of the induction of the mixed function oxidase enzymes in the liver. It is clear that rat, mouse and man may have these enzyme systems induced. However, the morphological sequelae of this induction is species specific. In the rat and mouse hepatic hypertrophy and hyperplasia are readily induced but such changes are not observed in man. The long-term effects of rat and mouse differ. In susceptible strains of mice a range of putative neoplastic lesions are induced by a wide range of compounds (Evans *et al.* 1992) but a similar range of neoplasia is not observed in the rat and no such lesions have been reported in man. It is these types of data that leads to the observation that in the U.S. National Toxicology Programme that the mouse liver is the commonest site for the induction of neoplasia. The converse of the observations of nodules in the mouse liver following enzyme induction is that in the rat liver, neoplasia are rarely observed while thyroid tumours are induced. This is now recognized to be secondary to the enhanced metabolism of T3 and T4 (Alison *et al.* 1994). In the mouse, thyroid tumours are not increased.

Tubular adenomas in mice arising following the sequence of ovarian atrophy, oestrogen depletion and enhanced FSH production is a further example of a species specific neoplasia secondary to endocrine dysfunction (Alison *et al.* 1994).

In way of introduction I have illustrated some general examples of species specificity of target organ toxicity. More specific examples will be given in this workshop showing how pathological, pharmacological and toxicokinetic and behavioural difference may be identified and the mechanisms elucidated. The recognition of these differences and the understanding of the mechanisms is essential for extrapolation such data from laboratory animals and man.

References

Alison, R.H., Capen, C.C. and Prentice, D.E. (1994). Neoplastic lesions of Questionable Significance to humans. Toxic. Path. **22**, p.179-186

Apperley, G.H., Brittain, R.T., Coleman, R.A., Kennedy I. and Levy, G.P. (1978). Characterization of the β adrenoceptor in the mesovarium of the rat. Br. J. Pharmac. **63**, p.345-346.

Bentley, P., Calder, I., Elcombe, C., Grasso, P., Struger, D. and Wiegand H.-J. (1993). Hepatic peroxisome proliferation in rodents and its significance for humans. Fd. Chem. Toxic. **31**, p.857-907.

Butler, W.H., Ford, G.P. and Newberne, J.W. (1987). A study of the effects of Vigabatrin on the central nervous system and retina of Sprague Dawley and Lister Hooded rats. Toxic. Path. **15**, p.143-148.

Dietrich, D.R. and Swenberg, J.A. (1991). Preneoplastic lesions in rodent kidney induced spontaneously or by non-genotoxic agents: predictive nature and comparison to lesions induced by genotoxic carcinogens. Mutation Res. **248**, p.239-260.

Evans, J.G., Collins, M.A., Lake, B.G. and Butler W.H. (1992). The histology and development of hepatic nodules and carcinoma in C3H/He and C57/BL/6 mice following chronic Phenobarbital Administration. Toxic. Path. **20**, p.585-594.

Gibson, J.P., Yarrington, J.T., Loudy, D.E., Gerbig C.G., Hurst, G.H. and Newberne J.W. (1990). Chronic toxicity studies with Vigabatrin, a GABA - Transaminase inhibitions. Toxic. Path. **18**, p.225-238.

Jack, D., Poynter, D. and Spurling N.W. (1983). Beta-adrenoceptor stimulants and mesovarian leiomyomas. Toxicology **27**, p.315-320.

Lake, B.S., Evans, J.G., Gray, T.J.B., Körösi, S.A. and North, C.J. (1989). Comparative studies on Nafenopin-induced Hepatic Peroxisome Proliferation in the rat, Syrian hamster, guinea pig and marmoset. Toxic. appl. Pharmac. **99**, p.148-160.

Noell, W.K., Walker, V.S., Kang, B.S. and Berman, S. (1966). Retinal damage by light in rats. Investve Ophthal. **55**, p.450-473.

Parzefall, W., Erter, E., Sedivy, R. and Schulte-Herman, R. (1991). Testing for induction of DNA synthesis in human hepatocyte primary cultures by rat liver tumour promoters. Cancer Res. **51**, p.1143-1147.

Sells, D.M. and Gibson, J.P. (1987). Carcinogenicity studies with Medroxalol Hydrochloride in rats and mice. Toxic. Path. **15**, p.457-467.

Weisse, I., Stotzer, H., Knappen, F. and Walland, A. (1971). The effect of chlonidine on the pupil diameter and the retina in rats, assessed in relation to the intensity of light. Arzneim Forsch Drug Res. **21**, p.821-825.

Species Specificity at the Molecular Level: the Case of Nitric Oxide Synthases

E. Felley-Bosco

Institute of Pharmacology and Toxicology, Rue du Bugnon 27, 1005 Lausanne, Switzerland.

Introduction

Nitric oxide (NO) is synthesized from the amino acid L-arginine (L-Arg) by two different types of NO synthase (NOS) (reviewed by Marletta, 1993, Knowles and Moncada 1994, Forstermann et al, 1994). The first type is represented by constitutive NOS, Ca^{++}/calmodulin dependent enzymes which deliver NO for short periods in response to receptor activation or physical stimulation. This type of NOS proteins is derived from two different genes, nNOS and eNOS, and are found in the brain, vascular endothelium, platelets and the cortex and medulla of adrenal gland in human and rodent species. The second type of NOS activity is called inducible (iNOS) and appears to be Ca^{++} independent. However, calmodulin is found tightly associated with the protein and a calcium concentration lower than the baseline level existing in the cytosol is sufficient to keep it active (Cho et al., 1992). Once expressed, iNOS synthesizes NO for long periods. The induction of iNOS is very different in human versus rodent species and will be discussed in detail below.

All NOS proteins are dimers under native conditions. They can be cytosolic or reversibly bound on membranes. NOS incorporates molecular oxygen into both NO and citrulline using L-Arg as a substrate. All isoforms require NADPH, flavins and 5,6,7,8 tetrahydrobiopterine (BH4) as cofactors. The latter is probably involved in the dimer stabilization (Baek et al, 1994). Flavins are arranged within NOS such that electrons are transferred: NADPH -->FAD-->FMN-->oxygenase component (Fig.1).

Recently NOS have been described to contain a heme (White and Marletta, 1992; McMillan et al, 1992). These functional and structural properties suggest that NOS belong to the cytochrome P450 (cytP450) family (Nelson et al, 1993). Gene analysis (see below) indicates that NOS might indeed represent a precursor of cytP450 enzymes. In that case, NOS would be the first of these enzymes located in the cytosol.

The aim of this chapter is to highlight some features of NOS proteins that present an interest from the point of view of a toxicologist with particular emphasis on iNOS since its activity is apparently totally different in human versus rodent species.

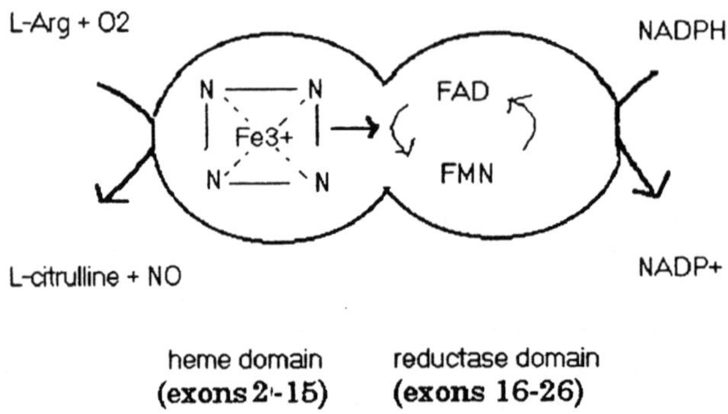

Fig.1 Schematic view of NOS: interaction of the flavin reductase and the heme oxygenase domain (adapted from Marletta, 1993; Marsden et al, 1993 and Chartrain et al, 1994).

NOS gene: coding sequence

The NOS comprise a family that can be distinguished on the basis of comparative DNA analysis. Sequences have been reported for the cDNAs encoding the constitutive neuronal NO synthases (nNOS) from rat and man, the constitutive endothelial NOS (eNOS) from bovine and human tissue, and the iNOS from a rodent macrophage line and human hepatocytes, chondrocytes and colon cell line.

A comparison of deduced protein sequences derived from the three classes of NOS cDNA show, overall, 50-60% similarity.

The C-terminal domain shows a high degree of similarity with the enzyme cytP450 reductase. Indeed it contains the sites for the flavins and NADPH cofactors and it is contained within the last eleven exons (see Fig.1).

In the N-terminal sequence (exons 2-15, Fig.1), there are the binding sites for the substrate, the heme and the tetrahydrobiopterine. These sites have not been fully identified yet. As mentioned above, NOS show convergent evolution with cytP450 family of enzymes. For this reason, heme binding sequence of the cyt P450 has been sought in iNOS. No match could be found (Nelson et al, 1993). Since in some cytP450 the heme binding site contains a cysteine-isoleucine (CI) sequence we sought such dipeptide in NOS and compared the consensus obtained for the three classes of NOS to protein sequences in a data bank (SwissProt). Surprisingly we found sequence similarity to heme binding sites in various proteins distinct from cytP450 (Fig.2). This confirms that if NOS is part of cytP450 family, it probably represents an ancestor gene.

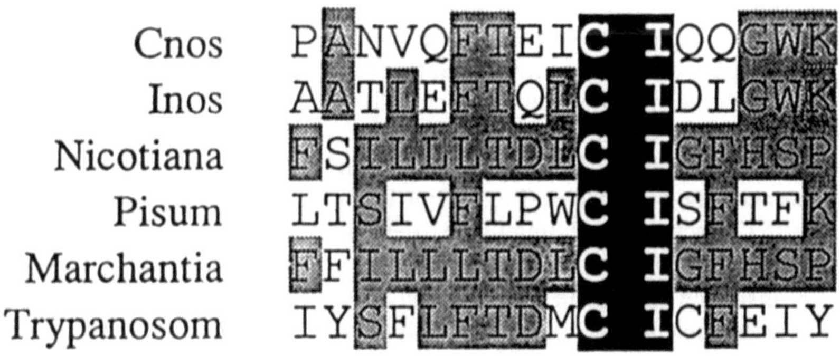

Fig.2 Alignment with the potential heme binding site of rat brain NOS (cNOS), mouse macrophage inducible NOS (iNOS) and the heme binding sites of common tobacco heme binding protein (Nicotiana), garden pea heme binding protein (Pisum), liver wort heme binding protein (Marchantia), and Trypanosoma brucei cytochrome c oxidase (Trypanosom).

The constitutive nNOS and eNOS seem to be highly conserved between species (93% amino acid identity between rat and human nNOS, 94% identity between bovine and human NOS). However, there is only 81% identity between human iNOS and rodent iNOS and the reason for such a difference is not clear at the present time.

NOS gene: promoter sequence

To date only the promoter sequences for the human eNOS and murine and human iNOS have been described (Marsden et al, 1993; Q. Xie et al, 1993, Lowenstein et al, 1993, Chartrain et al, 1994). The human eNOS has a « TATA-less » type promoter and the 5' flanking region contains proximal promoter elements consistent with a constitutively expressed gene that is found in endothelial cells, namely Spl and GATA motifs. In addition there are putative AP-1, AP-2, NF-1, heavy metal, acute-phase response, shear stress, and sterol-regulatory cis-elements (Marsden, 1993).

The inducible human and murine NOS promoters contain both a « TATA-box ». The murine iNOS promoter has been functionally divided in two domains: a) a downstream region mediating the lipopolysaccharide (LPS) induction and b) an upstream region crucial for response to interferon-γ. Since LPS is not able to induce iNOS in human cells, we performed a comparison of the available human and rodent iNOS promoter sequences, corresponding to the LPS sensitive region, using the Bestfit programm of the Genetics Computer Group, to investigate whether the difference might be justified by sequence divergence. This programm makes an optimal alignment of the best segment of similarity between two sequences using the local homology algorithm (Smith and Waterman, 1981). The analysis proved 72% identity between the two species (Fig.3). Furthermore, although LPS alone is not able to induce iNOS activity in human, the sites likely to be transactivated by factors stimulated by LPS (NF-κB and NF-IL6) are in the conserved regions and contain one base-pair mismatch in human and it would be interesting to check if this unique change is responsible for the different sensitivity to LPS.

Furthermore, stimulation by LPS alone or in addition with interferon-γ and/or TNFα and IL-1β is able to induce iNOS in a murine macrophage cell line but not in human macrophages. However, recent results (Kolb et al, 1994; Paul-Eugene et al, 1994) indicate that in the latter, IL-4 mediates the stimulation of iNOS activity by increasing the expression of the low-affinity receptor for immunoglobulin E (CD23). Subsequent ligation of CD23 with IgE/anti-IgE immune complexes induces the release of NO. It is known that IL-4 has different effects in murine versus human monocytes; for example in the latter it reduces their capacity to produce reactive oxygen species. Thus, IL-4 induced iNOS activity in human monocytes results in a totally different biological function than the one described for murine iNOS (Hibbs et al, 1990).

```
              .         .         .         .         .
-391 TGAGAGGATGGACAGGGATTAACCAGAGAGCTCCCTGCTGAGGAAAAAAT human
     |||||| |  |||||  |    |||||||||||| |||  || | |||||
-356 TGAGAGAACAGACAGAAA...GCCAGAGAGCTCCGTGCCCAGAACAAAAT murine

              .         .         .         .         .
     CTCCCAGATGCTGAAAGTGAGGCCATGTGGCTTGGCCAAATAAAACCTGG
     | | |||  |||| |||  |||  |||||| || || ||  ||| || |
     CCCTCAGCAGCTGCAAGCCAGGGTATGTGGTTTAGCTAAGAAAAGCCAGC

              .         .         .         .         .
     CTCC.......GTGGTGCCTCTATCTTAGCAGCCACCCTGCTGATGAACT
     ||||       |||   ||    | | |   | |  | |||||   |
     CTCCCTCCCTAGTGAGTCCCAGTTTTGAAGTGACTACGTGCTGCCTAGGG

              .         .         .         .         .
     GCCA...CCTTGGACTTGGGACCAGAAAGAGGTGG.GTTGGGTGAAGAGG
     ||||   ||||||||  | |||||| ||||| |||  |||  ||| ||
     GCCACTGCCTTGGACGGGCGACCAGGAAGAGATGGCCTTGCATGAGGATA

              .         .         .         .         .
     CACCACACAGAGTGATGTAACAGCAAGATCAGGTCACCCACAGGCCCTGG
     |||  |||||||||||||||             |||  ||||    || |
     CAC..CACAGAGTGATGTAA.............TCAAGCACACAGACTAG

                     TNF-RE                 NF-κB
              .         .         .         .         .
     CAGTCACAGTCATAAATTAGCTAACTGTACACA...AGCTGGGGACACTC
      |||  |  |||| ||| |||||||| | ||||   | |||||||| |||
     GAGTGTCCATCATGAATGAGCTAACT.TGCACACCCAACTGGGGACTCTC

       NF-IL6
              .         .         .         .         .
     CCTTTGGAAACCAAAAAAAAAAAAAAAAAAAAAGAGACCTTTATGCAAAAA
     ||||||| |                          ||  |||||||||
     CCTTTGGGA..........................ACAGTTATGCAAA..

                              TATA-Box
              .         .         .         .
     CAACTCTCTGGATGGCATGGGGTGAGTATAAATACTTCTTGGCTGC
      |   ||||| |  || ||| | | |||||||||| |  |||||||
     .ATAGCTCTGCAGAGCCTGGAG.GGGTATAAATACCTGATGGCTGC
```

Fig.3 Alignment of human (GenBank L26055) and murine (GenBank L23806) iNOS promoter sequences. The analysis shows that several transcription factors binding sites (TNF-resposive element, NF-κB binding site and NF-IL6 binding site) are conserved or contain only a single mismatch.

NOS activity: influence of cofactors and substrate

In the presence of L-Arg and BH4, the iNOS dimer produces citrullin and NO. The latter, in absence of superoxide anion, is readily transformed in nitrite which can be measured in the conditioned culture medium of cells expressing iNOS. This provides an easy way to investigate the effect of cofactors added directly to the cells on iNOS activity. Since it is known that BH4 synthesis varies within tissues (Werner-Felmayer et al, 1991), we tested the effect of BH4 addition on human bronchial epithelial cells transfected with the murine iNOS gene (2b-nos, Fig.4). The addition of BH4 slightly increases nitrite production in 2b-nos indicating that BH4 might act as a limiting factor if its synthesis is not proportional to the iNOS protein.

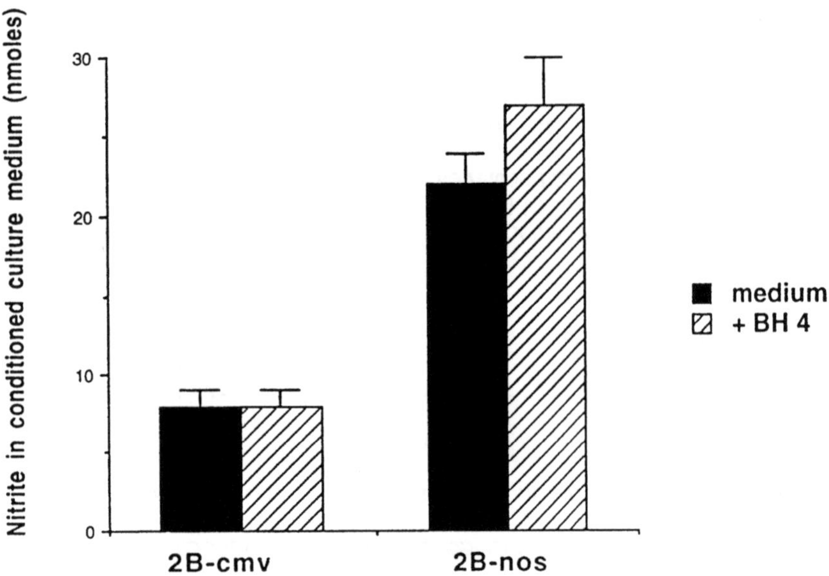

Fig. 4 Effect of BH4 addition on NO production, measured as nitrite in the conditioned culture medium, in human bronchial epithelial cells transfected with murine iNOS gene (2b-nos). No effect is observed in cells transfected with the control vector (2b-cmv).

Experiments *in vitro* have demonstrated that, in the absence of L-Arg, NOS induces uncoupled NADPH oxidation producing oxygen radicals (Abu-Soud and Stuehr, 1993, Baek et al, 1993, Klatt et al, 1994). *In vivo* L-Arg is supplied to cells by carriers that mediate facilitated diffusion and are specific for cationic amino acids (Ildicho Closs et al, 1993). This transport system, called system y^+, consists of at least three membrane proteins, CAT-1, CAT-2A and CAT-2B. CAT-1 is widely expressed in murine tissues and CAT-2A and CAT-2B are expressed in hepatocytes and stimulated macrophage, respectively, and result from differential splicing of transcripts from a single gene. CAT-2 has a 10-fold greater K_M for L-Arg, compared to CAT-1. It is not yet known how the expression of these transporters correlates with NOS activity but it is clear that they are essential for iNOS activity in stimulated macrophages (Hibbs et, 1990). In addition, the availaibility of L-Arg will depend on diet (Barbul, 1990) and arginase activity (Mills et al, 1992).

Cellular targets of NOS activity

The cellular targets of NOS activity will be different depending if L-Arg is present as substrate or not (see above). In the case of NO formation (Fig. 5) the targets are represented by a) the soluble guanylate cyclase (sGC), which mediates NO activity in signal transduction, b)[4Fe-4S]proteins, which are responsible for NO-induced cytostasis and c) nucleotides, which can be deaminated. On the other side, in the absence of the L-Arg, all known targets of oxygen radicals (Cerutti and Trump, 1991) are potentially involved.

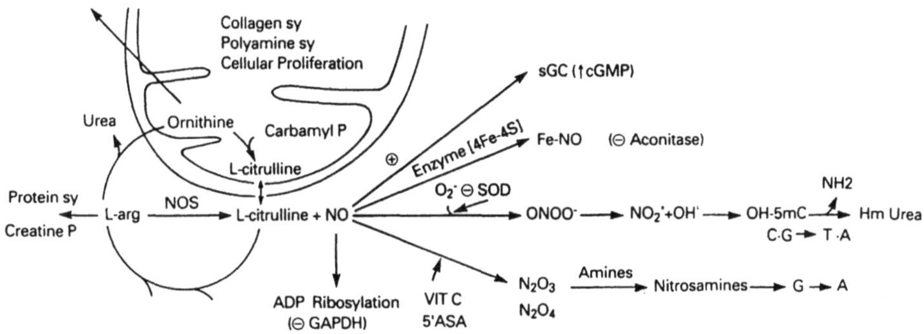

Fig.5 Cellular targets of NO include soluble guanylate cyclase (sGC), [4Fe-4S] enzymes, DNA and glyceraldehyde phosphate deshydrogenase (GAPDH).

The stimulation of sGC is apparently the main function of NO in human stimulated macrophages (Kolb et al, 1994), while in rodent macrophages NO is essentially cytostatic (Hibbs et al, 1990). The sGC is present not only in human macrophages but also in human neutrophils (Wyatt et al, 1993), human hepatocytes (Nussler et al, 1992) and bronchial epithelial cells (Felley-Bosco et al, 1994).

The interaction with [4Fe-4S]enzymes not only leads to cytostasis by the inhibition of aconitase and mitochondrial complex I (Hibbs et al, 1990) but also modulates iron metabolism. Indeed, when aconitase loses an atom of iron in its Fe-S cluster by Fe-NO reaction, it is transformed in the iron regulating factor (IRF) which maintains iron homeostasis (Drapier et al, 1993; Weiss et al, 1993).

NO is mutagenic to *S. thyphimurium* TA1535 (Isomura et al, 1984, Wink et al, 1991, Arroyo et al 1992,) and is able to deaminate deoxyribonucleosides, nucleotides, purine and calf thymus DNA in an aerobic environment *in vitro* (Wink et al 1991, Nguyen et al, 1992). NO was proved mutagenic in human TK6 lymphoblastoid cells (Nguyen et al, 1992). To our knowledge no study showing NO mutagenicity in rodent cells has been reported to date.

Conclusions

The analysis of the control and effects of NOS activity demonstrates a major variability between species in particular for iNOS. The difference is most evident between rodent and human cells. The iNOS coding sequence is less conserved between these two species, compared to the nNOS and eNOS. The absence of LPS induction of iNOS in human cells is apparently not due to the promoter sequence since the sites rensponsive to LPS activated factors are conserved. Finally NO synthesized from iNOS appears to have a different role in human versus rodent cells, since human macrophages iNOS is essentially involved in signal transduction.

These data indicate that it is necessary to make the appropriate species choice when the effects of iNOS activity are going to be investigated.

References

Abu-Soud H.H. and D.J. Stuehr (1993) Nitric oxyde synthases reveal a role for calmodulin in controlling electron transfer. Proc.Natl.Acad.Sci, USA, 90: 10769-10772.

Arroyo P.L., V. Hatch-Pigott, H.F. Mower, and R.V. Cooney (1992) Mutagenicity of nitric oxide and its inhibition by antioxidants. Mutat. Res. 281:193-202.

Baek K.J., B.A. Thiel, S. Lucas, and D. Stuehr (1994) Macrophage nitric oxide synthase subunits. J. Biol. Chem. 268: 21120-21129.

Barbul A. (1990) Physiology and pharmacology of arginine. In « Nitric oxide from L-arginine: a bioregulatory system » Moncada S and Higgs E.A. eds., Elsevier Science Publ., pp. 317-329.

Cerutti P.A. and B. F.Trump (1991) Inflammation and oxidative stress in carcinogenesis. Cancer Cells 3:1-7.

Chartrain N.A., Geller D.A., P.P. Koty, N.F. Sitrin, A. K. Nussler, E. P. Hoffman, T. R. Billiar, N. I. Hutchinson and J.S. Mudgett (1994) Molecular cloning, structure, and chromosomal localization of the human inducible nitric oxide sythase gene. J. Biol. Chem. 269: 6765-6772.

Cho H.J., Q. Xie, J. Calaycay, R. A. Mumford, K.M. Swidereck, T.D. Lee , and C. Nathan (1992) Calmodulin is a subunit of nitric oxide synthase from macrophages. J.Exp. Med. 176:599-604.

Drapier J-C, H. Hirling, J. Wietzerbin, P. Kaldy and L. C. Kuhn (1993) Biosynthesis of nitric oxide activates iron regulatory factor in macrophages. Embo J. 12: 3643-3649.
Felley-Bosco, E., Ambs, S., Lowenstein, C.J., Keefer, L.K., and Harris, C.C. Constitutive expression of inducible nitric oxide synthase in human bronchial epithelial cells induces c-fos and stimulates the cGMP pathway. Am.J. Respir. Cell Mol. Biol., in press.
Fostermann U., E.I. Closs, J.S. Pollock, M. Nakane, P. Schwarz, I. Gath and H. Kleinert (1994) Nitric oxide sythase isozymes. Hypertension 23: 1121-1131.
Hibbs J.B. , R.R. Taintor, Z. Vavrin, D.L. Granger, J-C Drapier, I.J. Amber and J.R. Lancaster Jr (1990) Synthesis of nitric oxide from a terminal guanidino nitrogen atom of L-arginine: a molecular mechanism regulating cellular proliferation that targets intracellular iron. In " Nitric oxide from L-arginine: a bioregulatory system. S. Moncada and E.A. Higgs ed., pp 189-223.
Ildicho Closs E., C.R. Lyons, C. Kelly and J.M. Cunningham (1993) Characterization of the third member of the MCAT family of cationic amino acid transporters. J. Biol. Chem. 268: 20796-20800.
Isomura, K., Chikahira M., Teranishi K., and Hamada K. (1984) Induction of mutations and chromosome aberrations in lung cells following in vivo exposure of rats to nitrogen oxides. Mut. Res., 136:119-125.
Klatt P., M. Schmid, E. Leopold, K. Schmidt, E.R. Werner and B. Mayer (1994) The pteridine binding site of brain nitric oxide synthase. J. Biol. Chem. 269: 13861-13866.
Knowles R.G. and S. Moncada (1994) Nitric oxide synthases in mammals. Biochem. J. 298: 249-258.
Kolb J.P., N. Paul-Eugene, C. Damais, K. Yamaoka, J.C. Drapier and B. Dugas (1994) interleukin-4 stimulates cGMP production by IFN-γ activated human monocytes. J. Biol. Chem. 269: 9811-9816.
Lowenstein C. J., E. W. Alley, P. Raval, A. M. Snowman, S.H. Snyder, S. W. Russell and W. J. Murphy (1993) Macrophage nitric oxide synthase gene: two upstream regions mediate induction by interferon γ and lipopolysaccharide. Proc. Natl. Acad. Sci. USA 90: 9730-9734.
Marletta M.A. (1993) Nitric oxide synthase structure and mechanism. J. Biol. Chem. 268: 12231-12234.
Marsden P.A., H.Q. Heng, S.W. Scherer, R.J. Stewart, A.V. Hall, X. Shi, L. Tsui, and K.T. Schappert (1993) Structure and chromosomal localization of the human constitutive endothelial nitric oxide synthase gene. J. Biol. Chem. 268: 17478-17488.
McMillan K., D.S.Bredt, D.J. Hirsch, S.H. Snyder, J.E. Clark, and B. S. Siler- Masters (1992) Cloned, expressed rat cerebellar nitric oxyde synthase contains stoichiometric amounts of heme, which binds carbon monoxide. Proc. Natl.Acad.Sci U.S.A. 89: 11141-11145.
Mills C.D., J. Shearer, R. Evans and M.D. Caldwell (1992) Macrophage arginine metabolism and the inhibition or stimulation of cancer. J. Immunol. 149: 2709-2714.
Nelson D.R., T. Kamataki, D.J. Waxman, F.P. Guemgerich, R. W. Estabrook, R. Feyereisen, F.J. Gonzales, M.J. Coon, I. C. Gunsalus, O. Gotoh, K. Okuda, and D.W. Nebert (1993) The P450 superfamily: update on new sequences, gene mapping, accession numbers, early trivial names of enzymes, and nomenclature. DNA and Cell Biol. 12: 1-51.
Nguyen T., Brunson D., Crespi C.L., Penman B.W., Wishnok J.S., and Tannenbaum S.R. (1992) DNA damage and mutation in human cells exposed to nitric oxide in vitro. Proc.Natl.Acad.Sci U.S.A 89:3030-3034.
Nussler A.K., Di Silvio M., Billiar T.R., Hoffman R.A., Geller D.A., Selby R., Madariaga J., and Simmons R.L. (1992) Stimulation of the nitric oxide synthase pathway in human hepatocytes by cytokines and endotoxin. J Exp.Med 176:261-264.
Paul-Eugene N., J.P. Kolb, C. Damais, A. Abadie, J.M. Mencia-Huerta, P. Braquet, J. Bousquet and B. Dugas (1994) β2-adrenoceptor agonists regulate IL-4-induced phenotypical changes and IgE-dependent functions in normal human monocytes. J. Leukoc. Biol. 55: 313-320.

Smith T.F. and M.S. Waterman (1981) Comparison of bio-sequences. Advances in applied mathematics 2: 482-489.

Weiss G., B. Goossen, W. Doppler, D. Fuchs, K. Pantopoulos, G. Werner-Felmayer, H. Wachter and M. W. Hentze (1993) Translation regulation via iron-responsive elements by the nitric oxide/NO-synthase pathway. Embo J. 12: 3651-3657.

Werner-Felmayer G., E.R. Werner, D. Fuchs, A. Hausen, G. Reibnegger and H. Wachter (1991) On multiple forms of NO synthase and their occurrence in human cells. Res. Immunol. 142: 555-561.

Wink D.A., Kasprzak K.S., Maragos C.M., Elespuru R.K., Misra M., Dunams T.M., Cebula T.A., Koch W.H., Andrews A.W., Allen J.S., and L.K. Keefer (1991) DNA deaminating ability and genotoxicity of nitric oxide and its progenitors. Science 254:1001-1003.

Wyatt T.A., T.M. Lincoln, and K.B. Prizwansky (1993) Regulation of human neutrophil degranulation by LY-83583 and l-arginine: role of cGMP-dependent protein kinase. Am. J. Physiol 265:C201-C211.

Xie Q., R. Whishnant and C. Nathan (1993) promoter of the mouse gene encoding calcium-independent nitric oxide synthase confers inducibility by interferon γ and bacterial lipopolysaccharide. J. Exp. Med. 177: 1779-1784.

Mechanisms Underlying Species-Specificity in Target Organ Toxicity.

Edward A Lock

Research Toxicology Section, Zeneca Central Toxicology Laboratory, Alderley Park, Cheshire, SK10 4TJ.

Introduction

One of the commonest reasons for a species difference in toxicity is a species difference in the route or rate of biotransformation of a chemical to its toxic metabolite. The biotransformation of foreign compounds usually occurs in two stages termed phase 1 and phase 2 reactions. The major phase 1 reactions include oxidation, reduction and hydrolysis, leading to the conversion of parent compound into substrate's suitable for phase 2 conjugation reactions. The interplay between the two phases is important for the biotransformation of foreign compounds, particularly with regard to the balance between activation and detoxification.

Activation of many chemicals is now widely accepted as the first step leading to toxic or carcinogenic effects. By far the most common mechanism for such activation is mixed-function oxidation by the haemoprotein cytochrome P450. A number of isozymic forms of this enzyme exist within each species, and their expression may be tissue-specific as well as species-specific. The different isozymic forms preferentially oxidise different substrates, and thus species differences in cytochrome P450 profile may contribute to species differences in toxicity. Many of the reactive electrophiles formed by cytochromes P450 mediated oxidation may undergo conjugation with glutathione and thereby inactivation. The formation of glutathione conjugates is catalysed by glutathione S-transferases, which also exist in multiple forms some being tissue-specific as well as species-specific analogous to the cytochromes P450.

The rate of metabolism of a chemical may also be a critical factor, certain metabolic routes can become saturated, thereby diverting metabolism via another route, which may lead to toxicity. Alternatively depletion of a cofactor such as glutathione may lead to an inability to protect against the generation of reactive electrophiles, and hence initiate the toxic process. Thus the balance between activation and detoxification is complex and in each case there is a need to establish the route and rate of metabolism of the particular chemical under study to help ascertain the basis for the mechanism of any toxicity.

However not all species differences in toxicity can be explained in terms of metabolism, different biochemical or physiological functions may also be important factors. With a chemical such as the herbicide paraquat, which undergoes little if any metabolism, the toxicity is due to the parent compound. Species differences in the toxicity of paraquat are primarily due to a species difference in (1) absorption of the dication across the gastrointestinal tract, (2) accumulation of the chemical via an energy-dependant polyamine transport system into alveolar type 1 and 2 epithelial cells and (3) the renal excretion and clearance of the chemical. In certain situations there may be a more profound species difference in response which may only be explained in terms of a major difference in biology. One such example is that of the male rat-specific protein nephropathy and carcinogenicity produced by a number of chemicals, which can be explained in terms of a male rat-specific protein, α2u-globulin.

It is not the intention to cover in this article all aspects of the metabolic basis for species-specific target organ toxicity, those interested in more detailed treatises are referred to the following books and the references therein (Hodgson and Guthrie, 1980; Timbrell, 1991) For this review I have selected two examples to illustrate some of the key points raised in the introduction.

Species Difference in the Route of Metabolism

The chemicals 2-[1-(ethoxyimino)propyl]-3-hydroxy-5-(2,4,6-trimethylphenyl)cyclohex-2-enone [ETC]) and 1-[4-(3-acetyl-2,4,6-trimethylphenyl)-2,6-cyclohexanedionyl]-O-ethyl propionaldehyde oxime [ATMP](Fig 1), both produce hepatic protoporphyria in the mouse following a single oral dose, while other species such as the rat, hamster or guinea pig are refactory to this injury (Brady et al. 1993; Frater et al. 1993).

Figure 1 ETC and ATMP

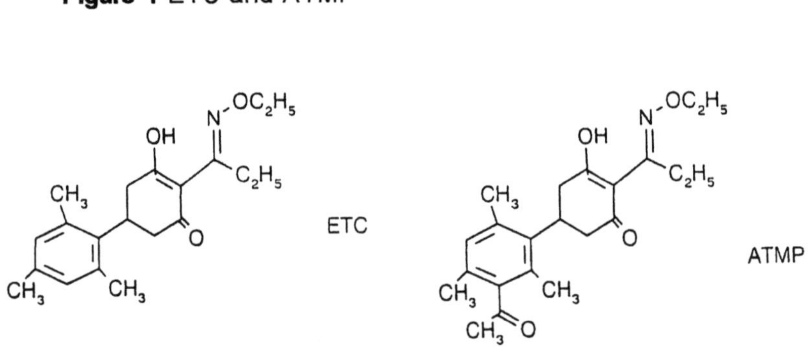

Hepatic protoporphyria is a disorder of liver haem metabolism characterised by inhibition of the enzyme ferrochelatase, leading to the accumulation of protoporphyrin IX, the substrate for the enzyme. Continuous administration of either ETC or ATMP to mice results in a marked protoporphyrin accumulation in the biliary system, leading to cholestasis and biliary cirrhosis. Other drugs have been reported to induce such a disorder, for example 3,5-diethoxycarbonyl-1,4- dihydrocollidine (DDC) and griseofulvin (GF). Both these drugs are thought to be metabolised in the liver by cytochromes P450 as suicide substrates, so that a fragment of the drug (a methyl group) becomes attached to the haem moiety of the enzyme, producing N-methylprotoporphyrin (De Matteis et al. 1980; Oritz deMontellano et al. 1981) a green pigment which accumulates in the liver and is a powerful inhibitor of ferrochelatase (DeMatteis et al. 1987; Marks et al. 1988). To ascertain the basis for the species difference in response to ETC and ATMP we investigated whether these compounds, which are structurally unrelated to DDC or GF, produced protoporphyria by a similar mechanism to these two drugs. A single dose of either ATMP or ETC to mice produced a marked accumulation of hepatic porphyrins within 24hr which was associated with a marked inhibition of liver ferrochelatase (Brady et al. 1993; Frater et al. 1993) and an accumulation in the liver of a green pigment which can inhibit ferrochelatase (Frater et al. 1993). The green pigment was isolated and identified as N-methylprotoporphyrin (Frater et al. 1993). Inhibitors of hepatic cytochrome P450, when given to mice before ATMP, afforded protection against ATMP-induced protoporphyria and the production of N-methylprotoporphyrin, suggesting a role for cytochrome P450 in the production of the metabolic disorder (Frater et al. 1993). Thus the mechanism of toxicity appears to be analogous to that reported for both DDC and GF. Studies with ETC have shown that the origin of the methyl group in N-methylprotoporphyrin isolated from the livers of treated mice is the drug.

Studies in other laboratory species with both ATMP and ETC showed no accumulation of porphyrins in the liver or inhibition of mitochondrial ferrochelatase (Brady et al. 1993; Frater et al. 1993). These findings indicate that the isozymic form of cytochrome P 450 that activates ATMP and ETC leading to the release of the methyl group which alkylates the haem moiety is either absent or only present at extremely low levels in rat, hamster or guinea pig liver. Hence the mechanistic basis for the species difference in protoporphyria is presumably due to a species difference in the route (and rate?) of formation of the reactive metabolite which can alkylate the haem moiety (Fig 2).

Figure 2 Probable basis for the species difference in hepatic protoporphyria produced by ATMP and ETC

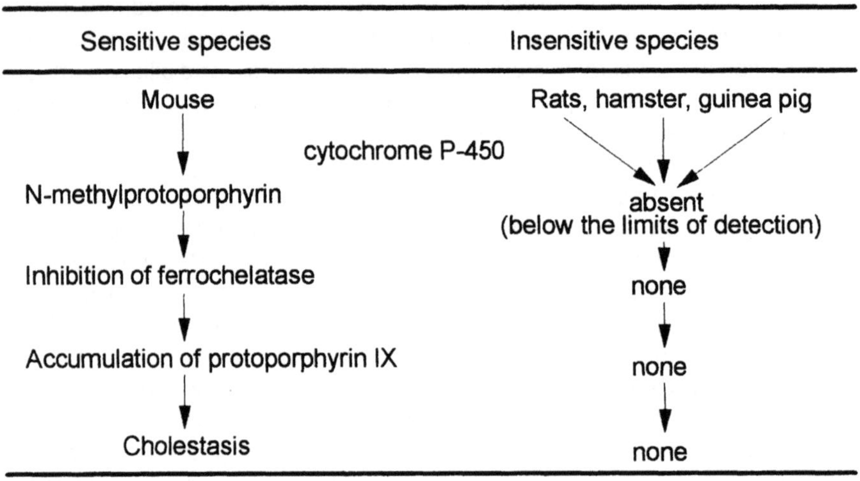

Species Difference in Proteins

A wide variety of organic chemicals have been reported to produce a male rat-specific nephropathy and, following chronic exposure, an increased incidence of renal tumours (Swenberg et al. 1989; Lock 1993; Hard et al. 1993). Studies to understand the basis for the acute renal injury and its subsequent progression showed that these chemicals all produced a marked increase in hyaline droplets in the proximal tubules of male rats. One of the most potent chemicals was the hydrocarbon 2,2,4-trimethylpentane (TMP) an important component in unleaded gasoline. The suggestion that the hydrocarbon nephropathy might be related to the male rat-specific protein α2u-globulin was first made by Alden and coworkers (1984), who identified α2u-globulin as the protein accumulating in the kidney of male rats treated with the hydrocarbon decalin. Since these initial studies many additional chemicals have been shown to produce α2u-globulin accumulation in male rats. None of these have caused a similar nephropathy in female rats or in either sex of any other species. Thus the α2u-globulin nephropathy appears to be a sex and species specific disease.

α2u-Globulin is a low molecular weight protein of 18,700D that is synthesised in the liver of male rats under androgenic control, secreted into the plasma and freely filtered by the glomerulus. About half of the α2u-globulin in the glomerular filtrate is reabsorbed by the cells lining the P2 segment of the proximal tubule, the remainder being excreted in the urine. Tubular reabsorption of α2u-globulin occurs via endocytosis. Whereas large amounts of protein are found in the urine of male rats, α2u-globulin is found in the urine of female and sexually immature rats at concentrations less than 1% of that found in young adult males. Females rats do not synthesis α2u-globulin in the liver, the small amounts observed being synthesised in the salivary gland.

A pivotal finding was that a metabolite of TMP became reversibly bound to α2u-globulin, accounting for its accumulation in the P2 segment of the proximal tubule of male but not female rats. Subsequent studies with other chemicals that produce a marked increase in protein droplets showed a similar interaction with α2u-globulin. Molecular modelling studies (Borghoff et al. 1991) have demonstrated that the chemicals fit into a hydrophobic pocket present in the three-dimensional structure of the protein. It is believed that this interaction between the chemical and the protein makes the protein less readily degradable by lysosomal enzymes in proximal tubular cells.
The accumulation of α2u-globulin is cytotoxic and results in single cell necrosis (Short et al. 1989). The increase in cell proliferation is localised in the P2 segment of the nephron and to a lesser extent in the P3 segment. Both the accumulation of α2u-globulin and the cell proliferation is dose-related and of considerable importance is the demonstration that kidneys from female rats that have been treated with identical doses to the male rats have no increase in the protein or cell proliferation.

The finding that the NBR strain of male rat does not synthesise the androgen-dependant form of α2u-globulin (Chatterjee et al. 1989) led to further testing of the hypothesis that it is an excessive accumulation of the protein that leads to the increased cell turnover. Studies with selected chemicals, such as d-limonene, showed that no protein accumulation or cell proliferation occurred in the NBR strain of rat (Dietrich and Swenberg 1991). This work and initiation-promotion studies using the renal carcinogen ethylhydroxyethylnitrosamine and d-limonene comparing α2u-globulin positive rats (F344) with the NBR strain showed that the promoting activity was entirely dependant on the presence of the protein. The following scheme (Fig 3) illustrates the key findings in this area of study, and the basis for the species difference in response (for a further details see the references).

Figure 3 Proposed basis for species difference in renal injury and carcinogenicity produced by chemicals that cause α2u-globulin accumulation.

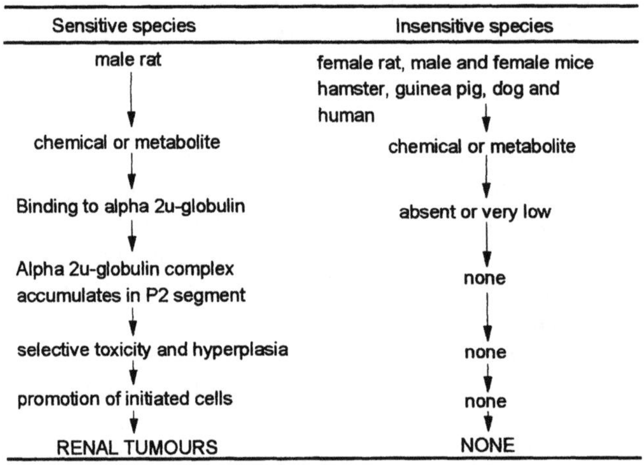

In this example the toxic and carcinogenic mechanism is clearly associated with the presence of a specific urinary protein α2u-globulin, that is not found in humans or other species. Several proteins are known to be members of this superfamily of low molecular weight proteins such as mouse urinary protein (MUP) which has about a 90% homology with the amino acid sequence of α2u-globulin. However, detailed studies with this protein established that the metabolites of d-limonene, although formed in the mouse, do not bind to the protein, and further that in the mouse kidney MUP does not appear to be reabsorbed by the cells of the proximal tubule (Lehmann-McKeeman and Caudill 1992a). This group also examined a number of other proteins in this super family including the human proteins α1-acid glycoprotein and retinol binding protein and β-lactoglobulin for their ability to bind these chemicals and concluded that the binding of these toxicants is unique to α2u-globulin (Lehmann-McKeeman and Caudill 1992b).

Conclusions

The two examples discussed illustrate how a detailed understanding of the mechanisms involved in the induction of hepatic protoporphyria and α2u-globulin nephropathy and carcinogenesis affords an explanation for the species difference in the toxicity and carcinogenicity. By understanding the mechanism of toxicity, which may or may not involve metabolism of the chemical, it is possible to make a realistic assessment of the risk from exposure to these chemicals in the workplace.

References

Alden CL, Kanerva RL, Ridder G and Stone LC (1984) The pathogenesis of the nephrotoxicity of volatile hydrocarbons in the male rat. In Renal effects of Petroleum Hydrocarbons. (MA Mehlman, GP Hemstreet, JJ Thorpe and NK Weaver Eds) pp107-120, Princeton Publ NJ.

Borghoff SJ, Miller AB, Bowen JP and Swenberg JA (1991) Characteristics of chemical binding to α2u-globulin in vitro-evaluating the structure-activity relationships.Toxicol Appl Pharmacol 107:228-238

Brady AM, Hasmall RL, Elcombe BM, Lock EA and DeMatteis F (1993) Effects of 2-[1-(ethoxyimino)propyl]-3-hydroxy-5- (2,4,6-trimethylphenyl)cyclohex-2-eone on hepatic haem biosynthesis: species differences in hepatic porphyria Toxicology 84: 199-215

Chatterjee B, Demyan WF, Song CS, Garg BD and Roy AK (1989) Loss of androgenic induction of α2u-globulin gene family in the liver of NIH black rats. Endocrinology 125: 1385-1388

DeMatteis F, Gibbs AH and Holley AE (1987) Occurrence and biological properties of N-methylprotoporphyrin. Ann NY Acad Sci 514:30-40

DeMatteis F,Gibbs AH, Jackson AH and Weersinghe S (1980) Conversion of liver haem into N-substituted porphyrins or green pigments: nature of the substituent at the pyrrole nitrogen atom. Febs Lett 119:109-112

Dietrich DR and Swenberg JA (1991) NCI-Black Reiter (NBR) male rats fail to develop renal disease following exposure to agents that induce α2u-globulin nephropathy. Fundam Appl Toxicol 16: 749-763

Frater Y, Brady AM, Lock EA and DeMatteis F (1993) Formation of N-methylprotoporphyrin in chemically induced protoporphyria. Arch Toxicol 67:179-185

Hard GC, Rodgers IS, Baetcke KP, Richards WL, McGaughy RE and Valcovic LR (1993) Hazard evaluation of chemicals that cause accumulation of α2u-globulin, hyaline droplet nephropathy, andtubule neoplasia in the kidneys of male rats. Environ Health Perspectives 99:313-349

Hodgson E and Guthrie FE (1980) Introduction to Biochemical Toxicology. Elsevier-North Holland

Lehmann-Mckeeman LD and Caudill D (1992a) Biochemical basis for mouse resistance to hyaline droplet neuropathy: lack of relevance of the α2u-globulin protein superfamily in this male rat-specific syndrome. Toxicol Appl Pharmacol 112: 214-221

Lehmann-McKeeman LD and Caudill D (1992b) α2u-globulin is the only member of the lipocalin protein superfamily that binds to hyaline droplets inducing agents. Toxicol Appl Pharmacol 116: 170-176

Lock EA (1993) α2u-globulin-related nephropathy. In Renal Disposition and Nephrotoxicity of Xenobiotics (MW Anders, W Dekant, D Henschler, H Oberleithner and S Silbernagl Eds) pp217-229 Academic Press London

Marks GS McCluskey SA, Mackie JE, Riddick DS and James CA (1988) Disruption of hepatic haem biosynthesis after interaction of xenobiotics with cytochrome P 450. Faseb 2 2774-2783

Oritz de Montellano, Beilan HS and Kunze KL (1981) N-methylprotoporphyrin IX: chemical synthesis and identification as the green pigment produced by 3,5-diethoxycarbonyl-1,4-dihydrocollidine treatment. Proc Natl Acad Sci USA 78:1490-1494

Short BG, Burnett VL and Swenberg JA (1989) Elevated proliferation of proximal tubular cells and the localisation of accumulated α2u-globulin in F344 rats during chronic exposure to unleaded gasoline or 2,2,4-trimethylpentane.Toxicol Appl Pharmacol 101: 414-431

Swenberg JA, Short BG, Borghoff SJ, Strasser J and Charbonneau M (1989) The comparative pathobiologgy of α2u-globulin nephropathy. Toxicol Appl Pharmacol 97: 35-46

Timbrell JA (1991) Principles of Biochemical Toxicology 2nd Edition Taylor and Francis, London

Species Specificity of Organ Toxicity: Behavioral Differences.

G. Bignami, G. Dell'Omo, and E. Alleva
Laboratory of Organ and System Pathophysiology, Istituto Superiore di Sanità, V.le Regina Elena 299, I-00161 Rome, Italy.

A contribution on behavioral differences in a discussion on species specificity of organ toxicity must first provide at least a tentative answer to a general question. This question is: what is the rationale for making recourse in toxicology to "soft" behavioral methods, in an age when "hard" methods (morphological, physiological, biochemical, and molecular) have reached a level of remarkable sophistication?

An anwer to this question must start by considering the status of behavioral responses or repertoires, which are not convenient dependent variables (McCleary 1961) and therefore need to be analyzed in their own right following rigorous rules (see e. g. Weiskrantz 1968), unless one accepts to miss a substantial part of the reality. In fact, behavioral repertoires are the main interface between the organism and the surrounding environment, including other organisms of the same and of different species, and constitute an important quota of the phenotypes on which selection forces have exerted their action.

Some of these repertoires can appear to be quite rigid, in the sense of being repeated with little variation between and within individuals of the same species in the face of specified internal and external stimulus conditions. It is now well established, however, that such invariance is not under exclusive genetic control, being the result of successive interactions between genetic and environmental factors during development (Lewontin 1974). Therefore, toxicologists must be ready to detect changes produced by treatments at successive developmental stages, as has been known for several decades by psychoendocrinologists involved in the study of changes produced by early exposure to sex hormones in apparently highly stereotyped sexually dimorphic behaviors (both reproductive and non-reproductive) (Naftolin and Butz 1981; Arnold and Gorski 1984).

On the other hand, a substantial portion of the behavioral repertoires continues to show considerable versatility ("behavioral plasticity") throughout the lifetime of the organism, which is a *qua non* for effective homeostasis, or even for survival, in a wide variety of different environmental conditions. Under less than extreme conditions, this behavioral variation is only partly amenable to an analysis aimed at defining different states of underlying regulatory mechanisms; hence the need of accurate behavioral assessments both to evaluate the effects of agents of toxicological interest and to identify the functional significance of changes in the CNS and elsewhere which are defined by physiological, biochemical and molecular methods. In fact, the latter changes are often quite heterogeneous after any given treatment, therefore not amenable to ranking according to relevance unless they can be confronted with a profile of behavioral effects.

The history of antidepressants shows the difficulties that one can encounter in this analysis; in fact, in the absence of a specific behavioral profile of antidepressant effects in laboratory animals, about fourty years have passed without our understanding which mechanism or mechanisms of action of these agents are critical for alleviating the symptoms of human depression. By contrast, the much better definition of the neuroleptic and anxiolytic syndromes in animals has made it possible to gain a more adequate understanding of the mechanisms of production of drug effects, including side effects of toxicological interest.

Significance and uses of behavioral data

When considering different types of agents endowed with noxious effects on the CNS and behavior, an important preliminary discriminant is whether or not these effects include clear-cut pathological changes which can be demonstrated by morphological and physiological-biochemical methods and are reflected in unmistakable neurological symptoms. Even when the pathological and clinical data apparently suffice to characterize the profile of a given agent, however, behavioral assessments are far from redundant. In fact, they can have considerable toxicological value, for example (i) for detecting changes in the initial stages of the intoxication, (ii) for assessing thresholds by the use of doses or concentrations lower than those that produce clear-cut pathological effects, and (iii) for identifying possible differences in vulnerability as a function of developmental stage.

On the other hand, a considerable literature concerning both humans and experimental animals shows that behavioral assessments are essential whenever an agent is suspected of producing harmful effects, but pathological changes are apparently absent or at least insufficient for an adequate toxicological evaluation. Two well-known examples of such a situation are the effects in children of extended exposure to low doses of lead (Needleman 1991) and those of prenatal exposure to low (subteratogenic) doses of alcohol (Barr et al. 1990; Fried and Watkinson 1990); a third example, concerning the effects in humans and in animals of prenatal exposure to benzodiazepines, will be dealt with in subsequent sections.

In cases like the ones just mentioned, studies on animals must carefully consider several methodological *caveats*. Some of the *caveats* are obvious, since they have to do with the risk of false negatives and with the well-known fact that an agreement has not yet been reached concerning test batteries suitable to minimize such a risk. In the case of motor activity, for example, several effects of considerable toxicological interest have been shown to vary from one test to another. A dose-dependent increase in preweaning locomotor activity was observed in rats after prenatal anticonvulsant (phenytoin) treatment when using a square open field, but failed to occur in the same laboratory when a circular field was used (Weisenburger et al. 1990). Neonatal antidepressant (clomipramine) treatment resulted in open-field hyperactivity at the adult stage whereas activity in a closed chamber was unaffected, which suggested that hyperreactivity to mild stress be the cause of the former change (Hartley et al. 1990). Haloperidol treatment from postnatal days 4 to 21 (a treatment known to produce considerable effects on CNS dopaminergic mechanisms) increased mouse activity in an open field but only with some combinations of animal strain, age, and time of testing; no changes in activity were found in a photocell activity cage, in a circular alley and in an operant chamber (Dallemagne and Weiss 1982).

These and other instances of test dependent effects show that negative data obtained by the use of only one test are at considerable risk of being false negatives, at least in the case of agents that can be reasonably suspected of being endowed with CNS effects. In addition, the literature contains several striking instances of negative behavioral data after treatments that produce widespread CNS effects, as for example after early developmental exposure to either alcohol (Nelson et al. 1988), parathion (Stamper et al. 1988), or methyl demeton (Meta-Systox) (Clemens et al. 1990).

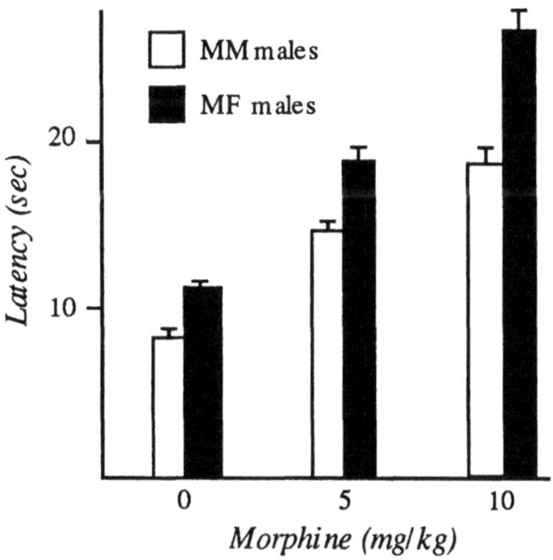

Fig. 1. Mean latency (+ S.E.M.) to first paw lick in a hot plate test (55 ± 1°C) in 40-day-old male CD-1 mice reared either in all male (MM) or male and female (MF) litters. MM males showed an enhanced reactivity to pain and a reduced effect of morphine given IP 15 min before testing. $N = 13$ in each group. (Reprinted with permission from Physiology & Behavior, Vol. 36, pp 779-781, Alleva E, Caprioli A, Laviola G, Postnatal social environment affects morphine analgesia in male mice, Fig. 1, Copyright 1986, Pergamon Press Ltd.)

At the opposite end, more subtle problems are encountered when dealing with the risk of false positives. A first *caveat* is that not any deviation from the behavioral profile of control animals is to be taken as a sign of toxicity. In fact, several types of isolated behavioral changes can be essentially neutral with respect to the adequate functioning of homeostatic mechanisms; this, for instance, is likely to be the case with changes in activity levels when the scores of all treated subjects remain within the range of the control population and other types of behavioral changes are consistently absent.

In addition, several behavioral end-points are characterized by a remarkable developmental plasticity; for example, pain reactivity (including analgesic drug responses) shows long-term changes not only after pre- and postnatal stress (Kinsley et al. 1988; Shimada et al. 1990), but also after non-stressful manipulation, such as variation of litter gender composition (for an example of the latter effect see Fig. 1) (Alleva et al. 1986; Laviola and Loggi 1992). Therefore, one should think twice before considering as signs of toxicity those behavioral changes that remain within the range of normal plasticity phenomena, at least in the absence of additional evidence pointing either to more widespread behavioral effects, pathological changes in the CNS, or both. The problem here is that there are several behavioral end-points with potential usefulness in toxicological assessments for which the range of normal plasticity phenomena is unknown or poorly known; therefore, there still is a scarcity of reference points suitable for separating the "pathological" from the "normal", a distinction which is far from being superposable to the one between the "significantly different" and the "not significantly different" from control level.

In any event, isolated behavioral changes of dubious functional significance should not be automatically discounted since they can provide cues for further verification of the treatment's profile, particularly when independent information — e.g., on CNS changes or possible effects in humans — suggests that further experimental efforts are advisable.

Specie-specific repertoires in toxicological assessments

In both pharmacology and toxicology, an increasing use has been made in recent years of observational methods aimed at studying well-defined components of species-specific behavioral repertoires (ethograms), often combined with the measure of responses with unmistakable functional significance, such as feeding and drinking. This approach, besides providing useful quantitative data, can help understanding whether treatment-induced changes are widespread or vice versa selective, affecting some components of the behavioral repertoire while sparing others. Such information can subsequently be used to orient studies on treatments' mechanisms of action, which would otherwise be like looking for a needle in a haystack.

The use of ethograms is greatly facilitated and made more reliable by videorecording and by the use of software evolved for the recording and processing of response scores (e. g. The Observer; Hile, 1991; Noldus, 1991). Specifically, videorecording minimizes disturbances to experimental animals and provides the experimenter with records that can not only be scored more reliably than the observers' visual input during test sessions (e.g. by exploiting replay), but also re-examined whenever there are good reasons to look for additional information. Moreover, scoring from videorecords can rely on a planning of observers' assignments which is less vulnerable to unforeseeable events (e.g., the sudden defection of an observer due to illness or other) than the one done when all scores must be collected during the test.

Therefore, the approach so far outlined fulfills some essential requirements such as those which go under the "3 R's" label (Gad 1990): *Refinement, Reduction* (of the number of animals, considering that one can do away with the redundancy needed in order to play safe in the absence of permanent records), and *Replacement* (in the sense that one or more additional experiments can often be spared by taking a harder look at the records of previous experiments).

In addition, this strategy of data collection and analysis is highly flexible; that is, it can easily be adapted to the needs of different types of first- and higher-tier assessments.

For example, the test can be performed in either a closed chamber or in an open arena, with either dim or bright illumination, and with either a low or a high level of background noise, so as to vary the type and intensity of environmental stimulation and therefore the amount of stress to which the subjects are exposed (see e.g. the recent analysis by Pinel et al. 1994 of rat defensive responses at different levels of illumination).

At a more sophisticated level, the familiarity or novelty of the situation can be manipulated in several ways, such as the use of different environments (e.g. home cage with own bedding, home cage with new bedding or soiled bedding from other animals' cages, and entirely novel situations) and of prolonged or repeated exposure to an originally novel environment. These manipulations can provide specific information concerning both the subjects' reactivity to environmental stimuli and habituation processes, which are deemed to be important components of the organism's adaptive capabilities, particularly since it has been realized that these functions are served by diversified mechanisms which can be brought into play separately from each other, depending on the type of experience (see e. g. the ontogenetic and pharmacological dissociation of within- and between-session habituation in Laviola et al. 1988).

Social (including sexual) interactions with either familiar or non-familiar partners of the same or the opposite sex are amenable to standardized assessments with minimal changes in the test set-up, provided of course that the observers are adequately trained and regularly checked for their reliability in score assignment (Caro et al. 1979; Martin and Bateson 1986; Alleva 1993).

Last, but not least, if there are controversies on the results, any one with adequate credentials can go back to treatment, video and score records and pick up any error the experimenters might have made, with the same confidence as when dealing with the records of physiological or biochemical experiments. This is obviously an important guarantee in the assessment of adverse effects, considering the liability of toxicologists whenever the results turn out to be false negatives or false positives with serious consequences on either human health and well-being, the finances of interested parties, the status and bargaining power of scientists relative to those of animal rights supporters, or all these things together.

The use of ethograms in the analysis of ozone (O_3) effects

The strategy outlined in the previous section can be further illustrated by some recent results concerning the behavioral effects of exposure to an environmental pollutant (ozone, O_3) at concentrations which are insufficient to produce clear-cut pathological effects (0.4 - 1.2 ppm) (Musi et al. 1994). The initiation of exposure to O_3 produced in adult CD1 mice several changes consisting of an increase of rearing, sniffing, grooming, feeding, and social contacts, while other responses such as digging and bar-holding were either unaffected or reduced. These changes, which were videorecorded without removing the animals from their home cages located in the exposure chambers, were markedly attenuated after about 30 min, resulting in a return to pre-exposure baselines within one hour. Such a profile is indicative of a disturbance like the one produced by several types of strong novel stimuli of various modalities, including typical habituation after marked initial effects.

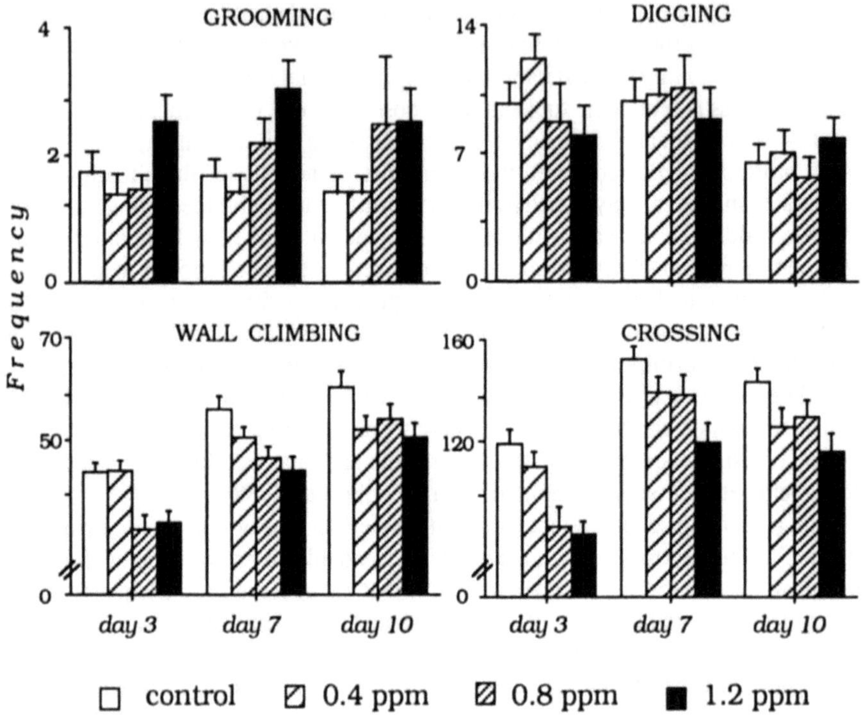

Fig. 2. Mean frequency (± S.E.M.) of selected responses of adult CD-1 mice at 3, 7, and 10 days after the start of continuous ozone (O_3) exposure. The data are from 5-min observation periods outside the exposure chambers, in cages identical to the home cages which were previously washed and furnished with clean bedding, thereby providing a novel environment. The graphs give examples of responses whose frequencies were either not significantly modified (digging), decreased (wall climbing, locomotion indicated by "crossing"), or increased (grooming) by O_3 exposure. $N = 20$ in each group (data for males and females pooled in the absence of significant sex effects on these measures). (Adapted with permission from NeuroToxicology, in press, Musi B, Dell'Omo G, Ricceri L, Santucci D, Laviola G, Bignami G, Alleva E, Effects of acute and continuous ozone (O_3) exposure on activity/exploration and social behavior of CD-1 mice, Fig. 2, Copyright 1994, Intox Press, Inc.)

Subacute effects of O3 were studied by videorecording the same animals at three successive stages (3, 7, and 10 days from the start of exposure) in a cage identical to the home cage, but outside the exposure chambers, so as to allow a more detailed analysis of their responses. These tests showed a concentration-dependent increase of grooming and digging paralleled by a reduction of locomotor activity (crossing) and of wall climbing (Fig. 2), a profile of effects similar to that produced by mild to moderate stress.

The parallel measure of food and water intake showed a concentration dependent reduction which was reflected in a decrease of body weight. These effects, however, were unlikely to be responsible for the behavioral changes just mentioned, since they showed a rapid tolerance; namely, O3 animals returned to control baselines in spite of continuing exposure after 3 days in the case of food intake, and after 7 days in the case of water intake and body weight, except for an additional 3 days of slight but significant depression of water intake at the highest (1.2 ppm) concentration.

These changes in feeding and drinking, which are essentially similar to previous findings in rats (Umezu et al. 1987), warn about possible confounding between indirect O_3 effects in the initial phase of exposure, which may be the consequence of a transient impairment of the nutritional state, and more direct effects, whose trend over time should be dissociated from that of the depression of consummatory responses and body weight.

Developmental O_3 effects in mice and rats: species differences or confounding of variables?

A detailed account of the more recent studies performed in our laboratory to assess the effects of prenatal exposure to O_3 cannot be given here. In brief, mid- and late gestational exposure (days 7-17 of pregnancy) with the same range of concentrations (0.4 - 1.2 ppm) failed to produce significant changes in postnatal somatic and neurobehavioral development as assessed both by appropriate test batteries and by studying the trend over age of specific responses, such as pups' ultrasonic vocalizations (Bignami et al. 1994). Somatic and neurobehavioral development was similarly unaffected in a subsequent experiment with more extended exposure (from before mating to pregnancy day 17); however, a more detailed study of possible long-term effects showed some "borderline" changes, namely, (i) a slight but significant modification of the ethogram at the prejuvenile and juvenile stages (increase of grooming and reduction of exploratory responses, again comparable to those produced by mild stress), and (ii) a limited deficit in the initial phase of radial maze learning at the young adult stage, in the absence of changes in activity and habituation (manuscript in preparation).

These data, whilst they confirm the circumscribed nature and the specificity of O_3 effects in mice, are apparently at variance with those of a previous study on rats (Kavlock et al. 1980). In fact, the latter showed reduced neonatal growth rates after exposure to O_3 (1.0 or 1.5 ppm) in mid- or late gestation (days 9-12 or 17-20) and also a delayed neurobehavioral development after late gestational treatment. This apparent discrepancy between rat and mouse data, although not yet subjected to more thorough investigation, is of potential methodological interest. In fact, it can be exploited to show how species differences and other possible sources of variance can be confounded with each other, unless experiments aimed at comparing treatment effects in different species are performed in otherwise identical conditions.

Specifically, our experiments used cross-fostering at birth of both treated and control litters to untreated-unhandled dams, whereas the litters in the rat experiment were nursed by their own mothers, which makes it possible that postnatal maternal effects (behavioral, nutritional, or other) played a role in the production of the developmental deficit. Secondly, the effects of the litter random factor were discounted in our experiments by the use of appropriate (mixed-model) analyses of variance (except in the case of non parametric data, for which equally effective methods are not available; Chiarotti et al. 1987), whereas litter and treatment effects were not separated from each other in data analyses of the rat experiment.

Thirdly, the fact that neurobehavioral impairment occurred in rats after late pregnancy but not after mid-pregnancy treatment is not sufficient to conclude that O_3 effects were direct and specific. On the one hand, these data point to a difference in vulnerability as a function of developmental stage, assuming for a moment that postnatal maternal effects or confounding of treatment and litter effects did not influence the results. On the other hand, however, the data cannot discriminate between a direct effect on vulnerable fetal targets and an indirect effect mediated by nutritional imbalance or stress in the period of maximal reduction of feeding and drinking by the dams after the onset of exposure.

The only safe inference at this point, in this and other similar cases, is that species factors should not be invoked as a *deus ex machina* in the face of a host of possible confounding variables which can mimic their effects; the correct strategy is rather to find ways and means for identifying the cause or causes of observed differences, which is an essential step in the identification of the treatment's mechanism of action.

Species-specific developmental time tables and treatment effects

Each species shows a specific pattern of neurobehavioral ontogenesis which has been shaped by evolutionary forces in relation to the different ecological niches occupied by the organism at successive stages of prenatal and postnatal development. In the case of the laboratory animals which are more widely used in toxicological studies, a fundamental distinction is the one between precocial rodent species (like the guinea pig, *Cavia porcellus*, and the spiny mouse *Acomys cahirinus*, of which more later), which have a relatively long pregnancy and give birth to young that are mature to the point of looking like "mini-adults", and altricial species with a much shorter pregnancy (about three weeks in the mouse and the rat), which give birth to young that are highly immature.

In the first three weeks of postnatal life, subjects of altricial species undergo a series of dramatic changes — both somatic (e.g. eye opening around the end of the second postnatal week) and neurobehavioral — which provide sensitive and specific end-points for the assessment of the effects of early treatments, either prenatal or postnatal. In the case of motor activity, for example, the maturation of monoaminergic regulatory mechanisms in the CNS is paralleled by a rapid increase of locomotor and other responses peaking at the time of eye opening (see e.g. Fibiger et al. 1970; Campbell and Mabry 1972; Alleva and Bignami 1985), and by the development of a sensitivity to the reinforcing properties of psychostimulants (such as amphetamine and cocaine) which is almost as marked as that of the adult (Laviola et al. 1992a; 1994). On the other hand, the functional control of cholinergic muscarinic mechanisms over locomotor activity and related motor responses achieves maturity only at a later time; that is, generally at the end of the third postnatal week, when the animals start showing adult-like habituation phenomena, but sometimes even later, depending jointly on animal strain and test factors (see e. g. Laviola et al. 1988).

In such a situation, the use of simple tests at successive ages, particularly when it includes the assessment of the effects of selected drug challenges such as amphetamine and scopolamine (Zenick 1983; Walsh and Tilson 1986; Bignami et al. 1992; Bignami 1994), can indicate whether an early (pre- and/or postnatal) treatment of toxicological interest can produce significant changes in the development of some important regulatory

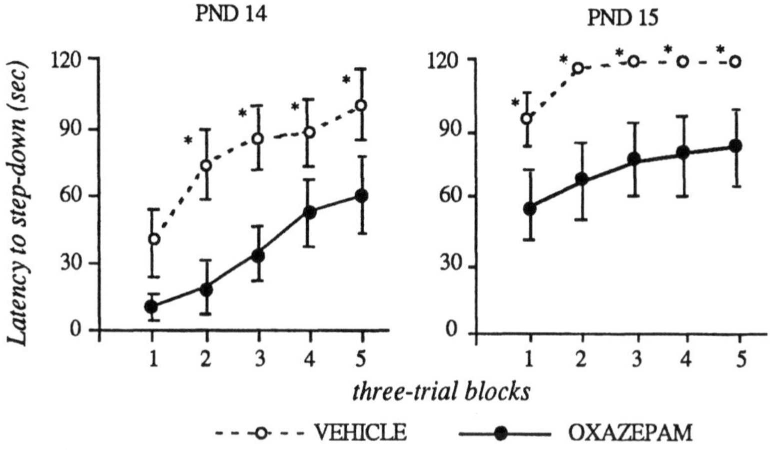

Fig. 3. Impairment of passive avoidance acquisition (step-down test, two sessions of 15 trials each on consecutive days) by two-week-old CD-1 male mice whose mothers had received a benzodiazepine treatment during pregnancy (oxazepam 15 mg/kg PO twice daily on pregnancy days 12-16). Mean (± S.E.M.) step-down latencies were lower than those of appropriate (vehicle) controls throughout the training period (N = 10 in each group; * P<0.01) (Adapted with permission from Brain Research Bulletin Vol. 33, pp 267-271, Ricceri L, Calamandrei G, Alleva E, Prenatal oxazepam affects passive avoidance performance of preweaning mice, Fig. 1, Copyright 1994, Pergamon Press Ltd.)

mechanisms. For example, several studies of prenatal benzodiazepine effects have shown both an attenuation of the increase in activity which normally occurs during the second week of life and a reduction of the activity enhancing effects of amphetamine, pointing to an interference with the development of monoaminergic regulatory mechanisms (Alleva et al. 1985; for review and additional references see Bignami et al. 1992). This effect is apparently a transient one, since one week later both activity levels and response to amphetamine were indistinguishable from those of controls, and also a selective one, since the hyperactivity response to scopolamine was not modified.

Data like those just mentioned have special relevance when it must be decided whether or not additional efforts need to be made in higher-tier studies with a variety of other response end-points, which are inevitably more burdensome and costly. In fact, studies on prenatal benzodiazepine effects were extended in our laboratory in different directions to assess possible changes in various behavioral functions.

One experiment was devoted to the study of social interactions of infant mice (Terranova et al. 1994). Prenatal treatment reduced active investigation and solicitation of littermates, increasing behavior related to the achievement and maintenance of a "passive" proximity with other subjects; in addition, normal differences between males and females were attenuated. Another experiment (see Fig. 3), also on preweanlings, showed a deficit in step-down passive avoidance acquisition (Ricceri et al. 1994). Both these studies, it should be noted, pointed to subtle changes at the same developmental stage when activity and response to amphetamine were affected in the experiments mentioned above.

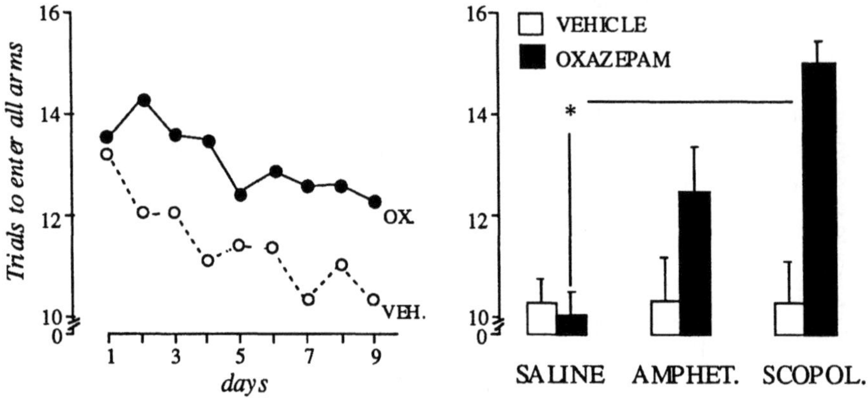

Fig. 4. Graph to the left: impairment of eight-arm radial maze learning (each arm baited with a drop of water before the start of each daily test) by CD-1 male mice exposed prenatally to oxazepam (same treatment as in Fig. 3; $N = 24$ in each group; overall significance of the prenatal treatment effect $P<0.02$). Graph to the right: disruptive effects of scopolamine (2 mg/kg IP 15 min before testing) on radial maze performance after the completion of the learning period in oxazepam but not in control mice. The effect of *d*-amphetamine (1 mg/kg), apparently in the same direction, did not reach statistical significance ($N = 8$ in each final group. * $P<0.01$). (Modified with permission from Brain Research Bulletin Vol. 29, pp 609-616, Laviola G, Pick CG, Yanai J, Alleva E, Eight-arm maze performance, neophobia, and hippocampal cholinergic alterations after prenatal oxazepam in mice, Figs. 1 and 2, Copyright 1992, Pergamon Press Ltd.)

Two other studies were devoted to the analysis of different types of learning abilities at the young adult stage. Both studies showed a learning deficit, which was mild and transient in a swimming navigation task (Dell'Omo et al. 1993) and more substantial in a radial maze task (see Fig. 4, graph to the left) (Laviola et al. 1992b). The latter study also gave information on the mechanisms which might be involved in the production of subtle long-term effects by prenatal benzodiazepine exposure. In fact, the animals of the treatment group showed a marked disruptive effect of scopolamine on maze performance after the completion of the learning period, which was absent in control mice. Moreover, when presented with a novel stimulus object in an open-field, they showed neophobia followed by perseverative responding with lack of habituation. Both Bmax and affinity (Kd) of cholinergic receptors were found to be increased in the hippocampal formation.

These and related data allow one to formulate some interesting working hypotheses concerning the mechanisms of action of the treatment in question, which cannot be discussed here. They acquire special significance when considering that the work in animals aimed at defining the profile of prenatal benzodiazepine effects has largely preceded, rather than followed, the first convincing demonstration of subtle neuropsychological deficits in children of mothers who had taken benzodiazepines during pregnancy (Viggedal et al. 1993). This shows that we may not be far from being able to move on to a genuinely previsional behavioral toxicology and teratology.

Several problems, however, remain to be solved, starting with the one created by the remarkable uncertainties about the rules in the game of cross-species comparisons.

Species differences in the patterns of behavioral development, which may be reflected in differences in treatment effects, are extremely difficult to assess when they are not as extreme as those between precocial and altricial species; in fact, they are inevitably confounded with the effects of several other factors, ranging from animal strain to a host of test variables. For example, there is no reliable criterion to adapt apparatus size to body size in an activity test aimed at comparing rats and mice at successive developmental stages. The same applies to the time parameters of test exposure, which are particularly important when studying the development of habituation, since the role of the species factor and that of the body size factor in determining the functioning of "internal clocks" cannot be disentangled from each other.

Of course, there might be extreme cases in which the effects profiles are totally different in two species across several test conditions, but no reliable examples of such a situation are apparently available. Therefore, at the present state of the art the attribution to species rather than to other factors of a particular profile of differences in treatment effects remains essentially a matter of opinion. Alternative avenues to the solution of this problem will be tentatively explored in the following sections.

"Developmental styles": the precocial spiny mouse (*Acomys cahirinus*) as a unique opportunity in comparative toxicology

"Developmental style" is an important variable in animal studies aimed at establishing behavioral toxicology models, since the neurobehavioral maturation of the human species is a rather peculiar case within the mammalian class. In fact, brain development in the majority of mammals is nearly completed at birth; sometimes, in Primates it continues during the early postnatal phase, but only in humans (a neotenic-type species) CNS maturation is delayed until later developmental phases (see critical presentation in Gould 1977, chapter 10). The problem then is to evolve a comparative approach taking into account the established fact that the human brain is a "mosaic" structure sharing both fast-developing and slowly-developing components, the latter sometimes reaching full maturation quite late in the postnatal period. In a study considering the first three weeks of rat postnatal life as comparable to middle and late prenatal periods in humans, the rat brain has been shown to approach the appearance of the adult stage within approximately six weeks after fertilization, while the human brain takes seven times more to achieve the same maturational stage (Bayer et al. 1993).

The history of the different developmental styles which have relevance in our case can be traced back to dinosaurs. In fact, dinosaur species, as well as their reptilian and bird survivor species, were and are characterized by two main styles of early postnatal development, the nidifugous and the nidicolous. Mammalian species also evolved similar "strategies" depending on the specific adaptive process leading to their radiation (Eisenberg 1981); among the species used in the laboratory, both styles are represented, the "precocial" one in the guinea pig and the "altricial" one in the rat and in the mouse, the use of which has progressively increased at the expense of other experimental animals (Beach 1950).

Today, rats (derived from wild populations of Norway rats) and mice (derived from the *Mus musculus* species "complex"; Festing and Lovell 1981) are the most commonly used laboratory animals for obvious practical/economic reasons (for their use in toxicology see Gad 1990; Lechat 1990); in particular, the near totality of behavioral toxicology studies are carried out on them. Such a situation has created a "subliminal consensus" that rats and mice represent useful and reliable models in the assessment of risks to the human health.

Rodents are born on a precocial-altricial continuum, with their relative independence from their dams varying greatly across species. *Acomys cahirinus* is unique among *Muridae* in that it is born furred, sighted and mobile after a relatively long pregnancy (39 days *versus* 22 and 20 days in the rat and in the mouse, respectively; Brunjes 1990). Norway rat (*Rattus norvegicus*) young are much more immature at birth; that is, they are born nude, sightless and unable either to thermoregulate effectively, or to move away from the confines of the maternal nest. Fur, thermoregulation capacity, hearing, vision, and mobility develop during the third week postpartum (Leon et al. 1984; for neurobehavioral differences in maturational pace between rats and mice see, e.g., Alleva and Calamandrei 1986). Even more immature at birth are Mongolian gerbils (*Meriones unguiculatus*), who also are born nude and blind, with poor thermoregulation, and unable to move away from the nest. The auditory, visual, and motor systems develop in this species even later than in Norway rats and in mice.

An important consideration at this point is that rats, mice and *Acomys* species all belong to the same *Murinae* sub-family; in other words, both the altricial and the precocial styles of development are represented within this same taxonomic sub-unit, which is a rare event that offers a unique opportunity for an effective comparative approach. The genus *Acomys* includes two precocial species, one of which (*A. cahirinus*) has been used successfully for laboratory studies (Leon et al. 1984; Birke et al. 1985; D'Udine and Alleva 1988; Brunjes et al. 1989; Brunjes 1990), although not yet in toxicological studies.

This fortunate occurrence allows *Acomys* species to be compared with the most extensively studied altricial species. At birth, for example, choline acetyltransferase, membrane-bound acetylcholinesterase levels, and density of muscarinic receptor sites are considerably higher in *A. cahirinus* than in rats; by contrast, adult levels are lower than in rats (Pintor et al. 1986). Moreover, the newborn *Acomys*, represents a suitable model for studies on early attachment and for the analysis of long-term effects on visual imprinting at the newborn stage (D'Udine and Alleva 1988). It appears therefore that a comparative analysis of early treatment effects using both precocial and altricial *Murinae* species might provide a wealth of useful information, reducing the present limitations created by the exclusive use of mice and rats.

Why semi-naturalistic settings?

Laboratory research is subtly naturalistic in intent, since its ultimate goal (in the present context, the goal of experimental neurobehavioral toxicology) is to provide appropriate animal models for evaluating human-centered problems occurring in the natural world.

The continuum ranging from laboratory studies to field research includes semi-naturalistic settings as intermediate situations in which methods developed in the laboratory are employed in more naturalistic contexts. This is the case, for example, with "mouse gardens" or wild rat studies carried out in semi-naturalistic settings ("visible burrow system" in Blanchard et al. 1993; 0.1 ha outdoor enclosures in Drickamer and Robinson 1993), both representing a common solution to reconcile rigorous laboratory practice with the need to assess the adaptive significance of the effects observed.

A semi-naturalistic setting can provide the opportunity to investigate components of the animal behavior repertoires which (i) are very difficult to study in the wild, and (ii) do not appear in current laboratory environments due to the scarcity of stimuli or the selection of anthropomorphic conditions. With field research, the problem of generalising to the natural life-style of animals is minimal; in this research, however, it is often difficult to perform the experimental manipulations which are necessary to reach the goals of a discipline like toxicology. Therefore, research conducted in semi-naturalistic settings appears appealing since it allows a validation of laboratory findings with respect to consequences on the life of animals in the wild, particularly when the focus is on the effects of toxicants on brain mechanisms, on their development, or behavioral adaptation. In fact, the behavior of animals in large outdoor enclosures is more similar to that occurring under natural conditions than it is in laboratory cages, where both quantity and quality of environmental stimuli are often dramatically curtailed.

For example, the study of behavioral and physiological effects of stress-inducing stimuli, including that of toxicant-induced changes, does not necessarily require the use of traditional laboratory methods such as exposure to electric shock or other physical stressors (cold water swimming, restraint, pulsating light or noise, air blast, etc). These analyses can also be performed following a more naturalistic approach, through the use of ethologically-relevant stimuli such as "fear-producing" cues emitted by natural predators. We recently characterized the reactions of laboratory mice to snake and fox scent, in order to provide an "alternative standard" for subsequent psychopharmacological or neurotoxicological evaluations aimed at assessing species-specific changes in emotional responding (Dell'Omo and Alleva 1994; Dell'Omo et al. 1994). We found that a predator cue elicits strain-dependent behavioral reactions, which appear to be amenable to the fear responses of the mouse populations from which the strains supposedly originated. The burying reaction of C57 mice is in fact reminiscent of a burrowing life style, such as the one of *Apodemus sylvaticus*, whereas CBA mice rather behave like surface living rodent species (Van Oortmerssen 1971).

Moreover, a natural setting may allow a better evaluation of subtle long-term effects of drugs or toxicants than that achieved by laboratory-based tests. In fact, subtle qualitative differences may only be detected by an ethogram taking into account a large number of components of the natural behavioral repertoire of the species. It appears therefore that behavioral effects of toxicants observed in the laboratory should be validated in naturalistic settings in order to verify their functional significance.

A first attempt to perform a semi-naturalistic study using the same prenatal oxazepam treatment as in an extended series of previous laboratory investigations (Alleva et al. 1985; see reviews in Alleva and Bignami 1986; Bignami et al. 1992; see also Section on *Species-specific developmental time tables and treatment effects*, particularly Figs. 3 and 4) was then conducted by comparing the behavior of CD-1 mice tested either in a classical open field or in the "real world", i.e., following release in the field (a meadow in the surroundings of Rome with no human artefacts within a radius of about 400 meters).

Specifically, adult male mice with a previous history of prenatal oxazepam treatment and appropriate controls were released in the field after being fitted with a radio collar, with the aim of sampling their behaviors over a period of 3 days. Unfortunately, the unexpectedly high local predatory pressure (all released mice were predated by nocturnal birds of prey within 48 hr) limited the observations to the first day after releasing (three periods of 30, 10, and 10 min, respectively, separated by 45 min intervals). The data (hitherto unpublished) showed a richer repertoire of behaviors than in the classical open-field test, including responses such as climbing, running, exploring, and sniffing; in particular, the field pattern of exploratory behavior was characterized by rapid and short runs separated by long-lasting sniffing bouts (the latter occurring in more protected areas such as grass spots). This pattern was substantially modified in oxazepam animals, due to an increase of locomotor activity and a reduction of feeding-like responses (e.g., manipulation and gnawing of twigs, grass blades, pieces of rubble, etc).

The results of this experiment suggest that the wide variety of environmental stimuli in the field can bring out subtle behavioral modifications which do not easily emerge in the standardized, stimuli-deprived situations currently used in laboratory studies. In particular, this approach seems to be suitable to separate major and widespread toxicant effects from more subtle situation-dependent effects revealed by exposure to a natural environment.

Natural populations as sentinels for environmental contaminants

Rodents are proverbial pests for humankind and have gained a bad reputation at least since the time of the Black Death in Europe. They represent one of the most successful "commensal-type" parasitic species, thriving on human products. Such commensal species are adapted to wide fluctuations of population size depending on the amount of available food. Accordingly, they have evolved in the direction of being highly prolific, which makes them so attractive for laboratory breeding and investigation. Most of the methods available today for neurobiology and behavior studies have been evolved in laboratory rodents, and the state-of-the-art in neurobehavioral pharmacology and toxicology reflects to a considerable extent a strict dependency on this rodent work (Cuomo et al. 1994).

On the other hand, biologists studying wild rodents have considered them mainly as pests to be put under control (health agencies, including World Health Organization, promote guide-lines for trapping them; Brooks and Rowe 1979). Yet, with our laboratory-derived knowledge of rodents it seems that we would all be well-served by eco-toxicological studies of natural populations aimed at monitoring CNS and behavioral alterations caused by exposure to chemicals dispersed in the environment (for a review of eco-toxical impacts of environmental contaminants on terrestrial small mammals see Shore 1994).

Rodents have been recommended for use in ecological monitoring at the population level since they are small and easy to handle, and since they spend their entire life cycle within a relatively restricted area, thereby providing highly specific information (Douglass

1984). Their high-density populations make them easily amenable to statistical analysis. In addition, rodents represent the main prey for terrestrial and avian predators (particularly raptors), which in turn are extremely sensitive bioindicators, due to their top position in trophic pyramids and to the spatial integration of their extended home ranges. However, the response of a given species to a given contaminant should not be used to predict the sensitivity of other species to the same contaminant, particularly in the case of pesticides (see e.g. the different sensitivities to four anticholinesterase insecticides found in the American kestrel, in the eastern screech-owl, and in the northern bobwhites by Wiemeyer and Sparling 1991). Another problem is that when behavioral changes have been identified, it is often difficult to relate them to exposure to a specific toxic compound. In fact, the changes observed in the laboratory following a treatment with a toxicant (e. g. an anticholinesterase) are not necessarily predictive of the effects in the wild, where behavior is adaptive and context-dependent, as remarked by Peakall (1994).

Furthermore, the behavioral alterations in many species living close to humans (including pets and farm animals) may be indicative of the presence and putative nature of toxicants which can affect neurobehavioral responses through CNS and other changes. To this purpose, species-specific ethograms aimed at bringing out abnormal responses or abnormal changes in selected patterns are needed.

Is such an eco/etho-toxicological approach a valuable investment for the future? Some preliminary suggestions can be made here in an attempt to answer this question. "Proactive" (previsional) studies should be based on a search for standardized "symptoms" established by both laboratory and semi-naturalistic studies carried out in selected sentinel species. On the other hand, a "reactive" multi-stage strategy, which is aimed at verifying the nature and relevance of effects already observed in humans, should

(i) use sentinel populations to evidence standard behavioral signs,

(ii) compare symptoms across species (including humans), and

(iii) use a convergent approach combining the above-mentioned information with that provided by tissue biochemical investigation and by environmental monitoring for a restricted range of suspected toxicants.

Data are available for some selected sentinel species that appear to be amenable to standardized behavioral toxicology assessments. Falcons, key bioindicators of environmental contamination, have already been used in toxicological laboratory studies in the past (Wiemeyer and Lincer 1987). However, eco-toxicological studies have not yet included behavioral indices of neurotoxicity, only reporting important effects on reproductive physiology, particularly eggshell reduction (Peakall and Kiff 1979). We recently initiated a pilot study on a bird species — the common swift, *Apus apus*, a strictly insectivorous bird which is seasonally exposed to polluted urban environments (from the beginning of April to the beginning of August in Rome) — to assess feather and organ contamination (by heavy metals, various pesticides, and policyclic aromatic hydrocarbons) in relation to possible signs of behavioral alteration. The data so far collected, however, are not yet sufficient to determine whether or not this bird, which is ubiquitous over most of the west Palearctic in the spring and summer period, can constitute an effective sentinel species.

Concluding remarks

The present overview of approaches, methods, and representative results in behavioral toxicology and teratology was not intended to provide a thorough survey of the present state of the art; the aim was rather to illustrate and discuss current problems and possible solutions, the latter often of a highly preliminary and tentative kind.

Two main provisional conclusions can be cautiously offered at this point. The first is that laboratory studies of toxicant effects should exploit more fully the considerable potential of species-specific repertoires, in the sense that choices should be based more often on rational criteria of a comparative-psychological and psychobiological kind, rather than sheer convenience. Of course, the emphasis on these criteria is not to de-evaluate the standardized assessments performed mainly in rats and mice, which have a highly useful function; the problem is rather that any effect of potential relevance should be analyzed in its own right by higher-tier studies aimed at maximizing both the specificity and the sensitivity of toxicological evaluation. In particular, the range of species and "developmental styles" brought under study should be extended before, rather than after, the limitations of present methodologies be revealed by some dramatic event like the thalidomide accident.

The second provisional conclusion is that there are now enough preliminary data to prompt a wider use of two types of assessments outside the laboratory scene, namely, experiments in semi-naturalistic settings and eco/etho-toxicological monitoring of natural populations. In fact, behavioral repertoires exploited in the laboratory are under severe constraints created by the fact that they represent an adaptation to a highly abnormal situation. Whether these repertoires tend to overreact, or viceversa to underreact, to the additional strain of chemical exposure of one or the other kind — due respectively to a summation of pathogenic influences or to the lack of sensitive components in impoverished environments ("zero-baseline effect") — is a matter that cannot be settled by arm-chair speculation, but rather requires adequately planned comparisons performed jointly by "traditional" toxicologists and by eco/etho-toxicologists.

In extreme conditions, obviously, laboratory data and "Silent Spring" types of effects can be easily reconciled with each other; their informative value, however, is overshadowed by that of clinical and epidemiological data providing strong evidence of human pathology. On the other hand, whenever less extreme conditions apply, the situation is often far from clear, as is shown by the skepticism expressed by some experts on much of the available evidence concerning toxic neurobehavioral effects in natural populations (see e.g. Peakall 1994). These and other related problems reinforce the notion that adequate efforts should now be made to extend and confront the information that can be obtained by following different approaches, with the aim of maximizing both the sensitivity and the specificity of toxicology assessments.

References

Alleva E (1993) Assessment of aggressive behavior in rodents. Methods Neurosci 14:111-137

Alleva E, Bignami G (1985) Development of mouse activity, stimulus reactivity, habituation, and response to amphetamine and scopolamine. Physiol Behav 34:519-523

Alleva E, Bignami G (1986) Prenatal benzodiazepine effects in mice: postnatal behavioral development, response to drug challenges, and adult discrimination learning. NeuroToxicology 7:303-318

Alleva E, Calamandrei G (1986) Odor-aversion learning and retention span in neonatal mouse pups. Behav Neur Biol 46:348-357

Alleva E, Caprioli A, Laviola G (1986) Postnatal social environment affects morphine analgesia in male mice. Physiol Behav 36:779-781

Alleva E, Laviola G, Tirelli E, Bignami G (1985) Short-, medium-, and long-term effects of prenatal oxazepam on neurobehavioural development of mice. Psychopharmacology 87:434-441

Arnold AP, Gorski RA (1984) Gonadal steroid induction of structural sex differences in the central nervous system. Annu Rev Neurosci 7:413-442

Barr HM, Streissguth AP, Darby BL, Sampson PD (1990) Prenatal exposure to alcohol, caffeine, tobacco, and aspirin: effects on fine and gross motor performance in 4-year-old children. Dev Psychol 26:339-348

Bayer SA, Altman J, Russo RJ, Zhang X (1993) Timetables of neurogenesis in the human brain based on experimentally determined patterns in the rat. NeuroToxicology 14:83-144

Beach FA (1950) The snark was a boojum. Amer Psychol 5:115-124

Bignami G (1994) Economical test methods for developmental neurobehavioral toxicity. Environ Health Persp, in press

Bignami G, Alleva E, Chiarotti F, Laviola G (1992) Selective changes in mouse behavioral development after prenatal benzodiazepine exposure: a progress report. Prog Neuro-Psychopharmacol Biol Psychiat 16:587-604

Bignami G, Musi B, Dell'Omo G, Laviola G, Alleva E (1994) Limited effects of ozone exposure during pregnancy on physical and neurobehavioral development of CD-1 mice. Toxicol Appl Pharmacol, in press

Birke LIA, D'Udine B, Albonetti ME (1985) Exploratory behavior of two species of Murid rodents, *Acomys cahirinus* and *Mus musculus*: a comparative study. Behav Neur Biol 43:143-161

Blanchard RJ, Yudko EB, Rodgers RJ, Blanchard DC (1993) Defense system psychopharmacology: an ethological approach to the pharmacology of fear and anxiety. Behav Brain Res 58:155-165

Brooks JE, Rowe FP (1979) Commensal rodent control. WHO, Vector Biology and Control Division, Geneva, paper VBC/79.726

Brunjes PC (1990) The precocial mouse, *Acomys cahirinus*. Psychobiology 18:339-350

Brunjes PC, Korol DL, Stern KG (1989) Prenatal neurogenesis in the telencephalon of the precocial mouse *Acomys cahirinus*. Neurosci Lett 107:114-119

Campbell BA, Mabry PD (1972) Ontogeny of behavioral arousal: a comparative study. J Comp Physiol Psychol 81:371-379

Caro TM, Roper R, Young M, Dank GR (1979) Inter-observer reliability. Behaviour 69:303-315

Chiarotti F, Alleva E, Bignami G (1987) Problems of test choice and data analysis in behavioral teratology: the case of prenatal benzodiazepines. Neurotoxicol Teratol 9:179-186

Clemens GR, Hartnagel RE, Bare JJ, Thyssen JH (1990) Teratological, neurochemical, and postnatal neurobehavioral assessment of METASYSTOX-R, an organophosphate pesticide in the rat. Fundam Appl Toxicol 14:131-143

Cuomo V, De Salvia MA, Petruzzi S, Alleva E (1994) Appropriate end points for the characterization of behavioral changes in developmental toxicology. Environ Health Persp, in press

Dallemagne G, Weiss B (1982) Altered adult behavior of mice following postnatal treatment with haloperidol. Pharmacol Biochem Behav 16:761-767

Dell'Omo G, Alleva E (1994) Snake odor alters behavior, but not pain sensitivity in mice. Physiol Behav 55:125-128
Dell'Omo G, Fiore M, Alleva E (1994) Strain differences in mouse response to odours of predators. Behav Proc, in press
Dell'Omo G, Wolfer D, Alleva E, Lipp H-P (1993) Impaired acquisition of swimming navigation in adult mice exposed prenatally to oxazepam. Psychopharmacology 111:33-38
Douglass RJ (1989) Assessment of the use of selected rodents in ecological monitoring. Environ Manag 13:355-363
Drickamer LC, Robinson AS (1993) Factors affecting reproductive success in house mouse. XXIII International Ethological Conference, 1-9 September 1993, Torremolinos, Spain, Abstracts, p 115
D'Udine B, Alleva E (1988) The *Acomys cahirinus* (Spiny mouse) as a new model for biological and neurobehavioral studies. Pol J Pharmacol Pharm 40:525-534
Eisenberg JF (1881) The mammalian radiation. The Atlone Press, London
Festing MFW, Lovell DP (1981) Domestication and development of the mouse as a laboratory animal. In: Berry RJ (ed) Biology of the House Mouse. Symposia of the Zoological Society of London, 47, Academic Press, London, pp 43-60
Fibiger HC, Lytle LD, Campbell BA (1970) Cholinergic modulation of adrenergic arousal in the developing rat. J Comp Physiol Psychol 72:384-389
Fried PA, Watkinson B (1990) 36- and 48-month neurobehavioral follow-up of children prenatally exposed to marijuana, cigarettes, and alcohol. J Dev Behav Ped 11:49-58
Gad SC (1990) Recent developments in replacing, reducing, and refining animal use in toxicologic research and testing. Fundam Appl Toxicol 15:8-16
Gould SJ (1977) Ontogeny and philogeny. The Belknap Press of the Harvard University Press, Cambridge, MA, USA
Hartley P, Neill D, Hagler M, Kors D, Vogel G (1990) Procedure- and age-dependent hyperactivity in a new animal model of endogenous depression. Neurosci Biobehav Rev 14:69-72
Hile MG (1991) Hand-held behavioral observations: The Observer. Behav Assess 13:187-196
Kavlock R, Meyer E, Grabowski CT (1980) Studies on the developmental toxicity of ozone: postnatal effects. Toxicol Lett 5:3-9
Kinsley CH, Mann PE, Bridges RS (1988) Prenatal stress alters morphine- and stress-induced analgesia in male and female rats. Pharmacol Biochem Behav 30:123-128
Laviola G, Dell'Omo G, Alleva E, Bignami G (1992a) Ontogeny of cocaine hyperactivity and conditioned place preference in mice. Psychopharmacology 107: 221-228
Laviola G, Dell'Omo G, Chiarotti F, Bignami G (1994) *d*-Amphetamine conditioned place preference in developing mice: relations with changes in activity and stereotypies. Behav Neurosci 108:514-524
Laviola G, Loggi G (1992) Sexual segregation in infancy and bi-directional benzodiazepine effects on hot-plate response and neophobia in adult mice. Pharmacol Biochem Behav 42:865-870
Laviola G, Pick CG, Yanai J, Alleva E (1992b) Eight-arm maze performance, neophobia, and hippocampal cholinergic alterations after prenatal oxazepam in mice. Brain Res Bull 29:609-616
Laviola G, Renna G, Bignami G, Cuomo V (1988) Ontogenetic and pharmacological dissociation of various components of locomotor activity and habituation in the rat. Int J Dev Neurosci 6:431-438
Lechat P (1990) Place de l'expérimentation animale en pharmacologie et en toxicologie. Thérapie 45:341-346
Leon M, Coopersmith R, Ulibarri C, Porter RH, Powers JB (1984) Development of olfactory bulb organization in precocial and altricial rodents. Dev Brain Res 12:45-53

Lewontin RC (1974) The genetic basis of evolutionary change. Columbia University Press, New York

Martin P, Bateson P (1986) Measuring behaviour. Cambridge University Press, Cambridge, UK

McCleary RA (1961) Response specificity in the behavioral effects of limbic system lesions in the cat. J Comp Physiol Psychol 54:605-613

Musi B, Dell'Omo G, Ricceri L, Santucci D, Laviola G, Bignami G, Alleva E (1994) Effects of acute and continuous ozone (O_3) exposure on activity/exploration and social behavior of CD-1 mice. NeuroToxicology, in press

Naftolin F, Butz E (eds) (1981) Sexual dimorphism. Science 211:1263-1324

Needleman HL (1991) The health effects of low level exposure to lead. Annu Rev Publ Health 12:111-140

Nelson BK, Brightwell WS, Mackenzie-Taylor DR, Burg JR, Massari VJ (1988) Neurochemical, but not behavioral, deviations in the offspring of rats following prenatal or paternal inhalation exposure to ethanol. Neurotoxicol Teratol 10:15-22

Noldus LPJJ (1991) The Observer: a software system for collection and analysis of observational data. Behav Res Meth Instr Comp 23:415-429

Peakall DB (1994) Disrupted patterns of behavior in natural population as an index of ecotoxicity. Environ Health Persp, in press

Peakall DB, Kiff LF (1979) Egg shell thinning and DDE residue levels among peregrine falcons *Falco peregrinus*: A global perspective. Ibis 121:200-204

Pinel JPJ, Mumby DG, Dastur FN, Pinel JG (1994) Rat (*Rattus norvegicus*) defensive behavior in total darkness: risk-assessment function of defensive burying. J Comp Psychol 108:140-147

Pintor A, Alleva E, Michalek H (1986) Postnatal maturation of brain cholinergic systems in the precocial murid *Acomys cahirinus*: comparison with the altricial rat. Int J Dev Neurosci 4:375-382

Ricceri L, Calamandrei G, Alleva E (1994) Prenatal oxazepam affects passive avoidance performance of preweaning mice. Brain Res Bull 33:267-271

Shimada C, Kurumiya S, Noguchi Y, Umemoto M (1990) The effects of neonatal exposure to chronic footshock on pain-responsiveness and sensitivity to morphine after maturation in the rat. Behav Brain Res 36:105-111

Shore RF, Douben PE (1994) Predicting ecotoxicological impacts of environmental contaminants on terrestrial small mammals. Rev Environ Contam Toxicol 134:49-89

Stamper CR, Balduini W, Murphy SD, Costa LG (1988) Behavioral and biochemical effects of postnatal parathion exposure in the rat. Neurotoxicol Teratol 10:261-266

Terranova ML, Alleva E, Laviola G (1994) Affiliation and neophobia in developing mice prenatally exposed to oxazepam. Behav Pharmacol 5:52-60

Umezu T, Shimojo N, Tsubone H, Suzuki AK, Kubota K, Shimizu A (1987) Effect of ozone toxicity on the drinking behavior of rats. Arch Environ Health 42:58-62

Van Oortmerssen GA (1971) Biological significance, genetics and evolutionary origin of variability in behaviour within and between inbred strains of mice (*Mus musculus*). Behaviour 38:1-92

Viggedal G, Hagberg BS, Laegreid L, Aronsson M (1993) Mental development in late infancy after prenatal exposure to benzodiazepines - a prospective study. J Child Psychol Psychiat 34:295-305

Walsh TJ, Tilson HA (1986) The use of pharmacological challenges. In: Annau Z (ed) Neurobehavioral toxicology. The Johns Hopkins University Press, Baltimore, pp 244-267

Weisenburger WP, Minck DR, Acuff KD, Vorhees CV (1990) Dose-response effects of prenatal phenytoin exposure in rats: effects on early locomotion, maze learning, and memory as a function of phenytoin-induced circling behavior. Neurotoxicol Teratol 12:145-152

Weiskrantz L (ed) (1968) Analysis of behavioral change. Harper and Row, New York
Wiemeyer SN, Lincer JL (1987) The use of kestrels in toxicology. Raptor Res Rep 6:165-178
Wiemeyer SN, Sparling DW (1991) Acute toxicity of four anticholinesterase insecticides to American kestrels, eastern screech-owls and northern bobwhites. Environ Toxicol Chem 10:1139-1148
Zenick H (1983) Use of pharmacological challenges to disclose neurobehavioral deficits. Feder Proc 42:3191-3195

Terrestrial ecotoxicity of chemical substances

(Chair: Dr. N. Grandy, France and
Dr. G. Karlaganis, Switzerland)

Development of OECD Test Guidelines and Hazard Assessment Procedures for Terrestrial Effects

N. J. Grandy
Organisation for Economic Co-operation and Development, 2 rue André Pascal, 75775 Paris, France

The initial step in the management of a chemical (including pesticides) is the identification of its intrinsic properties (i.e. hazard identification) which include its physical/chemical characteristics, its potential to degrade or accumulate and its toxicity to man and the environment. Hazard assessment (or risk characterisation) for chemicals integrates information on toxicity and potential exposure to determine the nature and magnitude of possible adverse effects that might result from their release into the environment.

An increasing number of countries now require, through legislation, that new and existing chemicals (including pesticides) are tested and assessed for potential risks for control purposes (e.g. notification, registration, classification and labelling). The availability of internationally harmonised and agreed testing methods and hazard assessment procedures increases the effectiveness and efficiency of chemicals control procedures and helps to 'share-the-burden' of assessments between countries. As part of its Chemicals Programme, OECD has, for some time, been working on the development of such harmonised methods and procedures with its activities on Test Guidelines and Hazard Assessment.

A particular driving force in the development of Test Guidelines and assessment procedures has been the introduction, in 1992, of the Pesticides Programme to the work of OECD, and these areas are two of the Programme's major activities.

This paper describes the OECD Test Guideline and Hazard Assessment Programmes, focusing particularly on methods and procedures applicable to the terrestrial environment.

The OECD Test Guidelines Programme

The 'OECD Guidelines for Testing of Chemicals' (OECD, 1993a) are internationally-agreed methods. They were first published in 1981 and cover testing in the areas of physical/chemical properties, effects on biotic systems health effects (toxicity). There are now over 70 OECD Test Guidelines which are available in English and French. The availability of internationally-agreed test methods helps to:

- o establish the validity and increase the international acceptance of data thus avoiding duplication of testing and hence the unnecessary use of laboratory animals;

- o facilitate international trade by avoiding or minimising barriers caused by differences in national legislation and thereby creating a level playing field for chemical industries in Member countries.

Mutual Acceptance of Data

In order to formalise the international acceptance of data, the OECD 'Decision concerning the Mutual Acceptance of Data in the Assessment of Chemicals' (i.e. the MAD Decision) was adopted in 1981. It is legally binding on all Member countries. The MAD Decision states that 'data generated in the testing of chemicals in an OECD Member country in accordance with OECD Test Guidelines and OECD Principles of Good Laboratory Practice shall be accepted in other Member countries for purposes of assessment and other uses relating to the protection of man and the environment'. Data generated in this way therefore have advantages over data generated using any other methods.

Development and Revision of OECD Test Guidelines

The development or revision of an OECD Test Guideline starts with a proposal from a Member country (or countries), the international scientific community (including industry), or the OECD Secretariat itself. It involves extensive peer review by experts (including the possibility for expert meetings or workshops) and the development of consensus in all Member countries.

A key role in Guideline development is played by the National Co-ordinators. Each Member country has a National Co-ordinator, designated by their government, who acts as a focal point for Test Guidelines work in their country. There is also a National Co-ordinator representing the Commission of the European Union (CEU). Each National Co-ordinator maintains an efficient network of technical experts (usually drawn from government, academia and industry) for consultation purposes. National Co-ordinators meet at least once a year to propose the work programme and the priorities and to decide on

Further details on Test Guideline development and revision procedures are described in OECD Environmental Monograph No. 76 (OECD, 1993b).

Guidelines for Terrestrial Effects Testing

Until recently, the focus of the Test Guidelines Programme with respect to environmental effects has been on the development of tests for the aquatic environment, although the following Guidelines for terrestrial organisms are available: 205 Avian Dietary Toxicity Test; 206 Avian Reproduction Toxicity Test; 207 Earthworm, Acute toxicity Tests; 208 Terrestrial Plants, Growth Test.

With the initiation of the Pesticide Programme however, a need for the revision and development of methods for assessing the effects of chemicals on terrestrial species as well as aquatic species has been identified (see Table 1).

Table 1. Results from a 1992 survey of Member countries and industry indicating the numbers of revisions and new Guidelines required for pesticide testing.

Test area	**Number of Guideline revisions proposed**	**Number of new Guidelines (or endpoints) proposed**
Physical-chemical properties and environmental fate	8	45
Ecotoxicology		
Aquatic	5	13
Terrestrial	**4**	**19**
Human health effects	27	9
Total	44	86

The outcome of the Test Guideline Survey indicated that a considerable amount of work was needed and that priorities would need to be set before work could begin. Three Task Forces (one on physical-chemical properties and environmental fate, one on ecotoxicology, and one on human health and occupational exposure) were established in June 1993. Each Task Force was comprised of pesticide experts from government and industry (including contract testing laboratories). Task Forces were requested to propose priorities for work (i.e. high, medium, low), taking into account the frequency with which data from a particular Guideline is requested for pesticide registration (i.e. using from results of a survey of Member country data requirements; OECD, 1994) and the availability of existing methods from other fora (e.g. ISO, US EPA, EPPO). They were also asked to make recommendations of how work should proceed.

The priorities proposed for terrestrial effects testing, and subsequently agreed by the Test Guideline National Co-ordinators, are given in Table 2 (high priorities) and Table 3 (medium and low priorities).

Table 2. High priorities for Guideline revision/development for terrestrial effects testing.

Test Area	Recommendation	Current status of work
Soil		
Soil microorganisms	new Test Guideline(s)	Working Group established
Earthworm reproduction	new Test Guideline	awaiting outcome of German ring test
Beneficial arthropods		
Honey bees - acute oral	new Test Guideline	developing drafts
- acute contact	new Test Guideline	developing drafts
- residues on foliage	new Test Guideline	developing drafts
Other arthropods	new Test Guideline(s)	no activity
Birds		
Acute oral	new Test Guideline	OECD/SETAC Workshop on Avian Toxicity Testing, Florida, USA, December 1994
Reproduction	revise Test Guideline 206	
Acceptability/palatability	new Test Guideline	

Table 3. Medium and low priorities for Guideline revision and development for terrestrial effects testing.

Medium priority	Low priority
- Earthworm acute	- Earthworm - field
- Other tests with soil organisms	- Avian - field
- Honey bees - field/semi-field	- Mammalian palatability
- Other beneficial arthropods - field/semi-field	- Wild mammal toxicity
- Avian dietary	- Mammalian - field
- Terrestrial plants	

As can be seen from Table 2, activities are now underway on most Guidelines given high priority. Work on the items given medium and low priority will be initiated as progress in the other areas allows.

In addition to the specific Guideline activities shown above and in a broader context than pesticides alone, the Netherlands are developing recommendations for balanced sets of tests for the terrestrial environment, which will take account of the need to include organisms from different functional groups (e.g. primary producers, decomposers, herbivores, detritivores, etc), different taxonomic groups (e.g. bacteria, plants, insects, spiders, birds, etc) and different routes of exposure (e.g. air, soil, water, direct contact).

OECD Work on Hazard Assessment

The overall objective of OECD's work on hazard assessment is, to the extent possible, to harmonise assessment procedures and reporting practices in order to assist Member countries in protecting human health and the environment from the potentially harmful effects of chemicals. Harmonisation of these procedures and reporting practices should enable countries to use part, or all, of each other's assessments in lieu of conducting assessments themselves. This would therefore help to 'share-the-burden' in, and reduce the costs of, the hazard/risk assessment of chemicals (including pesticides).

The work on hazard assessment is overseen by the Hazard Assessment Advisory Body (HAAB), which, like the group of National Co-ordinators of the Test Guideline Programme, consists of representatives from all OECD Member countries.

With respect to environmental hazard assessment, work to date, like the Test Guidelines Programme, has concentrated on the development of guidance for the aquatic environment, and guidance documents are available on aquatic effects assessment[1] (OECD, in preparation) and environmental exposure assessment[2] (OECD, 1993c).

For the terrestrial environment, work has been initiated with the development of a discussion document on terrestrial effects assessment by Denmark. In addition, a number of proposals for work in the terrestrial area were made at the recent OECD Workshop on Environmental Hazard/Risk Assessment[4] (London, May 1994) and are indicated in Table 4. These recommendations include proposals for work on Test Guideline development and hazard assessment procedures.

The proposal for the development of a set of standard terrestrial effects tests, (equivalent to the standard aquatic 'base set', comprising tests on algae, *Daphnia* and fish) is expanded on in Table 5. The Workshop proposed that the standard terrestrial tests should include representatives from four categories, i.e. soil organisms, plants, terrestrial invertebrates (above ground) and terrestrial vertebrates, and gave examples of the types of test species which could be included in each category. These recommendations are in agreement with the work already taking place, or proposed, in the Test Guidelines Programme and thus give added weight to the programme developed by the National Co-ordinators.

With respect to the other recommendations, OECD routinely co-ordinates its activities with other international organisations (i.e. recommendation 1) and work is already underway in OECD on the development of transparent, harmonised assessment reports (i.e. recommendation 2) and on the improvement of statistical data analysis (i.e. recommendation 5). The remaining recommendations will be considered by the HAAB at their next meeting in January 1995.

Table 4. Recommendations for further work on hazard assessment for the terrestrial environment made at the OECD Workshop for Environmental Hazard/Risk Assessment, London, May 1994.

Recommendations for further work	
General:	
1.	Integrate with other international organisations involved in development of risk assessment and test guidelines
2.	Encourage consistent transparent assessment reports and provide technical guidance to promote expert judgement
Specific:	
3.	Develop set of standard terrestrial effects tests
4.	Promote the improvement of extrapolation methods and uncertainty analysis, including assessment factors. Could involve the following steps: - review approaches taken in aquatic assessment and other types of risk analysis to identify suitable principles and approaches - construct database containing relevant published toxicity data for a wide range of terrestrial species - research project to derive new data for database using standard laboratory toxicity tests on plants and invertebrates - establish OECD working group to evaluate and make recommendations on how extrapolations should be handled for terrestrial effects
5.	Advance development of improved methods for assessing no-effect levels and other end-points. (Review range of parameters used as end-points and statistical methods for generating robust measures of toxicity).
6.	Develop better understanding of ecological significance of effects seen in the field on populations and ecosystems. Organise workshop to discuss and make recommendations for further work for OECD and other bodies?
7.	Provide criteria to judge validity of non-standard data.

Table 5. Proposal for a list of standard terrestrial tests made at the OECD Workshop on Environmental Hazard/Risk Assessment, London, May 1994. (see recommendation 3 in Table 4 above)

Category	**Types of test species**
Soil organisms	- one soil arthropod - earthworm - micro-organism function
Plants	- one species exposed through root uptake. - one species exposed through leaves
Terrestrial invertebrates (above ground)	- predatory mite - parasite wasp - honeybee - herbivore
Terrestrial vertebrates[5]	- bird

References

OECD (1993a) OECD Guidelines for Testing of Chemicals, continuing series, first published 1981.
OECD (1993b) Guidance Document for the Development of OECD Guidelines for Testing of Chemicals. OECD Environment Monographs No. 76.
OECD (1993c) Report on the OECD Workshop on the Application of Simple Models for Environmental Exposure Assessment. OECD Environment Monographs No. 69.
OECD (in preparation) Guidance Document for Aquatic Effects Assessment.

Notes

[1] effects assessment, i.e. the derivation of Predicted No Effects Concentrations (PNECs) by the use of extrapolation techniques such as assessment (application) factors, uncertainty factors and statistical approaches on the effects data available.

[2] exposure assessment, i.e. derivation of Predicted Environmental Concentrations (PECs) using data on physical/chemical properties, fate and behaviour, emission estimates, exposure models, etc.

[3] This Workshop covered procedures for effects and exposure assessments for the aquatic and terrestrial environments; recommendations relating to terrestrial effects are only detailed here.

[4] Mammalian toxicity is covered by tests for human health assessment. Amphibians are omitted because their most sensitive life-stages are aquatic.

Development of Guidance for Terrestrial Effects Assessment

L. Samsøe-Petersen and F. Pedersen
VKI, Water Quality Institute, Department of Ecotoxicology, Agern Allé 11, DK-2970 Hørsholm, Denmark

Within the OECD Chemicals Programme, priority has till recently been given to the development and harmonisation of strategies for the assessment of hazard and risk of chemicals in the aquatic environment, resulting in a Guidance Document for Aquatic Effects Assessment (Balk et al. 1993). The OECD Hazard Assessment Advisory Body (HAAB) has initiated the development of similar strategies for the terrestrial environment, and - with the aim of promoting discussions within the OECD member countries - a discussion document has recently been prepared (Samsøe-Petersen & Pedersen 1994).

Environmental effects assessment is part of environmental hazard and risk assessment of chemicals - the other part being exposure assessment. The aim of hazard and risk assessments is to protect the ecological structure and function of the environment as well as certain "key" species, which may be defined by ecological, commercial or conservational interests. In aquatic hazard and risk assessment, this aim is achieved by the calculation of a "Predicted No Effect Concentration" (PNEC) in the water, which is compared to the "Predicted Exposure Concentration" (PEC) in order to reach a "risk quotient" (PEC/PNEC), which can be used for an evaluation of the environmental concern for each chemical. The PNEC is most often derived from results of laboratory toxicity testing by extrapolation methods, based on empirical or statistical considerations. The PNEC is not a "level, safe for all organisms", but merely an estimation of the possibility of protecting most of the species in the environment.

The principles of environmental effects assessment for the aquatic environment are applicable for the terrestrial environment as well, but because of differences between aquatic and terrestrial environments, several issues need consideration before a guidance for terrestrial effects assessment can be fully developed. These issues are the topic of the present discussion.

1. Prerequisites for terrestrial effects assessment

In aquatic environments the water phase will disperse and dilute chemicals resulting in a relatively homogenous - compared to the terrestrial environment - exposure of the organisms to the chemicals. In terrestrial ecosystems non-volatile chemicals may stay - even in high concentrations - where they originally were released without being dispersed/diluted except through the action of wind, rain or living organisms. The fate and transport of volatile chemicals in terrestrial environments may be comparable to that of water soluble chemicals in aquatic environments.

While the primary exposure of aquatic organisms to chemicals is via the water, the exposure of terrestrial organisms is a combination of four routes: uptake from surfaces, (pore) water, (pore) air and food.

These differences imply that the foundation of procedures for effects assessment of chemicals in the terrestrial environment includes considerations concerning the relevance of terrestrial effects assessment for different types of chemicals as well as for different organisms. The relevance for chemicals is depending on the types of chemicals that will reach terrestrial environments, and which types and compartments of terrestrial environments they will reach. The relevance of terrestrial effects assessment for different organisms is depending on their place and way of living, which are determining their possibilities of exposure.

Furthermore, the choice and development of testing procedures must be considered, including the testing strategy, the choice of species and the standardisation of experimental conditions. Finally, the extrapolation methods for derivation of the PNEC must be considered.

2. The relevance of terrestrial effects assessment for different types of chemicals

2.1. Routes of emission of chemicals

The chemicals that may be released to and exert effects in terrestrial environments can follow aerial, aquatic or solid phase routes to the ecosystems. Most chemicals will arrive through several routes. Based on the character of the emission, the chemicals can be divided into two groups: those that are produced to be deliberately applied to the environment (the agricultural chemicals) and those that are not intended to reach the environment but are emitted as a consequence of different activities (the contaminants).

The primary differences between pesticides and other chemicals are basically due to differences in the patterns of use and release. Pesticides are designed to have toxic properties. They are produced and applied deliberately to the environment to exert toxic effects on living organisms, whereas industrial chemicals are designed, produced and used for quite different purposes and any toxic property they may have is - in most cases - unintended.

Chemicals contaminating the terrestrial environment via disposal of waste products or via deposition from the atmosphere have some features in common.

They have persisted through different kinds of treatment processes as e.g., waste water treatment, waste cleaning, or combustion/incineration. Water soluble and volatile chemicals may be transferred to the terrestrial environment via irrigation water. The water soluble chemicals are only expected to contaminate local areas on a temporary basis, while volatile chemicals are expected to evaporate during the irrigation process. Industrial chemicals, which are emitted to the atmosphere and to ground and surface water, are expected to be regulated by emission standards.

Thus, the chemicals for which terrestrial effects assessment are relevant are characterised by low (bio)degradability and relatively high adsorption to particles (see table 1). Chlorinated aromatics, and some PAHs belong to this group.

As to the toxicity, it is assumed that terrestrial effects assessment of industrial chemicals is preceded by the initial or refined aquatic testing stage, at which chemicals with no biological activity are sorted out. With pesticides, aquatic and terrestrial toxicity testing is initiated in parallel.

Table 1: Overview of main sources and chemicals released to the terrestrial environment. Shaded areas indicate chemicals for which terrestrial effects assessment is relevant.

Chemical / Source	Pesticides a.o. agricultural chemicals	Industrial chemicals which are toxic and not readily degradable/persistent		
		sorptive /lipophilic	water soluble	volatile
Direct, intended application as chemical	+	-	-	-
Direct, unintended application via sewage sludge	-/+	+	-	-
Direct, unintended application via solid waste (soil/compost)	-/+	+	-	-
Atmospheric deposition	+	+/-	+/-	+
Irrigation	-/+	-	+	+/-

2.2. Types and compartments of ecosystems

In intensively cultured areas, ecological diversity is intentionally reduced as certain species are regarded as pests and therefore reduced in number by e.g., the use of pesticides. In these areas ecological function is considered more important than the protection of a diverse ecological structure. The crop species are considered "key" species and ecological structure and function (including recycling of minerals, restriction of pests and pollination) are intended to allow for maximum yield of the crop. Thus part of the discussion on principles for terrestrial effects assessment should be oriented to protection levels and to which functions/species should be protected in the cultured areas. In non-cultured areas the protection should include as many functions/species as possible like in aquatic environments. This is however, predominantly a political issue and not fundamental to the principles of the effects assessment.

Terrestrial ecosystems are often considered as being composed of two compartments: one is the soil and the organisms living in it, the other is situated above the ground, and could be described as the aerial part of the ecosystem, including the organisms comprising and living in the canopy. In the following, this division of the terrestrial ecosystems shall be used, referring to soil dwelling organisms as *endogeic*, and to organisms living in the air or the canopy as *epigeic*.

The direct exposure of the two compartments of terrestrial ecosystems to different types of chemicals can be deduced from table 1. Combining exposure with the considerations regarding the relevance of terrestrial effects assessment for different types of chemicals, leads to the matrix of terrestrial compartments and chemicals for which effects assessment is relevant, presented in table 2.

Table 2: Relevance of terrestrial effects assessment for each compartment for different types of chemicals, considering direct exposure. Shaded areas indicate relevance of terrestrial effects assessment.

Chemical	Pesticides (treatment)		Toxic, not readily degradable/ persistent industrial chemicals		
Compartment	soil	crop	sorptive /lipophilic	water soluble	volatile
Endogeic	+	+	+	-	-
Epigeic	-	+	-	-	-

Table 2 shows that - considering direct exposure - terrestrial effects assessment is relevant for the endogeic compartment for agricultural pesticides and for toxic, non volatile, not readily degradable industrial chemicals, which are sorptive/lipophilic. For the epigeic compartment, sprayed pesticides is the only group of chemicals for which terrestrial effects assessment is relevant.

In terrestrial environments exposure via food may play a major role irrespective of the compartments considered. Thus, assessment of dietary exposure is relevant for both groups of chemicals in table 2 and should include the assessment of uptake of chemicals from air, plant surfaces and soil to assess the potential toxicity of food items.

3. Choice and development of testing procedures

3.1. Testing strategy

Aquatic effects assessment is described as a sequential process that may comprise three stages: initial, refined and comprehensive, where each stage depends on the type and quantity of information that is available. In initial or preliminary effects assessment the impact of the chemical is generally assessed from short term toxicity tests with only one or two representatives of primary producers (algae), primary consumers (daphnia) and predators (fish). The test design is based on the fundamental principle of the dose-response relationship and practised by the establishment of external concentration-response relations for the chemical. Based on the established relationship, the Lethal (or Effect) Concentrations (L(E)C50, L(E)C10 and L(E)C90) are estimated and the concentration expected not to be toxic to the end point(s) studied (the No Observed Effect Concentration (NOEC) or the Lowest Observed Effect Concentration (LOEC)) is assessed. Intermediate or refined effects assessments are based on chronic or semi-chronic tests, whereas (semi-) field studies may provide the basis for comprehensive effects assessments (Balk et al. 1993).

Terrestrial effects assessment should be based on a similar strategy, assuming preceding aquatic testing for industrial chemicals and parallel aquatic and terrestrial testing for pesticides, as described in section 2.1.

Consequently, Test Guidelines are needed for each of the testing stages in terrestrial effects assessment. For the initial testing stage, single species laboratory toxicity tests, using lethal and sublethal (L(E)C50 and NOEC) endpoints are needed. For the refined stage, (sub)chronic single species tests and two-several species toxicity tests, using species interactions as well as lethal and sublethal effects on each species as end points are relevant, and for the comprehensive stage, field (or semi-field) studies investigating structure and function of ecosystems are recommended.

For the testing strategy, the final question is, how many tests are needed at each stage to obtain enough results for the estimation of the PNEC. Statistical considerations (described in section 4) conclude that for the initial and/or the refined stage, the extrapolation should be based on results from tests with a minimum of five species, representing different groups of organisms. Furthermore, due to the compartmentalisation of the terrestrial environment and the exposure scenarios for different types of chemicals described previously, the Test Guidelines needed for terrestrial effects assessment at the initial stage should comprise organisms from both the endogeic and epigeic compartments. There is thus a need for Test Guidelines for five organisms from each of the two terrestrial compartments.

3.2. Choice of test species

The different groups of organisms to be represented, refer to the fact that terrestrial organisms differ basically with respect to (at least) exposure to chemicals, ecology and taxonomy.

To be representative with respect to the *route of exposure* at the individual level, the organisms should represent exposure via contact, via uptake from air/water and via food.

The *ecological criteria* are important for the identification of functional characteristics. They include the trophic level of organisms; primary producers, herbivores and carnivores as well as degraders - and for chemicals with a potential for biomagnification - predators at higher levels. Furthermore, the organisms of more importance in terrestrial ecosystems than in aquatic, need consideration. These include pollinators and - in cultured environments - natural enemies of pests.

To reflect and ensure the *taxonomical* complexity of ecosystems, members of major taxonomic groups of terrestrial ecosystems must be considered. The anatomy (e.g., respiratory mechanism, epidermal characteristics) and physiology (e.g., detoxification mechanisms) - and thereby the possibilities of uptake and intoxication - vary basically between taxonomic groups.

Thus, for the selection of organisms to be used for toxicity testing, the following aspects should be covered:

- exposure routes via water, air, contact and food
- ecological functions, i.e. primary producers, herbivores, carnivores, degraders and pollinators
- taxonomic groups, i.e. micro-organisms, plants and animals. Among the animals, the invertebrates should be represented by arthropods, preferably insects
- a wide array of sensitivity to chemicals
- "key" species selected from specific criteria.

There are a few OECD Test Guidelines for terrestrial organisms (OECD 1984), and an array of test methods have been developed, especially as a result of pesticide research. These are discussed in a Detailed Review Paper (DRP) (Stavola 1990) and a Discussion Paper prepared for the OECD (van Straalen & van Gestel 1993) as well as in the OECD "Status report on updating and development of terrestrial ecotoxicology test guidelines" (OECD 1991). At present, a project initiated by the OECD is being conducted in the Netherlands describing and using detailed criteria for the selection of test methods for the soil compartment (Leon & van Gestel 1994).

The choice of test species should of course be based on the recommendations from this project. However, tables 3 and 4 illustrate the selection of five test species for each of the two compartments, based on the above mentioned aspects.

Table 3: Example of existing test organisms (systems) to be used for chemicals reaching the endogeic compartment. Tests in (): to be considered for bioaccumulating chemicals. Shaded areas illustrate a choice of five tests for the "terrestrial base set of data".

Function / Taxonomy	Primary producer	Herbivore	Carnivore	Degrader
Micro-organisms				Functional endpoints
Plants	OECD 208			
Annelids				OECD 207
Arthropods		Budworm/Root fly	Beetle	Springtail
Birds		OECD 205/6		
Mammals		OECD 4xx	(OECD 4xx)	

OECD 4xx refers to the large array of tests with mammals using different routes of exposure. Depending on the properties of each chemical, the relevant test(s) could be chosen among these.

Table 4: Example of existing test organisms to be used for chemicals reaching the epigeic compartment, i.e. pesticides. Shaded areas illustrate a choice of five tests for the "terrestrial base set of data".

Function / Taxonomy	Primary producer	Herbivore	Carnivore	Degrader	Pollinator
Micro-organisms					
Plants	"Spray"				
Annelids					
Arthropods		Aphid/ mite/ weevil	Pa.wasp/ Pr.mite		Honey bee
Birds		OECD 205/6			
Mammals		OECD 4xx	OECD 4xx		

Pa. wasp = Parasitic wasp, Pr. mite = Predatory mite. "Spray" refers to the recommendation by Stavola (1990) that a Test Guideline exposing plants via the leaves should be developed.

3.3. Experimental conditions

Standardisation of the experimental conditions during the test is a prerequisite for obtaining reproducible and comparable results. As the development of several new

Test Guidelines with terrestrial organisms is expected to be initiated, it is very important that the principles for the standardisation are discussed. However, the design of tests with soil organisms poses a problem, as some of the organisms cannot be separated from the soil (e.g. micro-organisms), and compromises have to be made, which makes standardisation even more important. Design of *exposure systems* that allow for the investigation of separate routes of exposure, should also be given priority, even though this is not always possible.

In aquatic toxicity testing the *exposure concentration* is maintained at a constant level during the entire exposure period, e.g., by means of renewal of the water, in order to obtain reproducible results. This is not possible in tests performed with terrestrial substrates (except if the organisms are frequently moved to newly spiked substrates, which is a very laborious if not impossible procedure). Thus, administration of the chemicals in most terrestrial toxicity tests is performed once at the start of the test, resulting in possible declining exposure concentrations during the test[1]. For the development of standardised Test Guidelines, the experimental design of terrestrial tests should as far as possible allow for a constant exposure concentration. At least the exposure concentration should be measured by chemical analysis of samples collected during the entire exposure period.

The standardisation of the *substrate* to which the chemical is applied in contact tests is a major problem in terrestrial effects assessment. The need for standardisation is caused by the fact that the bioavailability of the chemicals - and thereby the actual exposure of organisms - in toxicity tests, is dependent on the substrate. Exposure in soil may be reduced by the sorption of chemicals to inorganic particles or organic molecules in the soil. The sorption is influenced by an array of factors e.g., the amount and character of organic matter and clay, the pH and the water content of the soil.

The best way of standardising this plethora of parameters is by the use of inert substrates, like sand or glass plates. However, again this is not possible - even for short term exposure - for testing with several soil organisms. Thus, one or a few standardised soils should be selected and defined. For this, an artificial soil as the one used in the OECD Test Guideline No. 207 with earthworms (OECD 1984) and standardised soils recommended for pesticide fate studies (the "Euro-soils" (Kuhnt & Muntau 1992)) are obvious candidates, but several problems have to be solved before a qualified choice can be made.

In conclusion, there is a need for research in the field of bioavailability of chemicals in soil before a final selection of standardised soil(s) can take place, and the recommendations are therefore preliminary. For the initial screening of the inherent toxicity of chemicals to endogeic organisms, sand should ideally be used

---

[1] Even though testing for terrestrial effects assessment of industrial chemicals is especially relevant for not readily degradable substances.

as the substrate. When this is not possible (e.g., for micro-organisms), standard soils should be used.

The last problem regarding comparable test designs for different test methods, is the selection of *end points*. Some of the tests developed for the testing of side effects of pesticides (e.g. Hassan 1992, EPPO 1993, 1994) are limit tests, excluding the estimation of the concentration-response relation. Thus the results cannot be used for a generic comparison of chemicals. Such tests should be further developed to include the use of dilution series of chemicals.

When the terrestrial "base set" of toxicity tests has been selected, the interpretation of the results from these laboratory toxicity tests should be *validated* by field studies. Initial validations could be based on existing data(bases).

4. Methods required for the calculation of PNEC

4.1. Prerequisites for the use of data from different tests

In aquatic effects assessment the calculation of the PNEC, with respect to direct exposure, of a chemical is often based on the results of a number of single species laboratory toxicity tests. The effect data (LC50s, EC50s and/or NOECs) are expressed by the concentration of the chemical in the water (e.g. mg/l). Terrestrial organisms are however exposed to chemicals via air, water, soil or food or combinations of these media and the units used for the description of dosage/concentration are thus not immediately comparable (e.g. μl/l, $\mu g/cm^2$, mg/kg, mg/day). Furthermore, the bioavailability and thus the toxicity of chemicals in soil depends on the soil type and test conditions as described above. Thus, the approach of calculating one PNEC is not immediately applicable to the terrestrial environment. However, the use of several PNECs would certainly impede the possibility of comparing and integrating results from ecotoxicity tests for effects assessment. Therefore, calculation methods for derivation of PNECs should include the possibilities of:

- normalising effect concentrations or NOEC-values estimated from experiments with different soils, to compensate for differences in bioavailability
- transforming effect data from tests employing different exposure methods (units) to effect concentrations for the same exposure situation / the same substrate
- extrapolating from single species laboratory test results to PNEC for the terrestrial environment.

Methods for *normalising* the results from tests using different soil types are based on theories concerning the distribution of chemicals between different phases; in soil this is mainly between the water phase (pore water) and the solid phase (particles, organic matter) and on the assumption that exposure of the organisms is mainly via the water phase, except for very lipophilic chemicals.

The parameters determining the partitioning of chemicals in soil include the pH and the contents of water, organic matter and clay particles. In general, organic compounds are sorbed to organic matter in the soil, depending on their lipophilicity. For polar organic compounds the soil pH is also important. For soils with low concentrations of organic matter, adsorption to inorganic particles (primarily clay particles) is also of importance.

Thus, based on more or less theoretical considerations, normalisation of effect concentrations obtained in different soils to one (standard) soil can be performed in order to enable the comparison of such data. However, as the calculation methods are based on several assumptions, validation of these methods by comparable toxicity testing experiments with different organisms, routes of exposure, substrates and chemicals should be initiated.

Methods for the *transformation* of results expressed in different units are almost lacking. In a proposal from pesticide research (Løkke 1994), all effect data are recalculated to area based dosages (i.e., kg/ha). This proposal is based on assumptions concerning the penetration of chemicals into soil. The method needs further development and validation before such calculations can be standardised. Prerequisites for this are that a standard scenario should be elaborated and that the estimation of the depth of penetration of chemicals into the soil must be based on characteristics of the chemical and soil type.

The extrapolation methods used for the derivation of the PNEC require that at least some data from ecotoxicity tests are available. For many existing chemicals the only data available are from tests with *aquatic organisms*. The use of aquatic data for the calculations of PNEC for the soil compartment may be recommended when terrestrial data are not available. Methods for the transformation of aquatic effect data to the terrestrial environment are based on assumptions concerning comparable sensitivity of aquatic and terrestrial organisms; exposure via pore water of soil organisms and bioavailability (e.g., Løkke 1994).

4.2. Exposure via food

In terrestrial environments it is necessary to include dietary exposure in the effects assessment. The derivation of a PNEC for soil, protecting species against exposure to toxic concentrations in food, i.e. secondary poisoning, calls for an evaluation of uptake of chemicals from soil, transport via the food chain and dietary toxicity studies.

For the terrestrial environment, this approach implies methods for the calculation of bioconcentration factors (BCFs) for plants and soil dwelling animal species. Effect data from tests exposing animals to chemicals in food, can then be transformed to concentrations in soil, provided that relationships between concentrations in soil and concentrations in plant and animal tissue are established. The literature addressing this issue is very limited, but Romijn et al. (1994)

suggested an algorithm for these calculations. The algorithm is based on the assumption that the daily food intake of the experimental bird or mammal is representative for the food intake in natural environments. The authors emphasise that this assumption could be an important source of uncertainty, as the food intake of different species in the environment is expected to vary with the lipid content of the food items, the body weight of the animal, and of course the availability of food.

Further development and validation of this algorithm is needed, including investigations of relationships between food consumption and food quality to be combined with existing knowledge concerning relationships between food consumption and body weight.

4.3. Extrapolation methods

After the normalisation and transformation of results of toxicity tests, the comparable effect concentrations for different species can be used for the calculation of the PNEC. In the following, methods for the derivation of a PNEC for the terrestrial soil compartment is described.

At an OECD workshop in Arlington in 1990 (OECD 1992), two different types of extrapolation methods were identified: the sensitivity distribution methods, and the application factor methods.

The *sensitivity distribution methods* assume that the species specific NOEC-values fit a specific distribution function (log-logistic, log-normal or log-triangular). By fitting the single species test results to the distribution function, a concentration can be calculated that is in theory hazardous only to a limited proportion of the total number of species in the ecosystem of concern. A PNEC-value for regulatory purposes is proposed to be the concentration that is calculated to be "safe" for 95% of the species (estimated with a possibility of 95%) and thus potentially hazardous to 5% of the species. This implies that chronic toxic effects to a small quantity of the species in the ecosystem of concern are accepted, although the calculated PNEC-value is merely an end-point for the calculations rather than a precise estimation of possible effects in the ecosystem (OECD 1992). For statistical reasons, it is recommended that the extrapolations are based on experimentally estimated NOEC-values for at least 5 species representing different taxonomic groups (see e.g., Balk et al. 1993). If NOEC-values for 5 species are not available, one of the application factor methods described below could be used instead.

Originally, the *application factor method* was developed by the US-EPA for the initial evaluation of chemicals for which only few toxicity data were available (EPA 1984). This method was further developed at the above mentioned OECD workshop (OECD 1992). The basic idea of this method is to divide the lowest available single species effect concentration or NOEC-value by an "uncertainty factor" or an "application factor" to derive an "Environmental Concern Level". This concentration may be used for initial effects assessment or for priority setting among an array of chemicals.

A similar method is used at present in the EU for the determination of Water Quality Objectives for List I substances in the context of Directive 76/464/EEC. The size of the application factor is however determined case by case - based on information regarding the number of test data, the slopes of toxicity curves, toxic activity, long term effects, genotoxicity etc (Bro-Rasmussen et al. 1994).

5. Scheme for terrestrial effects assessment

In summary, the different tools needed for terrestrial effects assessment (the estimation of PNEC for the terrestrial environment) may be amalgamated into a single flow scheme, as shown in figure 1.

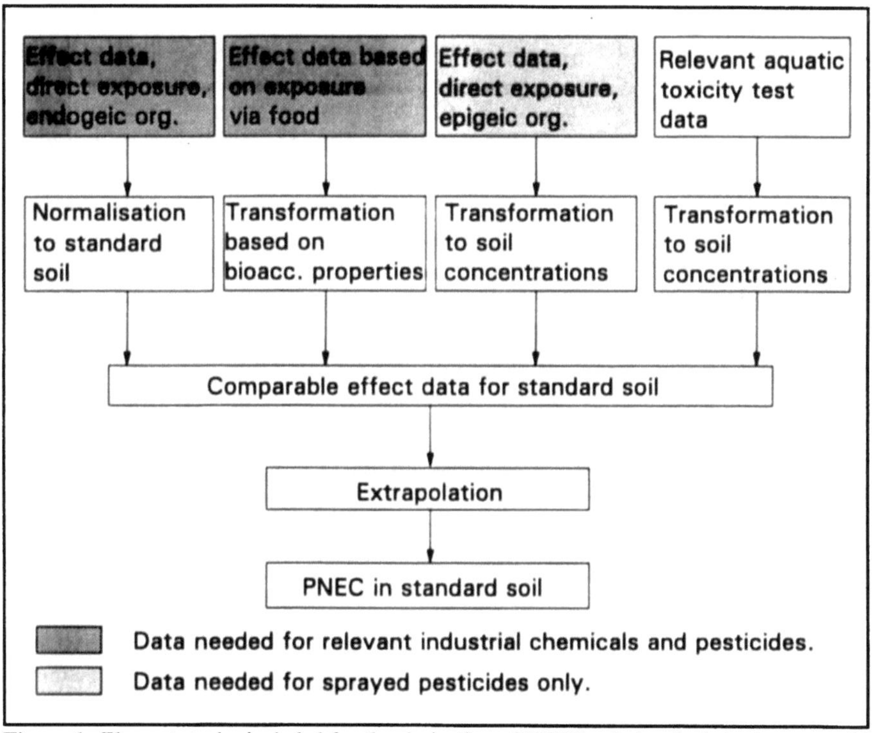

Figure 1. Elements to be included for the derivation of PNECs of chemicals in the terrestrial environment.

The scheme in figure 1 is suggested as a preliminary guidance for terrestrial effects assessment. As described in this paper, several of the components have not been fully developed and validated, and for the implementation of the strategy presented in the scheme, research and harmonisation is needed in most fields.

Acknowledgements

The work summarised in the present paper was initiated by the OECD Hazard Assessment Advisory Body (HAAB) and financed by the EC Commission, DG XI and the Danish Environmental Protection Agency. The project was supervised by G. Bennekou and C. Hansen (Danish EPA), H. Løkke and M. Bruus Pedersen (National Environmental Research Institute, Denmark) and P. Kristensen (VKI, Water Quality Institute, Denmark). Furthermore F. Bro-Rasmussen (Danish Technical University) gave valuable comments during the process. The authors wish to thank the institutions and persons who supported the work.

References

Balk F., P.C. Okkerman & S. Dogger (1993): Guidance document for aquatic effects assessment. Report prepared for the OECD Hazard Assessment Advisory Body, RIVM the Netherlands. 72 pp.

Bro-Rasmussen F., P. Calow, J.H. Canton, P.L. Chambers, A.S. Fernandes, L. Hoffmann, J.-M. Jouany, W. Klein, G. Persoone, M. Scoullos, J.V. Tarazona & M. Vighi (1994): EEC Water Quality Objectives for chemicals dangerous to aquatic environments (List 1), Reviews of Environmental Contamination and Toxicology, vol 137, in press.EPA (1984): Estimating "Concern Levels" for Concentrations of Chemical Substances in the Environment. Environmental Effect Branch, Health and Environmental Review Division.

EPPO (1993): Decision-making scheme for the environmental risk assessment of plant protection products. Bulletin OEPP/EPPO Bulletin, 23, 1, pp. 1-157.

EPPO (1994): Decision-making scheme for the environmental risk assessment of plant protection products. Bulletin OEPP/EPPO Bulletin, in press.

Hassan S.A. (ed.) (1992): Guidelines for testing the effects of pesticides on beneficial organisms: Description of test methods. IOBC/WPRS Bulletin, 1992/XV/3.

Kuhnt, G. & H. Muntau (1992): EURO-Soils: Identification, Collection, Treatment, Characterization. Joint Research Centre, Commission of the European Communities, ISPRA, Italy, 150pp.

Léon C.D. & C.A.M van Gestel (1994): Selection of a set of laboratory ecotoxicological test methods. Tests for the effects assessment of chemicals in terrestrial ecosystems. Vrije Universiteit, Amsterdam. In prep.

Løkke H. (1994): Ecotoxicological extrapolation - tool or toy? In: Donker, M., Ejisackers H., & F. Heimbach (eds.): Ecotoxicology of soil organisms. SETAC Special Publication series, Lewis Publishers, Chelsea, USA, pp. 411-426.

OECD (1984): Guidelines for testing of chemicals. OECD, Paris.

OECD (1989): Report of the OECD workshop on ecological effects assessment. OECD Environment Monographs, No. 26. OECD, Paris.

OECD (1991): Status Report on updating and development of terrestrial toxicology test guidelines. OECD, Paris.

OECD (1992): Report of the OECD workshop on the extrapolation of laboratory aquatic toxicity data to the real environment. OECD Environment Monographs No. 59, OECD, Paris.

Romijn C.A.F.M., R. Luttik & J.H. Canton (1994): Presentation of a general algorithm to include effect assessment on secondary poisoning in the derivation of environmental quality criteria. Part 2. Terrestrial food chains. Ecotoxicology and Environmental Safety, 27, pp. 107-127.

Samsøe-Petersen, L. & F. Pedersen (1994): Discussion paper regarding terrestrial effects assessment. Prepared for the Organisation for Economic Cooperation and Development (OECD). VKI, Water Quality Institute, Denmark. 64pp.

Stavola (1990): Detailed review paper on terrestrial ecotoxicology test guidelines (May 1990), OECD, Paris.

van Straalen N.M. & C.A.M. van Gestel (1993): Ecotoxicological test methods using terrestrial arthropods. Discussion Paper for the OECD Test Guidelines Programme, Vrije Universiteit, Amsterdam, the Netherlands.

Application of Quick and Simple Plant Bioassays to Assess the Genotoxicity of Environmental Pollutants - Detection of Potential Health Hazards of Air, Water and Soil contaminants

Te-Hsiu Ma
Laboratory of Environmental Mutagenesis, Department of Biological Sciences, Western Illinois University, Macomb, IL 61455 USA

Pollution has been created by human activities for the fulfillment of the greedy desires of human beings for comfort, convenience and enjoyment ever since the invention of the wheel. However, the intensity and severity of pollution has emerged as a serious and grave problem only in the last few decades. In order to reduce pollution to a tolerable level, and thus avoid the eventual total destruction of the ecosystem and at the same time to minimize human affliction and misery of human suffering, the pollution problem should be faced in realistic terms. Based upon this concept and the diversity of the pollution problems, pollution should be discussed under the following generic definition:

THE PRESENCE OF ANY ENTITY IN AN EXCESSIVE QUANTITY AT THE WRONG PLACE AND/OR AT THE WRONG TIME

Only under this broad sense, might it be possible to find a partial solution of this complex environmental problem. It is up to us, humans, to begin to turn around the present course of action, and bring about a long lasting stability and harmony with the nature, and our ecosystem, the vital units of the biosphere of this planet.

Perhaps the most noticeable pollution under this generic definition is the human pollution or over-population. One can see the consequence and feel the pressure of this increasing source of pollution. Air, water and soil pollution may be visible in some cases. but the most dangerous kind are those which are invisible. After a short 10 years practice in chemical pest control in the field, more than 50% of the shallow wells in the farm in some mid-western states of America, can no longer provide drinkable water, and this water is even unsafe for cattle consumption. The invisible and the odorless kind of pollution in the environment, especially in mixture forms, require urgent attention. A first alert system, such as the canary carried over the miner's hat, or the smoke alarm installed in the common house hold, is needed to insure the safety of the air we breath, water we drink, and the soil in which our crops are raised. This is the kind of bioassays/biomonitors that will be addressed in the discussion.

For more than 3 decades, since the awareness of the pollution problem, chemical analysis has been the predominant role in the environmental protection program. A small fraction of analysis relies on biological test results. The major reason for this reliance of chemical analyses are as follows: Based upon the evolution of sciences, the latter portion of the 20th century is still in the chemistry era whereas the early part of the century was dominated by the science of physics. Therefore, chemists rule the world, with such slogans as "better living through chemistry". It was precisely, the better living which has created most of the pollutants. And yet the mighty chemists believe firmly that they are the only ones who can solve the pollution problem. They have the intricate analytical instruments, and they can detect parts per trillion concentration of the pollutants in the air and in the water and portable Mass Spectrometers can be used to monitor pollutants in the field. Lengthy reports are complied in the authoritative agencies and further funding of such activities have escalated year after year. Our policy makers fail to understand that pollutants always exist in mixtures of chemical and the conditions for their existence are variable and constantly changing. Their effects on the biological system on a singular basis may have nothing to do with the effects of the chemicals when they are in mixed forms under real world circumstances. And yet the majority of the biological laboratories test chemical agents singularly, especially emphasize the *in vitro* carcinogenic effects.

There are more than 200 bioassays (Water et al. 1990) available at the present time. Most of them are only good for *in vitro* testing under laboratory conditions, especially those using animal tissues or cells. Some use the whole animal, such as dead fish in a river as the signal of water pollution, or decline of certain protozoan populations in a lake as indicators of environmental degradation. These bioassays are too insensitive and too slow. A number of sensitive detectors of water and air pollutants have been available since the early 1930s. Hardly any of them have been funded by the federal or state agencies for the practical use of environmental monitoring.

Most of the sensitive, quick and simple bioassays comprise plant systems. The sensitive target cells in the biological system are germ cells and especially those of plants. The germ cells of higher plants such as *Tradescantia* are located in the flower buds which have four layers of leafy structures which are known to be very porous and allow gaseous agents to diffuse readily. The next layer of protection is the anther wall or the ovary and ovule wall. The other route of entry of polluting agents are the vascular tubules from the stem up to the flower parts.

Based upon radioisotope labelling experiments conducted in the *Tradescantia* plant (Nauman et al. 1979), agents in the air can diffuse into the flower buds within a very short time. In contrast, the target cells (namely germ cells) of a whole animal are well shielded from toxic agents. Generally, the polluting agents that are either in the air, water, or any food stuff have to gain their entry into the animal body through the circulatory (blood) system, and eventually reach the testis or ovary after several circulating routes through the lung, liver and other accessory organs of the circulatory system. The polluting agents would be diluted through these pathways and often metabolized and become harmless by-products.

Tradescantia, the generic name of a group of plants in the family of *Commelinaceae*, is a close relative of the Wondering Jew, and the Spider Plant. There are more than 20 species found in North and South America. The extensively cultivated clones for experimental and genotoxicity studies are clone 02 and 4430 which are hybrids and heterozygous for the blue (dominant) and pink (recessive) alleles. A special clone 03 derived from (*T. paludosa*), has been used in cytogenetic studies since the early 1930s by Karl Sax and his students. The microspore chromosomes of clone 03 are relatively large and only 6 in its haploid genome. Most of the principles of chromosome/chromatid aberrations were developed by Prof. Sax and his co-workers with the help of X-irradiation. The stamen-hair-mutation and meiotic chromosome abnormalities are used as the cytogenetic endpoints of the current bioassays induced by environmental agents.

The Standard Protocol of the Tradescantia-Stamen-Hair-Mutation (Trad-SHM) Bioassays

The Tradescantia-Stamen-Hair-Mutation (Trad-SHM) bioassay was developed in the early 1960s by Dr. Arnold H. Sparrow and his co-workers at the Biology Division of the Brookhaven National Laboratory, Long Island, NY (Underbrink, A.G. et al. 1973). The target cells are the mitotic cells of the stamen-hair of clone 4430 or 02 which are heterozygous for the blue and pink alleles. The induced change represents a somatic mutation changing from the dominant blue to the recessive pink (Emmerling-Thompson and Nawrocky, 1982). This somatic mutation, although occurring in the cells of stamen hairs, also may take place in the microspore mother cells which are the germ cells in the anthers of the flower. For this reason, mutations which happened in the somatic cells of the stamen hair may also happen in the germ cells and become a heritable change and be transmitted to future generations.

A standard protocol for the Trad-SHM assay is currently in print in Mutation Research (Ma et al. 1994a). This standard protocol was validated under the

International Programme on Chemical Safety, World Health Organization, The United Nations and the United Nations Environment Programme. The detailed procedure of this Trad-SHM bioassay was described earlier by Underbrink et al. (1973).

The major steps of the Trad-SHM bioassay are outlined below:

1. Raising plants of clone 4430 or 02 in the greenhouse, growth chamber or in the field - Plants grow best in small clones of 10 - 25 individual plants in a pot of loose soil or patches of plants in the field. In order to stimulate their flowering process, plants should be grown under a 16/8 hr light/dark photoperiodic cycle.

2. Collecting plant cuttings for *in situ* monitoring or in laboratory testing of chemical or physical agents - At least 20 plant cuttings which bear at least 9 exposed flower buds in the young inflorescences are necessary to conduct an experiment.

3. Recovery duration - generally, 7 to 11 days are required in order to capture the cells of peak sensitivity.

4. Scoring of pink mutation events - Flowers in full bloom of the treated and control plant samples are collected daily after the recovery period, and cooled in a refrigerator to facilitate the combing of the usually tangled hairs on the stamen. After the hairs are combed on each of the 6 stamens of a flower (the sample unit for scoring), pink cells amongst the blue cells of each of the stamen hairs are identified under a dissecting microscope. Pink cell or cells intercepted by blue cells within the hair or at the terminal position of a hair are considered to be pink mutation events. Since the mutated cells in the earlier stage of development may divide to give rise to more pink cells, thus each pink cell is not considered as a single mutational event. Pink events are scored from 5 to 12 flowers for each of the experimental groups for 5 consecutive days during the peak sensitivity period (7th - 11th days), and only the high scores from 3 consecutive days are used to derive the means and standard deviations. The mutation rates of a given experimental group are expressed in terms of number of mutation event per 1000 hairs.

5. Statistical analyses of the mutation rates are done by a Student-t analysis if only one treated and one control group were used. ANOVA analysis of variation and Donnett-t statistics are used to determine the significance at the 0.01 or 0.05 level.

A sample of the data from the mutagenic test of Tordon (a systemic herbicide) is given in Table 1. (Table 1)

Table 1. Tradescantia-Stamen-Hair-Mutation Assay on the mutagenicity of Tordon*

Date	Expt. group	Flower numbers 1	2	3	4	5	6	7	8	9	10	11	12	Pink event	M/1000 hairs
7th	C	0	0	2	0	0								2	1.33
	T-1	2	0	0	2	0	1	0	1	0				6	2.22
	T-2	0	1	1	1	1								4	2.67
	T-3	2	3	0	1	4								5	6.67
8th	C	2	0	2	0	0	1	1	0	0	0			6	2.00
	T-1	0	1	2	3	1	1	0	1	0	1			10	3.33
	T-2	0	0	2	3	5	0	0						10	4.76
	T-3	0	4	0	2	4								10	6.67
10th	C	2	1	1	0	0	0	1	0	0	0			5	1.67
	T-1	1	0	0	1	2	0	0	0	0	2			6	1.82
	T-2	1	0	1	0	0	2	2	1	0	2	0	0	9	2.50
	T-3	1	0	2	0	4	2							9	5.00

Statistical Analysis with ANOVA and Dunnett-t statistic

Expt. group	Dose ppm	n	Mutation/1000 hairs Mean	+/- S. D.	Signif. (0.05)	Remarks
Control	0	3	1.67	0.33		
Treated-1	1	3	2.46	0.78		
Treated-2	5	3	3.31	1.26	+	
Treated-3	10	3	6.11	0.96	+	lower vigor

*Tordon: 4-amino-3,5,6-trichloropicolinic acid, a systemic herbicide against deep-rooted herbaceous weeds and woody plants.

The database of this bioassay was presented in a recent review (Ma, 1994b) in the Proceedings of the Conference of the International Association for Great Lake Research and Estuarine Research Federation. The database covers the areas of studies on pesticides (Mohammad and Ma 1983), *in situ* air monitoring (Schairer et al., 1982, Schairer 1983), *in situ* water monitoring in Canada (Grant et al., 1992), radiobiological tests (Sparrow et al., 1973, Underbrink et al., 1973) and radiation monitoring around nuclear power plants (Ichikawa, 1981), water and sediment tests (Underbrink et al., 1984), gaseous mutagens (Van't Hof et al., 1982) and known mutagens in liquid forms (Ma et al., 1994b).

The Standard Protocol of the Tradescantia-Micronucleus (Trad-MCN) Bioassay:

The standard procedure for this bioassay was established under the validation study of the International Program on Chemical Safety, World Health Organization The United Nations and the United Nations Environment Program is published in Mutation Research (Ma et al. 1994c). A detailed description of this bioassay was published earlier (Ma, 1983, Ma et al., 1978).

The major steps of the Trad-MCN protocol are outline below:

1. Raising plants of clone 4430 or 03 is the same as that for the Trad-SHM bioassay.
2. Collecting plant cuttings for treatment - Around 15 plant cuttings which bear at least 9 exposed flower buds on the young inflorescences for each experimental group are required to conduct the experiment. They are maintained in tapwater or nutrient solution for *in situ* exposure, or for laboratory testing of environmental agents similar to that for the Trad-SHM bioassays.
3. The treated and control inflorescences are fixed in Carnoy's fluid for 24 hr and then transferred into 70% ethanol for storage. Slide preparation for MCN scoring can be done any time later.
4. Slide Preparation: The fixed inflorescences are coded before the slides are made. Only the early tetrad stage of the meiotic microspore mother cells were used to prepare slides for scoring of the micronuclei.
5. The MCN are scored from the intact tetrads (4 cells attached together as one unit) and enter the data into various columns of the scoring sheet. The total number of MCNs are tallied from tetrads which contain 1, 2, 3, 4, 5, or more MCNs. The means and standard deviations of each experimental group are derived from 5 sample populations. The MCN frequencies are expressed in terms of number of MCN per 100 tetrads.
6. Statistical Analyses are done in the same manner as in the Trad-SHM bioassay.

A sample of the data for the genotoxicity test of Primaquine (a new anti-malaria drug) is shown in Table 2.

Table 2. Sample Data Score: Tradescantia-Micronucleus Bioassay Dose Related MCN frequencies induced by Primaquine*

Expt. group	N 0	Number of Micronuclei I	II	III	IV	V	Sum	MCN/100 tetrads	Remarks
1	296	4					300		Control
		4					4	1.3	(0 ppm)
2	295	2	3				300		
		2	6				8	2.7	
3	291	9					300		
		9					9	3.0	
4	300	2	1	1			304		
		2	2	3			7	2.3	
5	300	4	1	1			306		
		4	2	3			9	2.9	
							Mean & SD	2.44 +/- 0.69	
1	352	7	1				360		T-1
		7	2				9	2.5	(100 ppm)
2	300	11	3	1			315		
		11	6	3			20	6.3	
3	300	4	2	1		1	308		
		4	4	3		5	16	5.3	
4	300	2	2	3	1		309		
		2	4	9	4		19	6.2	
5	300	4	3				307		
		4	6				10	3.2	
							Mean & SD	4.7 +/- 1.7	
1	296	3		2			300		T-2
		3		6			9	3.0	(200 ppm)
2	302	12	2				316		
		12	4				16	5.1	
3	296	6	2	1			305		
		6	4	3			13	4.3	
4	300	5	5	3			313		
		5	10	9			24	7.7	
5	300	9	2				311		
		9	4				13	4.2	
							Mean & SD	4.86 +/- 1.76	

*Primaquine: A new drug for the new strain of sporozoa of malaria.

Table 2 continued

Expt. group	0 N	Number of micronuclei I	II	III	IV	V	Sum	MCN/100 tetrads	Remarks
1	284	13	3				300		T-3
		13	6				19	6.3	(300 ppm)
2	293	4	2	1			300		
		4	4	3			11	3.7	
3	278	20	2				300		
		20	4			1	24	8.0	
4	290	8	1				300		
		8	2			5	15	5.0	
5	294	2	3	1			300		
		2	6	3			11	3.7	
							Mean & SD		5.3 +/- 1.87
1	300	6	4	1	1		312		T-4
		6	8	3	4		21	6.7	(400 ppm)
2	300	7	6	1	1		315		
		7	12	3	4		26	8.3	
3	300	36	6	3			345		
		36	12	9			57	16.5	
4	302	2	3			1	308		
		2	6			5	13	4.2	
5	300	14	2				316		
		14	4				18	5.7	
							Mean & SD		8.3 +/- 4.3
1	288	10	2				300		T-5
		10	4				14	4.7	(600 ppm)
2	292	4	1	2		1	300		
		4	2	6		5	17	5.7	
3	271	18	6	5			300		
		18	12	15			45	15.0	
4	272	25	1	2			300		
		25	2	6			33	11.0	
5	294	3	1	1	1		300		
		3	2	3	4		12	4.0	
							Mean & SD		8.1 +/- 4.7

Statistical analysis with ANOVA and Dunnett-t statistic

Expt. group	Dose (ppm)	n	MCN/100 tetrads Mean	+/- S. D.	Signif. (0.05)	Remarks
Control	0	5	2.4	0.7		
Treated-1	100	5	4.7	1.7		
Treated-2	300	5	4.9	1.8	+	Positive
Treated-3	300	5	5.3	1.9	+	positive
Treated-4	400	5	8.3	4.3	+	Positive
Treated-5	600	5	8.1	4.7	+/-	over dose

The database of the Trad-MCN for various categories is partially presented in the review paper published in 1990 (Ma 1990) and the revised database in the review of 1994 (Ma 1994b). Selected references for various categories of studies under this bioassay are listed as follows: Anderson and Ma (1982) studied the effect of low dose gamma radiation. Genotoxicity of brackish water (Chen and Zhang, 1982), surface water, polluted air (Fang, 1981) and contaminated soil (Ho et al., 1983) were studied in china. The laboratory tests of the Chicago municipal sludge samples (Hopke et al., 1982), the *in situ* monitoring of a lead smelter in St. Louis (lower et al., 1983), as well as the gaseous emissions and leachate from a landfill site in Chicago (Ma, 1993) have been conducted. Results of two year studies of drinking water from a reservoir have been published (Ma et al., 1984a). More than 100 chemicals in gaseous or liquid forms including known carcinogens, mutagens, and clastogens have been tested with the Trad-MCN bioassay (Ma et al., 1984b). A variety of *in situ* air monitoring data, such as industrial complexes, parking garages, truck stops and bus terminals as well as indoor air qualities have been carried out (Ma et al., 1982, 1984b). The contaminated wastewater and urban air of the city of Queretaro, Mexico (Ruiz et al., 1992), and polluted water of Lake Superior of Canada (Grant et al., 1992) have been monitored. The genotoxicity of creosote-contaminate soil have been detected by the Trad-MCN bioassay (Baud-Grasset et al., 1993).

Numerous conferences related to many facets of environmental protection or pollution take place every month of the year somewhere in the world. Most of these important conferences have brought up problems, identified new agents as the susplcious carcinogens, but few deal with remediation, and the reduction of the pollutants or relate social ethics to the pollution problem.

In the early 1970s, the slogan for environmental problems was:

"THE SOLUTION OF POLLUTION IS DILUTION"

Not until the 80s, the slogan changed into:

"THE SOLUTION OF POLLUTION IS REDUCTION".

Reduction here in a broad sense, means reduce our human desire, the GNP, the new gadgets, the demand of energy consumption. Recycling is one of the essential ways of reduction. For more than 2000 years of civilization in the old world, there was no solid waste problem. In less than 200 years in the American society, we have solid waste problems not only in the urban metropoly, but also in the rural community of 20,000 in Macomb, IL, a small college town. Reduction and recycling can not be accomplished by law but rely only on social ethics or morality. The magnitude of accomplishment toward reduction in a tiny fraction of improvement in morality may exceed a thousand times that of legislation of new laws and regulations. This can be illustrated by the driving habits on a freeway. The conscientious ethical driver needs no speed regulations, and on the contrary, the law offenders install radar detectors, just to defeat the law in order to speed up to 10 - 20 Km above the speed limit.

mm MORALITY, Km LEGALITY

When the law prohibits the illegal dumping of hazardous wastes in a municipal landfill, the next thing that happens is the discovery of the hazardous waste on the roadside. When scientific evidence is overturned by million dollar law suits, laboratories abandoned the traditional animal tests.

The policy makers create new laws and regulations, and hire lawyers to protect new laws. Ironically, the law protecting lawyers happens to be confronted with more powerful lawyers for the law breakers. Such consequences are not surprising since legality is the counterpart of morality. That is to say, morality is the foundation of social order and community harmony. Laws are created by humans when morality is in decline. Consequently, more laws in the book would reduce the level of social morality, although the rate of moral decline is a much slower process than the establishment of laws. A good example of this is the American Social Welfare system. By law, the low income mother legally obtains child support while the recipients of the welfare money intentional reproduce more children against the social norm of the moral person.

Is there a solution for environmental pollution? Should we be the problem presenters like those that have appeared in the numerous conferences which allegedly dealt with environmental problems? It is about time to offer a solution.

THE SOLUTION OF POLLUTION IS MODERATION.

Moderation in all thing is the essence of the Doctrine of the Mean, one of the books of Confucius (English translation by Legge and Ride, 1991). Moderation should be a morality guided upward movement in every aspect of life.

Moderation, in reference to pollution problem, is an action under a conscientious effort to reduce the production and consumption of materials or energy in every day life. It can not and should not be accomplish overnight to the degree that the environment is totally pollution free. It is a gradual but consistent strive toward perfection. It may take decades or centuries to see the improvement. A drastic

change overnight by the regulatory powers of law only bring disaster. Moral improvement is a slow process attainable only through education.
Education in the current discussion means family, school, and social education processes through a life time and through generations, especially family education. Education in reference to the pollution problem means to demonstrate the harmful effects of pollution at the grassroots level. Quick and Simple plant bioassays are excellent tools for environmental education. It is up to us human beings to speed up this life long process or leave it to Gaia. If so, nature will determine the fate of all living beings by its cruel ways. The so-called *Homo sapiens*, the Wise Men species may be headed towards extinction.

Moderation is a moral based movement - Confucius said in His The Doctrine of the Men (Legge and Ride 1991) (Fig. 1)

施诸己而不愿亦勿施于人

WHEN YOU DO NOT LIKE WHEN DONE TO YOURSELF,
DO NOT DO TO OTHERS

Moderation can be carried out within a social group, and among different social groups within different geographic boundaries. Only when moderation is carried out on the global level, the environmental problem can be lessened. Moderation is a morality base upward adjustment at the individual level, the family level, the social level, and necessitates nation-wide practice, international movement, and global action.

Let us reemphasize the underlying cause of pollution, stated at the beginning of this discussion. A movement in moderation should be focused on our current and future scientific activities. If Confucius were alive today, he would say:

THERE ARE THREE SOURCES OF POLLUTION
THOSE GENERATED BY APPLIED PHYSICS DID A LITTLE HARM.
THOSE PRODUCED BY APPLIED CHEMISTRY GAVE US DEEP TROUBLE
THOSE CREATED BY APPLIED BIOLOGY WILL LEAVE ENDLESS
TURMOIL TO MANKIND.

Reference

Anderson, V. A. and T. H. Ma (1982) Micronuclei induced by low-dose cobalt-60 gamma irradiation in *Tradescantia* pollen mother cells. Environ. Mutagenesis, 4:348.

Baud-Grasset, A. and F. Baud-Grasset, J. M. Bifulco, J. R. Meirer and T. H. Ma (1993) Reduction of genotoxicity of a creosote-contaminated soil after fungal treatment determined by the Tradescantia-Micronucleus test, Mutation Res., 303:77-82.

Chen, D. and Y. Zhang (1982) Tradescantia-Micronucleus tests on water samples from several seashore areas of Qingdao and Jiaoshouwan, J. Environ. Sci., 3:35-37. (in Chinese with English abstract)

Emmerling-Thompson,M. and M. M. Nawrocky (1982) Genetic basis for using *Tradescantia* clone 4430 as an environmental monitor of mutagens, J. Hered., 71:261-265.

Ichikawa, S. (1981) *In situ* monitoring with *Tradescantia* around nuclear power plants, Environ. Health Persp., 37:145-164.

Fang, T. (1981) A report on the studies of effects of environmental pollutants on chromosomes - A Sino-American colleborated research on environmental mutagen in the air and water samples from some industrial areas of Qingdao, PRC and on the pesticide-DDV. J. Shandong Coll. Oceanology, 11:9-11.

Grant, W. F., H. G. Lee, D. M. Logan and M. F. Salamone (1992) The use of *Tradescantia* and *Vicia faba* bioassays for the *in situ* detection of mutagens in an aquatic environment, Mutation Res., 270:53-64.

Ho, J., R. Zhou, T. Fang (1983) Tradescantia-Micronucleus tests on fluoride contaminated soil. Chinese Soil Utilization and Conservation. In: Proc. 5th Ann. Meet. Chinese Soil Sci. Soc. 2:325-326. (in Chinese with English Abstract)

Hopke, P. K., M. J. Plewa, J. B. Johnson, D. Weaver, S. G. Wood, R. A. Larson and T. Hindley (1982) Multitechnique screening of Chicago municipal sludge mutagenicity, Environ. Sci. Technol., 16:140-147.

Legge, J. and L. Ride (1991) Translation of Chinese Classics, 3rd ed. SMC Publ. Inc. Hong Kong.

Lower, W. R., W. A. Thompson, V. K. Drobney and A. F. Yanders (1983) Mutagenicity in the vincinity of a lead smelter, Teratog. Carcinog. Mutagen., 3:231-243.

Ma, T. H.(1983) Tradescantia-Micronculeus (Trad-MCN) test for environmental clastogents, In: *In vitro* toxicity testing of environmental agents: current and Future Possibilities, Pt. A., Kolber, Wong, Grant, DeWoskin and Hughes (Eds.)

Plenum Publ. Corp. NY pp. 191-214.

Ma, T. H. (1990) Tradescantia-Micronucleus test on clastogens and *in situ* monitoring, In: Mutation and Environment, Part E. Environmental Genotoxicity, Risk, and Modulation, M. L. Mendelsohn and R. J. Albertini (Eds), Wiley-Liss, NY pp.83-90.

Ma,T. H. (1993) Genotoxicity of gaseous agents and leachates from a landfill and the air around an incinerator, The Final Report to: Office of Salid Waste Research, Center for Solid Waste Management and Research, Institute for Environmental Studies, University of Illinois, 34 pages.

Ma, T. H., A. H. Sparrow, L. A. Schairer and A. F. Nauman (1978) Effect of 1,2-dibromoethane (DBE) on meiotic chromosomes of *Tradescantia*, Mutation Res., 58:251-258.

Ma, T. H., V. A. Anderson, S. S. Sandhu (1982) A preliminary study of the clastogenic effects of diesel exhaust fumes using Tradescantia-Micronucleus bioassays, In; Short-term Bioassays in the Analysis of Complex Environmental Mixtures II. Waters, Sandhu, Huisingh, Claxton and Nesnow (Eds.) Plenum Publ. Corp., NY pp. 352-358.

Ma, T. H., V. A. Anderson, M. M. Harris, R. E. Neas and T. S. Lee (1984a) Mutagenicity of drinking water detected by the Tradescantia-Micronucleus test. Can. J. Genet. Cytol. 27:143- 150.

Ma, T. H., M. M. Harris, V. A. Anderson, I. Ahmed, K. Mohammad, J. L. Bare and G. Lin (1984b) Tradescantia-Micronucleus (Tra-MCN) tests on 140 health related agents. Mutation Res., 138: 157-167.

Ma, T. H., G. L. Cabrera, A. Cebulska-Wasilewska, R. Chen, F. Loarca, A. L. Vandenberg and M. F. Salamone (1994a) Tradescantia-Stamen-Hair-Mutation bioassay, Mutation Res., (in press).

Ma, T. H. (1994b) *Tradescantia* (spiderwort) plants as biomonitors for the genotoxicity of environmental pollutants, In: Proc. Annual Conference of International Association for Great Lake Research and Esturaine Reseaerch Federation, Symposium: Biomonitors and Biomarkers as Indicators of Environmental Change in the Great Lake Basin, F. Butterworth (ed), Plenum Publ. Corp. June 5 - 9, 1994, (in press)

Ma, T. H., G. L. Cabrera, R.Chen, B.S. Gill, S. S. Sandhu, A. L. Vandenberg, M. F. Salamone (1994c) Tradescantia-Micronucleus bioassay, Mutation Res. (in press)

Mohammad, K., T. H. Ma (1983) Tradescantia-Micronculeus (Trad-MCN) and Tradescantia-Stamen-Hair-Mutation (Trad-SHM) tests on common pesticides, Environ. Mutagenesis, 5:65.

Nauman, C. H., P. J. Klotz and L. A. Schairer (1979) Uptake of tritiated 1,2-

dibromoethane by *Tradescantia* floral tissue. Relation to induced mutation frequency in stamen hair cells. Environ. Health Perspect., 19:201-215.

Ruiz, E. F., V. M. E. Rabago, S. U. Lecona, A. B. Perez and T. H. Ma (1992) Tradescantia-Micronucleus(Trad-MCN)bioassayonclastogenicityofwastewater and *in situ* monitoring, Mutation Res., 270:45-51.

Schairer, L. A. (1983) Mutagenicity of ambient air at selected sites in the United States using *Tradescantia* as a monitor, In *In vitro* Toxicity Testing of Environmental Agents, Current and Future Possibilities, Part A. Kolber, Wong, Grant, DeWoskin and Hughes (Eds) Plenum Press, NY pp.

Schairer, L. A., R. C. Saukulis and N. R. Tempel (1982) Monitoring ambient air for mutagenicity using the higher plant *Tradescantia*, In Genotoxic Effects of Airborne Agents, R. R. Tice, D. L. Costa and K. M. Schaich (Eds) Plenum Press, NY pp. 123-140.

Sparrow, A. H., L. A. Schairer and R. Villalobos-Pietrini (1973) Comparison of somatic mutation rates induced in *Tradescantia* by chemical and physical mutagens, Mutation Res., 21:238-239.

Underbrink, A. G., L. A. Schairer and A. H. Sparrow (1973) *Tradescantia* stamen hairs: a radiobiological test system applicable to chemical mutagenesis, In: Chemical Mutagens: Principle and Methods for Their Detection, Vol 3, A. Hollaender (Ed.) Plenum Press, NY pp. 171-207.

Underbrink, A. G., W. R. Lower, A. F. Yanders and T. K. Ranney (1984) New methodology for assessing the toxicity of water and related sediments, In: Proc. 18th Ann. Conf. Trace Substances in Environmental Health, Univ. Missouri, Columbia MO, pp. 351-355.

Van't Hof, J. and L. A. Schairer (1982) *Tradescantia* assay system for gaseous mutagens, A Report of the U. s. Environmental Protection Agency, Gene-Tox Program, Mutation Res., 99:301-315.

Waters, M. D., A. L. Brady, H. F. Stack and H. E. Brockman (1990) Antimugenicity profiles for some model compounds, Mutation Res., 238:57-85.

Comparison of susceptibilities of species used in toxicological and environmental risk assessment to chemicals (pesticides)

W. Pflüger

Institute of Environmental Biology, Bayer AG, Crop Protection, Research Centre Monheim, D - 51368 Leverkusen

Toxicology and Ecotoxicology

In making a comparison between ecotoxicology and toxicology one must keep in mind the major differences in the goals of these disciplines. Although both deal with potentially toxic substances, they have different methods, endpoints, intensities of individual tests, and different risk assessment procedures and evaluations. Toxicology aims at the best possible hazard prediction and protection of individuals, whereas ecotoxicology looks at populations, species, ecosystems and the sustainability of ecological functions. The purpose of the comparison made in this discussion is not to suggest further species and tests for better predictions in human toxicology. Human toxicology is too advanced and too well-founded to expect that essential contributions could be made from the comparatively few and rough observations made in ecotoxicology. The search for very rare individual risks, e.g. hyperallergic reactions, and the development of special animal tests for their detection is a typical example for toxicological research (e.g. Zbinden, 1991); such risks are hardly important in ecology, because a very low incidence reflects by definition a very low risk for individuals and practically no risk for a population or species.

Why extrapolation?

Toxicologists know the difficulties of extrapolating results from rats and mice to humans, which for a general biologist are quite closely related species. In ecotoxicology, where only a few representative species can be tested, the problem of extrapolation to other species and taxonomic categories in the environment appears even more difficult than in toxicology. Nevertheless, there is a certain need and a great deal of political pressure for such extrapolations aimed at establishing, for example, safety standards or quality criteria covering all biota, even if the data base is small, such as it is for many "industrial chemicals". Pesticides have a comparatively large environmental data base. I will therefore present data from a few widely used pesticides, and show some features of one of the currently proposed extrapolation systems for determination of protective quality criteria in the environment.

Finally, we will have a look at the behaviour of highly mobile terrestrial animals, which in ecological assessments can be as important as the intrinsic toxicity of a substance. This aspect of exposure presents a particularly difficult problem in the terrestrial environment, and conclusions on risk drawn without considering such aspects may be highly erroneous. Bird toxicity and risk assessment of pesticides will be given as an example.

A. Comparison of 5 active ingredients

For simplification, some typical ecotoxicological endpoints are presented for the two herbicides, Atrazine and Metamitron, as well as for three insecticides, Azinphos-methyl, Cyfluthrin, and Parathion-ethyl. Except for some aquatic organisms, the results of the chronic tests are not presented.

The following overview of short term test methods and endpoints shows that there are great differences in methods, exposure and endpoints (Table 1, long term and chronic tests not included).

Table 1: Typical Short-Term Tests and Endpoints

Category	Species	Application/ Exposure	Carrier / Medium	Duration/ Frequence	Acute (a)/ Chronic (c)	Endpoint
Mammals	rat	per os	gavage	1 x	a	LD 50
Birds	quail quail / duck	per os in food	capsule/gavage food	1 x 5 d	a subacute	LD 50 LC 50
Fish	trout / bluegill, orfe	static / flow through	water	96 h	a	LC 50
Water flea	Daphnia magna	static / flow through	water	48 h	a	EC 50 (= LC 50)
Algae	Scenedesmus sp./ Selenastrum sp.	static	water	72 h (8 generations)	c	EC 50 / NOEC (cell division)
Bacteria	Pseudomonas activat. sludge	static static	water water	16 h 0,5 - 3 h	c a - c	EC 50 (EC 0) EC 50
Earthworms	Eisenia fetida	whole body	soil	14 d	a	LC 50
Bees	honey bee	topical contact oral in food	water / solvent sucrose solution	1 x 1 x (1-2h)	a a	LD 50 LD 50

It is obvious that such different endpoint values cannot be directly compared. The test concentration or dose is a parameter measured or applied outside the body. Effects such as LC_{50} and LD_{50} values depend on concentrations inside the body. Equal amounts in surrounding or applied media such as water, soil, solvent, capsules, food etc. do not mean similar concentrations in body liquid or tissues. Finally, the variations in adsorption, cell permeability, endogenous transport, metabolism, enzymes and receptors will not only contribute to species variability, but also differ in their relative importance between various compounds.

Taking this into consideration, it would be of quite limited value to make highly sophisticated and detailed comparisons in such an overview, even if the differences in exposure were partly compensated for by a method of normalisation. It seems sufficient to look for characteristic features of the substances and to roughly compare the endpoints in relative dimensions of orders of magnitude. The values presented in Figures 1 to 5 are derived from standard tests for registration and in some cases from literature studies of apparently high quality. If several values were available, the geometrical mean is presented.

Atrazine (Fig. 1) has a low acute mammalian and bird toxicity, with all values falling within the same order of magnitude. The toxicity for fish and Daphnia is low, falling in the range of 20 to 100 mg/l. As to be expected for a herbicide, the algae are highly sensitive with a NOEC of 11 µg/l. As is found for almost all pesticides, toxicity to bacteria is low, particularly if the test is not performed on an individual species like Pseudomonas putida, but on the functionally relevant "activated sludge", which represents the community responsible for biological purification in sewage plants. In activated sludge the EC_{50} values of most pesticides exceed 10000 mg/l.
The earthworms, with an LC_{50} of about 80 mg/kg dry weight of soil, are in a range where no effects are to be expected in the field. The value of more than about 100 µg per bee in topical application confirms the lack of insecticidal activity.

Concerning the extrapolation from laboratory to field, a comparison of many pesticides from several different chemical classes has shown, that the effects in aquatic ecosystem tests, even with long term exposure, has generally been covered by the lowest chronic NOEC value from simple aquatic standard tests. The latter have even frequently been more sensitive. These findings are supported by an increasing number of studies (see references under "conclusions" at the end of chapter A). They are contrary to the wide spread belief that ecosystems are always more sensitive than single species and

Fig. 1 Atrazine

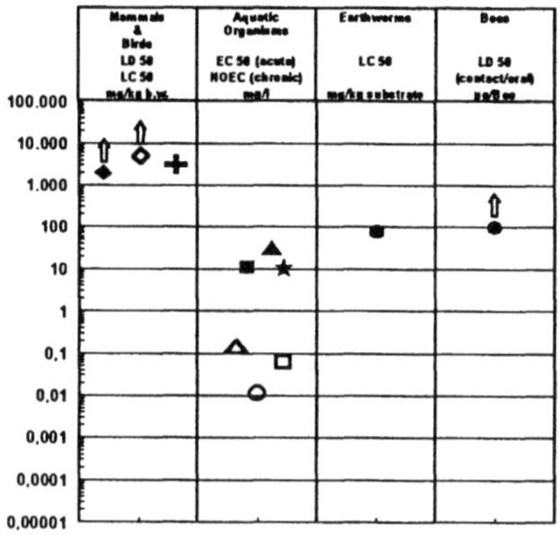

Fig. 2 Metamitron

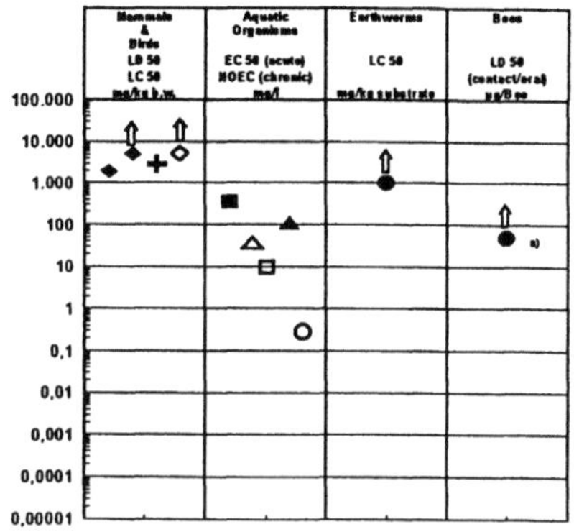

a) oral in herbicidal mixture

Fig. 3 Azinphos-methyl

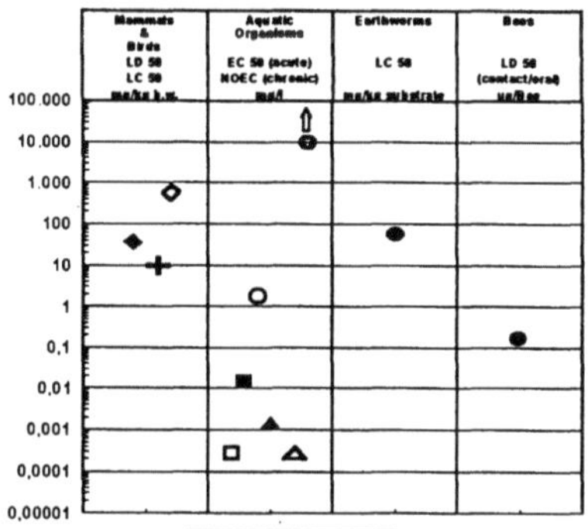

Fig. 4 Parathion-ethyl

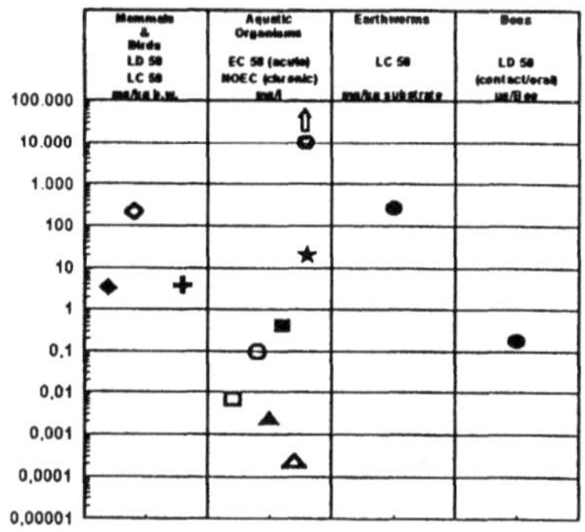

Fig. 5 Cyfluthrin

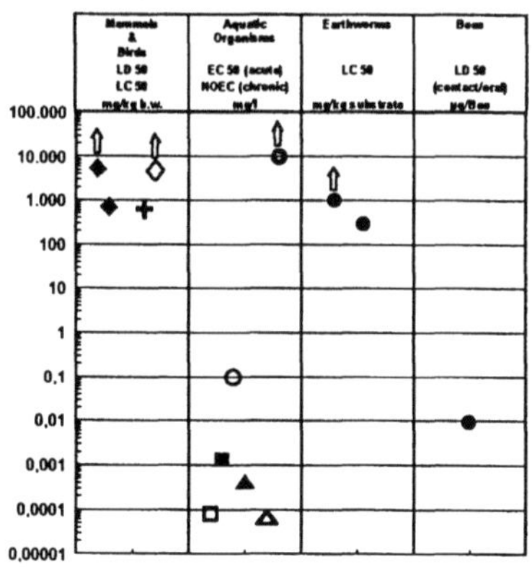

Legend:

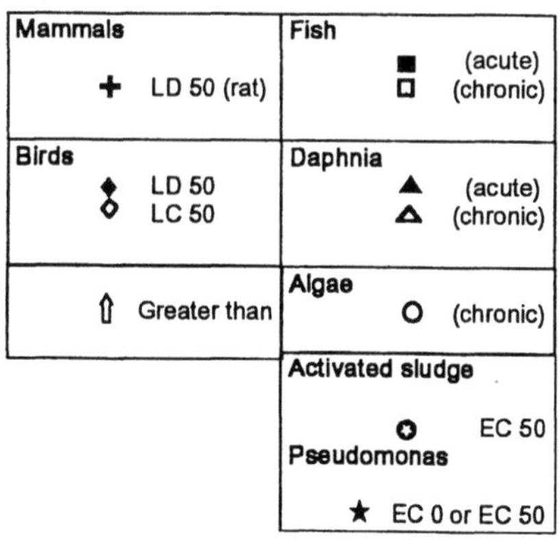

the effects principally unpredictable; this assumption is one of the reasons for safety factors of considerable range. For Atrazine, the lowest standard laboratory NOEC value for unicellular green algae was 11 µg/l, whereas in numerous field tests, noticeable effects did not appear below about 20 µg/l. Even considerably higher field NOEC's have been reported, particularly in flowing water systems (Huber 1994).

The second herbicide Metamitron (Fig. 2), which is widely used in sugar beets, shows a similar pattern of test results at a low level of toxicity for vertebrates and aquatic organisms in acute studies. The aquatic long term toxicity is very low for fish and Daphnia; the specificity of the high algal toxicity is even much more pronounced than in the case of Atrazine. The compound is not toxic to earthworms and bees.
Azinphos and Parathion (Fig. 3, 4) are "classical" organophosphorous insecticides with a high intrinsic toxicity of the active ingredient to vertebrates, aquatic animals and, of course, bees. As is to be expected, the algal toxicity is relatively low, in these cases about 3 orders of magnitude lower than the toxicity for aquatic animals.

Cyfluthrin (Fig. 5) is a pyrethroid and reveals the typical properties of this class of insecticides: very low toxicity for mammals and birds and very high toxicity for insects and aquatic invertebrates. The toxicity to algae is about 3 orders of magnitude lower than for aquatic animals. In laboratory experiments, pyrethroids show a high toxicity for fish. This is in contrast to their activity under practical field conditions. The very low solubility of pyrethroids in water, and their extreme adsorptivity to material suspended in natural waters, drastically decreases their bioavailability to fish. Thus, pyrethroids are good examples of how intrinsic physical properties of compounds can have pronounced effects on their bioavailability and consequent toxicity.
The relative differences in toxicity values, given in orders of magnitude, are summarised in Table 2. Since the different test methods have not been compensated by an attempt of normalisation, the minimum difference has been substracted from the highest, because this part could be caused by a systematic difference in methods, exposures, and measuring units.

This method of analysis is without a doubt crude, but it does allow confirmation of some findings which are already known from much more extensive comparisons in the literature. I believe that the following conclusions can be

Tab.2 **Relations of acute (short term) toxicity-values (orders of magnitude)**

	Atrazine	Metamitron	Azinphosmethyl	Parathionethyl	Cyfluthrin	Range of differences in magnitude a)	b)
bird LD_{50} / rat LD_{50}	0	0	0 - 1	0 - 1	0 - 1	1	+
rat LD_{50} / fish EC_{50}	2	2	2 - 3	1	5 - 6	4 - 5	-
rat LD_{50} / earthworm LC_{50}	1 - 2	1	0 - 1	1 - 2	0 - 1	1 - 2	?
bird LC_{50} / earthworm LC_{50}	1 - ?	0 - 1 (?)	1	0	0 - 1 (?)	(1 - ?)	?
rat LD_{50} / bee LD_{50}	1	1	1	1	4 - 5	3 - 4	-
rat LD_{50} / algae NOEC	4 - 5	4	0 - 1	2	3	4 - 5	-
fish EC_{50} / daphnia EC_{50}	0	0	1 - 2	1	0	1 - 2	+
fish EC_{50} / earthworm LC_{50}	0	0 - 1 (?)	3	2 - 3 (?)	5	5	-
earthworm EC_{50} / bee LD_{50}	0	1 (?)	2	3	4	4	-

0 = same order of magnitude   1 = difference exceeding..orders of magnitude
a) = highest minus lowest relation   b) = comparability yes (+) / no (-) / questionable (?)

drawn, even if there are always exceptions (some supportive references are given in brackets):

- The efficacy of substances with specific biological activity, such as pesticides, focus on the species for which they were selected (e.g. herbicides on plants, insecticides on insects). They are predictable from basic laboratory tests.
- Acute and chronic effects are correlated, within a certain range of magnitude (Kenaga 1982; Hollihn et al. 1991).

- Acute and chronic laboratory results allow a good prediction of short and long term aquatic ecosystem effects, particularly if the fate of the substance is known (Emans et al. 1993, CEC-Research-Project EV 4-0110, 1993, Persoone & Janssen, 1994; Pflüger & Heimbach 1993, Crossland et al. 1994; Hill et al. 1994; Belanger 1994; Leeuwangh 1994).
- Warm-blooded vertebrates such as birds and mammals are comparable in sensitivity.
  Warm-blooded vertebrates should not be compared to fish (Tucker & Leitzke, 1979).
- Aquatic animals like fish and Daphnia are often similar in susceptibility (Thurston et al., 1985, Maki, 1979, Mayer & Ellersieck, 1988).
  Plants are not comparable.
- Toxicities for terrestrial vertebrates and invertebrates are not readily comparable.

B. Extrapolation methods intending to protect 95 % of species

There have been a variety of attempts to use ecotoxicological data from individual substances to define "safe concentrations" for the environment (cf. Van Straalen et al., 1994). The most simple method is the application of "safety factors" which depend on the amount of data available, i.e., the degree of uncertainty. For example, if there was only one acute toxicity value available, the environmentally tolerable concentration would be set at, for instance, 1/1000 of this value. If several chronic NOEC values were available, this "safety factor" could be lowered to e.g. 1/100 or 1/10. Such rough procedures for estimation are understandable, if very few data are available for a substance. The system is clear and takes precautionary aspects into account. The values derived, however, are not necessarily correct in the sense that they closely approach the "true" value of the ecological effect threshold in the ecosystem.

Efforts have been made to develop more scientific methods aimed at protecting, for example, at least 95 % of the species in an ecosystem. The basic assumption for these methods is a log normal or log logistic distribution of toxicities over the whole range of organisms. The NOEC's (or LC/EC50-values) from various test species are then considered representative. The distribution curve derived from these values can be used to calculate the so-called "Maximum Tolerable (or Permissible) Concentration" (MTC or MPC), with theoretically 5 % of the NOEC's being below this value. Stated differently, this means up to 5 % of the species could theoretically be affected by the substance (for further details see e.g. OECD 1992, Emans et al., 1993).

As an example, I refer to a recent presentation of Løkke (1994) on the insecticide Dimethoate. The author collected a relatively large number of test results on a high variety of species and taxonomical groups. For better comparison, he tried to normalise the test data by calculating the amount of active ingredient per hectar which would be necessary to reach the respective NOEC. I will not discuss the particular ways of conversion, they have been described by the author.

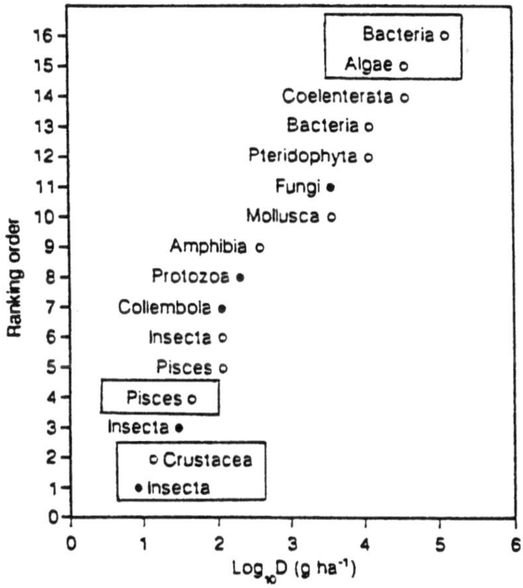

Fig. 6:
Normalized aquatic and terrestrial toxicity data on dimethoate in ranking order, as presented by Løkke 1994 (Fig. 4). Only the lowest and the highest values are results from species usually tested for registration (inside rectangles, added as a modification of the author's original figure).

When he ranked his results, he found a quite continuous distribution of toxicity, thus seemingly justifying the basic assumption of "normal distribution" of toxicity throughout the biota (Fig. 6). This as well as the assumption of random sampling do not apply to active ingredients. Typically, a toxicity distribution with two maxima has to be admitted (e.g. in fig. 6: fish, crustaceans and insects as opposed to bacteria and algae). The test species for risk assessment have not been selected at random. In consequence, the "MRC"-values calculated from all test results reflect the variability of (non-random) test samples and not the actual toxicity to the most susceptible species, i.e. they will, for statistical reasons, be considerably lower than ecologically justified.

Moreover, in practice the number of tested species, at least from normalised and validated studies submitted for registration, are much less numerous than those presented by Løkke. For reasons of animal protection and resource limitations, the problem can not be solved by drastically increasing the number of species tested. This would not even significantly increase safety, because traditional risk assessment is already based on highly susceptible organisms, and there is no need to generate a pool of "intermediate" toxicity values just for statistical reasons. I therefore do not believe, that this - theoretically interesting - concept will be largly applied in practice.

Rather, future ecological risk assessment may give more weight to probabilistic aspects of risk, thus allowing us to focus on major risks of a significant frequency.

C. Behavioural aspects

The importance of behaviour for hazard assessment is particularly apparent with mobile species like birds. They feed on fields or crops which have been treated with potentially toxic substances, for instance insecticidal seed treatments or granules. Many birds feed on seeds, and birds may also ingest granules as pieces of "grit" (stones helping to grind seeds in the stomach). In certain cases, a bird can take up toxic amounts of a pesticide. From the basic LD50 and LC50 studies, the toxic quantities can easily be calculated. However, in practice, intoxications are rare, even in cases where a high or very high potential hazard for birds are predicted. This is due to:

1. repellency, which is more or less pronounced with many active ingredients, and
2. the learning capabilities and the resulting avoidance responses of the birds.

In practice, a compound that is potentially toxic to birds will be safe if its palatability is low, the toxic dose is distributed over a sufficient number of items (e.g. seeds, granules) and the perception of non-palatability is rapid enough to induce timely avoidance behaviour.

The formulation has an important function in helping the birds to learn the discrimination of unpalatable, treated seeds, e.g. by adding a suitable colour. The support of the colour in improving the learning process was for example demonstrated in studies with Methiocarb, a compound which is known for its exellent bird "repellency" (Greig-Smith 1987, Grau et al 1992, Schafer et al 1983). Before such formulations are commercialised, certain tests are performed under more or less "worst case" conditions. These must demonstrate an effective avoidance behaviour (e.g. following guideline 25-1 of the BBA). According to this guideline, starved birds are released into a pen where, for example, treated and untreated seeds are spread on the bottom. The stress conditions are selected in a way that there is pressure for the birds to take up toxic amounts of treated food. If the birds are able to differentiate between the two types of food and to avoid the treated seeds before taking up a toxic amount, the formulation can be expected to be safe in practice (Grau et al., 1992).

Without taking the avoidance behaviour of the animals into consideration, the application of many formulations, which are safe in practice, would be classified as lethal or highly hazardous to birds.

Conclusions for risk assessment and regulation:

- The risk of substances must be evaluated on a case by case basis, taking into account the specific properties of each compound, its particular function and its use pattern in the environment.
- Similar test results from substances can lead to very different results in risk evaluations, depending on the use of the substance and the goals of protection.
- "Quality objectives" (such as "acceptable environmental concentrations") cannot be derived for all groups of substances in the same schematic way. If the acceptability of certain concentrations has been demonstrated following specific guidelines or regulations (e.g. registration of pesticides or various types of "biocides"), the same substances can not be regulated under another scheme for the same environmental compartment with different rules.

Literature:

Belanger SE (1994) Review of experimental microcosm, mesocosm, and field tests used to evaluate the potential hazard of surfactants to aquatic life and the relation to single species data. In: Freshwater field test for hazard assessment of chemicals. (Eds. Hill, I.R. et al.), Lewis Publ., Boca Raton, 287-314

Biologische Bundesanstalt für Land- und Forstwirtschaft (1993) Prüfung von Ködermitteln, Granulaten und behandeltem Saatgut auf Vogelgefährdung - Annahmeversuche. Richtlinien BBA, Teil VI, 25-1, 8pp., Braunschweig

Commission of the European Communities, DG XII (1983) Development and validation of methods for evaluating chronic toxicity to freshwater ecosystems. Report summary CEC contract No. EV 4V-0110-UK(BA), CEC Brussels

Crossland N O; Heimbach F; Hill I R ; Boudou A ; Leeuwangh, P; Matthiessen P; Persoone G (1994) Summary and recommendations of the European Workshop on Freshwater Field Tests (EWOFFT). IN: Freshwater field test for hazard assessment of chemicals. (Eds. Hill, I.R. et al.), Lewis Publ., Boca Raton, xxv-xxxvii

Emans HJB; Plassche EJ v.d.; Canton JH; Okkerman PC; Sparenburg PM (1993) Validation of some extrapolation methods used for effect assessment. Env. Tox. Chem. 12, 2139-2154

Grau R; Pflüger W; Schmuck R (1992) Acceptance studies to assess the hazard of pesticides formulated as dressings, baits and granules, to birds. Brighton Crop Protection Conference, Proceedings 7c-4, 787-792.

Greig-Smith PW, (1987) Hazards to wildlife from pesticide seed treatments. BCPC Mono No 39, Application to seeds and soil (Ed. T. Martin)

Hill IR; Shaw JL; Maund SJ (1994) Review of aquatic field tests with pyrethroid insecticides. In: Freshwater field tests for hazard assessment of chemicals. (eds. Hill IR et al.), Lewis Publ., Boca Raton, 249-271

Hollihn KU; Fischer R; Fritsch H; Guth JA; Johnen BG; Knauf W; Pflüger W; Säle M; Voget M (1991) Pflanzenschutzmittel und Naturhaushalt - Prüfung, Gefährdungsabschätzung und Risikobeurteilung, Teil III: Stufenkonzept für Prüfschemata als Grundlage der Gefährdungsabschätzung und der Risikobeurteilung. Gesunde Pflanzen, 43(7), 221-231

Huber W (1994): Atrazine in aquatic test systems: An evaluation of ecotoxicological risks. In: freshwater field tests for hazard assessment of chemicals (eds. Hill IR et al.), Lewis Publ., Boca Raton, 273-286

Kenaga EE (1982) Predictability of chronic toxicity from acute toxicity of chemicals in fish and aquatic invertebrates. Env. Tox. Chem. 1, 347-358

Leeuwangh P (1994) Comparison of chlorpyrifos fate and effects in outdoor aquatic micro- and mesocosms of various scale and construction. In: Freshwater field tests for hazard assessment of chemicals. (eds. Hill IR et al.), Lewis Publ., Boca Raton, 217-248

Løkke H (1994) Ecotoxicological extrapolation: Tool or toy? In: Donker MH et al. (ed.), Ecotoxicology of soil organisms, Lewis Publ., Boca Raton, 411-425

Maki AW (1979) Correlations between Daphnia magna and fathead minnow (Pimephales promelas) chronic toxicity values for several classes of test substances. J. Fish. Res. Board Can. 36, 411-421

Mayer FL; Ellersieck MR (1988) Manual of acute toxicity: interpretation and data base for 410 chemicals and 66 species of freshwater animals. US Dept. Inter. Fish and Wildlife Service, Resource Publ. 160

Organisation for Economic Cooperation and Development (1992) Report of the OECD workshop on the extrapolation of laboratory aquatic toxicity data to the real environment. OED Environm. Monogr. No. 59, Paris, 43 pp.

Persoone G; Janssen CR (1994) Field validation of predictions based on laboratory toxicity tests. In: Freshwater field tests for hazard assessment of chemicals. (eds. Hill IR et al.) 379-397

Pflüger W; Heimbach F (1993) Can laboratory toxicity data be used to predict effects in the field? In : Environmental criteria for assessing agricultural pesticides (eds. Rijtema PE et al.), DLO Winand Staring Centre, Wageningen/Netherlands, Seminar Report 73, pp. 59 - 60

Schafer EW jr, Bowles WA, Hurlbut J (1983) The acute oral toxicity, repellency and hazard potential of 998 chemicals to one or more species of wild and domestic birds. Arch. Environm. Contam. Tox 12, 355-382

Thurston RV; Gilfoil TA; Meyn EL; Zajdel RK; Aoki TI; Veith GD (1985) Comparative toxicity of ten organic chemicals to ten common aquatic species. Water Research 19(9), 1145-1155

Tucker RK; Leitzke JS (1979) Comparative toxicology of insecticides for vertebrate wildldife and fish. Pharmac. Ther. 6, 167-220

Van Straalen NM, Leeuwangh P, Stortelder PBM (1994) Progressing limits for soil ecotoxicological risk assessment. In: Ecotoxicology of soil organisms (eds. Donker MH et al.), Lewis Publ., Boca Raton, 397-409

Zbinden G (1991) Predictive value of animal studies in toxicology. Regulat. Toxicol. Pharmacol. 14, 167-177

Toxicological Evaluation of Pollutants in Soil - Concept of the AGU* and Assessment of Tolerable Concentrations for Metals

B. Liebl, U. Schwegler, E. Roscher, G. Kerscher

*Arbeitsgemeinschaft Umwelthygiene der Akademie für das öffentliche Gesundheitswesen im Bayerischen Staatsministerium für Arbeit und Sozialordnung, Familie, Frauen und Gesundheit; Landesuntersuchungsamt für das Gesundheitswesen Südbayern, D-85762 Oberschleißheim, FRG

Introduction

In view of the high number of waste sites in the industrialized world, health authorities are more and more faced with the problem having to evaluate health risk posed by contaminated soil. Especially sites where children might play are often a matter of concern. In particular, small children may be endangered because of their high susceptibility and potentially high exposure due to direct oral intake of soil. Guideline values set for this population group are usually also applicable to less susceptible or less exposed persons (older children, adults). Playground values have been derived using various methods, but to date no generally accepted procedure for hazard assessment has been established. We report here on a health-based approach for setting values of tolerable soil concentrations above which from a toxicological point of view remediation is indicated. The proposed method was applied to a series of metals often involved at hazardous waste sites.

Concept of hazard assessment

Apart from exposure resulting from contaminated soil, in most cases "background" exposure to the chemicals of concern must be considered, due to their ubiquitous existence in the environment from either natural and/or anthropogenic sources. The first step in the present concept is an assessment of the average exposure to potential contaminants through air, drinking water and food, independent of possible exposure caused by soil. Intake via food, mostly being the main route of exposure, is preferably derived from duplicate portion studies. These are based on the analysis of exact duplicates of all foods (and beverages) as consumed by a selected group of individuals thus taking into account the addition and loss of substances during food processing and preparation (e.g. cleaning, cooking) and providing most realistic and reliable information about actual dietary intake.

Estimated average intake via air, drinking water and food is subtracted from a generally approved tolerable dose likely to be without an appreciable health risk during a lifetime, e.g. PTWI (provisional tolerable weekly intake; WHO 1987b). If no PTWI has been established, PMTDI (provisional maximum tolerable daily intake), ADI (acceptable daily intake), RfD (reference dose) or comparable values are used. The resulting difference represents acceptable intake by the additional pathway of incorporation via polluted soil:

$$\textit{tolerable intake via soil} = \textit{tolerable intake}[e.g. PTWI] - \textit{average intake (air, drinking water, food)}$$

Based on this estimation, tolerable soil concentration can be calculated:

$$mg\ agent / kg\ soil = \frac{\textit{tolerable intake via soil}[mg\ agent / kgBW\ x\ day] \times \textit{body weight}[kg]}{\textit{intake of soil}\ [g / day] \times \textit{absorption rate}[\%]}$$

Maximum oral exposure to surface soil is assumed for children in the age group between two and four years, due to their playing behavior (hand-soil-mouth-contacts) and the average time spent playing in the open air (Kaiser and Roßkamp 1992). Accordingly, about three-year old children (body weight: 15 kg, respiratory volume: 6 m^3/day, water consumption: 1 liter/day, caloric intake: 420 kJ/kg body weight) are taken as a basis for all calculations. Estimates on oral intake of soil by children ranging from 0.05 to 10 g/ day have been reported in the literature (Eikmann et al. 1989). We use a value of 0.5 g/ day derived from the work of Ruck et al. (1989) who compared the results of several national and international studies and noticed an intake of this amount by 95 % of all children up to 6 years of age when playing in the open air. In cases where tolerable intakes already consider bioavailability issues (e.g. PTWI, ADI), absorption rates are set to 100 %. Similarly, an absorption rate of 100 % is used, if no reliable data are available.

The described model calculations are meant as a tool in the process of defining remediation guideline values for contaminated soil. However, they should not be applied schematically. Finally recommended values may differ from calculated ones, e.g. due to uncertain toxicological or exposure data, or simply because of reasonable achievability.

Assessment of tolerable soil concentrations for metals

Table 1 gives a list of several relevant metals together with their tolerable intakes as defined by internationally approved expert committees. A survey of average human exposure to these elements through air, drinking water and food, derived from actual exposure data for European residents, is shown in Table 2. Data on

Table 1. Tolerable intake of metals

Element	Value	Name	Committee
As	15 µg/kgBW x week[a]	PTWI	JECFA
Cd	7 µg/kgBW x week	PTWI	JECFA
Cr	4.8 µg/kgBW x day	RfD	USEPA
Cu	500 µg/kgBW x day	PMTDI	FAO/WHO
Hg	5 µg/kgBW x week[b]	PTWI	JECFA
Pb	25 µg/kgBW x week[c]	PTWI	JECFA
Ni	16 µg/kgBW x day	RfD	USEPA
Zn	0.3 - 1 mg/kgBW x day	PMTDI	WHO

Abbreviations: BW, body weight; PMTDI, Provisional Maximum Tolerable Daily Intake; PTWI, Provisional Tolerable Weekly Intake; RfD, Reference Dose; USEPA, U.S. Environmental Protection Agency; JECFA, Joint FAO/WHO Expert Committee on Food Additives; FAO, Food and Agriculture Organization of the United Nations; WHO, World Health Organization

[a] for inorganic arsenic
[b] of which not more than 3.3 µg/kgBW x week should be present as methylmercury
[c] for infants and small children

exposure via food represent the range of mean values from various duplicate portion studies performed in Switzerland, Belgium, The Netherlands and Scandinavia, including results from a most recent German investigation carried out in connection with the second "National Environmental Survey" (Becker et al. 1994). Exposure data were transformed to intakes of 3-year old children using the parameters given above. Data from dietary studies performed on adults were used in half of their amount. According to Kaiser and Roßkamp (1992), this approximation is possible, provided that food is similarly composed, because coloric intake of children amounts to about half the value of adults. Based on the calculated ranges listed in Table 2, average total intakes via air, drinking water and food, expressed as percentage of tolerable intakes, were estimated as shown in Table 3. Based on these data, average tolerable intake via soil and tolerable soil concentrations (Table 3) were calculated as described. The resulting contents correspond to the proposed guideline values (Table 3), except for Cu and Pb. In the case of Cu an additional safety factor of 10 was used to account for uncertainty in assessing health hazard for small children. The value calculated for Pb is in the range of

Table 2. Average intake of metals via air, drinking water and food by children

	Source	Average exposure	Absorption rate [%]	Calculated intake by children [µg/kgBW x day]	[% of tolerable intake]
As	Air	Rural areas 1 - 10 ng/m^3 [a]	30 [a]	< 0.01	< 0.2
		Urban areas 5 - 20 ng/m^3 [a]			
	Drinking water	< 1 µg/l [b]	-	< 0.07	< 3.1
	Food	9 - 21 µg/person x day [c,d]	-	0.30 - 0.70	14.0 - 32.6
				Σ 0.38 - 0.78	17.3 - 35.9
Cd	Air	Rural areas 1- 5 ng/m^3 [a]	25 [d]	< 0.01	< 0.1
		Urban areas 10- 20 ng/m^3 [a]			
	Drinking water	0.07 µg/l [d]	-	< 0.01	< 0.5
	Food	7 - 18 µg/person x day [d,e,h]	-	0.23 - 0.60	23.0 - 60.0
				Σ 0.25 - 0.62	23.6 - 60.6
Cr	Air	Rural areas 5 - 20 ng/m^3 [i]	100	< 0.01	< 0.6
		Urban areas 4 - 70 ng/m^3 [i]			
	Drinking water	< 10 µg/l [b]	-	< 0.67	< 14.0
	Food	22 - 30 µg/person x day [j]	-	0.73 - 1.00	15.2 - 20.8
				Σ 1.41 - 1.68	29.8 - 35.4
Cu	Air	Rural areas 10- 60 ng/m^3 [i]	20 [d]	< 1	< 0.1
		Urban areas 20-500 ng/m^3 [i]			
	Drinking water	< 100 µg/l [b]	-	< 7	< 1.4
	Food	0.75-1.4 mg/person x day [k,l,m]	-	25 - 47	5.0 - 9.4
				Σ 33 - 55	6.5 - 10.9
Hg	Air	Rural areas 2 ng/m^3 [a]	80 [a]	< 0.01	< 0.5
		Urban areas 10 ng/m^3 [a]			
	Drinking water	0.01 µg/l [d]	-	< 0.01	< 0.1
	Food	6 - 14 µg/person x day [c,d,e]	-	0.20 - 0.47	28.0 - 65.8
				Σ 0.22 - 0.49	28.6 - 66.4
Pb	Air	1 - 2 µg/m^3 [d]	30 [f]	0.12 - 0.24	3.4 - 6.7
	Drinking water	2 - 5 µg/l [g]	-	0.13 - 0.33	3.6 - 9.2
	Food	22 - 100 µg/person x day [c,d,e]	-	0.71 - 3.29	19.9 - 92.1
				Σ 0.96 - 3.86	26.9 - 108.0
Ni	Air	Rural areas 1 - 10 ng/m^3 [a]	100	< 0.01	< 0.1
		Urban areas 20 ng/m^3 [a]			
	Drinking water	< 10 µg/l [n]	-	< 0.66	< 4.1
	Food	80 - 150 µg/person x day [c]	-	2.66 - 5.00	16.6 - 31.3
				Σ 3.33 - 5.67	20.8 - 35.5
Zn	Air	Rural areas 60 - 250 ng/m^3 [i]	100	< 1	< 0.3
		Urban areas 50 - 2000 ng/m^3 [i]			
	Drinking water	< 100 µg/l [b]	-	< 7	< 2.3
	Food	8.4 -11.7mg/pers.x day [e,k,m,o]	-	280 - 390	28.0 - 130.0
				Σ288 - 398	30.6 - 132.6

[a]Air quality guidelines for Europe (WHO 1987a); [b]Aurand et al. 1991; [c]Knutti and Zimmerli 1985; [d] Kaiser and Roßkamp 1992; [e]Becker et al. 1994; [f]Cumbrowski 1991; [g]Müller and Weigert 1990; [h]Zimmerli et al. 1989; [i]Lahmann 1990; [j]Kumpulainen 1992; [k]Bro et al. 1990; [l]Buchet et al. 1983; [m]Ellen et al. 1990; [n]WHO 1991; [o]van Dokkum et al. 1989

Table 3. Assessment of tolerable soil concentrations and comparison with background contents

	Estimated average intake via air, drinking water and food	Estimated tolerable intake via soil		Tolerable soil concentration [mg/kg soil]		Background soil concentration [a] [mg/kg soil]
	[% of tolerable intake]	[% of tolerable intake]	[µg/kgBW x day]	Calculated Value	Proposed guideline value	
As	30	70	1.5	45	50	29
Cd	50	50	0.5	15	15	0.8
Cr [b]	30	70	3.4	102	100	100
Cu	10	90	450	13500	1500	36
Hg	50	50	0.36	11	10	0.3
Pb	50	50	1.79	54	300	85
Ni	35	65	10.4	312	300	35
Zn	75	25	75 [c]	2250	2000	140

[a] Standard reference value A (Holländische Liste 1988)
[b] Cr(VI)
[c] using the lower PMTDI-margin (0.3 mg/kgBW x day)

usually found background concentrations and therefore not suitable to decide on need for remediation. In agreement with other working groups we propose a tolerable concentration of 300 mg Pb/kg soil, because it has been shown that soil concentrations up to 500 - 1000 mg Pb/kg soil do not lead to any measurable elevation of blood Pb levels.

Discussion

The present approach was designed to establish conservative but realistic remediation guideline values for contaminated soil. The model calculations are based on a most realistic evaluation of "background" exposure caused by air, drinking water and food. Safety is given, because exposure is assessed for small children. This population group tends to ingest relatively large amounts of soil. However, the tolerable intakes used for hazard assessment include safety factors and are defined for lifetime exposure, while only a limited span of life is spent playing in the open air. Moreover, children usually dont play on bare soil for 365 days per year as suggested in the present calculations. Overestimation may also result from ignoring bioavailability issues due to missing data or by using tolerable intakes actually defined for intake through food thus presuming same bioavailability from soil as from food. Many pollutants (e.g. Pb) tend to bind tightly to soils, particularly if they have been in contact with the soils for a considerable length of time (LaGoy and Quirk 1994).

On the other hand, it must be emphasised that the proposed guideline values have preliminary character and should be treated with care. In the individual case, several factors not considered may substantially impact hazard estimation. For example, average intake via air is mostly considered less relevant (see Table 2). It can become important, however, if larger amounts of dust stirred up from contaminated bare soil are inhaled. The suggested exposure assessment does not take account of dermal absorption which might contribute to total exposure (e.g. pollution by organic Hg compounds) but can hardly be quantified. Besides systemic distribution, dermal exposure may lead to sensitisation (e.g. Cr(VI), Ni). An interesting approach for establishing soil concentrations based on the development of allergic contact dermatitis has recently been reported (Horowitz and Finely 1994). In this context it is relevant to note that, especially in the case of some metals (e.g. As, Hg, Cr), speciation (state of oxidation, organic, inorganic compounds) is desirable but often not practicable. Using average exposure via drinking water may lead to understimation, for example if natural levels are high (e.g. As) or pollution caused by drinking water pipes occurs (e.g. Cu, Zn). Intake through food may vary considerably depending on individual dietary habits (e.g. intake of As or Hg through fish). Additional exposure may result from widespread contamination (e.g. Cd, Hg) if significant quantities of food grown on polluted soil (gardens, local agriculture) are consumed.

In summary, the present approach provides remediation guideline values for polluted soil, based on a realistic evaluation of health hazard for young children playing on the ground. Nevertheless, the acceptability of certain concentrations in soil depends on the local situation requiring further validation considering the individual case.

References

Anderson RA, Bryden NA, Polansky MM (1992) Dietary chromium intake. Freely chosen diets, institutional diets, and individual foods. Biol. Trace Element Res. 32: 117-121

Aurand K, Hässelbarth U, Lange-Asschenfeldt H, Steuer W (eds) (1991) Die Trinkwasserverordnung: Einführung und Erläuterungen für Wasserversorgungsunternehmen und Überwachungsbehörden, 3rd ed. Erich Schmidt Verlag, Berlin

Becker K, Krause C, Nöllke P, Schulz C (1994) Dietary intake of some trace and toxic elements in the FRG - A 24h duplicate study. Presentation at the 5th Nordic Symposium on Trace Elements in Human Health and Disease: Abstract O-4

Bro S, Sandström B, Heydorn K (1990) Intake of essential and toxic trace elements in a random sample of Danish men as determined by the duplicate portion sampling technique. J. Trace Electrolytes Health Dis. 4: 147-155

Buchet JP, Lauwerys R, Vandevoorde A, Pycke JM (1983) Oral daily intake of cadmium, lead, manganese, copper, chromium, mercury, calcium, zinc and arsenic in Belgium: A duplicate meal study. Fd. Chem. Toxicol. 21: 19-24

Cumbrowski J (1991) Die Schwermetallbelastung des Menschen durch die Umwelt. Forum Städte-Hyg. 42: 134-142

LaGoy PK, Quirk TC (1994) Establishing generic remediation goals for the polycyclic aromatic hydrocarbons: critical issues. Environ. Health Persp. 102: 348-352

Müller J, Weigert P (1990) Bleigehalte in und auf Lebensmitteln. ZEBS-Hefte 2/1990.

Eikmann T, Michels S, Krieger T, Weber H, Einbrodt HJ (1989) Umwelthygienische Grundlagen und Problematik der Richtwertfestsetzung für Schadstoffe in Böden. Forum Städte-Hyg. 40: 333-337

Ellen E, Egmond E, van Loon JW, Sahertian ET, Tolsma K (1990) Dietary intakes of some essential and non-essential trace elements, nitrate, nitrite and N-nitrosamines, by dutch adults: Estimated via a 24-hour duplicate portion study. Fd. Add. Contam. 7: 207-221

Holländische Liste (1988) Leidraad Bodemsanering. Staatsuitgeverij, s'Gravenhage, Niederlande

Horowitz SB, Finley BL (1994) Setting health-protective soil concentrations for dermal contact allergens: A proposed methodology. Regul. Toxicol. Pharmacol. 19: 31-47

Kaiser U, Roßkamp E (1992) Gesamtbelastung des kindlichen Organismus mit Schwermetallen. Bilanzierungsversuch unter Berücksichtigung der Aufnahme über Spielplatzsand und Bodenpartikel. Bundesgesundhbl. 35: 197-203

Knutti R, Zimmerli B (1985) Untersuchungen von Tagesrationen aus schweizerischen Verpflegungsbetrieben. Mitt. Gebiete Lebensm. Hyg. 76: 206-232

Kumpulainen JT (1992) Chromium content of foods and diets. Biol. Trace Element Res. 32: 9-18

Lahmann E (1990) Luftverunreinigung-Luftreinhaltung. Erich Schmidt Verlag, Berlin

Ruck A (1988) Bodenaufnahme durch Kinder - Abschätzungen, Annahmen. In: Rosenkranz D, Einsele G, Harreß HM (eds) Bodenschutz - Ergänzbares Handbuch, BoS 5. Lfg. V/90. Erich Schmidt Verlag, Berlin

Ruppert H, Joneck M (1988) Anthropogene Schwermetallanreicherungen in bayerischen Böden vor dem Hintergrund der natürlichen Grundgehalte. Bayerisches Geologisches Landesamt (ed) Reihe "Materialien" des Bayerischen Staatsministeriums für Landesentwicklung und Umweltfragen, Nr. 54

USEPA (U.S. Environmental Protection Agency) (1988) Chromium. Rev. Environ. Contam. Toxicol. 107: 39-52

USEPA (1988) Nickel. Rev. Environ. Contam. Toxicol. 107 103-115

van Dokkum W, de Vos RH, Muys T, Wesstra JA (1989) Minerals and trace elements in total diets in the Netherlands. Brit. J. Nutr. 61: 7-15

Viereck L, Kramer M, Eikmann T, König W, Bertges WD, Gableske R, Krieger T, Michels S, Exner M, Weber H (1991) Ableitung von Richtwerten für Metalle auf Kinderspielplätzen in Nordrhein-Westfalen. Öff. Gesundh.-Wes. 53: 7-15

WHO (1987a) Air quality guidelines for Europe. WHO Regional Publications. European Series No. 23, Kopenhagen

WHO (1987b) Principles for the safety of food additives and contaminants in food. Environ. Health Criteria 70, Geneva

WHO (1991) Nickel. Environmental Health Criteria 108. Geneva

Zimmerli B, Bosshard E, Knutti R (1989) Vorkommen und gesundheitliche Bewertung toxischer Spurenelemente in der Nahrung. Mitt. Gebiete Lebensm. Hyg. 80: 490-518

Immunodeficient mouse systems

(Chair: Dr. M. Lövik, Norway)

The SCID (Severe Combined Immunodeficiency) Mouse - Its Biology and Use in Immunotoxicological Research

Martinus Løvik
Section for Environmental Immunology, Department of Environmental Medicine, National Institute of Public Health, Oslo, Norway.

In medical research, there is a need to bridge the gap between *in vivo* animal experimentation and studies on human cells and tissues *in vitro*. Two clefts have to be bridged - one is the species difference, the other is the *in vivo - in vitro* difference. The SCID mouse reconstituted with human cells may represent an important tool for building the bridges needed. Models based on reconstituted SCID mice make experimentation with human cells in an *in vivo* situation possible without the ethical and practical constraints that limit studies in man.

This review is focused on the use of SCID mice reconstituted with human cells in immunotoxicology. The review gives a brief overview of the biology of the SCID mouse, and a description of the main variants of human immune cell reconstituted SCID mouse models. Thereafter follows a discussion of problems connected with the different models, and possible approaches for improving the reconstituted SCID mouse as a tool in immunotoxicological research. The review, which is extensively referenced, ends with an overview over reports on the use of SCID mouse models in immunotoxicology.

Biology of the SCID mouse

Gene defect. SCID (severe combined immunodeficiency) mice were first described in 1983 in C.B-17 mice (Bosma et al. 1983). C.B-17 mice are an immunoglobulin allotype (IgH) variant of BALB/c mice. SCID mice have a spontaneous autosomal recessive mutation on chromosome 16 in a homozygous state, causing a defect VDJ recombinase system. This system is used to rearrange the genes for the antigen-specific receptors of B and T lymphocytes (Lieber et al. 1988). SCID mice also have defects in a double-stranded (ds) DNA break repair system, leading to reduced ability to repair dsDNA breaks occurring after exposure to ionizing radiation or chemical agents (Fulop and Phillips 1990). Antigen receptor gene rearrangements may involve dsDNA breaks, and the dsDNA repair defect is likely to underlie both the radiation sensitivity and the failure to productively rearrange antigen receptor genes (Hendrickson 1993).

Phenotype. Because of the defect in antigen receptor gene rearrangement, SCID mice lack functional B and T lymphocytes. The mice are profoundly lymphopenic, and their thymus size is only between 1% and 10% of normal. The SCID thymocytes are CD3-, CD4-, and CD8-negative. Lymphoid tissues, such as Peyer's patches and lymph nodes, are hypoplastic and contain only few cells that can be recognized as lymphocytes. There are few T cells, most of which are probably not functional (Custer et al. 1985). Pre-B lymphocytes carrying the B lymphocyte marker B220 and cytoplasmic IgM are not detectable (Bosma et al. 1983). Serum contains no or only very low levels of immunoglobulin. SCID mice are unable to mount a specific immune response to T dependent as well as T independent antigens, and they are not capable of immunologically rejecting allogeneic cells. However, NK-cell mediated rejection is possible in SCID mice. In contrast to the lymphoid lineage (B and T lymphocytes), the development and function of cells of the myeloid lineage apparently is not affected by the *scid* mutation. The bone marrow of SCID mice appears normal. Granulocyte, erythrocyte, megakaryocyte, monocyte/macrophage and NK-cell differentiation and function is intact (Bosma et al. 1991).

Husbandry. Because of their immunodeficiency, SCID mice are highly suceptible to a number of viral and bacterial pathogens, and to *Pneumocystis carinii* infection. They must be kept in a protected environment, either in a plastic isolator or a laminar air flow hood. Food, water and bedding should be sterilized. All handling and experimental procedures are optimally performed under aseptic conditions. Some investigators use drinking water supplemented with antibiotics; however, it has been reported that mice given antibiotics have a reduced immune function after reconstitution with lymphocytes compared to antibiotics-free mice (Ashany et al. 1992). -The price of a SCID mouse obtained from a commercial breeder is about 5 times that of a standard inbred mouse, and in addition comes the cost of screening for leakiness (see below). Breeding of SCID mice requires housing facilities with advanced equipment and strict precautions for the maintenance of a clean colony. The cost of buying as well as breeding SCID mice is, therefore, high.

SCID mouse models

SCID mouse models are based on the pecularity of SCID mice that they are unable to mount a specific immune response. One type of model use the SCID mouse as a 'test-tube' in which microorganisms and cells can grow without the interference of immune mediated phenomena, or it can be passively immunized with selected components of the immune system, like specific antibodies. SCID mice have no immune system of their own that can make the interpretation of passive immunization experiments difficult, in contrast to mice that have some immunological competence of their own. One of the most successful areas for the use of SCID mouse models so far has been in tumor biology, ranging from tumor development to therapy. SCID mice are permissive to some infections that immunologically competent mice would not be susceptible to. Also, the fact that the SCID mouse has a defect that for practical purposes is limited to the specific immune system, makes it a good tool for the study of the contribution of the natural immune system (macrophages, granulocytes, NK cells) to the resistance to certain infections and tumors. There are a number of reviews of the use of SCID mice in the study of tumor biology and infectious diseases (Mueller and Reisfeld 1991; Cesano et al. 1992; Stanley and Virgin 1993), in addition to the discussion of these topics in general reviews of SCID mouse models.

It may be of interest also for immunotoxicological applications that not only human tumors but also normal liver tissue (Sacci et al. 1992), skin (Kawamura et al. 1992, Yan et al. 1993), thyroid tissue (Martin et al. 1993), synovium (Rendt et al. 1993), and endometrium (Aoki et al. 1994) have been transplanted into SCID mice.

Reconstitution of SCID mice. SCID mice may be reconstituted with congenic, allogeneic or xenogeneic cells. Most work has been done with mouse, rat, bovine, equine, or human cells. After reconstitution by transfer of congenic murine lymphoid tissue, bone marrow or fetal liver cells, donor-derived lymphocytes can be detected in the lymphoid tissues. Serum immunoglobulins become detectable, and both the antibody response to T-dependent antigens and cytotoxic T cell activity are restored. Mature murine B and T cells and cloned lymphoid cell lines have been successfully transferred to SCID mice (Bosma and Carroll 1991; Sprent et al. 1991). The ability to reconstitute SCID mice with selected populations of murine lymphoid cells provides a powerful tool for analysing the role of a given lymphoid-cell type in the immune response. Also, genetically manipulated or otherwise altered cells can be used for reconstitution. SCID mice have been found useful for the growth of hybridomas (Ware et al. 1985; McKnight et al. 1991; Pistillo et al. 1992). However, we will focus here on reconstitution of SCID mice with elements of the human immune system. This reconstitution can be accomplished by two different procedures, described below.

Reconstitution with postnatal human lymphocytes. SCID mice can be reconstituted with human peripheral blood mononuclear cells (and are then by some investigators denominated hu-PBL-SCID mice). Using this approach, human PBLs are injected into the peritoneal cavity (i.p.)(Mosier et al. 1988). Reconstitution by the intravenous route is seldom successful (Mosier et al. 1988). As an alternative to PBLs, human tonsillar lymphocytes (Nadal et al. 1991), splenocytes (Thirdborough et al. 1993), cerebrospinal fluid cells (Saeki et al. 1992), thymocytes (Schönbeck et al. 1993), bone marrow (Kumpel et al. 1992), synovial lymphocytes (Tighe et al. 1990) and umbilical cord blood cells (Kumpel et al. 1992) have been used. After i.p. transplantation with cell suspensions, the activation and possible expansion of B lymphocytes result in the production of human IgG, IgM, IgA and IgE immunoglobulins (Smith et al. 1991; Nadal et al. 1991; Hesselton et al. 1993; Kilchherr et al. 1993; Steinsvik et al. 1994). Further, production of many human autoimmune antibodies present in the donor has been achieved (Tighe et al. 1990; Ashany et al. 1992; Duchosal 1992; Saeki et al. 1992; Macht et al. 1993;

Schönbeck et al. 1993; Dickey et al. 1994). Human antibody levels in reconstituted mice peaks most often between 1 and 3 months, but may be detectable for a year after transfer of cells. It has been reported that preformed plasma cells that continue to produce Ig have a life span between 3 days and 3 months (MacLennan and Gray 1986), but there may be some proliferation of B lymphocytes in SCID mice. The immunoglobulin response shows a degree of oligoclonality (Saxon et al. 1991; Aaberge et al. 1992), which may be caused by abnormal immunoregulatory conditions and stimuli for proliferation. Human immunoglobulin production in hu-PBL-SCID mice has been shown to depend on the presence of T lymphocytes (Ueno et al. 1992; Macht et al. 1993).

There are reports of specific antibody production in response to immunization of mice transplanted with cell suspensions from adult donors (Mosier et al. 1988; Nadal et al. 1991; Kumpel et al. 1992; Carlsson et al. 1992; Duchosal et al. 1992; Aaberge et al. 1992; Dickey et al. 1994; Steinsvik et al. 1994). The most consistent response may be to T independent antigens (Aaberge et al. 1992). There is considerable variability in most of the antibody responses, which may possibly in some instances be caused by a low number of precursor cells for the given specificity. Antibodies induced by immunization with a pneumococcal polysaccharide vaccine were shown to offer a degree of protection to challenge of transplanted and vaccinated mice with virulent pneumococci (Aaberge et al. 1992). Attempts to elicit primary immune responses in hu-PBL-SCID mice have generally been unsuccessful, except for immunization with T independent antigens (Aaberge et al. 1992). A recent report (Sandhu et al. 1994) indicates that pretreatment of recipient SCID mice with anti-asialo GM-1 antibody and irradiation facilitates the induction of a primary immune response to environmental antigens.

The transplanted human cells remain to a large extent in the peritoneal cavity (Hoffmann-Fezer et al. 1992), but human cells can also be found in peripheral blood, thymus, lung, liver, spleen and lymph nodes (Tary-Lehmann and Saxon 1992; Murphy et al. 1992a, 1992b; Hoffman-Fezer et al. 1992, 1993; Martino et al. 1993; Ladel et al. 1993; Hesselton et al. 1993). There are some rather divergent results in the literature as to the extent of human cell colonization of organs outside the peritoneal cavity. The human cell reconstitution appears to be difficult to quantify, to vary considerably between animals, and to vary with cells from different sources and with cells from different donors.

In mice injected with single cell suspensions, the normal microenvironment required for an immune response is largely lacking. The human cells reside within the murine tissues and are likely to interact with murine cells. Human cells may form aggregates (Hoffmann-Fezer et al. 1992) that facilitate their interaction. However, the human-human interactions may be different from those in normal human organs and also from those in fetal tissue transplanted to SCID mice (see below). Human cell interactions may therefore be suboptimal or at worst abnormal, and there may be human-murine interactions of unknown significance. The importance of the micro-environment is illustrated by experiments reported by Schönbeck et al. (1993). In parallell experiments with i.p. injection of thymocyte suspension and transplantation of small pieces of intact myasthenia gravis thymus, it took 100 times more suspended cells than intact tissue cells to obtain similar levels of specific autoantibodies. -Not only do the anatomic distribution of human cells in murine organs differs from normal human anatomy, the cells also appear to be in a different functional state. Human T cells in the hu-PBL-SCID mouse have been reported to be refractory to *ex vivo in vitro* stimulation (Tary-Lehmann and Saxon 1992), and in graft-versus-host reactions (see below) they express activation markers like MHC class II antigens (Hoffmann-Fezer et al. 1992, 1993).

Some other limitations to the use of hu-PBL-SCID mice should be mentioned. First, when Epstein-Barr-virus (EBV) positive donors are used, EVB-transformed lymphocytes and even aggressive B-cell lymphomas may arise in transplanted mice (Okano et al. 1990). This B lymphocyte tumor generation has been shown to be T lymphocyte dependent (Veronese et al. 1992). Second, depending on the number of cells transplanted, some hu-PBL-SCID mice develop signs of graft-versus-host disease which may range from subclinical to lethal (Murphy et al. 1992a; Hoffmann-Fezer et al. 1993). The severity of the graft-versus-host reaction has been reported to correlate with the number of human cells in the peripheral circulation of the animals, and to be reduced if natural killer cells are removed from the human cells before transplantation (Xun et al. 1993). A peculiar form of graft-versus-host reaction that may result in severe anemia is the production of human antibodies against mouse erythrocytes (Williams et al. 1992; Pirrucello et al. 1992).

Reconstitution with human fetal tissues. SCID mice may be surgically transplanted under the kidney capsule with small fragments of fetal human hematolymphoid organs, e.g. thymus, liver, bone marrow and peripheral lymphoid organs. These mice are often referred to as SCID-hu mice. Transplantation of tissue fragments may be combined with injection of cell suspensions (McCune et al. 1988). The idea is to transplant a normal human microenvironment with human progenitor cells. Reconstitution with these procedures is in some ways more complete than with the single-cell suspension procedure, because the cells are supported by a more normal microenvironment. SCID-hu mice transplanted with fetal thymus and liver show continued T lymphopoiesis and production of human B lymphocytes for half a year or more after transplantation (Vandekerckhove et al. 1993). Cells expressing T lymphocyte markers are detected in peripheral blood and are functionally similar to normal human T cells (Krowka et al. 1991; Vandekerckhove et al. 1991). However, few or no human cells are detected in SCID thymus or lymph nodes, and few human cells are detectable in the SCID spleen. Circulating human cells other than T lymphocytes are generally not detectable. Therefore, the primary use of thymus/liver SCID-hu mice is in the study of human T cell development within the transplanted tissue. SCID-hu mice have become an important model for the study of human hematopoiesis and lymphopoiesis (McCune et al. 1991) and for the study of certain human infections, in particular HIV infection and cytomegalovirus infection (McCune et al. 1990; Mockarski et al. 1993; Stanley et al. 1993). Reports on antibody production in SCID mice reconstituted with human fetal tissues are few (Vandekerckhove et al. 1993), compared to reports on human antibodies in mice transplanted with mononuclear cells from peripheral blood. Antibody production is enhanced if fetal lymph node fragments are transplanted, and the engrafted human lymph nodes are reported to be both macroscopically and microscopically similar to intact human lymph nodes (McCune et al. 1988, 1991). An advantage of the fetal tissue model with regard to functional studies is that it may be possible to achieve a primary immune response in this model (McCune et al. 1991). However, the usefulness of the SCID-hu model for studies on human immune system function is limited by the fact that full reconstitution is not achieved with regard to morphology or lymphocyte distribution in the mice, and apparently the generation of immune responses is somewhat difficult to achieve. Ethical considerations will limit the use of fetal tissues in some countries, and practical problems connected with obtaining and preparing material for transplantation is larger with fetal tissues compared to peripheral blood from adults. EBV transformants are not observed with the fetal tissue fragment procedure, and graft-versus-host reactions are rarely seen (McCune et al. 1991). The engraftment of postnatal thymus has also been studied, but the organ assumes an abnormal histologic appearance and function (Barry et al. 1991).

Human fetal bone may be transplanted subcutaneously into SCID mice to obtain a human multilineage hematopoietic compartment. Active human multilineage hematopoiesis has been observed for as long as 5 months in this model (Kyoizumi et al. 1992). Low numbers of human B cells and myeloid cells were found in the peripheral circulation in less than half the animals, and T cell lymphopoiesis was not observed.

Problems, possibilites and further development of the SCID mouse models

Biological incompatibilities. Biological incompatibilities may exist that will limit the reconstitution of SCID mice with human cells even under the most optimal conditions. There are or may be species differences in lymphocyte size, surface charge and glycosylation. Species divergence in specific receptors and their ligands that mediate communication between immune system cells and for example endothelial cells may affect localization of lymphocytes in lymphoid organs. One may attempt to overcome these problems by using SCID mice transgenic for adhesion molecules and other molecules.

Reversion of the scid phenotype - leakiness. The absence of host lymphocyte differentiation and B and T lymphocyte function is not absolute in SCID mice. On the contrary, between 2% and 23% of young adult scid mice have detectable serum immunoglobulin levels, reaching 30-47% of the mice after one year (Bosma et al. 1988). Most old SCID mice have some lymphocyte function. SCID mice with a certain level of lymphocyte function are called leaky (Bosma 1992), and SCID mice must be screened for leakiness before use. B and T lymphocyte function in leaky SCID mice is more or less oligoclonal. In addition to age, the development of leakiness has been reported to depend on how clean the conditions are under which the animals are kept and on mouse strain background (the *scid* gene has been crossed into C3H and C57BL mice). Activated components of the human immune system may induce leakiness across the species barrier (Donjon et al. 1993). It has been reported that the tendency to leakiness can be reduced by selective inbreeding of SCID mice (Inohara et al. 1993). Large inter-mouse variability is a problem that affects all parameters in SCID mice reconstituted with human cells. Variability can only to a limited extent be reduced by transplanting cells from the same donor to all mice in an experiment. Some of the variability may be due to low degrees of developing leakiness. *Scid* phenotype mice created by inactivating the recombinase-activating genes (RAG-deficient mice) do not exhibit leakiness (Mombaerts et al. 1992), and the defect in RAG knock-out mice supposedly is less pleiotropic than the similar defect caused by the spontaneous mutation. RAG knock-out mice transplanted with human cells may therefore give improved reproducibility of human immune cell function compared to SCID mice.

Pretreatment of recipient mice. It has been reported that pretreatment of recipient mice with the anti-NK-cell antibody anti-asialo GM-1 and irradiation enhances function of transplanted cells and tissues in SCID mice (Barry et al. 1991; Sandhu et al. 1994).

Mediator treatment and prestimulation of cells. It has been reported that human growth hormone treatment (Murphy et al. 1992b) enhances human T cell engraftment in the periphery of SCID mice, and that *in vitro* preactivation of the human cells to be transplanted have a similar enhancing effect (Armstrong et al. 1992; Murphy et al. 1993). Cytokine treatment may enhance human cell engraftment and function (Spiegelberg et al. 1994).

Mouse strain background. The *scid* gene has been crossed into C3H and C57BL/6 mice, and into NK-cell deficient *beige/beige* mice (Nonoyama et al. 1993a; Mosier et al. 1993). Each of these variants may have certain advantages, but have not so far proved to represent major improvements of the SCID models. An interesting approach is the crossing of the *scid* gene into NOD mice. According to unpublished data, the SCID/NOD mouse appears promising with regard to better and more consistent human immune cell function.

Assays for human immune cell function. The low number of assays available for the measurement of human immune cell fuction in reconstituted SCID mice is an important limitation to the use of SCID mice in immunotoxicology. Spontaneous IgG production is regularly obtained and is used also as a general measure of transplant function. Human IgG has a half-life of about 10 days in SCID mice. However, because of active secretion into the bile, human IgA (and also to some degree IgM) reach only low levels in SCID mice (Smith et al. 1991). Assays for class-specific immunoglobulin producing cells (e.g. ELISPOT) or immunohistochemistry can be used to quantify IgA- and IgM-producing cells (Nadal et al. 1991). Measureable IgE levels are regularly achieved in reconstituted SCID mice (Kilchherr et al. 1993). A number of specific antibody response models have been described in hu-PBL-SCID mice (Mosier et al. 1988; Nadal et al. 1991;

Carlsson et al. 1992; Duchosal et al. 1992; Kumpel et al. 1992; Markham et al. 1992; Nonoyama et al. 1993b), but large variation in the response is a problem (Pollock et al. 1994). Also, whereas a secondary antibody response can be obtained, a primary antibody response has until recently (Sandhu et al. 1994) been only rarely obtained with T dependent antigens. In contrast, a response to T independent antigens is regularly achieved in mice transplanted with peripheral blood mononuclear cells from unimmunized human donors (Aaberge et al. 1992). The mixed lymphocyte reaction and cytotoxicity assays and assays for proliferative response of transplanted cells to mitogen and antigen specific stimulation have been used only to a quite limited extent.

Morphological studies can be done on the transplanted thymus (rat, man) and to a certain extent on bone marrow or bone marrow-like structures and transplanted fetal lymph nodes. Organ architecture, cell morphology and membrane marker expression can be analysed, and absolute and relative numbers of different cell types in different localizations can be studied.

Reports on the use of SCID mouse models in immunotoxicology

An interesting approach to increase the relevance of the SCID models is to use mice transplanted with rat thymus and liver (SCID-ra mice) as a xenotransplant model in parallell with SCID mice reconstituted with human cells or tissues. Used in parallell with SCID-hu mice, SCID-ra mice will to some degree constitute a control system that allows experimental *in vivo* effects in the normal intact animal to be compared directly with effects observed in a xenotransplant SCID model with the same species (De Heer et al. 1993). Also, it may be relevant to use SCID mice reconstituted with murine cells as controls, in particular when there are available data on immunotoxic effects previously obtained *in vivo* or *in vitro* in mouse systems (Pollock et al. 1994).

SCID mice reconstituted with human peripheral blood mononuclear cells (hu-PBL-SCID mice) or mouse spleen cells (mo-SPL-SCID mice) were administered either 2,3,7,8-tetrachlorodibenzo-p-dioxin or cyclosporin A (Pollock et al. 1994). Comparison of the chemical-related changes in immunoglobulin and tetanus antibody concentrations revealed some similarities between control mice and mo-SPL-SCID or hu-PBL-SCID mice, particularly in the cyclosporin A groups. However, the anti-tetanus response was relatively poor and experimental variation was large. SCID mice have been successfully used to study the effect of cytokines on the hematopoietic and immune systems (Lapidot et al. 1992; Kyoizumi et al. 1993; Spiegelberg et al. 1994), to study the effect of microbial superantigens on T cell development (Waller et al. 1992; Baccala et al. 1993), and the toxicity of irradiation on human hematopoietic progenitor cells (Kyoizumi et al. 1994).

References

Aaberge IS, Michaelsen TE, Rolstad AK, Groeng E-C, Solberg P, Løvik M. SCID-hu mice immunized with a pneumococcal vaccine produce specific human antibodies and show increased resistance to infection. Infect Immun 1992;60:4146-4153.

Aoki D, Katsuki Y, Shimizu A, Kakinuma C, Nozawa S. Successful heterotransplantation of human endometrium in SCID mice. Obstet Gynecol 1994;83:220-228.

Armstrong N, Cigel F, Borcherding W, Hong R, Malkovska V. *In vitro* preactivated human T cells engraft in SCID mice and migrate to murine lymphoid tissues. Clin Exp Immunol 1992:90:476-482.

Ashany D, Hines J, Gharavi A, Mouradian J, Elkon KB. Analysis of autoantibody production in SCID-systemic lupus erythematosus (SLE) chimeras. Clin Exp Immunol 1992;88:84-90.

Baccala R, Vandekerckhove BAE, Jones D, Kono DH, Roncarolo M-G, Theofilopoulos AN. Bacterial superantigens mediate T cell deletions in the mouse severe combined immunodeficiency-human liver/thymus model. J Exp Med 1993;177:1481-1485.

Barry TS, Jones DM, Richter CB, Haynes BF. Successful engraftment of human postnatal thymus in severe combined immune deficient (SCID) mice: differential engraftment of thymic components with irradiation versus anti-asialo GM-1 immunosuppressive regimens. J Exp Med 1991;173:167-180.

Bosma GC, Custer RP, Bosma MJ. A severe combined immunodeficiency mutation in the mouse. Nature 1983;301:527-530.

Bosma GC, Fried M, Custer RP, Carroll A, Gibson DM, Bosma MJ. Evidence of functional lymphocytes in some (leaky) *scid* mice. J Exp Med 1988;167:1016-1033.

Bosma MJ, Carroll AM. The SCID mouse mutant: definition, characterization, and potential uses. Ann Rev Immunol 1991;9:323-350.

Bosma MJ. B and T cell leakiness in the SCID mouse mutant. Immunodeficiency Rev 1992;3:261-276.

Carlsson R, Mårtensson C, Kalliomäki S, Ohlin M, Borrebaeck CAK. Human peripheral blood lymphocytes transplanted into SCID mice constitute an in vivo culture system exhibiting several parameters found in a normal humoral immune response and are a source of immunocytes for the production of human monoclonal antibodies. J Immunol 1992;148:1065-1071.

Cesano A, Hoxie JA, Lange B, Nowell PC, Bishop J, Santoli D. The severe combined immunodeficient (SCID) mouse as a model for human myeloid leukemias. Oncogene 1992;7:827-836.

Custer RP, Bosma GC, Bosma MJ. Severe combined immunodeficiency (SCID) in the mouse. Pathology, reconstitution, neoplasms. Am J Pathol 1985;120:464-477.

De Heer C, Verlaan APJ, Penninks AH, Schuurman H-J, Van Loveren H. The SCID-ra mouse: rat T-cell differentiation in the severe combined immunodeficient mouse. APMIS 1993;101:467-479.

Dickey WD, Harley JB, Scofield RH. Human autoantibody production in the severe combined immunodeficiency (*scid*) mouse. Clin Exp Rheumatol 1994;12:157-161.

Donjon CM, Morkowski H, Cheung RK, Leeder JS, Dosch H-M. Reversion of the SCID phenotype by human T cell grafts. Development of cross-species immunocompetence. J Immunol 1993;151:5948-5954.

Duchosal MA. SCID mice in the study of human autoimmune diseases. Springer Semin Immunopathol 1992;14:159-177.

Duchosal MA, Eming SA, Fischer P, Leturcq D, Barbas CF III, McConahey PJ, Caothien RH, Thornton GB, Dixon FJ, Burton DR. Immunization of hu-PBL-SCID mice and the rescue of human monoclonal Fab fragments through combinatorial libraries. Nature 1992;355:258-262.

Fulop GM, Phillips RA. The *scid* mutation in mice causes a general defect in DNA repair. Nature 1990; 347:479-482.

Hendrickson, EA. The SCID mouse: relevance as an animal model system for studying human disease. Am J Pathol 1993;143:1511-1522.

Hesselton RM, Koup RA, Cromwell MA, Graham BS, Johns M, Sullivan JL. Human peripheral blood xenografts in the SCID mouse: characterization of immunologic reconstitution. J Inf Dis 1993;168:630-640.

Hoffmann-Fezer G, Kranz B, Gall C, Thierfelder S. Peritoneal sanctuary for human lymphopoiesis in SCID mice injected with human peripheral blood lymphocytes from Epstein-Barr virus-negative donors. Eur J Immunol 1992;22:3161-3166.

Hoffmann-Fezer G, Gall C, Zengerle U, Kranz B, Thierfelder S. Immunohistology and immunocytology of human T-cell chimerism and graft-versus-host disease in SCID mice. Blood 1993;12:3440-3448.

Inohara H, Nomura T, Hongyo T, Nakajima H, Kawaguchi T, Fukuda K, Sutoh K, Iwasa T, Matsunaga T. Reduction of leaky lymphocyte clones producing immunoglobulins and thymic lymphocytic leukemia by selective inbreeding of SCID (severe combined immunodeficiency) mice. Acta Otolaryngol 1993;Suppl 501:107-110.

Kawamura T, Niguma T, Fechner JH, Wolber R, Beeskau MA, Hullett DA, Sollinger HW, Burlingham WJ. Chronic human skin graft rejection in severe combined immunodeficient mice engrated with human PBL from an HLA-presensitized donor. Transplantation 1992;53:659-665.

Kilchherr E, Mandak V, Wagner K, Heusser CH. Regulation of human IgE response in hu-PBL-SCID mice. Cell Immunol 1993;151:241-256.

Krowka JF, Sarin S, Namikawa R, McCune JM, Kaneshima H. Human T cells in the SCID-hu mouse are phenotypically normal and functionally competent. J Immunol 1991;146:3751-3756.

Kumpel BA, Poole GD, Jackson DJ. Specific anti-A responses of SCID mice populated with human lymphoid cells from peripheral blood, umbilical cord, bone marrow and spleen after immunization with group A erythrocytes. Immunology Letters 1992;33:163-168.

Kyoizumi S, Baum CM, Kaneshima H, McCune JM, Yee EJ, Namikawa R. Implantation and maintenance of functional human bone marrow in *SCID*-hu mice. Blood 1992;79:1704-1711.

Kyoizumi S, Murray LJ, Namikawa R. Preclinical analysis of cytokine therapy in the SCID-hu mouse. Blood 1993;81:1479-1488.

Kyoizumi S, McCune JM, Namikawa R. Direct evaluation of radiation damage in human hematopoietic progenitor cells *in vivo*. Radiation Res 1994;137:76-83.

Ladel CH, Kaufmann SHE, Bamberger U. Localization of human peripheral blood leukocytes after transfer to C.B-17 scid/scid mice. Immunology Letters 1993;38:63-68.

Lapidot T, Pflumio F, Doedens M, Murdoch B, Williams DE, Dick JE. Cytokine stimulation of multilineage hematopoiesis from immature human cells engrafted in SCID mice. Science 1992;255:1137-1141.

Lieber MR, Hesse JE, Lewis S, Bosma GC, Rosenberg N, Mizuuchi K, Bosma MJ, Gellert M. The defect in murine severe combined immune deficiency: joining of signal sequences but not coding segments in V(D)J recombination. Cell 1988;55:7-16.

Macht LM, Corrall RJ, Banga JP, Elso CJ. Control of human thyroid autoantibody production in SCID mice. Clin Exp Immunol 1993;91:390-396.

MacLennan ICM, Gray D. Antigen-driven selection of virgin and memory B cells. Immunol Rev 1986;91:61-85.
Markham RB, Donnenberg AD. Effect of donor and recipient immunization protocols on primary and secondary human antibody responses in SCID mice reconstituted with human peripheral blood mononuclear cells. Infect Immun 1992;60:2305-2308.
Martin A, Valentine M, Unger P, Lichtenstein C, Schwartz AE, Friedman EW, Shultz LD, Davies TF. Preservation of functioning human thyroid organoids in the *scid* mouse: 1. System characterization. J Clin Endocrinol Metab 1993;77:305-310.
Martino G, Anastasi J, Feng J, Mc Shan C, DeGroot L, Quintas J, Grimaldi LME. The fate of human peripheral blood lymphocytes after transplantation into SCID mice. Eur J Immunol 1993;23:1023-1028.
McCune JM, Namikawa R, Kaneshima H, Shultz LD, Lieberman M, Weissman IL. The SCID-hu mouse: murine model for the analysis of human hematolymphoid differentiation and function. Science 1988;241:1632-1639.
McCune JM, Namikawa R, Shih C-C, Rabin L, Kaneshima H. Suppression of HIV infection in AZT-treated SCID-hu mice. Science 1990;247:564-566.
McCune JM, Péault B, Streeter PR, Rabin L. Preclinical evaluation of human hematolymphoid function in the *SCID-hu* mouse. Immunol Rev 1991;124:45-62.
McKnight ME, Prather K, Koda K, DeBoer KF, Glassy MC. Use of severe combined immunodeficient (SCID) mice to produce human hybridoma ascites. Hum Antibod Hybridomas 1991;2:190-193.
Mocarski ES, Bonyhadi M, Salimi S, McCune JM, Kaneshima H. Human cytomegalovirus in a SCID-hu mouse: thymic epithelial cells are prominent targets of viral replication. Proc Natl Acad Sci USA 1993;90:104-108.
Mombaerts P, Iacomini J, Johnson RS, Herrup K, Tonegawa S, Papaioannou VE. RAG-1-deficient mice have no mature B and T lymphocytes. Cell 1992;68:869-877.
Mosier DE, Gulizia RJ, Baird SM, Wilson DB. Transfer of a functional human immune system to mice with severe combined immunodeficiency. Nature 1988;335:256-259.
Mosier DE, Stell KL, Gulizia RJ, Torbett BE, Gilmore GL. Homozygous *scid/scid;beige/beige* mice have low levels of spontaneous or neonatal T cell-induced B cell generation. J Exp Med 1993;177:191-194.
Mueller BM, Reisfeld RA. Potential of the scid mouse as a host for human tumors. Cancer Metastasis Rev 1991;10:193-200.
Murphy WJ, Bennett M, Anver MR, Baseler M, Longo DL. Human-mouse lymphoid chimeras: host-*vs*-graft and graft-*vs*-host reactions. Eur J Immunol 1992a;22:1421-1427.
Murphy WJ, Durum SK, Longo DL. Human growth hormone promotes engraftment of murine or human T cells in severe combined immunodeficient mice. Proc Natl Acad Sci USA 1992b;89:4481-4485.
Murphy WJ, Conlon KC, Sayers TJ, Wiltrout RH, Back TC, Ortaldo JR, Longo DL. Engraftment and activity of anti-CD3-activated human peripheral blood lymphocytes transferred into mice with severe combined immune deficiency. J Immunol 1993;150:3634-3642.
Nadal D, Albini B, Scläpfer E, Chen C, Brodsky L, Ogra PL. Tissue distribution of mucosal antibody-producing cells specific for respiratory syncytial virus in severe combined immune deficiency (SCID) mice engrafted with human tonsils. Clin Exp Immunol 1991;85:358-364.
Nonoyama S, Smith FO, Bernstein ID, Ochs HD. Strain-dependent leakiness of mice with severe combined immune deficiency. J Immunol 1993a;150:3817-3824.

Nonoyama S, Smith FO, Ochs HD. Specific antibody production to a recall or a neoantigen by SCID mice reconstituted with human peripheral blood lymphocytes. J Immunol 1993b;151:3894-3901.

Okano M, Taguchi Y, Nakamine H, Pirruccello SJ, Davis JR, Beisel KW, Kleveland KL, Sanger WG, Fordyce RR, Purtilo DT. Characterization of Epstein-Barr virus-induced lymphoproliferation derived from human peripheral blood mononuclear cells transferred to severe combined immunodeficient mice Am J Pathol 1990;137:517-522.

Pirruccello SJ, Nakamine H, Beisel KW, Kleveland KL, Okano M, Taguchi Y, Davis JR, Mahloch ML, Purtilo DT. Hemagglutination and graft-versus-host disease in the severe combined immunodeficiency mouse lymphoproliferative disese model. Am J Pathol 1992;140:1187-1194.

Pistillo MP, Sguerso V, Ferrara GB. High yields of anti-HLA human monoclonal antibodies can be provided by SCID mice. Human Immunology 1992;35:256-259.

Pollock PL, Germolec DR, Comment CE, Rosenthal GJ, Luster MI. Development of human lymphocyte-engrafted SCID mice as a model for immunotoxicity assessment. Fundam Appl Toxicol 1994;22:130-138.

Rendt KE, Barry TS, Jones DM, Richter CB, McCachren SS, Haynes BF. Engraftment of human synovium into severe combined immune defcient mice. Migration of human peripheral blood T cells to engrafted human synovium and to mouse lymph nodes. J Immunol 1993;151:7324-7336.

Sacci JB, Schriefer ME, Resau JH, Wirtz RA, Detolla LJ, Markham RB, Azad AF. Mouse model for exoerythrocytic stages of *Plasmodium falciparum* malaria parasite. Proc Natl Acad Sci USA 1992;89:3701-3705.

Saeki Y, Mima T, Sakoda S, Fujimura H, Arita N, Nomura T, Kishimoto T. Transfer of multiple sclerosis into severe combined immunodeficiency mice by mononuclear cells from cerebrospinal fluid of the patients. Proc Natl Acad Sci USA 1992;89:6157-6161.

Sandhu J, Shpitz B, Gallinger S, Hozumi N. Human primary immune response in SCID mice engrafted with human peripheral blood lymphocytes. J Immunol 1994;152:3806-3813.

Saxon A, Macy E, Denis K, Tary-Lehmann M, Witte O, Braun J. Limited B cell repertoire in severe combined immunodeficient mice engrafted with peripheral blood mononuclear cells derived from immunodeficient or normal humans. J Clin Invest 1991;87:658-665.

Schönbeck S, Padberg F, Marx A, Hohlfeld R, Wekerle H. Transplantation of myasthenia gravis thymus to SCID mice. Ann NY Acad Sci 1993;681:66-78.

Smith CIE, Abedi MR, Islam KB, Johansson MEB, Christensson B, Hammarström L. Humoral immunity in *scid* mice reconstituted with cells from immunoglobulin-deficient or normal humans. Immunol Rev 1991;124:1-26.

Spiegelberg HL, Beck L, Kocher HP, Fanslow WC, Lucas AH. Role of interleukin-4 in human immunoglobulin E formation in hu-PBL-SCID mice. J Clin Invest 1994;93:711-717.

Sprent J, Schaefer M, Hurd M, Surh CD, Ron Y. Mature murine B and T cells transferred to SCID mice can survive indefinitely and many maintain a virgin phenotype. J Exp Med 1991;174:717-728.

Stanley SK, McCune JM, Kaneshima H, Justement JS, Sullivan M, Boone E, Baseler M, Adelsberger J, Bonyhadi M, Orenstein J, Fox CH, Fauci AS. Human immunodeficiency virus infection of the human thymus and disruption of the thymic microenvironment in the SCID-hu mouse. J Exp Med 1993;178:1151-1163.

Stanley SL, Virgin HW IV. SCID mice as models for parasitic infections. Parasitology Today 1993:9:406-411.

Steinsvik TE, Aaberge IS, Belsnes D, Gaarder PI, Løvik M. Specific human IgE response after immunization with birch pollen in *scid* mice transplanted with human peripheral blood mononuclear cells - a model for human allergy? Toxicology In Vitro 1994;8:xxx (In press).

Tary-Lehmann M, Saxon A. Human mature T cells that are anergic in vivo prevail in SCID mice reconstituted with human peripheral blood. J Exp Med 1992;175:503-516.

Thirdborough SM, Chamberlain JC, Swift SM, James RFL, London NJM. The kinetics of SCID-hu reconstitution: a comparison between human splenocytes and peripheral blood leukocytes. Cell Transplant 1993;2:339-343.

Tighe H, Silverman GJ, Kozin F, Tucker R, Gulizia R, Peebles C, Lotz M, Rhodes G, Machold K, Mosier DE, Carson DA. Autoantibody production by severe combined immunodeficient mice reconstituted with synovial cells from rheumatoid arthritis patients. Eur J Immunol 1990;20:1843-1848.

Ueno Y, Ichihara T, Hasui M, Maruyama H, Miyawaki T, Taniguchi N, Komiyama A. T-cell-dependent production of IgG by human cord blood B cells in reconstituted SCID mice. Scand J Immunol 1992;35:415-419.

Vandekerckhove BAE, Krowka JF, McCune JM, De Vries JE, Spits H, Roncarolo MG. Clonal analysis of the peripheral T cell compartment of the SCID-hu mouse. J Immunol 1991;146:4173-4179.

Vandekerckhove BAE, Jones D, Punnonen J, Schols D, Lin H-C, Duncan B, Bacchetta R, De Vries JE, Roncarolo M-G. Human Ig production and isotype switching in severe combined immunodeficient-human mice. J Immunol 1993;151:128-137.

Veronese ML, Veronesi A, D'Andrea E, Del Mistro A, Indraccolo S, Mazza MR, Mion M, Zamarchi R, Menin C, Panozzo M, Amadori A, Chieco-Bianchi L. Lymphoproliferative disease in human peripheral blood mononuclear cell-injected SCID mice. I. T lymphocyte requirement for B cell tumor generation. J Exp Med 1992;176:1763-1767.

Waller EK, Sen-Majumdar A, Kamel OW, Hansteen GA, Schick MR, Weissman IL. Human T-cell development in SCID-hu mice: staphylococcal enterotoxins induce specific clonal deletions, proliferation, and anergy. Blood 1992;80:3144-3156.

Ware CF, Donato NJ, Dorshkind K. Human, rat or mouse hybridomas secrete high levels of monoclonal antibodies following transplantation into mice with severe combined immunodeficiency disease (SCID). J Immunol Methods 1985;85:353-361.

Williams SS, Umemoto T, Kida H, Repasky EA, Bankert RB. Engraftment of human peripheral blood leukocytes into severe combined immunodeficient mice results in the long term and dynamic production of human xenoreactive antibodies. J Immunol 1992;149:2830-2836.

Xun C, Brown SA, Jennings CD, Henslee-Downey PJ, Thompson JS. Acute graft-versus-host-like disease induced by transplantation of human activated natural killer cells into SCID mice. Transplantation 1993;56:409-417.

Yan H-C, Juhasz I, Pilewski J, Murphy GF, Herlyn M, Albelda SM. Human/severe combined immunodeficient mouse chimeras. An experimental in vivo model system to study the regulation of human endothelial cell-leukocyte adhesion molecules. J Clin Invest 1993;91:986-996.

THE SCID MOUSE AS A TOOL TO BRIDGE THE GAP BETWEEN ANIMAL AND HUMAN RESPONSES

Henk Van Loveren and Cees De Heer
National Institute of Public Health and Environmental Protection
PO Box 1, 3720 BA, Bilthoven

Direct immunotoxicity can express itself in many different forms. Many effects are associated with morphologic changes, such as the effect of TCDD on the weight and histology of the thymus. Also changes in peripheral lymphoid tissues or lymphoid cells in the respiratory tract or the skin are often indicative of immunotoxicity. Associated with such morphological changes are often functional changes. These may range from altered antibody responses to specific antigens, e.g. tetanus toxoid or many others, to alterations in resistance to infections or tumors. Such effects of exposure to immunotoxic agents are usually readily detected with in vivo studies using experimental animals. Usually, such experiments are carried out in mice or rats. It is not easy to predict, on the basis of such studies in experimental animals, what the risk of encountering adverse effects of exposure to immunotoxicants for man would be. If human exposure does take place, the actual exposure dose is often difficult to establish, and there is an obvious restriction in potential effects that may be studied.

Current research activities are devoted to bridge the gap between the animal experiments and the human system. One such effort focuses on in vitro systems, in which relevant cell types of animal and human origin are exposed to immunotoxicants, and relevant read outs are used to investigate possible differences or similarities in sensitivity and mechanism of immunotoxicity between animal cells and those of man. Such information may form the basis for better extrapolation of the larger data base on adverse effects of exposure to the immunotoxicant as it exists in experimenal animals. Another model that is currently under investigation, and that is aimed at serving the same goal, is the Severe Combined Immune Deficient (SCID) mouse that is reconstituted with a human immune system, or at least certain aspects thereof.

The SCID mouse has a abberant gene on chromosome 16, that results in malfunctioning of a DNA recombinase enzyme. This enzyme sustains the genomic rearrangement in lymphocytes during maturation that is required for generation of the specific T cell receptor and antibody repertoire on T and on B lymphocytes, respectively. Because of this defect, SCID mice lack functionally active T and B lymphocytes, and are therefore excellent hosts for receiving xenografts, including grafts comprising lymphoid tissue.

There are two models where SCID mice are reconstituted with human lymphoid tissue. One model is constituted by injecting human peripheral blood lymphocytes intraperitoneally. In the other model thymus tissue is grafted under the kidney capsule. Effects of exposure on these human grafts is compared to effects of similar grafts, but, then of mouse or rat origin. Comparison of effects on these grafts may provide valuable information that can be used to extrapolate the animal data to man.

The model in which peripheral blood lymphocytes (PBL) are injected is studied by Pollock et al. (1994). These authors injected PBLs from human individuals that were sensitized to tetanus. Such animals were immunized further after grafting with tetanus toxoid. Total (human) immunoglobulin levels as well as specific antibody responses to tetanus toxoid were used as measures of functionality. As control model, SCID mice reconstituted with spleen cells from mice sensitized to tetanus toxoid, or normal, sensitized mice were used. Dosing these animals with cyclosporin induced in sensitized mice and also in SCID mice reconstituted with mouse cells effects on total immunoglobulin level as well as on specific antibody responses. The SCID mice reconstituted with human lymphoid cells showed altered immunoglobulin levels after dosing with cyclosporine. This data, which of course is in line with known effects of cyclosporin in man, indicates that the model is suitable for detecting effects on human lymphocyte function. Unfortunately, specific antibody responses were not affected in the SCID mouse reconstituted with human cells, in contrast to intact mice or mice reconstituted with mouse cells. This finding is at odds with what one would have expected from what is known of cyclosporin. It should be noted that the specific anti-tetanus toxoid antibody titers decreased in time after grafting, regardless of exposure to cyclosporin. This indicates that the functionality of the transferred lymphoid cells was poor, and it illustrates the limitations of this system. Exposure to TCDD, as it was able to alter total immunoglobulin levels and specific antibody titers in both intact mice as well as in SCID mice reconstituted with mouse splenic cells, failed to affect such values in the SCID mouse with human cells. This may be taken as an indication of relative insensitivity of the humoral immune system of man for TCDD, and this would corroborate that no effects in man on humoral responses after accidental exposure to TCDD have been observed. However, the relatively poor specific antibody response to tetanus toxoid, and the absence of any effect of cyclosporin in these mice, should be taken as a caution to interprete data gained with this model with the greatest of care.

In the second model in which SCID mice are used, fetal thymus tissue is transplanted, in conjunction with fetal liver tissue from the same donor, under the kidney capsule. The liver tissue serves as a source for precursor thymocytes, since fetal liver is a hemopoietic organ. The material should be derived from human fetus of 15-20 gestational weeks; in younger fetuses it is not possible to find thymus tissue; older material fails to grow out and results in abnormal thymic architecture. Histopathologically, transplantation of fetal thymus and liver results in thymus transplants with an architecture that is strikingly similar as compared to a normal thymus. This is underlined by virtually identical composition of thymocyte phenotypes, i.e. double negative cells, CD4/CD8 double positive cells, and CD4 and CD8 single positive cells. De Heer et al (1993), described the control model for this construct, i.e. SCID mice grafted with thymus and liver tissue from rat fetuses from gestational day 17. These authors used normal rats treated with one single dose of TCDD, and also treated SCID-ra and SCID-hu mice, i.e. SCID mice reconstituted with either rat or human fetal liver and thymus tissue. Using morphometrical analysis, preliminary data indicated that the decreased cortex to medulla ratio as it was observed after TCDD treatment, was parallelled by a decreased cortex to medulla ratio in the human transplants. So this information indicates for the first time, that human thymus tissue is sensitive to TCDD exposure at similar doses at which rat thymuses are affected by TCDD.

As is true for the model in which lymphoid cells are injected, also this model in which tissue transplants are carried out does have definite drawbacks. For one, in contrast to the SCID-hu, in the SCID-ra model rat lymphocytes gain access to lymphoid organs of the host, possibly as a consequence of their evolutionary closeness. This may in return affect the thymus, since it is likely that if feed back mechanisms are operational in this system, in the SCID-ra model there may be a more pronounced drain of cells from the thymus graft, as compared to the SCID-hu. Moreover, in some SCID-ra mice, as opposed to SCID-hu mice, signs of graft versus host responses in the skin are observed. This indicates the functionality of the donor cells within the host, but also indicates a further difference between the SCID-ra and the SCID-hu model. These differences must be taken into account when comparing effects of exposure to immunotoxicants.

In sum, both models do have some potential for comparing immunotoxicity in man and animals, and may therefore be valuable for extrapolating the larger data base as it exists in experimental animals to man. There may be some doubts on the functionality of the lymphocytes after injection into the SCID mice, and this model therefore presents serious limitations. The model using the transplants seem to have less of this problem, since the grafts do have an almost normal appearance, but this model may be restricted to only analysing effects on the thymus, and hence investigation of chemicals that have the thymus as their main target. It should also be noted that for this model, obtaining human material is often very difficult, and in some countries perhaps impossible. In any case, neither of these models is suited to be applied for screening purposes. They may only be valuable if animal research indicates effects of the chemical on the specific end point to be measured in the particular model. However, before these models may be applied on a somewhat more routine basis, more compounds need to tested using these models.

REFERENCES

1. Pollock PL, Germolec DR, Comment CE, Rosenthal GJ, Luster MI
   Development of human lymphocyte-engrafted SCID mice as a model for immunotoxicity assessment.
   Fundamental and Applied Toxicology 1994; **22**: 130-138.

2. Heer C de, Verlaan APJ, Penninks AH, Schuurman HJ, Van Loveren H.
   The SCID-RA mouse: rat T cell differentiation in the severe combined immunodeficient mouse.
   APMIS 101: 467-479.

The SCID Mouse and its use for Immunopharmacological and Immunotoxicological Investigations

C. H. Ladel[1], H. Püschner[2], and U. Bamberger[2]
1 Department of Immunology, University of Ulm, 89070 Ulm/Donau, Germany
2 Department of Experimental Pathology and Toxicology, Dr. Karl Thomae GmbH, 88400 Biberach/Riss, Germany

Differentiation of early B- and T-lymphocytes is disturbed in SCID (*S*evere *C*ombined *I*mmuno*d*eficiency) mice due to defects in the system of recombinases responsible for the synthesis of antigen-specific receptors on these cells, i.e. immunoglobulines and T-cell receptors (Bosma et al., 1983; Schuler and Bosma, 1989; Yancopoulos et al., 1986).

Thus, SCID mice lack functional lymphocytes (Malynn et al., 1988) and are unable to reject allo- as well as xenotransplants from different tissues. They therefore allow to establish model systems to study murine or human tissues in a defined in vivo setting. In particular, studies with human peripheral blood mononuclear lymphocytes (huPBL) transferred to SCID by intraperitoneal injection have been reported (Mosier et al. 1988). In this model, generally known under the name SCID-hu, transferred huPBL have been observed to traffic from the peritoneum to the lymphoid organs of the host and to remain functionally active, as shown by the secretion of human immunoglobulin (huIg) and antigen-specific humoral immune responses occuring after immunization with recall antigens (Mosier et al., 1988; Duchosal et al., 1992).

In addition, they allow to study human autoimmune diseases (Duchosal et al, 1990; Tighe et al.,1990) and to develop an in vivo model for the infection with HIV (Mosier et al., 1989; Rowe et al., 1991; Torbett et al., 1991).

Our goal was to establish and characterize the SCID-hu system and to use it for studies of the effects of compounds with known immunomodulatory activity in humans, in particular recombinant human interleukin-2 (IL-2), Cyclosporin A, *S*taphylococcal *E*nterotoxin *B* (SEB) and Bacterial *L*ipo*p*oly*s*accaride (LPS).

Description of the SCID-huPBL system

For breeding as well as for the performance of experiments, CB17 scid/scid mice derived from the original FoxChase strain and obtained via BomMice, Bomholtgard, Denmark, were kept under laminar air flow conditions in the barrier-sheltered specific pathogen-free area of the animal facilities of Dr. Karl Thomae GmbH.

To determine serum levels of murine immunoglobulin (muIg), an *e*nzyme-*l*inked *i*mmuno*s*orbant *a*ssay (ELISA) was established in microplates. The assay was performed as a sandwich ELISA, using polyclonal rabbit anti-muIg antibodies (Dako, Hamburg, Germany) bound to the solid phase, peroxidase-labelled polyclonal rabbit anti-muIg antibodies (Dako) to detect bound muIg and a standard muIg preparation (Immunotech SA, Marseille, France) for internal control. A detection limit of 6 ng muIg per ml serum was determined. Only CB17 scid/scid mice with a serum level of muIg below 50 ng per ml serum at the age of 4 to 6 weeks were used for experiments and breeding. Mice with higher muIg serum concentration were looked upon as "leaky". Leakyness is known to occur in the offspring of CB17 scid/scid with a rate of 10 - 20 % and is supposed to depend on spontaneous recombination of Ig genes (Carroll et al., 1988; Carroll et al., 1989; Ferrier et al., 1990; Petrini et al., 1990).

Density centrifugation on Ficoll-Hypaque (Boehringer Mannheim, Mannheim, Germany) was used to isolate PBL from citrate-treated blood taken from healthy volunteers. Subsequently, PBL were washed three times in *p*hosphate-*b*uffered *s*aline (PBS) and resuspended in this buffer at a concentration of 10^8 cells per ml. The transfer of huPBL to CB17 scid/scid was performed by intraperitoneal injection of $5x10^7$ cells in 0.5ml PBS.

To detect huPBL in murine tissues, they were stained in vitro before transfer to CB 17 scid/scid with the fluorochrome PKH 26 GL (Zynaxis, Malvern, PA, USA), known to be stable in vivo. This technical approach was shown to be suitable for both, flow-cytometric and histological detection (for details, see: Ladel et al., 1993).

Using flow-cytometry, we were able to detect huPBL at low frequency in blood, spleen and thymus of CB17 scid/scid up to 4 weeks after transfer, due to the red light emitted after excitation in the laser light of 499 nm wavelength (Table 1).

Table 1. Cytofluorimetric detection of PKH 26 GL-labelled huPBL in lymphoid organs of SCID-hu

	% positive cells at different times after transfer[a]			
Lymphoid Organ	1 week	2 weeks	3 weeks	4 weeks
Peripheral Blood	2.7 ± 1.3	4.1 ± 1.3	3.0 ± 0.9	3.9 ± 1.2
Spleen	5.6 ± 2.9	4.0 ± 0.4	2.7 ± 0.9	2.7 ± 0.4
Thymus	5.9 ± 0.1	9.2 ± 3.1	9.4 ± 3.9	4.2 ± 0.3

[a] data represent mean ± SD of 3 cytofluorimetric analysis. "Leaky" control was gated on 2% positive cells.

Histological investigations confirmed the presence of stained huPBL in the spleen, with evidence of clustering around vessels, and their distribution over the whole thymus.

Cytofluorometric detection of distinct subsets of huPBL in blood or preparations of lymphoid tissues of CB17 scid/scid was performed usind fluorochrome-labelled murine monoclonal antibodies with specificity for surface markers of huPBL (for details, see: Ladel et al., 1992).

A sandwich ELISA was established to detect huIg in the serum of CB17 scid/scid, using solid-phase-bound polyclonal rabbit anti-huIg antibodies (Dako) and peroxidase-labelled polyclonal rabbit anti-huIg antibodies (Dako) to detect bound huIg. A standard preparation of huIg (Immunotech SA, Marseille, France) was used as internal control. Low cross reactivity to muIg and a detection limit of 10 ng per ml serum could be shown.

The concentration of huIg in the serum of CB17 scid/scid increased exponentially until week 4 after transfer of huPBL, reached a plateau of about 2,000 to 3,000 µg per ml serum and remained stable until week 12 before slowly declining (Fig. 1).

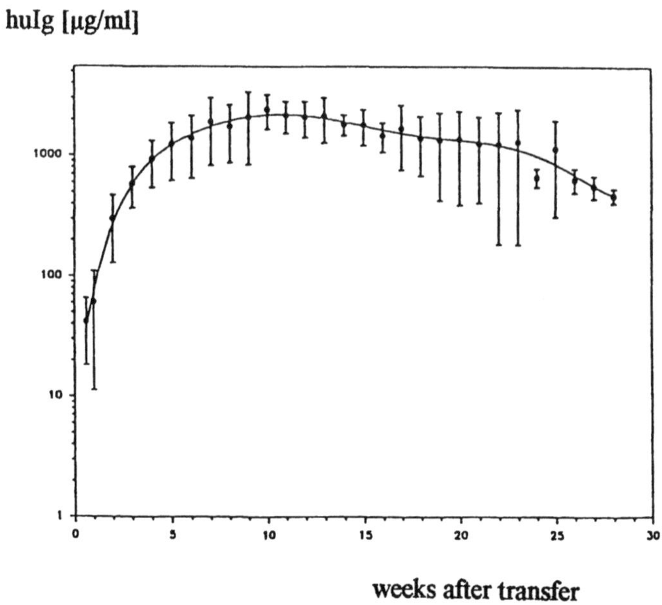

Fig. 1. Time course of huIg level in serum of SCID-hu. Data represent mean ± SD of 10 analysis.

Using peroxidase-labelled rabbit polyclonal antibodies with specificity for human IgM, human IgG and human IgA (all obtained from Dako), respectively, to detect bound huIg in the sandwich ELISA described above, the determination of class-specific huIg from serum of mice was possible.

Antibodies of all Ig classes could be detected in the serum of CB17 scid/scid from day 17 to day 184 after transfer of huPBL with a distribution comparable to the one in human blood (Tab. 2).

Table 2. Human immunoglobulin classes in the serum of SCID-hu

	Concentration in µg/ml at different times after transfer of huPBL[a]							
Immunoglobulin class	d17	d31	d52	d101	d115	d142	d158	d184
Human IgM	5	135	105	188	131	109	68	8
Human IgG	425	875	1864	1826	1538	1200	701	522
Human IgA	2	3	3	4	4	4	4	1

[a]data represent two experiments with different donors

Half-life of huIg injected intravenously into CB17 scid/scid resembled the one determined in the immunocompetent wild-type strain CB17 +/+ and was depending on the Ig class, with about 110 h for IgG, 48 h for IgM and 30 h for IgA. Thus, the Ig levels determined in CB17 scid/scid after transfer of huPBL cannot be explained by passive transfer.

4 to 6 weeks after the transfer of huPBL, CB17 scid/scid were immunized with 75 I.U. tetanus toxoid (TT; Tetanol[R], Behringwerke AG, Marburg/Lahn, Germany). The resulting humoral immune response was measured using an ELISA with TT directly bound to the solid phase. Binding antibodies were detected with peroxidase-labelled polyclonal rabbit anti-huIg. A standard human anti-TT preparation (Tetagam[R], Behringwerke AG) was used as internal standard and a detection limit of 0.002 I.U. TT was determined.

A TT-specific humoral immune response could be detected in all SCID-hu studied. The resulting anti-TT titer was shown to depend on the TT-specific immune status of the donor. A typical example of a strong immune response is shown in Fig. 2, occuring in a CB17 scid/scid bearing huPBL from a donor immunized with TT within the last 12 months.

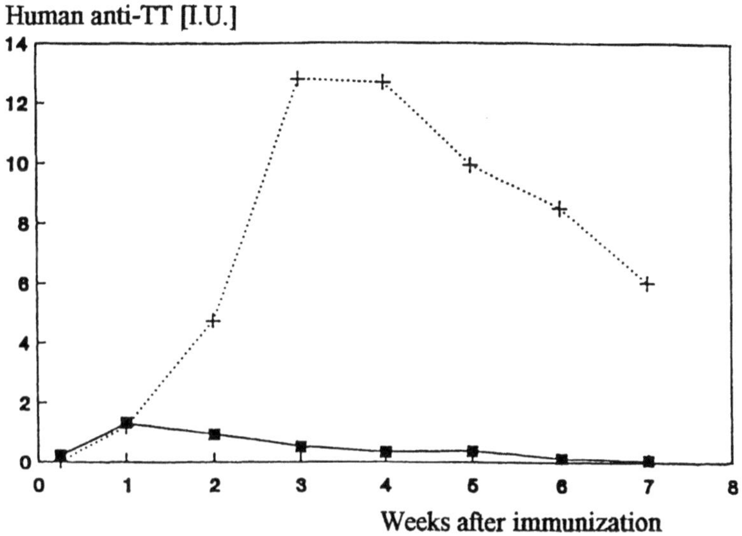

Fig. 2. Human anti-TT in SCID-hu. Last immunization of the donor of huPBL was within the last 12 months (■ = unimmunized control SCID-hu; + = SCID-hu immunized with 75 I.U. TT)

In conclusion, we were able to establish a SCID-hu model by intraperitoneal injection of $5x10^7$ huPBL into CB17 scid/scid kept under laminar air flow conditions, which were characterized by a serum level below 50 ng muIg per ml at the age of 4 to 6 weeks. Transferred huPBL could be shown to populate the lymphoid organs of the mice, to produce huIg for a period of over 28 weeks after transfer and to mount a humoral immune response to the recall antigen TT.

Effects of recombinant human interleukin-2

1,000 I.U. IL-2 (Boehringer Mannheim) were administered intravenously into CB17 scid/scid 6 to 8 weeks after the transfer of huPBL. This treatment resulted in an increase in the plasma level of huIg, which was observed from week 1 to week 4 after injection of IL-2 (Fig. 3).

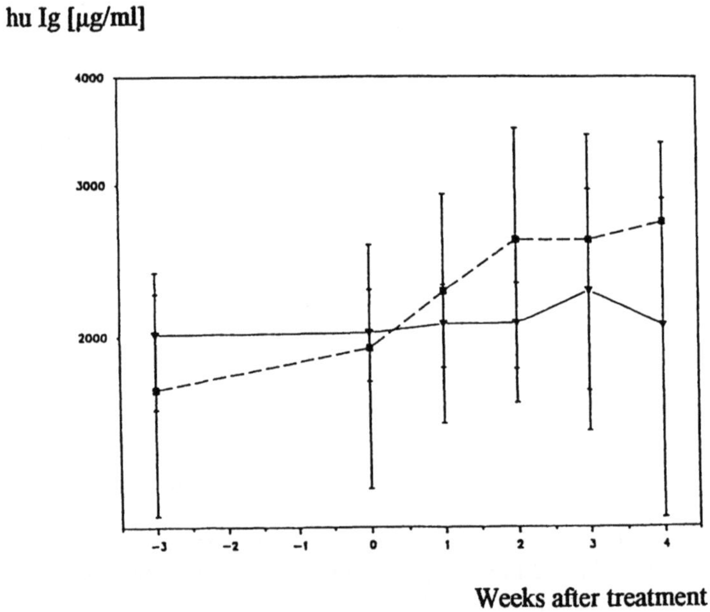

Fig. 3. Influence of IL-2 treatment on human Ig level in SCID-hu. Data represent mean ± SD of 5 different experiments (▼ = untreated control SCID-hu, ■ = SCID-hu treated with 1,000 I.U. IL-2 on day 0)

Furthermore, IL-2 treatment induced an about fivefold increase in the maximum humoral immune response to TT, when mice were immunized 1 week after treatment (Fig. 4).

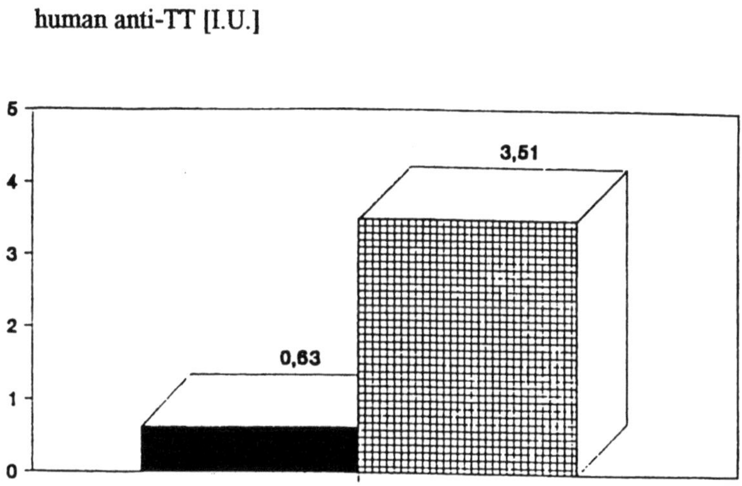

Fig. 4. Influence of IL-2 on the human humoral immune response to TT in SCID-hu. Animals were treated with 1,000 I.U. IL-2 (light column) or buffer (dark column) and one week later immunized with 75 I.U. TT.

In addition, IL-2-induced alterations in the distribution of huPBL subsets in peripheral blood, spleen and thymus of the murine host were observed.

In summary, increased concentrations of huIg in the plasma and a stimulated secondary humoral immune response to TT were observed subsequent to a single intravenous administration of 1,000 I.U. IL-2.

Effects of Cyclosporin A

Cyclosporin A (Sandimmun[R], Sandoz, Basel, Switzerland) at doses up to 25 mg per kg bodyweight was administered intravenously once weekly over a period of 4 weeks into CB17 scid/scid 6 to 8 weeks after the transfer of huPBL.

A transient decrease of huIg in the murine serum was observed over the time of treatment. About 2 weeks after the end of Cyclosporin A administration, huIg levels returned to control values (Fig. 5).

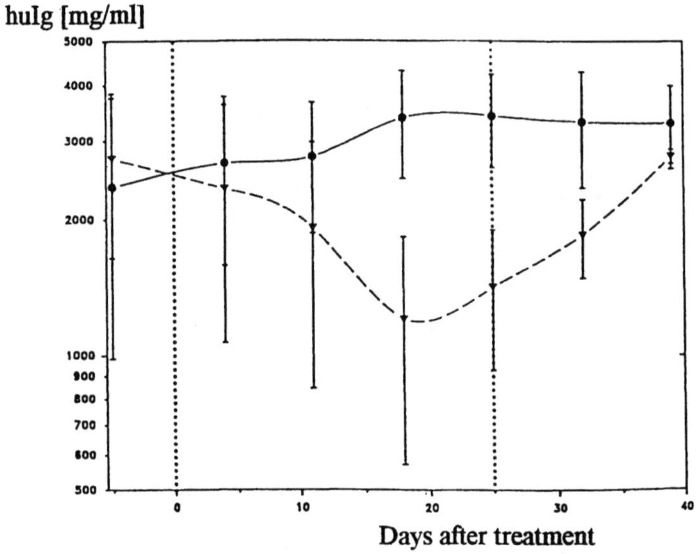

Fig. 5. Effect of treatment with Cyclosporin A on human Ig level in serum of SCID-hu. Data represent mean ± SD of 4 untreated control SCID-hu (●) and 6 SCID-hu (▼) treated with 25 mg Cyclosporin A per kg bodyweigh once weekly over 4 weeks.

When SCID-hu were immunized with TT (75 I.U.) one week after the start of treatment with Cyclosporin A, the resulting humoral immune response was clearly reduced (Fig. 6).

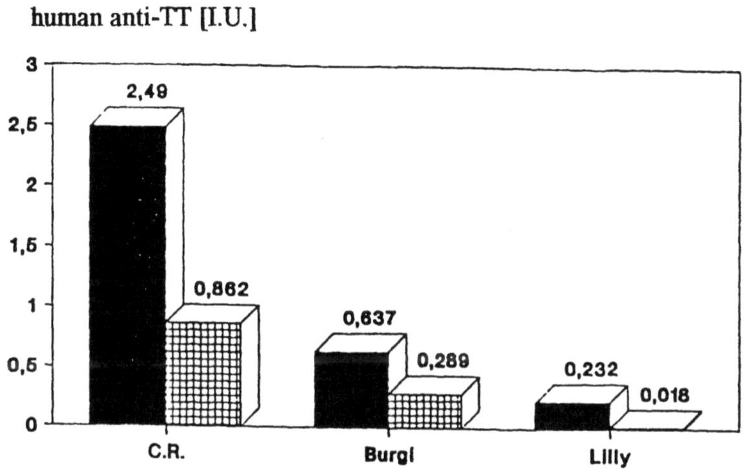

Fig. 6. Influence of treatment with Cyclosporin A on human humoral immune response to TT in SCID-hu. Maximum of immune response in SCID-hu bearing huPBL from 3 different donors are shown. Dark colums represent values of untreated controls, light ones those of mice treated once weekly over 4 weeks with 25 mg Cyclosporin A per mg bodyweight, starting one week before immunization.

The strength of this reduction was shown to be dose-dependent, with a 70 % decrease of the maximum titer at 5 and a more than 90 % decrease at 25 mg per kg bodyweight in the case of a low responder (Fig. 7).

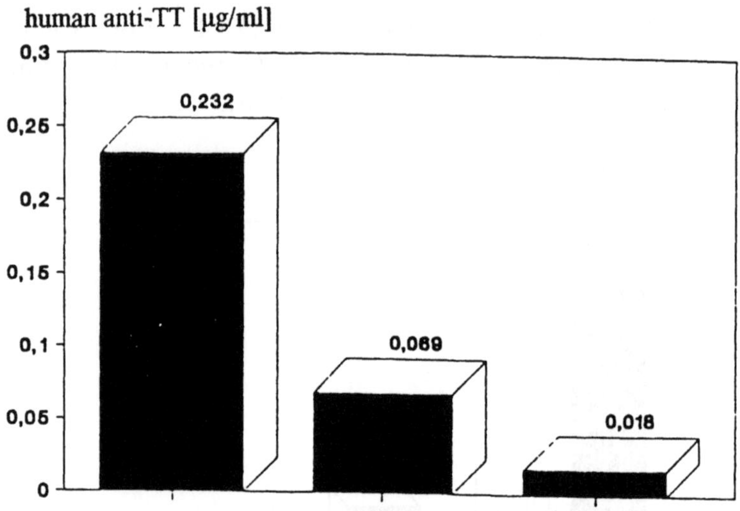

Fig. 7. Influence of treatment with different doses of Cyclosporin A on human humoral immune response to TT in SCID-hu. Data represent maximum immune response of untreated control SCID-hu (first column), SCID-hu treated with 5 mg (second column) or 25 mg Cyclosporin A per kg bodyweight once weekly over 4 weeks, starting one week before immunization.

When huPBL from a known EBV-positive donor were transferred to CB17 scid/scid, huIg in serum increased faster and to much higher concentrations than in animals bearing huPBL from EBV-negative donors and mice died from week 6 to week 12 after transfer of huPBL. The histological examination revealed a strong infiltration of murine lymphoid tissues with human lymphoid cells. When CB17 scid/scid bearing huPBL from EBV-positive donors were treated with Cyclosporin A (25 mg/kg), huIg in the serum was increased compared to untreated controls. However, when those mice were immunized with TT (75 I.U.) subsequent to treatment with Cyclosporin A, the expected result of an about 90 % reduced maximum of the humoral immune response was observed. This is in accordance with reports on the induction of clinical mononucleosis in EBV-positive patients after treatment with Cyclosporine A (Shapiro et al., 1988).

Thus, in SCID-hu, immunopharmacological as well as immuno-toxicological effects of Cyclosporin A treatment expected from the experience in the human situation could be detected.

Effects of Staphylococcal Enterotoxin B

SEB as a superantigen binds to MHC gene products and induces immunomodulation by a combination of stimulatory and cytotoxic effects (White et al., 1989; Dohlsten et al.,1991). In mice, systemic immunosuppression has been shown in vivo (Moseley et al., 1991).

A single intravenous injection of 10 µg SEB (Sigma, St. Louis, Mo, USA) resulted in a rapid occuring and long lasting suppression of huIg levels in the serum of SCID-hu (Fig. 8).

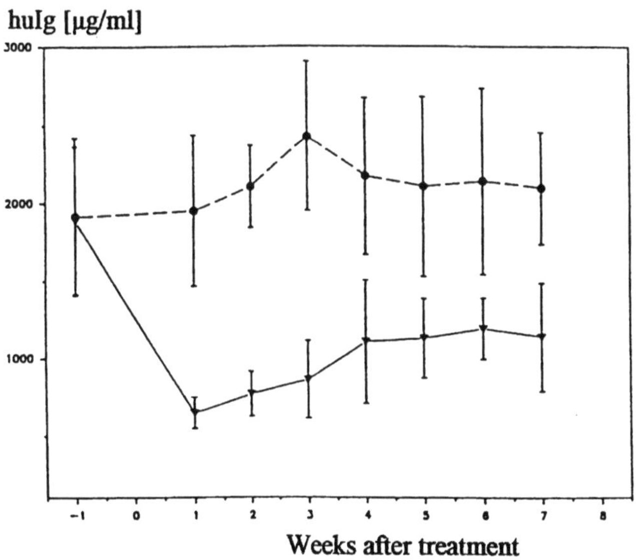

Fig. 8. Influence of intravenously administerd SEB on human Ihg in serum of SCID-hu. Data represent mean ± SD of 6 different experiments. (● = untreated control SCID-hu, ▼= SCID-hu treated with 10 µg SEB per animal on day 0)

In addition, the tissue distribution of huPBL in CB17 scid/scid was altered, decreased frequency of CD3-positive as well as CD4-positive and CD8-positive cells was most obvious.

The SCID-hu model was thus suitable to demonstrate immunomodulatory effects resulting from a single administration of 10 µg SEB per mouse.

Effects of Bacterial Lipopolysaccaride

LPS is known to exert a number of effects on the human immune system, affecting mostly B-lymphocytes and cells of the monocyte/macrophage lineage (Richardson et al., 1989). 6 to 8 weeks after transfer of huPBL, 5 µg LPS from E. coli, Serotype O26:B6 (Sigma) was administered intravenously into CB17 scid/scid, resulting in a fast occuring decrease of huIg levels in the serum. This decrease lasted for 4 weeks and serum concentrations of huIg returned to control values about 11 weeks after administration (Fig. 9).

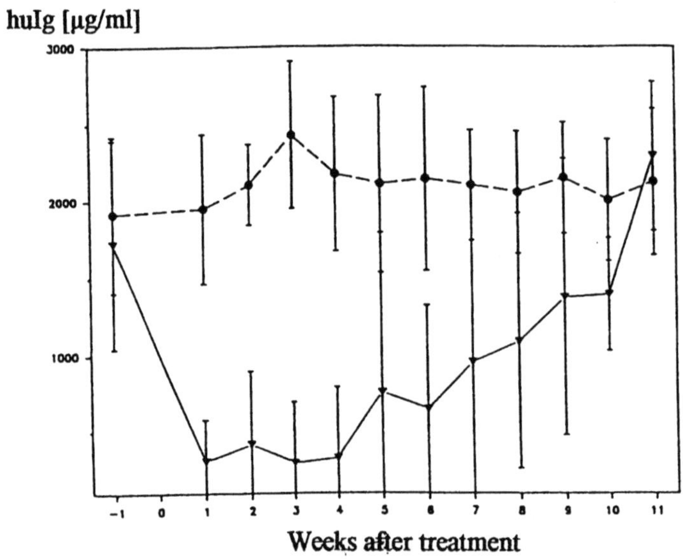

Fig. 9. Influence of intravenous administration of LPS on human Ig in serum of SCID-hu.Data represent mean ± SD of 4 different experiments. (● = untreated control SCID-hu, ▼= SCID-hu treated with 5 µg LPS per animal on day 0)

When SCID-hu were immunized with TT (75 I.U.) 11 weeks after the treatment with LPS, i.e. at the time when serum levels of huIg had returned to normal, a clearly enhanced humoral immune response was observed (Fig. 10).

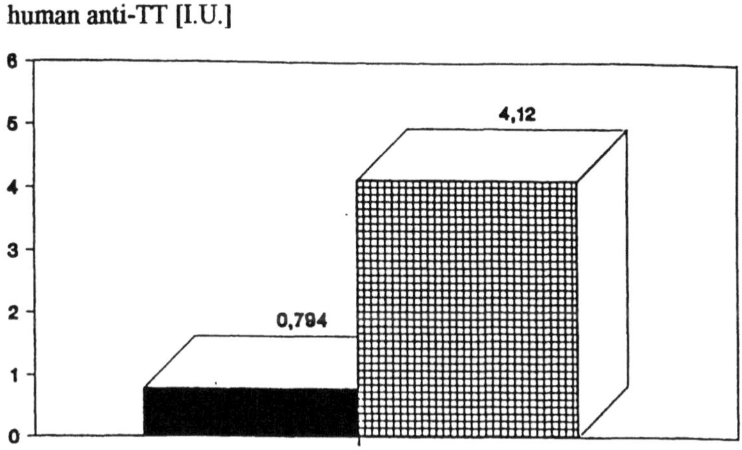

Fig. 10. Influence of intravenous administration of LPS on human humoral immune response to TT. Data represent maximum immune response of untreated control SCID-hu (dark column) and SCID-hu treated with 5 µg LPS per animal 11 weeks before immunization with 75 I.U. TT.

In summary, the SCID-hu model could detect immunomodulatory effects resulting from a single intravenous administration of 5 µg LPS per mouse, which could be expected from the toxin's known effects on human immunocompetent cells.

Conclusion

The SCID-hu model was established and characterized in our laboratory. When IL-2, Cyclosporin A, SEB or LPS were tested in this model, immunomodulatory effects were seen, which could be explained by known effects of these substances on human immunocompetent cells. It thus seems worthwile to further study the use of this model for immunopharmacological as well as immunotoxicological purposes.

References

Bosma GC, Custer RP, Bosma MJ (1983) A severe combined immunodeficiency mutation in the mouse. Nature 301:527-530

Carroll AM, Bosma MJ (1988) Detection and characterization of functional T cells in mice with severe combined immune deficiency. Eur J Immunol 18:1965-1971

Carroll AM, Hardy RR, Bosma MJ (1989) Occurence of mature B (IgM^+,$B220^+$) and T ($CD3^+$) lymphocytes in scid mice. J Immunol 143:1087-1093

Dohlsten M, Hedlund G, Kalland T (1991) Staphylococcal-enterotoxin-dependent cell-mediated cytotoxicity. Immunol Today 12:147-150 (1991)

Duchosal MA, McConahey PJ, Robinson CA, Dixon FJ (1990) Transfer of human systemic lupus erythematosus in severe combined immunodeficiencent (scid) mice. J Exp Med 172:985-988

Duchosal MA, Eming SA, Fischer P, Leturcq D, Barbas CF, McConahey PJ, Caothien RH, Thornton GB, Dixon FJ, Burton DR (1992) Immunization of hu-PBL-scid mice and the rescue of human monoclonal Fab fragments through combinatorial libraries. Nature 355:258-262

Ferrier P, Covey LR, Li SC, Suh H, Malynn BA, Blackwell TK, Morrow MA, Alt FW (1990) Normal recombination substrate V_H to DJ_H rearrangements in pre-B cell lines from scid mice. J Exp Med 171:1909-1918

Ladel CH, Püschner H, Kaufmann SHE, Bamberger U (1992) Human peripheral blood leukocytes transplanted on CB17 scid-scid mice are transferred to their offspring. Eur J Immunol 22:1735-1740

Ladel CH, Kaufmann SHE, Bamberger U (1993) Localisation of human peripheral blood leukocytes after transfer to C.B-17 scid/scid mice. Immunol Lett 38:63-68

Malynn BA, Blackwell TK, Fulop GM, Rathbun GA, Furley AJW, Ferrier P, Heinke LB, Phillips RA, Yancopoulos GD, Alt FW (1988) The scid defect affects the final step of the immunoglobuline VDJ recombinase mechanism. Cell 54:453-460

Moseley AB, Huston DP (1991) Mechanisms of Staphylococcus aureus exotoxin A inhibition of Ig production by human B cells. J Immunol 146:826-832.

Mosier DE, Gulizia RJ, Baird SM, Wilson DB (1988) Transfer of a functional human immune system to mice with severe combined immunodeficiency. Nature 335:256-259

Mosier DE, Gulizia RJ, Baird SM, Spector S, Spector D, Kipps TJ, Fox RI, Carson DA, Cooper N, Richman DD, Wilson DB (1989) Studies of HIV infection and the development of Epstein-Barr-Virus-related B cell lymphomas following transfer of human lymphocytes to mice with severe combined immunodeficiency Curr Top Microbiol 152:195-200

Petrini J.H.-J, Carroll AM, Bosma MJ (1990) T-cell receptor gene rearrangements in functional T-cell clones from severe combined immune deficient (scid) mice: reversion of the scid phenotype in individual lymphocyte progenitors. Proc Natl Acad Sci USA 87:3450-3453

Richardson RP, Rhyne CG, Fong Y, Hesse DG, Tracey KJ, Marano MA, Lowry SF, Antonacci AC, Calvano SE (1989) Peripheral blood leukocyte kinetics following in vivo lipopolysaccaride (LPS) administration to normal human subjects. Ann. Surg. 210:239-245

Rowe M, Young LS, Crooker J, Stokes H, Henderson S, Rickinson AB (1991) Epstein-Barr virus (EBV)-associated lymphoproliferative disease in the scid mouse model: implications for the pathogenesis of EBV-positive lymphomas in man. J Exp Med 173:147-158

Schuler W, Bosma MJ (1989) Nature of the scid defect: a defective VDJ recombinase system. Curr Top Microbiol 152:55-62

Shapiro RS, McClain K, Frizzera G, Gajl-Peczalska KJ, Kersey JH, Blazer BR, Arthur DC, Patton DF, Greenberg JS, Burke B, Ramsay NKC, McGlave P, Filipovich AH (1988) Epstein-Barr virus associated B cell lymphoproliferative disorders following bone marrow transplantation. Blood 71:1234-1241

Tighe H, Silverman GJ, Kozin F, Tucker R, Gulizia R, Peebles C, Lotz M, Rhodes G, Machold K, Mosier DE, Carson DA (1990) Autoantibody production by severe combined immunodeficient mice reconstituted with synovial cells from rheumatoid arthritis patients. Eur J Immunol 20:1843-1848 (1990)

Torbett BE, Picchio G, Mosier DE (1991) Hu-PBL-scid mice: a model for human immune function, AIDS and lymphomagenesis. Immunol Rev 124:139-164

White J, Herman A, Pullen AM, Kubo R, Kappler JW, Marrack P (1989) The Vß-specific superantigen staphylococcal enterotoxin B: stimulation of mature T cells and clonal deletion in neonatal mice. Cell 56:27-35 (1989)

Yancopoulos GD, Blackwell TK, Suh H, Hood L, Alt FW (1986) Introduced T cell receptor variable region gene segments in pre-B cells: evidence that B and T cells use a common recombinase. Cell 44:251-259

Stereochemical effects

(Chair: Dr. J.P. Seiler, Switzerland and
Prof. Dr. J. Diezi, Switzerland)

Chirality - from Molecules to Organisms

J.P. Seiler
Intercantonal Office for the Control of Medicaments
Erlachstrasse 8, CH-3000 Bern (Switzerland)

Chirality, the "handedness" of molecules, is a stereochemical property which may exert profound influences on the biological, i.e. pharmacodynamic, pharmacokinetic, metabolic or toxicologic, properties of those compounds that can exist as mirror images of themselves.

For decades after the initial discovery and separation of the two chiral forms of tartaric acid in the form of their respective mirror image crystals by Louis Pasteur, such substances could be prepared synthetically only as racemates, and only under fortunate circumstances could their single enantiomers be separated and analyzed. On the other hand, biosynthetic pathways had forever produced pure enantiomers, from relatively simple molecules like L-amino acids and D-carbohydrates, to the most complex ones, e.g. antibiotics, steroid hormones, and fungal or marine toxins. It was, however, necessary to have developed first the analytical and synthetical-chemical tools, before, in the last two decades, chirality could really begin to occupy its considered place and role in the fields of pharmacology and toxicology.

A very great input into the scientific dissection of chirality effects first came from pharmacology. This development was certainly supported by the wide array of pharmacologically active natural substances extractable from their plant or animal sources in pure enantiomeric form. But also the relatively modest amounts necessary to perform *in vitro* tests on tissues or on cellular receptors compared with the amounts needed to perform e.g. chronic toxicity studies may be regarded as having been helpful in this respect. One example of this kind of investigations may be sufficient to be mentionned here. The bronchodilating agent Formoterol, a drug used by inhalation by asthmatic patients, acts as an agonist at the β_2-adrenoceptor. The molecule has two chiral centers and can exist therefore in four enantiomeric forms (the drug Formoterol actually is just one of the two diastereomers, being the racemate of the (R,R)- and the (S,S)-forms, resp.). Agonist activity is obviously connected with the R-configuration at both chiral centers, as the (R,R)-enantiomer is by far the most potent, with the (S,S)-one being practically without activity (Trofast et al. 1991).

However, the extrapolation of such *in vitro* data to the *in vivo* effects is not so straightforward, as a number of possible difficulties and confounding factors have to be considered; to name just a few of them: Enantiomeric purity; receptor specificity and distribution; stereospecific pharmacokinetics and metabolism; and *in vivo* chiral inversion. But let us quit the field of pharmacodynamics here and enter the one of toxicology.

One such example of a stereospecific toxic effect is the induction of hepatic microgranulomata in rats by only one of the four stereoisomers of the insecticide fenvalerate. Mechanistical studies could demonstrate that these lesions were produced not by the substance itself, but by a cholesteryl-ester, which, because of steric constraints, could only be formed by one of the enantiomers (Okuno et al. 1986). Other toxicopathological effects that have been observed to be induced stereospecifically include the induction of hepatic peroxisome proliferation by the enantiomers of 2-ethylhexanoic acid (Macherey et al. 1992), by clofibrate analogues (Esbenshade et al. 1990), and by the leukotriene antagonist MK-0571 (Grossmann et al. 1992). In this latter case it could be shown that this effect is due to the enantioselective activation of a specific peroxisome proliferator-activated receptor (Boie et al. 1993).

Differences in toxicity may be mechanistically related to pharmacodynamic effects, as well as to differences in pharmacokinetics and metabolism. There may, however, also exist "intrinsic" differences in the toxicity of enantiomeric pairs. The one area of toxicology, where such "intrinsic" effects and enantiomeric specificities are probably best investigated is the field of genotoxicity. In a sense, the genotoxic interaction of a chiral molecule with its target, DNA, can be compared to the pharmacodynamic interaction of a chiral molecule with its receptor: Both are in a fixed spatial conformation, enabling only one enantiomeric form of the molecule to fully interact with the receptor. In the case of genotoxicity the "receptor" is DNA with its fixed three-dimensional, helical structure, making it easy to visualize the possibility of enantiomer-specificity in such interactions.

Already 20 years ago, this "intrinsic" property of enantiomeric interactions became known through the investigations on the mutagenic and carcinogenic mode of action of benzo(a)pyrene, as reviewed by Conney et al. (1994). The ultimate mutagenic entity of this polycyclic aromatic hydrocarbon is the dihydro-diol-epoxide which exists in four diastereomeric forms. Of these four stereochemically different molecules three do not exhibit very strong activity neither in the induction of lung tumors (Buening et al. 1978) nor in the initiation of skin carcinogenesis (Slaga et al. 1979), and it is the (7R,8S,9S,10R)-form which can be seen as the main culprit not only of benzo(a)pyrene carcinogenicity, but also of its mutagenicity (Wood et al. 1977).

Binding of a chiral molecule to DNA can itself be affected by the enantiomeric form of the reactant, and the resulting adduct can then affect subsequent reactions at this site in a number of ways, depending on the stereochemical conditions and circumstances. A first such possibility could be the differential reactivity of the respective enantiomers with DNA bases, as has been shown for benzo(c)phenanthrene dihydrodiol epoxide enantiomers, where the

(+)-1-isomer is reacting almost exclusively with adenine, in preference over guanine, as an adduct ratio of 8.4 A to 1 G has been determined. The other isomers are exhibiting a lesser degree of preference in this respect, with the (-)-2-isomer reacting in a ratio of about 2 A to 1 G and with the remaining two isomers yielding ratios of 1.4 A to 1 G (Dipple et al. 1987).

Another possible effect, a bulky residue like a PAH bound to DNA can have, may depend on the direction the adduct is pointing to with respect to the polarity of the DNA strand and subsequently to the direction in which enzymatic repair or replication is proceeding. A recent paper by Mao et al. (1993) has indeed shown that if the benzo(a)pyrene residue bound to the only guanine in the model oligonucleotide used in this investigation is pointing towards the phosphodiesterase as this enzyme proceeds along the DNA, it tends to block hydrolysis completely at the bond before the adduct, whereas, if the residue is pointing away, the enzyme may proceed in the hydrolytic process further along the oligonucleotide.

As a further parameter, not only chemical reactivity can be shown to play a role, but also the local sequence of the DNA itself can be important in differentially determining the fate of stereospecific lesions. Styrene oxide is a rather simple molecule with just one chiral center and thus existing in two enantiomeric forms. Chemical reactivity, as evidenced by their respective interactions with model nucleophiles, is equal for both enantiomers, as are the respective values of total bound styrene residues on DNA. However, mutagenic activity is different for the two enantiomers, with the R-enantiomer being roughly twice as active as the S-one in bacterial assays (Seiler 1990), while the S-oxide produced more chromosomal aberrations and SCEs than the R-compound (Sinsheimer et al. 1993). Obviously other factors than the purely "intrinsic" mutagenic activity of the two enantiomers have to be responsible for this divergence in biological effects, factors that may determine the effective target concentrations of the respective enantiomers. In the context of styrene oxide it is known that in mammalian cells R-styrene oxide is more rapidly detoxified by glutathione transferases and by epoxide hydratase than the corresponding S-form, thus profoundly influencing the actual reactant concentrations at the DNA.

A facet of the pure, intrinsic mutagenic activity of this molecule may be glimpsed at through the investigations of Latham et al. (1993). In this paper it is quite exciting to see, what profound differences are possible with respect to mutagenicity and to "toxicity" (i.e. replication arrest) depending on both the chirality of the reacting molecule as well as on the local sequence context. In these experiments R- and S-styrene oxide was reacted with the two adenine residues in the codon 61 of a human ras-gene construct. The four different nucleotides thus obtained were then assayed on the one hand for the extent of A -> G transitions and on the other hand for their respective replication competence. The four adenine N^6-adducts proved to behave very differently in both respects. Point mutations, i.e. A -> G transitions, could only be observed to an appreciable extent with the S-modification at the codon position 2, while the R-enantiomer bound at this same position did to a great extent abolish the replication competence of the gene construct, in other words, it would become a cytotoxic lesion *in vivo*. Such a difference in effects can now be said with confidence to truly represent an intrinsic property of the molecule's chirality. Of course, treatment of whole DNA with styrene oxide does not only lead to adenine N^6-adducts, but also to a number of other, potentially mutagenic adducts as well, and it will be the sum of total adduct formation, together with the local sequence context and the chirality-dependent lesion expression of the different adducts which will, in the end, determine the relative mutagenic activity of the two enantiomers.

Doubtlessly, mutations have their influence on the organism as a whole, but another in toxicological discipline, effects on the organism may be seen in a more direct way, as it is only in this respect that lesions will become manifest. Thus, a few remarks on chirality effects in teratogenicity shall conclude this introduction. Valproic acid is an achiral molecule, containing, however, a prochiral centre; it is a very valuable antiepileptic drug, but it is also a recognized animal and human teratogen, producing mainly neural tube defects (spina bifida in humans and exencephaly in mice). Several derivatives of valproic acid have been tested for a possible - and very much desirable - separation of teratogenic and antiepileptic effects of the valproate basic structure, and in the course of this work, chiral substances have also been investigated. The compound 2-n-propyl-4-pentynoic acid thereby proved to be about twice as potent a teratogen in animal studies as the parent substance valproic acid, and furthermore, its S-(-)-enantiomer was shown to be far more potent in this respect than the R-(+)-one (Nau et al. 1991). This enantiomeric specificity of action again might have been due to stereoselective pharmacokinetics, either in terms of metabolism or clearance on the maternal side, or of uptake rates or accumulation on the fetal side of this experimental system, or it might have been due to a true intrinsic enantioselectivity of teratogenic action. In a paper by Hauck et al. (1992) it could indeed be shown that differences in pharmacokinetics do not exist between the two enantiomers, and that thus the vast difference in teratogenic activity observed must be due to true intrinsic enantioselectivity of action.

Having arrived now at effects on whole organisms and thus at the end of this journey from molecules to organism, I will finish my introduction by a summary in two or three sentences: Life is built on asymmetry, on chirality, and this is true from the blueprint of atomic structures to the form of our hands, which have given the term "chirality" its name. Handedness of molecules may profoundly influence their effects on chemical and biological reactions. Enzymatic reactions *in vitro*, metabolism *in vivo* as well as pharmacodynamic responses, and also toxicity, may mirror the images of chiral substances as we have seen in the examples on genotoxicity and teratogenicity. Future research in this fascinating area will certainly allow us further and additional insights into mechanisms and modes of action due to this specific aspect of molecular, and biological, asymmetry called chirality.

References

Boie Y, Adam M, Rushmore TH, Kennedy BP (1993) Enantioselective Activation of the Peroxisome Proliferator-activated Receptor. J Biol Chem 268:5530-5534

Buening MK, Wislocki PG, Levin W, Yagi H, Thakker DR, Akagi H, Koreeda M, Jerina DM, Conney AH (1978) Tumorigenicity of the optical enantiomers of the diastereomeric benzo(a)pyrene 7,8-diol-9,10-epoxides in newborn mice: Exceptional activity of (+)-7ß,8α-dihydroxy-9α,10α-epoxy-7,8,9,10-tetrahydrobenzo(a)-pyrene. Proc Natl Acad Sci (USA) 75:5358-5361

Conney AH, Chang RL, Jerina DM, Wei S-JC (1994) Studies on the metabolism of benzo(a)pyrene and dose-dependent differences in the mutagenic profile of its ultimate carcinogenic metabolite. Drug Metabolism Reviews 26:125-163

Dipple A, Pigott MA, Agarwal SK, Yagi H, Sayer JM, Jerina DM (1987) Optically active benzo(c)phenanthrene diol epoxides bind extensively to adenine in DNA. Nature 327:535-536

Esbenshade TA, Kamanna VS, Newman HAI, Tortorella V, Witiak DT, Feller DR (1990) *In vivo* and *in vitro* peroxisome proliferation properties of selected clofibrate analogues in the rat. Biochem Pharmacol 40:1263-1274

Grossmann SJ, DeLuca JG, Zamboni RJ, Keenan KP, Patrick DH, Herold EG, VanZwieten MJ, Zacchei AG (1992) Enantioselective Induction of Peroxisomal Proliferation in CD-1 Mice by Leukotriene Antagonists. Toxicol Appl Pharmacol 116:217-224

Hauck R-S, Elmazar MMA, Plum C, Nau H (1992) The enantioselective teratogenicity of 2-*n*-propyl-4-pentynoic acid (4-yn-VPA) is due to stereoselective intrinsic activity and not differences in pharmacokinetics. Toxicology letters 60:145-153

Latham GJ, Zhou L, Harris CM, Harris TM, Lloyd RS (1993) The Replication Fate of R- and S-Styrene Oxide Adducts on Adenine N^6 Is Dependent on Both the Chirality of the Lesion and the Local Sequence Context. J Biol Chem 268:23427-23434

Macherey AC, Grégoire S, Tainturier G, Lhuguenot JC (1992) Enantioselectivity in the Induction of Peroxisome Proliferation by 2-Ethylhexanoic acid. Chirality 4:478-483

Mao B, Li B, Amin S, Cosman M, Geacintov NE (1993) Opposite Stereoselective Resistance to Digestion by Phosphodiesterase I and II of Benzo(a)pyrene Diol Epoxide-Modified Oligonucleotide Adducts. Biochemistry 32:11785-11793

Nau H, Hauck R-S, Ehlers K (1991) Valproic Acid-Induced Neural Tube Defects in Mouse and Human: Aspects of Chirality, Alternative Drug Development, Pharmacokinetics and Possible Mechanisms. Pharmacol Toxicol 69:310-321

Okuno Y, Seki T, Ito S, Kaneko T, Watanabe T Yamada T, Miyamoto J (1986) Differential metabolism of fenvalerate and granuloma formation. II. Toxicological significance of a lipophilic conjugate from fenvalerate. Toxicol Appl Pharmacol 83:157-169

Seiler JP (1990) Chirality-dependent DNA reactivity as the possible cause of the differential mutagenicity of the two components in an enantiomeric pair of epoxides. Mutation Res 245:165-169

Sinsheimer JE, Chen R, Das SK, Hooberman BH, Osorio S, You Z (1993) The genotoxicity of enantiomeric aliphatic epoxides. Mutation Res 298:197-206

Slaga TJ, Bracken WJ, Gleason G, Levin W, Yagi H, Jerina DM, Conney AH (1979) Marked differences in the skin tumor-initiating activities of the optical enantiomers of the diastereomeric benzo(a)pyrene 7,8-diol-9,10-epoxides. Cancer Res 39:67-71

Trofast J, Oesterberg K, Källströ B-L, Waldeck B (1991) Steric Aspects of Agonism and Antagonism at ß-Adrenoceptors: Synthesis of and Pharmacological Experiments with the Enantiomers of Formoterol and Their Diastereomers. Chirality 3:443-450

Wood AW, Chang RL, Levin W, Yagi H, Thakker DR, Jerina DM, Conney AH (1977) Differences in mutagenicity of the optical enantiomers of the diastereomeric benzo(a)pyrene 7,8-diol-9,10-epoxides. Biochem Biophys Res Commun 77:1389-1396

Interactions between the in vitro metabolism of xenobiotics and fatty acids. The case of ibuprofen and other chiral profens.

J.M. Mayer, B. Testa, M. Roy-de Vos, C. Audergon and J.C. Etter
Ecole de Pharmacie, Université de Lausanne, BEP, CH-1015 Lausanne, Suisse

Xenobiotic carboxylic acids can act as substrates for the pathways of fatty acid biotransformation. Thus, they can form abnormal fatty acids by addition of 2-carbon units, or be conjugated with cholesterol, bile acids, carnitine and other amino acids. Finally, xenobiotic carboxylic acids can substitute endogeneous fatty acids in glycerolipid synthesis producing hybrid triacylglycerols and phospholipids (Caldwell, 1985; Caldwell and Marsh, 1983; Huston, 1982). In order to enter the routes of lipid biochemistry, xenobiotic acids like natural fatty acids must be activated by coenzyme A. The acyl-CoA thioesters thus formed are pivotal in the metabolism of xenobiotic acids (Fig. 1).

2-Arylpropionic acids (profens) are a major group of non-steroidal anti-inflammatory drugs (NSAIDs). These compounds exist in two enantiomeric forms due to the presence of an asymmetric carbon atom alpha to the carboxyl function (Fig. 2). *In vitro* tests have shown that the anti-prostaglandin synthase activity of profens resides almost exclusively with the (+)-(*S*)-enantiomers (Shen, 1981). Other pharmacological aspects seem to be non-stereoselective. Both enantiomers of ibuprofen inhibited equally neutrophil cells (Villanueva et al., 1993) and displayed the same antinociceptive effect (Brune et al., 1993). All profens except naproxen (Fig. 2) are marketed as racemates.

Profens exhibit enantioselective pharmacokinetics, the most intriguing aspect of which is a unidirectional chiral inversion from the (-)-(*R*)- to the (+)-(*S*)-enantiomer (Fig. 3) (Hutt and Caldwell, 1983; Mayer, 1990; Testa, 1986; Testa and Mayer, 1988). The reaction is documented by a large number of *in vivo* and *in vitro* studies using a variety of animal species and profens. Thus, when *rac*-ibuprofen was incubated in isolated hepatocytes the *S*-enantiomer was eliminated at a much lower rate than its *R*-antipode. During the incubation of (*R*)-ibuprofen an important amount of the *S*-enantiomer was formed, bringing direct proof that metabolic chiral inversion had occurred (Müller et al., 1990). The absence of formation of (*R*)-ibuprofen during the incubation of (*S*)-ibuprofen demonstrates the unidirectionality of the process.

In the present paper, we discuss recent findings, both published and unpublished, on three aspects of profen metabolism. First, the variability and mechanism of their chiral inversion is considered. This is followed by a focus on the pivotal role of the profenoyl-CoA thioesters and the influence of coenzyme A levels on chiral inversion. These findings then lead to a preliminary understanding of the mechanisms underlying the interaction of profens with lipid biochemistry, e.g. β-oxidation, formation of hybrid triglycerides and hypolipidemic activity.

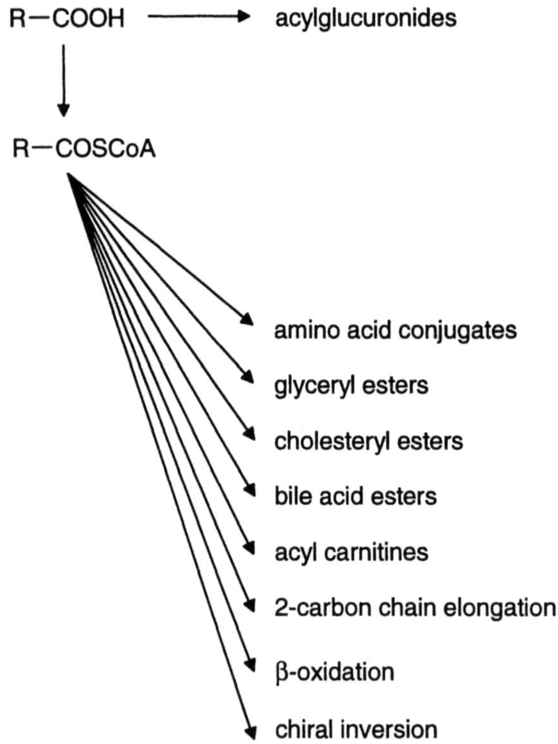

Fig. 1. Pathways in the metabolism of xenobiotic acids

Variabilities in chiral inversion

Not all profens undergo metabolic chiral inversion. Only two other profens, fenoprofen (Fig. 2C) (Rubin et al., 1985) and benoxaprofen (Fig. 2D) (Bopp et al., 1979) have revealed significant inversion in humans. The stereoselective disposition of profens in laboratory animals does not always correspond to that observed in humans. Thus, the inversion half-life of (*R*)- to (*S*)-benaxoprofen in rat is about 2.6 h whereas the conversion in humans is much slower (about 108 h) (Simmonds et al., 1980). Flurbiprofen (Fig. 2E) (Jamali et al., 1988), indoprofen (Fig. 2F) (Buttinoni et al., 1983) and carprofen (Fig. 2G) (Kemmerer et al., 1979) are not inverted in rats and in humans. However, profens such as flunaxoprofen, tiaprofenic acid (Fig. 2H) and ketoprofen (Fig. 2I) undergo significant inversion in rats but are not inverted in man (Foster and Jamali, 1987; Jamali, 1988; Pedrazzini

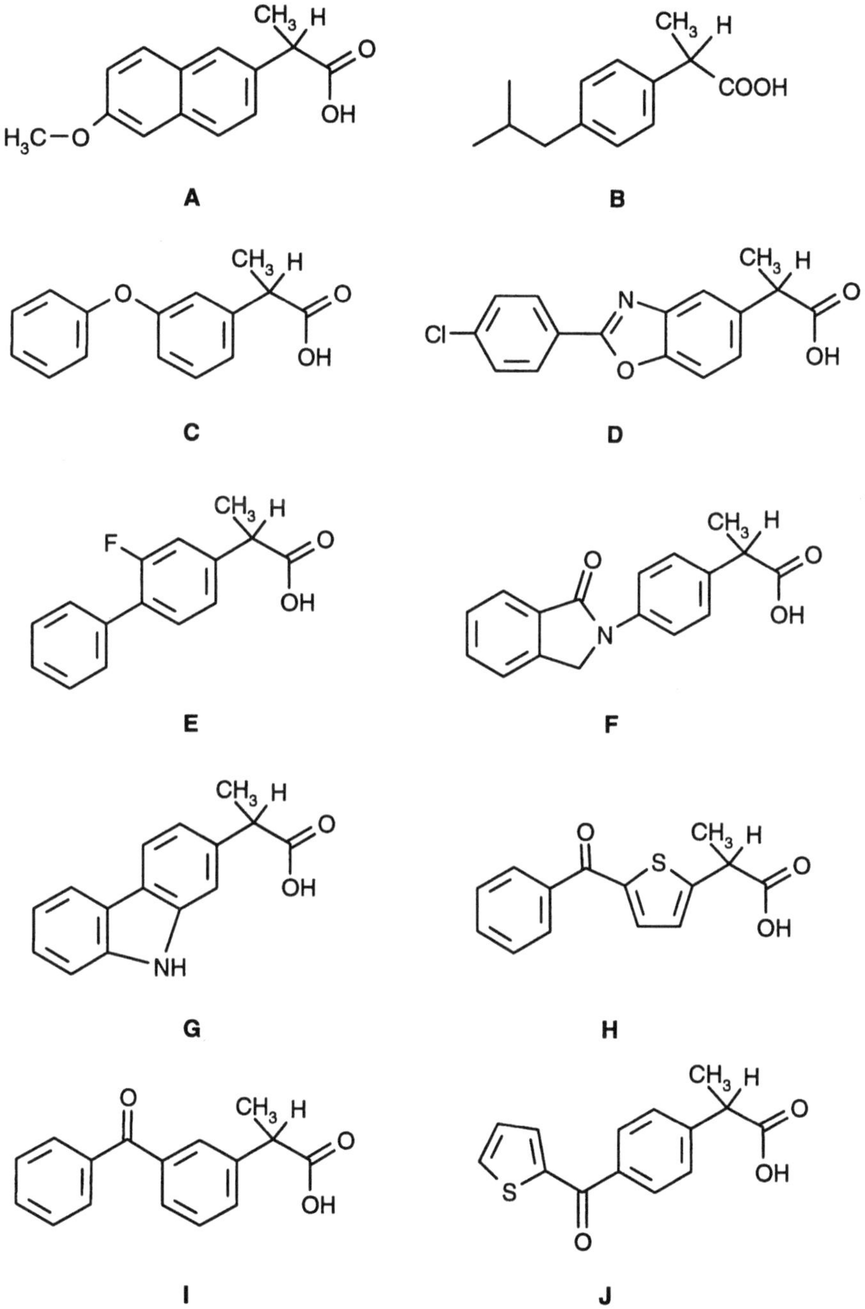

Fig. 2. Chemical structure of some profens: naproxen (**A**), ibuprofen (**B**), fenoprofen (**C**), benaxoprofen (**D**), flurbiprofen (**E**), indoprofen (**F**), carprofen (**G**), tiaprofenic acid (**H**), ketoprofen (**I**), suprofen (**J**).

et al., 1988). There are other examples for important interspecies differences. Thus, ketoprofen which is extensively inverted in the rat shows less than 10% inversion in the rabbit (Abas and Meffin, 1987). No inversion occurs for 2-phenylpropionic acid in the mouse in contrast to rat and rabbit (Fournel and Caldwell, 1986).

There are some exceptions to the rule of unidirectional inversion. Thus, both enantiomers of 2-phenylpropionic acid and ibuprofen were inverted with nearly the same efficiency in dogs and guinea-pigs, respectively (Tanaka et al., 1992; Chen et al., 1991). It appears from this brief compilation that the rate of configurational inversion depends on both animal species and nature of substrate, but no clear generalization emerges from available data.

Mechanism of inversion

The mechanism of chiral inversion involves activation of the 2-arylpropionic acid by formation of the acyl-coenzyme A thioester via an AMP-intermediate (Fig. 3)(Tracy and Hall, 1992; Knihinicky et al., 1989; Knights et al., 1988; Lee et al., 1985; Fears et al., 1978; Nakamura et al., 1981; Wechter et al., 1974). The unidirectional inversion of configuration is explained in terms of the acyl-CoA formation being substrate enantioselective for the *R*-enantiomer, while the *S*-enantiomer is not a substrate for the acyl-CoA synthetase(s) catalyzing the reaction. Recently, it has been established that the formation of the AMP-intermediate is the stereoselective step in the CoA thioester formation (Geißlinger, 1993)(Fig. 3). The *R*-thioester may then epimerize to the *S*-thioester. The enzyme catalyzing this reaction, a "2-arylpropionyl-CoA epimerase", has recently been purified to homogeneity from rat liver cytosol and mitochondria (Shieh and Chen, 1993). The last step in the sequence is hydrolysis of the thioester (Fig. 3). Since CoA thioesters are stable in plasma and in buffer solutions under physiological conditions, their hydrolysis must be mediated by hydrolases. Studies with chemically synthetized profenoyl-CoA thioesters revealed that neither the epimerization nor the hydrolytic step are enantioselective (Knihinicki et al., 1991).

The CoA thioester intermediate

The study of arylpropionyl-CoA thioester formation in rat liver tissue homogenates and subcellular fractions have confirmed that thioesters are formed only from the *R*-enantiomer (Knadler and Hall, 1990). Flurbiprofen (Fig. 2C), a profen which does not undergo inversion, did not form acyl-CoA conjugates. The enzyme catalyzing stereoselectively the formation of (*R*)-fenoprofenoyl-CoA has been identified as the microsomal long-chain fatty acyl-CoA synthetase (EC 6.2.1.3) (Knights et al. 1988).

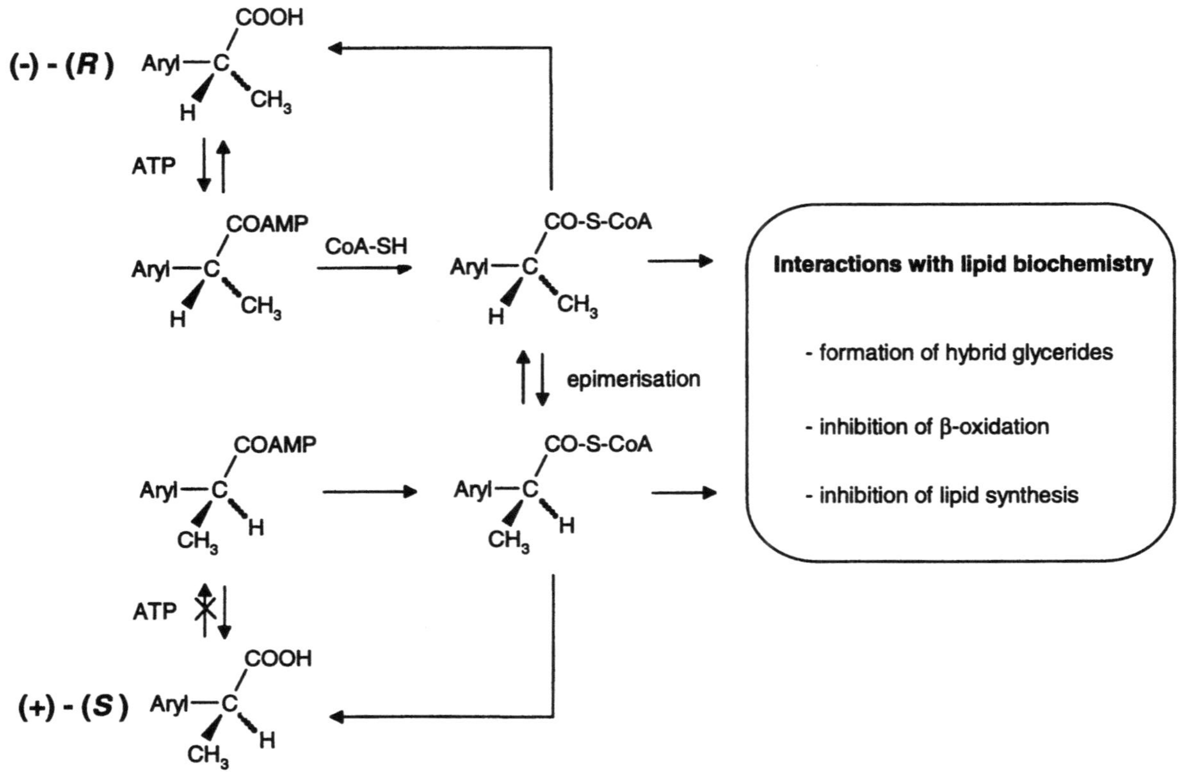

Fig. 3. Mechanism for the unidirectional chiral inversion of profens via epimerisation of 2-arylpropanoyl-coenzyme A thioesters and their possible interactions with lipid biochemistry.

Influence of coenzyme A levels on chiral inversion of profens

When we incubated (*R*)-ibuprofen with palmitic acid, its concentration remained constant (Fig. 4A). Thereafter, the *R*-isomer concentration decreased slowly and paralleled formation of the antipode. The onset of the inversion was shown to depend on the concentrations of palmitic acid in the incubation medium. Octanoic acid blocked the chiral inversion of (*R*)-ibuprofen in rat hepatocytes to a similar extent as did palmitic acid (Fig. 4B). This observation may suggest that both long- and medium-chain fatty acyl-CoA synthetases are involved in the formation of ibuprofenoyl-CoA. However, investigations with rat liver homogenates (Tracy et al., 1993) revealed that palmitic acid is a much more potent inhibitor (apparent K_i = 0.005 mM) of ibuprofenoyl-CoA formation than octanoic acid (apparent K_i = 0.19 mM). The effect of octanoic acid may therefore be due not only to an interaction at the catalytic site but also to the depletion of the cosubstrate (i.e. CoA).

In order to verify this hypothesis we measured intracellular CoA levels in isolated rat hepatocytes. The evolution of CoA concentrations during the incubation period in hepatocytes is shown in Figures 5 and 6. (*R*)-Ibuprofen reduced transiently CoA levels whereas (*S*)-ibuprofen was without effect (Fig. 5). Palmitic as well as octanoic acid reduced CoA concentrations followed by a return to normal levels (Fig. 6). Concomitant incubation of palmitic or octanoic acid with (*R*)-ibuprofen produced a increased and prolongued depletion of CoA. The inhibition of the chiral inversion of (*R*)-ibuprofen by palmitic and octanoic acid can therefore be explained by the competition of these acids for CoA..

In order to investigate further the influence of cellular CoA levels we also studied the effect of clofibric acid on the inversion of (*R*)-Ibuprofen. Clofibric acid induces a number of enzymes, in particular, long-chain acyl-CoA synthetase and acyl-CoA hydrolases. Furthermore, clofibric acid treatment increases the amount of total hepatic CoA. The effect of clofibric acid in the incubation medium on the metabolism of (*R*)-ibuprofen in isolated rat hepatocytes is shown in Figure 7. It appears that in the presence of clofibric acid (*R*)-ibuprofen is eliminated more rapidly and more (*S*)-ibuprofen is formed than in control incubations. It can therefore be concluded that clofibric acid enhances the chiral inversion of (*R*)-ibuprofen. Since the effect of clofibric acid is immediate, it is unlikely that the induction of acyl-CoA synthetase(s) is the cause of increased inversion. Measurements of the CoA-levels revealed that clofibric acid produces a transitory but substantial increase in intracellular CoA concentrations (Fig. 8).

We conclude therefore that the increased inversion of (*R*)-ibuprofen is the result of the greater availability of CoA resulting from the effect of clofibric acid. Increased inversion of (*R*)-ibuprofen has also been observed in liver homogenates of rats pretreated with clofibric acid (Knights et al., 1991).

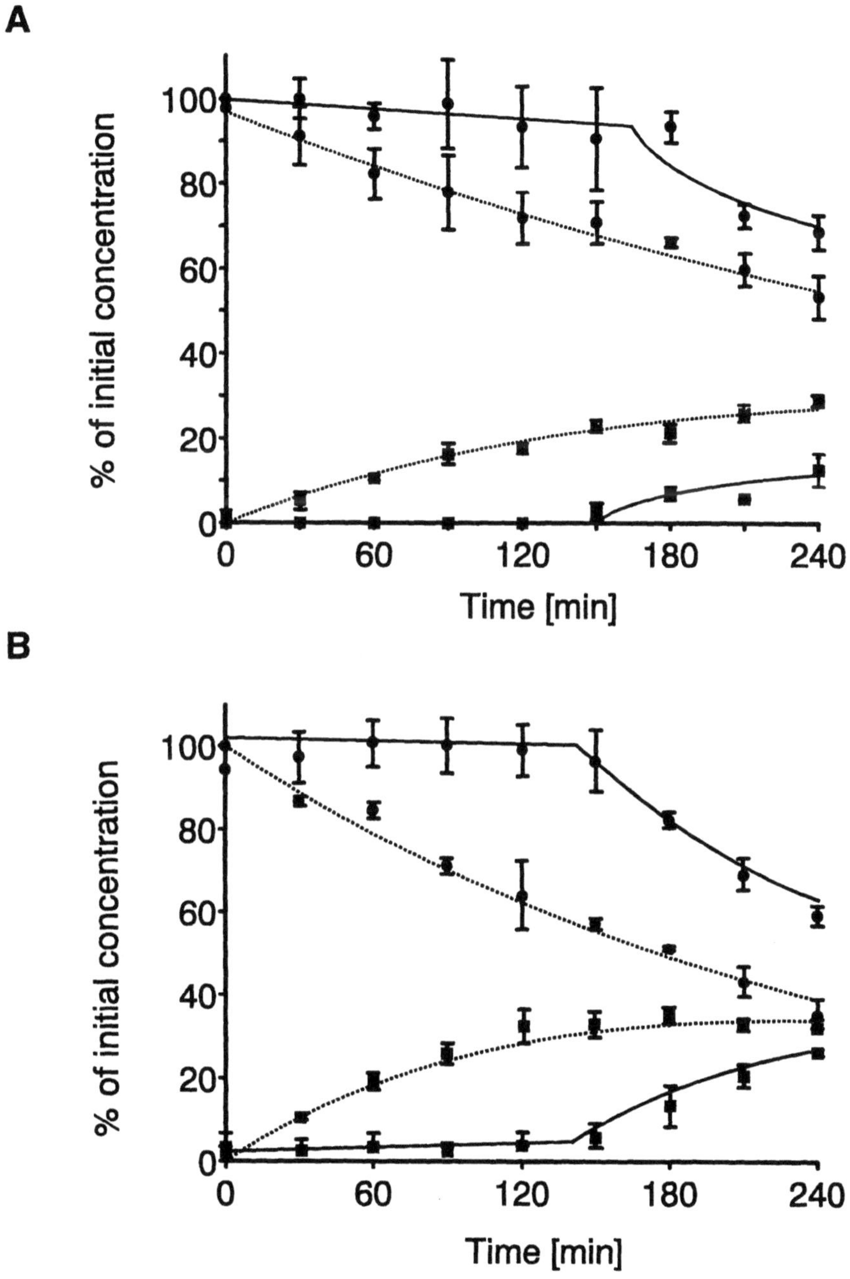

Fig. 4. Effect of fatty acids on the chiral inversion of (-)-(R)-ibuprofen (5mg/l) in rat hepatocyte suspensions (5x10^5 cells/ml); ------ control without fatty acids, 0.5% BSA. (**A**) incubation medium containing 1.45 mM palmitic acid bound to 0.5% BSA. (**B**) incubation medium containing 1.45 mM octanoic acid bound to 0.5% BSA. (-)-(*R*)-ibuprofen (●); (+)-(*S*)-ibuprofen (■).

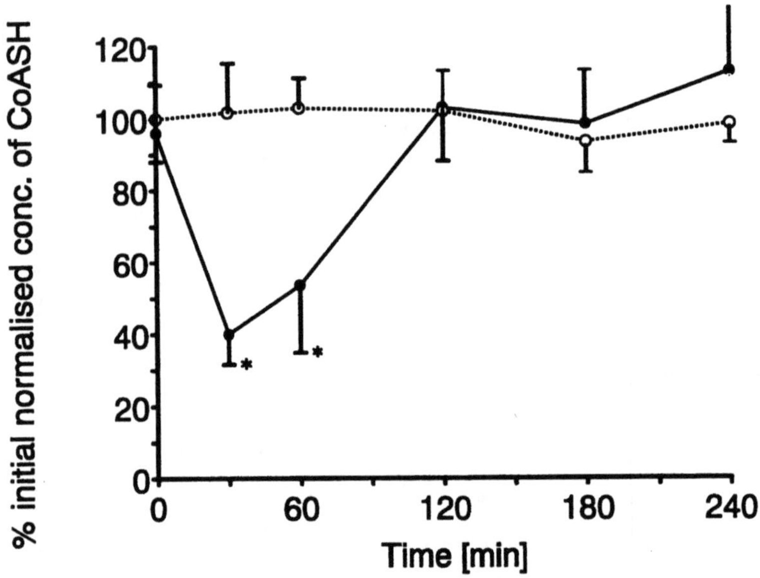

Fig. 5. CoA levels in rat hepatocyte suspensions (5x10^5 cells/ml) containing 0.5% BSA during the incubation of: —— (*R*)-ibuprofen (24.2 μM) and .... (*S*)-ibuprofen (24.2 μM). Mean of three incubations per rat. Student *t*-test (*): $P < 0.05$.

Interaction with β-oxidation

Another possible point of attack of profens on lipid biochemistry is the β-oxidation of fatty acids. Indeed, pirprofen has been found to elicit a concentration-dependent inhibition of mitochondrial β-oxidation of fatty acids, a finding postulated to account for the microvesicular steatosis produced by this drug (Geneve et al., 1987). Ibuprofen and tiaprofenic acid similarly inhibited β-oxidation, but (*S*)-naproxen was inactive (Geneve et al., 1987). Freneaux et al. (1990) investigated the stereoselectivity of the inhibition of β-oxidation caused by ibuprofen enantiomers. They assessed β-oxidation by the formation of acid-soluble products during the incubation of mouse liver mitochondria with labeled palmitic acid and the separated enantiomers of ibuprofen. In order to examine the potential role of CoA sequestration by the *R*-enantiomer, incubations were conducted with various concentrations of exogenous CoA. It was found that at the highest CoA concentration used (50 μM), a concentration naturally occurring in liver cytosol, both (*S*)- and (*R*)-ibuprofen inhibited the formation of β-oxidation products. At lower CoA concentrations, however, the effect varied with the enantiomer. (*R*)-Ibuprofen increasingly inhibited β-oxidation with decreasing CoA concerntration whereas the effect of (*S*)-ibuprofen was independent on CoA

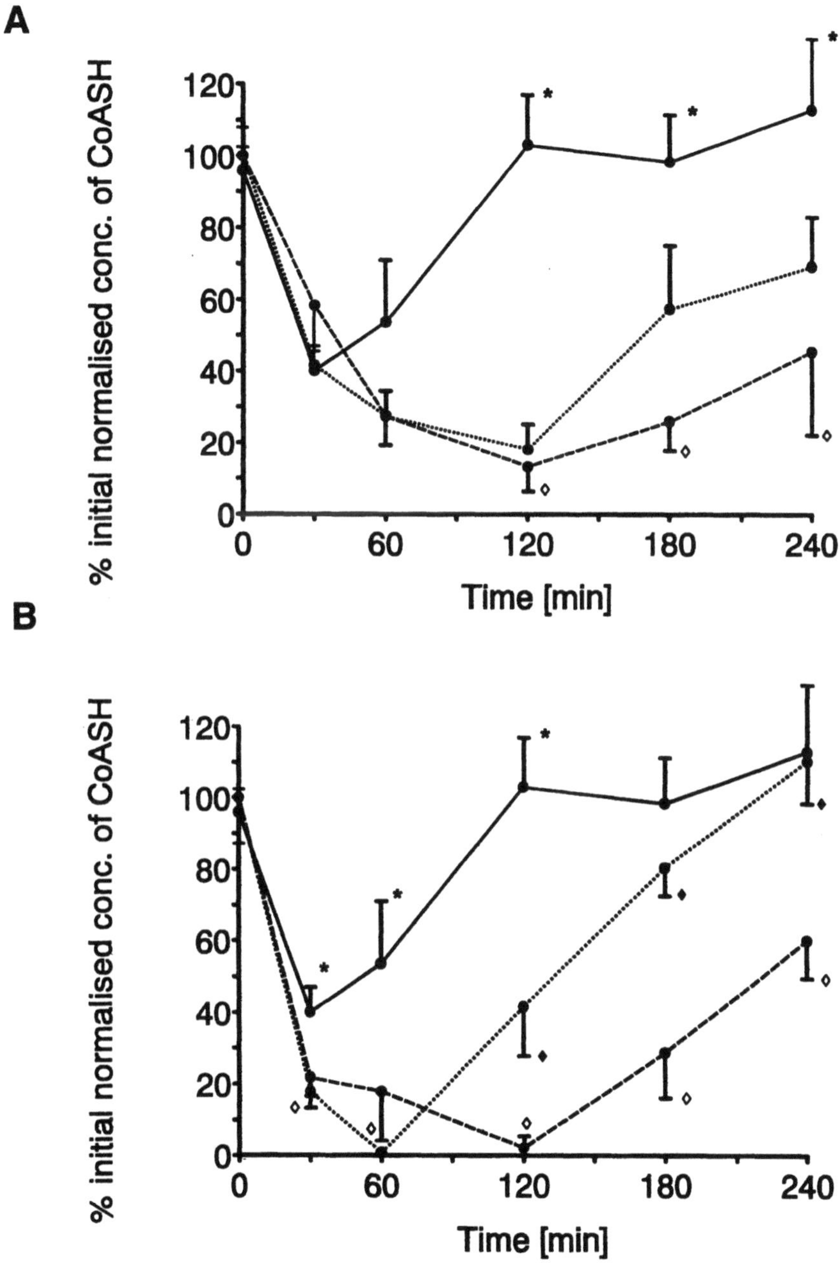

Fig. 6. CoA levels in rat hepatocyte suspensions ($5x10^5$ cells/ml) containing 0.5% BSA during the incubation of (**A**): —— (*R*)-ibuprofen (24.2 μM), ……. palmitic acid (1.45 mM), --------- (*R*)-ibuprofen and palmitic acid. (**B**): —— (*R*)-ibuprofen (24.2 μM), ……. octanoic acid (1.45 mM), --------- (*R*)-ibuprofen and octanoic acid. Mean of three incubations per rat. Student *t*-test (*),(◊): $P < 0.05$.

concentration. A different result was observed *in vivo*. After administration of labeled palmitic acid to mice, β-oxidation assessed by exhalation of labelled CO_2 was inhibited by both enantiomers to nearly the same extent (Freneaux et al., 1990). Based on these results it was concluded that the inhibition of β-oxidation apparently results from two different effects, namely a non-stereoselective effect independent of CoA levels and of unknown mechanism, and a stereoselective effect dependent on CoA levels which can be explained by the sequestration of CoA by the *R*-enantiomer.

We investigated the effect of (*R*)-ibuprofen on the rate of elimination of palmitic acid in rat hepatocytes. To this end, we developped a HPLC method which allows a more direct assessment of the fate of palmitic acid by measuring its disappearance from the incubation medium. It was found that the half-life of palmitic acid increases with increasing (*R*)-ibuprofen concentrations. Since (*S*)-ibuprofen did not modify the elimination of palmitic acid, we conclude that the inhibition of palmitate metabolism is stereoselective for (*R*)-ibuprofen (Müller et al., 1992).

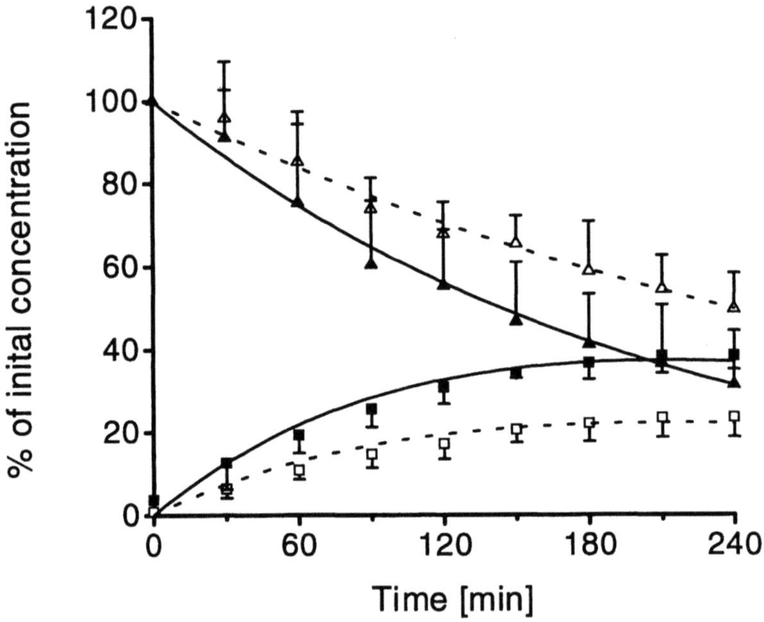

Fig. 7. Influence of clofibric acid in the incubation medium on the chiral inversion of (*R*)-ibuprofen in isolated rat hepatocytes. --------- incubation of (*R*)-ibuprofen (24.2 μM); ——— simultaneous incubation of (*R*)-ibuprofen (24.2 μM) and clofibric acid (96.8 μM). (-)-(*R*)-ibuprofen (Δ); (+)-(*S*)-ibuprofen (■).

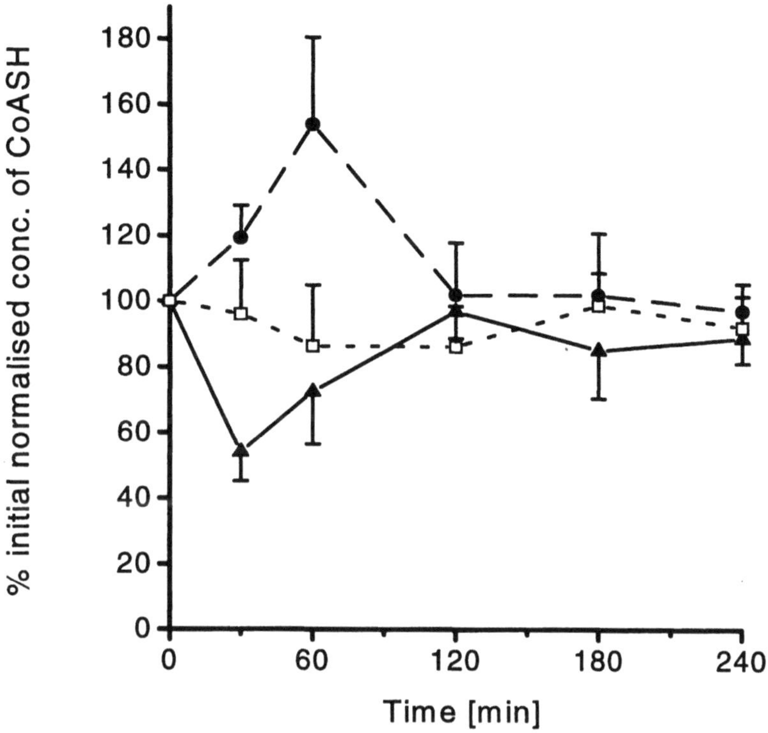

Fig. 8. CoA levels in rat hepatocyte suspensions (5x10^5 cells/ml) containing 0.5% BSA during the incubation of: —— (*R*)-ibuprofen (24.2 μM), — — clofibric acid (96.8 μM), ……. (*R*)-ibuprofen and clofibric acid. Mean of three incubations per rat.

Formation of 'hybrid' triglycerides

Acyl-CoA thioesters are a pivotal point in the metabolism of xenobiotic acids (Fig. 1). In particular, xenobiotic acids activated by coenzyme A can enter the routes of lipid biochemistry. Using glycerol and palmitate as precursors, Fears et al. (1978) showed that ibuprofen, ketoprofen and fenoprofen are incorporated into 'hybrid' triglycerides by rat liver slices. Of particular relevance is the stereoselectivity of profen incorporation into lipids. Following chronic treatment of rats with (*S*)-ibuprofen, only negligible amounts of the drug were incorporated into adipose tissues whereas after the administration of (*R*)-ibuprofen both enantiomers were detected at significant levels in lipids (Williams et al., 1986). The synthesis of "hybrid" triglycerides from (*R*)-fenoprofen in isolated hepatocytes and adipocytes was shown to follow Michaelis-Menten kinetics, with (*S*)-fenoprofen reacting very slowly (Sallustio et al., 1988). Ibuprofen accumulates as hybrid triglycerides in adipose tissue, since it is very slowly eliminated (the half-life in adipose tissue >100 hours) (Adams et al., 1969). It has been suggested that the elimination of ibuprofen from adipose tissues is stereoselective, favoring removal of (*S*)-ibuprofen (Williams et al., 1986).

Hypolipidemic activity

Various 2-arylpropionates have been shown to decrease serum triacylglycerol levels *in vivo* in rats (Fears et al., 1981). Sallustio et al. (1990) demonstrated the stereoselectivity of this interaction since only the *R*-enantiomer of fenoprofen inhibits endogenous triacylglycerol synthesis in adipocytes and hepatocytes. The hypolipidemic activity appears to be correlated with the capacity of profens to form hybrid triglycerides. However, it is probably not the production of a hybrid triglycerid as such that interferes with lipid synthesis, but rather the accumulation of the activated CoA intermediate.

The hypolipidemic action of these profens is probably due to the inhibition of acetyl-CoA carboxylase, the rate-limiting enzyme in fatty acid biosynthesis. Kemal and Casida (1992) investigated the inhibition of acetyl-CoA carboxylase by aryloxyphenoxypropionate herbicides (diclofop, haloxyfop and fluazifop), ibuprofen, fenoprofen and their synthetic CoA thioesters. They showed that the inhibitory effect resides essentially in the CoA thioesters, the parent compounds being far less active (Kemal and Casida, 1992; Smith and Stern, 1983). The inhibition of acetyl-CoA carboxylase appeared to be stereoselective for the *R*-enantiomer of diclofopyl-CoA. (*S*)-Ibuprofenoyl-CoA, chemically synthesized, has the same absolute configuration as (*R*)-diclofopyl-CoA and was found to be a better inhibitor of acetyl-CoA carboxylase than the *R*-enantiomer. It must be noted that (*S*)-ibuprofenoyl-CoA is only formed after administration of (*R*)-ibuprofen. The hypolipidemic effect of ibuprofen is thus absent when the pure *S*-enantiomer is administrated.

Conclusion

The published and as yet unpublished results briefly reviewed here show conclusively that xenobiotic acids may enter physiological pathways of lipid biochemistry. In most if not all cases, the first step in these effects is the formation of an acyl-CoA conjugate leading to "hybrid" triglycerides and inhibition of lipid β-oxidation. "Hybrid" triglycerides have the potential to form long-lasting residues in adipose tissues and to be incorporated into membranes.

Because acyl-CoA conjugates of profens are also the intermediates in inversion of inactive (*R*)-profens to active (*S*)-profens (Caldwell, 1984; Caldwell and Marsh,

1983), they can be viewed as a pivotal point leading to bioactivation and to pathways of potential toxication. Such considerations should be taken into account when debating the relative merits of racemic versus eutomeric drugs (Testa and Trager, 1990).

References

Abas A, Meffin PJ (1987). Disposition of 2-arylpropionic acid nonsteroidal anti-inflammatory drugs. IV Ketoprofen disposition. J Pharmacol Exp Ther 240:637-641

Bopp RJ, Nash JF, Ridolfo AS, Shepard ER (1979). Stereoselective inversion of (R)-(-)-benoxaprofen to the (S)-(+)-enantiomer in humans. Drug Metab Disposit 7:356-359

Buttinoni A, Ferrari M, Colombo M, Cesarini R (1983). Biological activity of indoprofen and its optical isomers. J Pharm Pharmacol 35:603-604

Caldwell J (1984). Xenobiotic acyl-coenzyme A: critical intermediates in the biochemical pharmacology and toxicology of carboxylic acids. Biochem Soc Trans. 12:9-11

Caldwell J, Marsh MV (1983). Interrelationships between xenobiotic metabolism and lipid biosynthesis, Biochem Pharmacol 32:1667-1672

Cox JW , Cox SR, van Giessen G, Ruwart MJ (1985). Ibuprofen stereoisomer hepatic clearance and distribution in normal and fatty *in situ* perfused rat liver. J Pharmacol Exp Ther 232:636-643

Fears R, Richards DH (1981). Association between lipid-lowering activity of aryl-substituted carboxylic acids and formation of substituted glycerolipids in rats. Biochem Soc Trans 9:572-573

Fears R, Baggaley KH, Alexander R, Morgan B, Hindles RM (1978). The participation of ethyl 4-benzoyloxybenzoate (BRL 10894) and other aryl-substituted acids in glycerolipid metabolism. J Lip Res 19:3-11

Foster RT, Jamali F (1987). Stereoselective pharmacokinetics of ketoprofen in rat: influence of route of administration. Pharm Res 4:s-117

Fournel S, Caldwell J (1986). The metabolic chiral inversion of 2-phenyl propionic acid in rat, mouse and rabbit. Biochem Pharmacol 35:4153-4159

Freneaux E, Fromenty B, Berson A, Labbe G, Degott C, Letteron P, Larrey D, Passayre D (1990). Stereoselective and nonstereoselective effects of ibuprofen enantiomers on mitochondrial β-oxidation of fatty acids. J Pharmacol Exp Ther 255:529-535

Geißlinger G (1993). Reiner Enantiomer von 2-Arylpropionsäuren: Ein therapeutischer Fortschritt?. Habilitationsschrift, Universität Erlangen-Nürnberg

Geneve J, Hayat-Bonan B, Labbe G, Degott C, Letteron P, Freneaux E, Le Dinh T, Larry D, Pessayre D (1987). Inhibition of mitochondrial β-oxidation of fatty acids by pirprofen. Role in microvesicular steatosis due to this nonsteroidal anti-inflammatory drug. J Pharmacol Exp Ther 242:1133-1137

Hutt AJ, Caldwell J (1983). The metabolic chiral inversion of 2-arylpropionic acids - a nouvel route with pharmacological consequences. J Pharm Pharmacol 35:693-704

Jamali F (1988). Pharmacokinetics of enantiomers of chiral non-steroidal anti-inflammatory drugs. Eur J Drug Metab Pharmackin 13:1-9

Jamali F, Berry BW, Tehrani MR, Russell AS (1988). Stereoselective pharmacokinetics of flurbiprofen in humans and rats. J Pharm Sci 77:666-669

Kemmerer JM, Rubio FA, McClain RM, Koechlin BA (1979). Stereospecific assay and stereospecific disposition of racemic carprofen in rats. J Pharm Sci 68:1274-1280

Knights KM, Drew R, Meffin PJ (1988). Enantiospecific formation of fenoprofen coenzyme A thioester in vitro. Biochem. Pharmacol. 37:3539-3542

Knights KM, Addinall TF, Roberts BJ (1991). Enhanced chiral inversion of *R*-ibuprofen in livers from rats treated with clofibric acid. Biochem Pharmacol 41:1775-1777

Lee EJD, Williams K, Day R, Graham G, Champion D (1985). Stereoselective disposition of ibuprofen enantiomers in man. Br J Clin Pharmacol 19:669-674

Lombard A, Rossetti V, Buffa M, Gabriel L, Miglietta (1988). In vitro studies on the stereoselective inversion of R(-) to S(+)-ketoprofen. IRSC Med. Sci. 13:1025

Mayer JM, Stereoselective metabolism of anti-inflammatory 2-arylpropionates (1990). Acta Pharm Nord 2:197-216

Mayer JM, Bartolucci C, Maître JM, Testa B (1988). Metabolic chiral inversion of anti-inflammatory 2-arylpropionates: lack of reaction in liver homogenates, and study of methine proton acidity. Xenobiotica 18:533-543

Müller S, Mayer JM, Etter JC, Testa B (1990). Metabolic chiral inversion of ibuprofen in isolated rat hepatocytes. Chirality 2:74-78

Müller S, Mayer JM, Etter JC, Testa B (1992). Influence of palmitate and benzoate on the unidirectional chiral inversion of ibuprofen in isolated rat hepatocytes. Biochem Pharmacol 44:1468-1470

Nakamura Y, Yamaguchi T (1987). Stereoselective metabolism of 2-phenylpropionic acid in rat. I. In vitro studies on the stereoselective isomerization and glucuronidation of 2-phenylpropionic acid. Drug Metab Disposit 15:529-534

Nakamura Y, Yamaguchi T, Takahashi S, Hashimoto S, Iwatani K, Nakagawa Y (1981). Optical isomerisation mechanism of R(-)-hydratropic acid derivatives. J Pharmacobio Dyn 4:s-1

Pedrazzini S, De Angelis M, Zanoboni MW, Furgione A (1988). Stereochemical pharmacokinetics of the 2-arylpropionic acid non-steroidal antiinflammatory drug flunoxaprofen in rats and in man. Arzneim Forsch 38:1170-1175

Rubin A, Knadler MP, Ho PPK, Bechtol LD, Wolen RL (1985). Stereoselective inversion of R-fenoprofen to S-fenoprofen in humans. J Pharm Sci 74:82-84

Sallustio BC, Meffin PJ, Knights KM (1988). The stereospecific incorporation of fenoprofen into rat hepatocyte and adipocyte triacylglycerols. Biochem Pharmacol 37:1919-1923

Sallustio BC, Knights KM, Meffin PJ (1990). The stereoselective inhibition of endogeneous triacylglycerol synthesis by fenoprofen in rat isolated adipocytes and hepatocytes. Biochem Pharmacol 40:1414-1417

Sanins SM, Adams WJ, Kaiser DG, Halstead GW, Baillie TA (1990). Studies on the metabolism and chiral inversion of ibuprofen in isolated rat hepatocytes, Drug Metab Disposit 18:527-533

Shen TY (1981). Nonsteroidal anti-inflammatory agents. In: ME Wolff (ed) Burger's Medical Chemistry 4th edn. Wiley, New York, Part III, p 1205-1261

Shieh WR, Chen CS(1993).. Purification and characterization of novel "2-aryl-propionyl-CoA" epimerases from rat liver cytosol and mitochondria. J Biol Chem 268:3487

Simmonds RG, Woodage TJ, Duff SM, Green JN (1980). Stereospecific inversion of (R)-(-)-benoxaprofen in rat and man. Eur J Metab Pharmacokin 5:169-172

Testa B (1986). Chiral aspects of drug metabolism. Trends Pharmacol. Sci. 7:60-64

Testa B, Mayer JM (1988). Stereoselective drug metabolism and its significance in drug research. In: Jucker E (ed) Progress in Drug Research, Vol. 32. Birkhäuser, Basel, p 249-303

Testa B, Trager WF (1990). Racemates versus enantiomers in drug development: Dogmatism or pragmatism? Chirality 2:129-133

Wechter WJ, Loughhead DG, Reischer RJ, van Giessen GJ, Kaiser DG (1974). Enzymatic inversion at saturated carbon: nature and mechanism of the inversion of R(-) p-iso-butylhydratropic acid. Biochem Biophys Res Comm 61:833-837

Williams K, Day R, Knihinicki R, Duffield A (1986). The stereoselective uptake of ibuprofen enantiomers into adipose tissue. Biochem Pharmacol 35:3403-3405

Side effects and toxic reactions of chiral drugs: A clinical perspective

M. Eichelbaum
Dr. Margarete Fischer-Bosch-Institut für Klinische Pharmakologie,
Auerbachstr. 112, 70376 Stuttgart

A survey of 1850 drugs regularly prescribed reveal that 1045 are chiral. 517 of the 1045 chiral drugs are derived from natural sources or are modifications of naturally occurring compounds. Due to the stereospecificity of biological synthesis 509 (99 %) of these naturally occurring drugs are obtained in a stereochemically pure form which is the biological active isomer. By contrast, 467 (88 %) of the 528 drugs with a chiral center produced by chemical synthesis are used as racemates since chemical synthesis usually leads to racemic compounds unless stereospecific synthesis is employed, and only 61 are stereochemically pure drugs (Ariens 1990, Kleeman und Engel 1982, Bailey 1986). Chemically, and to an even greater extent biologically, a racemic drug is not a single compound, but a 50:50 mixture of 2 stereoisomers. In view of the widespread therapeutic use of racemic drugs and the pronounced differences that can exist both in the activity and the disposition of the enantiomers contained in racemates, interest in the stereochemical aspects of action, disposition, side effects and toxicity of racemic drugs and the implications for drug use has grown in recent years. No generalizations can be made concerning whether and to what extent the activity, in either qualitative or quantitative terms, differs between stereoisomers. It is not unusual for the stereoisomer of a drug to have a high degree of stereoselectivity for one action, but no stereoselectivity for another action. The more active enantiomer is designated as eutomer, whereas the enantiomer with the lesser pharmacological activity is the distomer (Lehmann et al. 1976). The situation that both enantiomers have the same therapeutic potency, side effects and toxicity is rather seldomly encountered.

For most racemic drugs there are differences in the activity between enantiomers both in quantitative or qualitative terms, where

i) the enantiomers have the same pharmacological activity but differ in potency,
ii) the activity is mainly associated with one enantiomer, and
iii) both enantiomers are active but have qualitatively different actions, e.g. the stereoisomers are agonists or antagonists at different receptors.

Whereas differences in activity, side effects and toxicity between stereoisomers have been known for many years, only recently has the importance of stereoselectivity in drug disposition been realized and the possible consequences of the phenomenon been appreciated. Since all the processes associated with the absorption, protein binding, tissue binding, metabolism, renal and biliary excretion of racemic drugs involve an interaction between chiral drug molecules and chiral biological macromolecules drug dispostion can be stereoselective. Differences in protein binding, tissue binding and renal clearance of enantiomers is usually modest. It is the biotransformation by the drug metabolizing enzyme which exhibits the greatest degree of stereoselectivity of all the processes contributing to drug disposition. In general there is a two to five-fold difference in metabolic clearance between stereoisomers, but for some drugs greater differences in clearance are observed. As a consequence of stereoselective drug disposition the concentrations of the stereoisomers in the organism will be different from the 50:50 ratio in the drug preparation. If there are differences in potency depending on whether the active or less active enantiomer is preferentially affected, there may be amplification or attenuation of in vivo as compared to in vitro drug potency. For some racemic drugs chiral inversion has been reported which is the enzyme-catalyzed metabolic conversion of an enantiomer into the opposite enantiomer. If the less active enantiomer is converted into the more active enantiomer, this leads to drastic attenuation of in vivo drug potency.

In this context another aspect of chirality deserves attention. If a chiral drug is metabolized to an active chiral metabolite where differences in potency exist between the enantiomers of the parent drug similar differences in potency can exist between the enantiomers of the active metabolites. Therefore the contribution of a metabolite to the pharmacological and toxicological effect will depend on the stereochemical configuration of the metabolite formed, since there can be preferential formation of a metabolite from one enantiomer. Data on the contribution of active chiral metabolites to therapeutic or toxic effects are often not meaningful since they have been obtained with the racemate of metabolite not taking into account that differences can exist in potency for the enantiomers. In contrast to drug side effects which are an extension of the desired pharmacological action due to high drug concentrations, drug toxicity such as rash or hepatotoxicity is usually not related to the dose/concentration and is unpredictable. This type of toxicity is often not mediated by the parent drug but rather by biologically reactive intermediates which are formed during metabolism. Since metabolism of enantiomers exhibits selectivity with regard to rate and route of biotransformation, only one enantiomer may form the reactive intermediate and be responsible for this type of toxicity.

In view of the differences in activity, disposition, side effects and toxicity which can exist between enantiomers, two drugs are in fact being administered when a racemic drug is prescribed. Recognition of this aspect has stimulated a lively debate as to whether the racemic drugs in therapeutic use should still be used as racemates or whether such drugs should be replaced by drug preparations containing only the active stereoisomers. Several arguments favoring the use of the pure enantiomer had been put forward. There can be no doubt that the use of stereochemically pure drugs will result in less complex pharmacokinetics, less complex concentration-effect relationship, less complex drug interaction, a decreased burden of xenobiotic material, and in some instances there will be an improved pharmacological profile. However, in the clinical context one is not so much concerned with the differences that might exist in the activity and the pharmacokinetics of enantiomers contained in racemic drug preparations, but rather with the benefits to be gained in terms of therapeutic efficacy and safety if a stereochemically pure drug instead of a racemic drug were to be used. In evaluating the benefits to develop a stereochemically pure drug from an existing racemate one should apply the same criteria as in the case of drug combinations, since racemates are a fixed 50:50 combination of two drugs. In the case of a drug combination both components should contribute to the therapeutic efficacy. If we have a drug combination where one drug of such a combination is mainly responsible for the side effects and toxicity and not contributing to the therapeutic efficacy, such a combination would not be approved. If we apply these criteria to racemic drugs presently in the therapeutic use, there are only a few convincing examples for which this scenario is valid, e.g. one enantiomer is active, the other enantiomer therapeutically inactive, but contributes mainly side effects and toxicity. The decision whether to use a single enantiomer or the racemate should be made on a case by case basis, and various scenarios can be envisaged concerning the relative merits of developing stereochemically pure drugs or continuing to use the racemate.

1. The stereoisomers produce similar pharmacological effects, however, one stereoisomer is more potent than the other in producing the desired pharmacological effect. Therefore only the more potent stereoisomer is the main active therapeutic moiety. The other stereoisomer does not contribute significantly to the therapeutic efficacy of the drug. However, for other pharmacological effects which contribute to therapeutic efficacy of the drug for different indications there may be less or no difference in potency between enantiomers. Hence, it is doubtful that the use of the enantiomer will improve the therapeutic index and decrease the incidence of side effects.

   A high degree of stereoselectivity is observed for the ß-adrenoceptor antagonism. The eutomer being at least 10 to 100 times more potent than the distomer (Walle *et al.*, 1988). It has, therefore, been suggested that for indications where β-blocking activity is desired, enantiomerically pure dosage forms containing only the S-enantiomer should be used. The therapeutic effect as well as most of the adverse reactions to these compounds are due to their antagonism of norepinephrine at β-adrenoceptors and thus it is doubtful that removal of the R-enantiomer, which makes little contribution to these pharmacological effects, would reduce the incidence and severity of side effects while treating coronary heart disease and hypertension. Indeed the profile and incidence of adverse drug reactions to timolol, used as a pure S- enantiomer, are similar to those of racemic β-adrenergic antagonists (Reidenberg, 1988). In contrast to β-adrenergic receptor antagonism, other pharmacological effects elicited by these compounds do not exhibit stereoselectivity. For example R- and S-propranolol are equipotent in their membrane stabilising effects (Jaillon *et al.*, 1980), efficacy in migraine (Stensrud *et al.*, 1976) and inhibition of thyroxine metabolism (Heyma *et al.*, 1980). The R-enantiomers may thus be useful for indications independent of β-adrenergic blockade where these actions may be unwanted side effects. S-timolol applied topically to the eye not only lowers intraocular pressure but also elicits systemic ß-blockade and therefore should not be used in those patients for whom ß-adrenergic antagonists are contraindicated (112). However, lack of knowledge of this fact has caused a substantial number of severe adverse cardiovascular, pulmonary and CNS drug reactions which, according to a recent survey of 450 case reports, has resulted in the death of at least 32 patients (113). Since R-timolol has 1/4th the activity of S-timolol with regard to lowering ocular pressure but only 1/50th to 1/90th the activity of S-timolol with regard to systemic ß-blockade, R-timolol may be expected to be superior to S-timolol for the treatment of wide angle glaucoma (114). Therefore, by using R- instead of S-timolol it may be postulated that serious adverse effects could be avoided and the drug might even be able to be used in patients with wide angle glaucoma and concomitant diseases which contraindicate the use of S-timolol (115). However, in a recent study in patients with asthma, the doses of R-timolol required to lower intraocular pressure to the same extent as S-timolol caused an increase in airway resistance (116). Thus, in patients with asthma R-timolol may not be superior to S-timolol for the treatment of wide angle glaucoma as had been predicted on the basis of data obtained in various animal models and normal subjects whether R-timolol is in fact superior to S-timolol in terms of cardiovascular and CNS side effects is not yet known (Nelson *et al.*, 1986).

2. Both stereoisomers are pharmacologically active, but have qualitatively different actions, e.g. they are agonists or antagonists at different receptors. Administration of a racemate will therefore be analogous to giving a combination of two different drugs. Depending on the pharmacological profile the racemate or enantiomer is superior.

In the case dobutamine we have the situation where both enantiomers are active having qualitative different action and hence a combination of two drugs is given. Both enantiomers contribute to the therapeutic effect and the racemate is superior in therapeutic efficacy to the administration of a single enantiomer.

The enantiomers of dobutamine are agonists at different receptors. (+)-dobutamine has β_1- and β_2-adrenoceptor agonist activities which result in peripheral vasodilation in addition to positive inotropic and chronotropic effects. (-)-dobutamine, however, has α_1-adrenoceptor agonist activity which produces a positive inotropic effect with concomitant vasoconstriction (Ruffolo *et al.*, 1985). Racemic dobutamine thus stimulates both the α_1- and β_1-adrenoceptors in the myocardium, resulting in a greater inotropic response than after administration of the individual enantiomers. When the racemate is administered this therapeutic effect is accompanied by little net change in blood pressure, since the α-adrenoceptor mediated vasoconstriction of the (-)-enantiomer is counteracted by the β_2-adrenoceptor mediated vasodilation caused by the (+)-enantiomer.

3. One stereoisomer has less of the therapeutic effect than the other but has similar or greater side effects, resulting in the less potent isomer contributing little to the desired therapeutic effect while making a disproportionate contribution to the undesirable effects. Therefore the use of enantiomers will improve the therapeutic index.

There are several interesting examples among these drugs where one enantiomer contributes little to the therapeutic effect but is mainly responsible for the undesirable effects.

The intravenous anaesthetic ketamine is administered as a racemic mixture and its use has been associated with distressful emergence reactions. Equianaesthetic doses of S-ketamine (1 mg/kg), R-ketamine (3 mg/kg) and racemic ketamine (2 mg/kg) have been compared in patients undergoing elective surgery. Both quantitative and qualitative differences between the enantiomers were noted. S-Ketamine produced less agitated behaviour, less perioperative pain (White *et al.*, 1980), more rapid recovery of psychomotor skills (White *et al.*, 1985) and was more acceptable to patients than R- or racemic- ketamine. R-ketamine did not produce EEG slowing comparable to that of S-ketamine or the racemic mixture, and thus it would be more difficult to achieve an adequate depth of anaesthesia with R-ketamine (White *et al.*, 1985). Administration of an enantiomerically pure dosage form containing S-ketamine may, therefore, improve the margin of safety and patient acceptance of ketamine anaesthesia.

4. The therapeutic activity resides exclusively with one enantiomer, the other enantiomer, which does not contribute to therapeutic efficacy, is responsible for the toxicity. In such a situation the use of the enantiomer is mandatory. L-Dopa and D-penicillamine are drugs which were originally introduced as racemates. Later on the racemate was replaced by an enantiomerically pure drug preparation because of severe adverse drug reaction caused by the enantiomer which did not contribute to therapeutic efficacy. Thus by removing the distomer from the drug preparations toxicity could be avoided or drastically reduced.

D-penicillamine is used in the treatment of Wilson's disease, cystinuria and rheumatoid arthritis. The severe adverse reaction such as optic neuritis are caused mainly by the L-enantiomer and therefore only the D-enantiomer is used. As the mechanism underlying the toxicity its is assumed that L-penicillamine is incorporated into proteins. D-penicillamine could not undergo such a reaction because proteins are made up of only L-aminoacids (Kean et al., 1991).

Recognition that Parkinsons disease is associated with diminished brain L-dopamine concentrations resulted in the trial of oral dopa for the treatment of Parkinsons disease. It was hypothesised that *in vivo* L-dopa would be metabolised to L-dopamine and thus be active. However, as costs associated with production of the pure enantiomer L-dopa were prohibitive, initial studies in humans were conducted using a racemic mixture of D- and L- dopa. Dopa was efficacous, however, the incidence of side-effects, in particular granulocytopenia, was unacceptably high (Cotzias *et al.*, 1967). By administering L-dopa alone efficacy was main tained while the incidence of side effects decreased (Cotzias *et al.*, 1969).

5. It is very important to consider the optical purity of the isomers used in assessing stereoselectivity in drug action and side effects. In many studies these data are not provided. If the optical purity of a less active enantiomer is 99.5%, the presence of 0.5% of the active enantiomer that is 1000 times more potent than the distomer will result in activity which could be attributable to the less active enantiomer, but which actually should be attributed to the small quantity of the active enantiomer present. This oversight has resulted in the use of the distomer with deleterious consequences for the patients.

The critical importance of the optical purity for saftey and efficacy of a enantiomerically pure drug can be illustrated using D-thyroxine. D-thyroxine was used in the late sixties to treat patients with hyperlipidemia. L-thyroxine which is the natural occuring stereoisomer of this thyroid hormone is at least 15-40 times more potent. Compared to L-thyroxine D-thyroxine induces a greater decrease in serum cholesterol per increase in oxygen consumption and heart rate indicating that D-thyroxine had a preferential effect on cholesterol metabolism. However a double blind trial of d-thyroxine (6 mg) in patients with chronic heart disease and previous myocardial infarction was halted prematurely as the mortality associated with D-thyroxine was greater than that of the placebo group (Coronary Drug Research Group, 1972). It had been assumed that the D-thyroxine tablets used contained negligible amounts of L-thyroxine. Later on it was reported that available pharmaceutic preparation of D-thyroxine contained between 0.5 - 2.3 % L-thyroxine. This seems to be a small contamination. However, at D-thyroxine dose of 8 mg/day administered this corresponds to a L-thyroxine dose ranging from 40 to 180 μg/day. Treatment of athyrotic patients with these D-thyroxine tablets in doses of 8 mg/day resulted in physiological L-thyroxine levels and L-thyroxine levels were proportional to the degree of L-thyroxine contamination of the two lots of D-thyroxine tablets. Therefore it is quite likely that all the reported D-thyroxine action and side effects are due to L-thyroxine contamination (Young et al. 1984).

Conclusions

The very existence of stereoselectivity in drug action and disposition does not necessarily mean that in the future only stereochemically pure compounds instead of racemates should be used. For old drugs and compounds being introduced in the near future it should be considered on a case by case basis whether a single enantiomer may offer a therapeutic advantage. If there are indications that a single enantiomer may offer an advantage in terms of efficacy or safety, then the development of an enantiomerically pure drug should be considered. For new drugs it should be established that the racemate is superior in terms of safety and efficacy to the eutomer before use of the racemate is approved.

References

Ariens EJ. Racemic Therapeutics - Problems all along the line. In: Chirality in Drug Design and Synthesis, Academic Press Ltd. London 1990; 29-43.

Bailey DM (ed). Annual Reports in Medicinal Chemistry, Academic Press London. (1983-1986); 19-21.

Cotzias GC, Van Woert MH, Schiffer LM. Aromatic amino acids and modification of Parkinsonism. N Eng J Med 1967; 276: 374-379.

Cotzias GC, Papavasiliou PS, Gellene R. Modification of Parkinsonism. Chronic treatment with L-dopa New Enl J Med 1969; 280: 337-345.

Heyma P, Larkins RG, Higgonbotham L, Ng KW. D-Propranolol and DL-propranolol both decrease conversion of L-thyroxine to L-triiodothyronine. Br Med J 1980; 24-25.

Jaillon P, Heckle J, Weissenburger J, Cheymol G. Cardiac electrophysiologic properties of dl-propranolol, d-propranolol, l-propranolol and dl-pindolol in anesthetized dogs. J Pharmacol Exp Ther 1980; 212: 347-353.

Kean WF, Lock CJL, Howard-Lock HE. Chirality in antirheumatic drugs. The Lancet 1991; 338: 1565-1568.

Kleeman A. and Engel J. Pharmzeutische Wirkstoffe. Thieme Verlag. Stuttgart 1982.

Lehmann FPA, Rodrigues De Miranda JF and Ariens EJ. Stereoselectivity and affinity in molecular pharmacology. In Progress Research (ed.) E. Jucker, Birkhauser Verlag, Basel, 1976; 20: 101-142.

Nelson WL, Fraunfelder FT, Sills JM, Arrowsmith JB, Kuritsky JN. Adverse respiratory and cardiovascular events attributed to timolol ophthalmic solution 1978-1985. Am J Ophthalmol 1986; 102: 606-611.

Reidenberg M. The therapeutic use of stereochemically pure drugs - a pragmatic point of view. In Wainer IW, Drayer DE ed. Drug Stereochemistry. New York: Marcel Dekker Inc, 1988; 365-369.

Ruffolo RR, Messick K. Effects of dopamine, (±)- dobutamine and the (+)- and (-)-enantiomers of dobutamine on cardiac function in pithed rats. J Pharmacol Exp Ther 1985; 235: 558-565.

Stensrud P, Sjaastad O. Short-term clinical trial of propranolol in racemic form (Inderal), d-propranolol and placebo in migraine. Acta Neurol Scandinav 1976; 53: 229-232.

Walle T, Webb JG, Bagwell EE, Walle UK, Daniell HB, Gaffney TE. Stereoselective delivery and actions of β-receptor antagonists. Biochem Pharmacol 1988; 37: 115-124.

White PF, Ham J, Way WL, Trevor AJ. Pharmacology of ketamine isomers in surgical patients. Anesthesiology 1980; 52: 231-239.

White PF, Schüttler J, Shafer A, Stanski DR, Horai Y, Trevor AJ. Comparative pharmacology of the ketamine isomers. Br J Anaesth 1985; 57: 197-203.

Young WF, Gorman CA, Jiang NS, Machacek D, Hay ID. L-Thyroxine contamination of pharmaceutical D-thyroxine: probable cause of therapeutic effect. Clin Pharmacol Ther 1984; 36: 781-787.

Analysis of (S)- and (R)-Propranolol in Human Plasma and Urine by a Specific Immunoenzymatic Assay *versus* HPLC

C. Pham-Huy [1], A. Sahui-Gnassi [1], F. Massicot [1], H. Galons [2], H. Dutertre-Catella [1], B. Radenen [1], J.-M. Warnet J[1], H.T.Duc [3] & J.R. Claude J[1]

[1]Laboratoire de Toxicologie (EA 207) & [2]Laboratoire de Chimie Organique, Faculté de Pharmacie, Université René-Descartes , 75270 Paris Cedex 06, [3]INSERM U-267, Laboratoire d'Immunogénétique, Hôpital P. Brousse, 94800 Villejuif, France

Propranolol, a racemic β-adrenergic blocking agent, is widely used in the treatment of cardiovascular diseases (hypertension, angina pectoris and cardiac arrythmia).

In spite of an identical structure, (S)- and (R)-propranolol enantiomers have different phamacological and pharmacokinetic properties : (S)-propranolol is about 100 times more potent than its (R)-antipode upon β-receptors (Barrett & Cullum, 1968) and has a longer plasma half-life (Kawashima *et al.*, 1976). Thus, the determination of propranolol enantiomers in biological fluids is of great interest for therapeutic monitoring as well as for pharmacokinetic and toxicological studies. There are two different approaches to the enantioselective determination of racemic propranolol : immunochemical assays and physicochemical assays. Most of the immunoassays cannot distinguish the two enantiomers, except the two methods using specific antibodies : one of them uses radioimmunoassay (Kawashima *et al.*, 1976) and the other, that we recently developed, consists in an enzyme-linked immunoassay (ELISA) (Sahui-Gnassi *et al.*, 1993). Most of the physicochemical assays are achieved by HPLC using chiral stationary phase or chiral derivatization.

The aim of this study was to compare the selective determination of these 2 enantiomers in human plasma and urine by an immunoenzymatic assay (ELISA) using specific antibodies and by an HPLC method after chiral derivatization.

Materials and Methods

Biological fluid sampling : 12 plasma and 12 urine samples collected at different times from 4 healthy volunteers given orally 40 or 80 mg of the racemic drug were comparatively analyzed by ELISA and HPLC for the enantioselective determination of (S)- and (R)-propranolol.

ELISA assay : A sequential competitive and sandwich technique was performed as previously described (Sahui-Gnassi *et al.*, 1993). Briefly, for the analysis of (S)- and (R)-propranolol, two antibodies were used : a specificic antibody to (S)-propranolol and an another antibody to (R,S)-propranolol. The levels of (R)-enantiomer in biological fluids were calculated by difference between those of the racemic and of (S)-propranolol. For the determination of (S)- or (R,S)-propranolol, diluted rabbit antiserum to (S)- or to (R,S)-propranolol in phosphate buffer was mixed with human plasma or urine samples or standard calibration solutions. After incubation, absorbed antiserum was tested with (S)- or (R,S)-conjugate previously fixed on the wells of an immunoplate. After a second incubation and washing, a diluted peroxidase antirabbit IgG conjugate solution was added. After addition of the substrate of peroxidase, the plates were read at 492 nm (Multiskan Plus spectrophotometer). The sample results were reported on a standard calibration curve set up under the same conditions.

HPLC analysis : After chiral derivatization with (R)-phenylethylisocyanate (PEIC), HPLC was performed as previously described (Pham-Huy *et al.*, 1994). Briefly, after extraction of human plasma or urine sample or standard calibration solution in presence of 4-methylpropranolol as internal standard (IS) and ammonium hydroxide, the residue was derivatized with PEIC. After evaporation, the derivatized compound was dissolved in the mobile phase (methanol/water, 72.5/27.5, v/v), then separated on a Novapak C18 column and detected with a fluorescence detector.

Results

The results of 12 plasma (S)- or (R,S)-propranolol concentrations determined by ELISA compared to those of the corresponding enantiomer or racemic levels obtained with HPLC are shown in Figures 1 and 2. For the determination of (S)- and (R,S)-propranolol in plasma, the correlation coefficients r were respectively 0.994 and 0.996 and the regression lines were respectively : $y = 1.03\ x + 1.06$ and $y = 1.06\ x + 1.18$ where y = ELISA and x =HPLC.

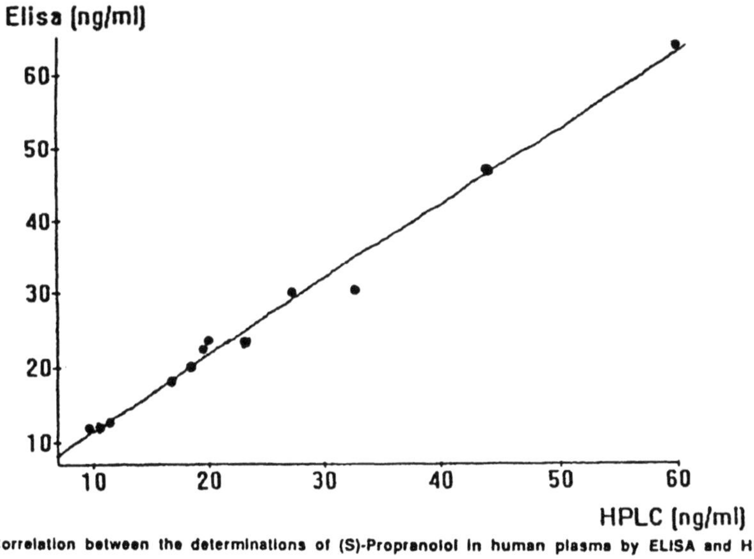

Fig 1. Correlation between the determinations of (S)-Propranolol in human plasma by ELISA and HPLC :

$$Y_{ELISA} = 1.03\ X_{HPLC} + 1.06$$
$$r = 0.994 \quad (n = 12)$$

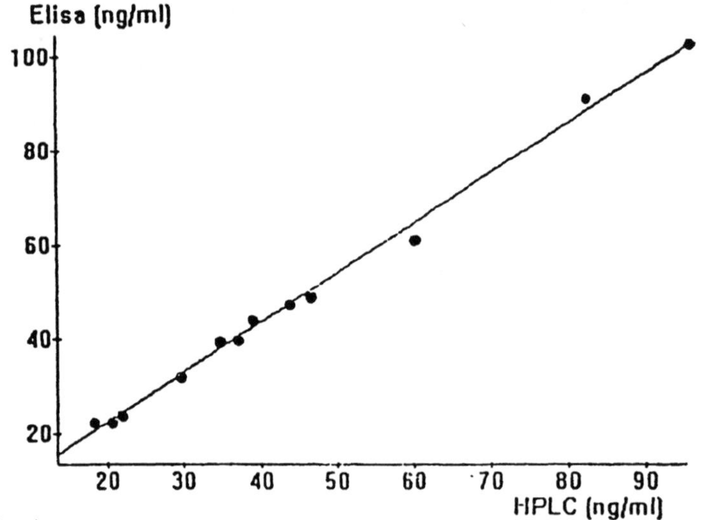

Fig 2. Correlation between the determinations of (RS)-Propranolol in human plasma by ELISA and HPLC :

YELISA = 1.06 xHPLC + 1.18
r = 0.996 (n = 12)

The results of 12 urine (S) or (R,S)-propranolol concentrations determined by ELISA compared to those of corresponding enantiomer levels obtained with HPLC are shown in Figures 3 and 4. In these cases, the statistical parameters were :
r= 0.993 and y = 0.98 x + 6.3 for (S)-propranolol
r= 0.990 and y = 0.99 x + 5.3 for (R,S)-propranolol where y= ELISA and x= HPLC.

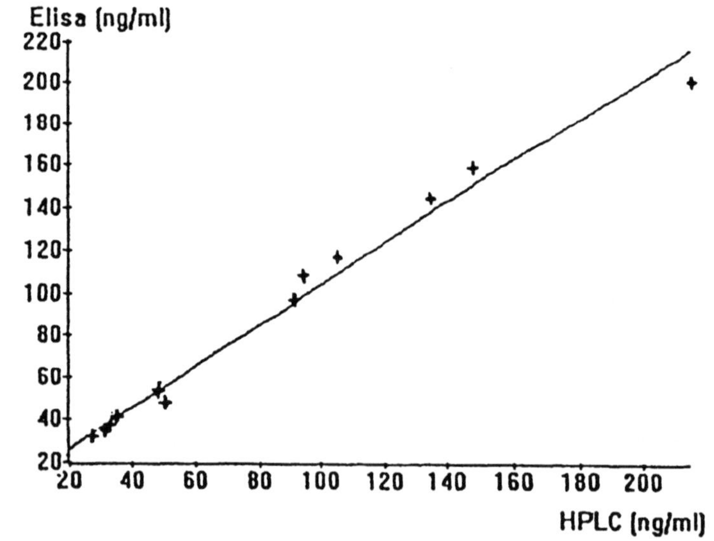

Fig 3. Correlation between the determinations of (S)-Propranolol in human urine by ELISA and HPLC :

YELISA = 0.98 xHPLC + 6.3
r = 0.993 (n = 12)

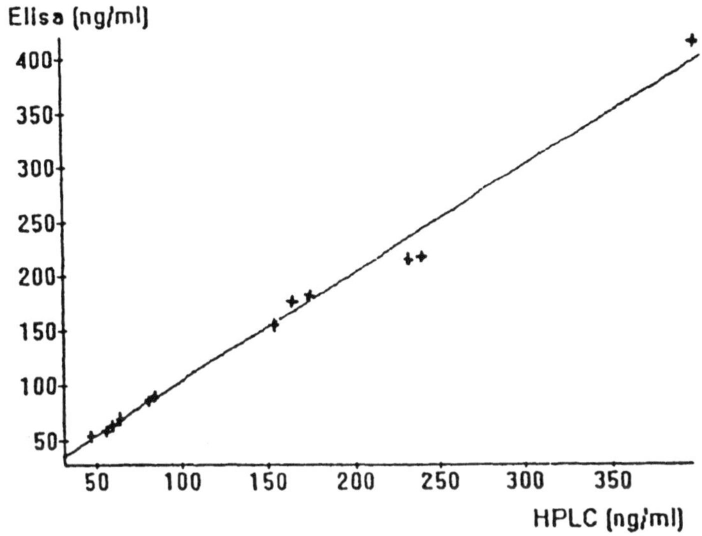

Fig 4. Correlation between the determinations of (RS)-Propranolol in human urine by ELISA and HPLC :

$$y_{ELISA} = 0.99\ x_{HPLC} + 5.3$$
$$r = 0.990 \quad (n = 12)$$

Figures 5 and 6 show the levels of (S)-and (R)-propranolol obtained by these 2 methods in the plasma and urine samples of one healthy volunteer collected at different times after oral absorption of 80 mg racemic propranolol.

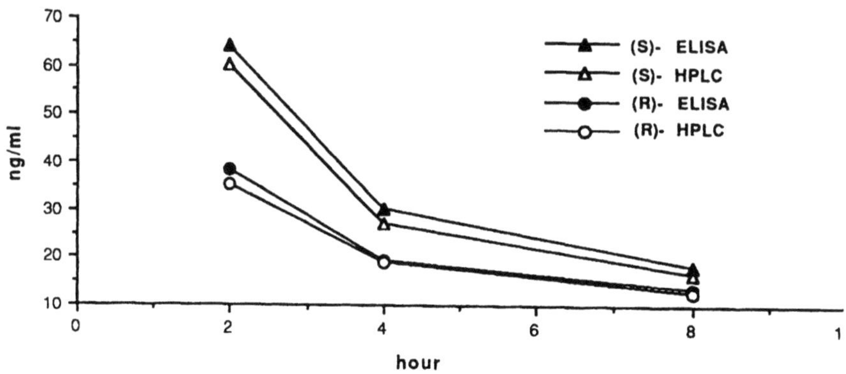

Fig. 5 : ELISA and HPLC determinations of (S)- and (R)-propranolol levels in plasma of one healthy volunteer collected at different times after oral administration of 80 mg racemic drug

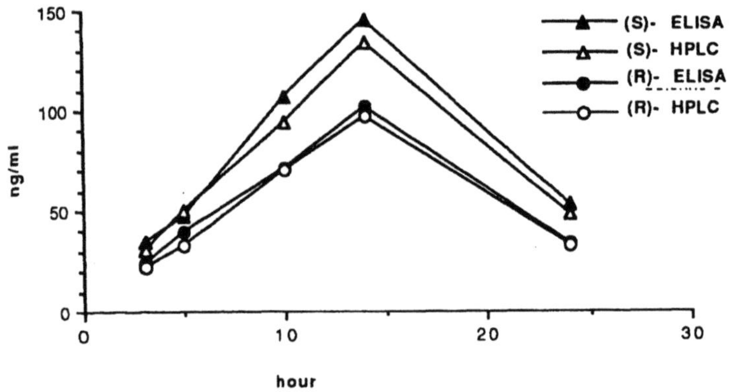

Fig. 6 : ELISA and HPLC determinations of (S)- and (R)-propranolol levels in urine of one healthy volunteer collected at different times after oral administration of 80 racemic drug

Discussion

The results show a good correlation between the two selective methods (r # 0.99) for plasma as well as for urine. However, the levels obtained with ELISA are slightly higher than those from HPLC (about 5-10%) (Figures 5 and 6). This is due to a weak interference about 5% of anti-(S)-propranolol antibody with its antipode (R). The advantages of the immunoenzymatic assay are its high sensitivity (0.7 ng/ml *versus* 1.3 ng/ml by HPLC), the lack of need for extraction and a possible automatization.

The preliminary pharmacokinetic study in 4 healthy volunteers showed a maximum concentration of both (S)- and (R)-propranolol at 2 h, then a parallel decrease at 4 and 8 h after oral administration. The levels of both enantiomers in plasma and urine were also dose-dependent. Moreover, both methods lead to assume that the two propranolol enantiomers have a different metabolism, as (S)-propranolol levels are always higher than those of the (R)-isomer both in plasma and urine. The (S)/(R) ratios observed in the 4 healthy volunteers varied between 1.1-1.5 for plasma and 1.2-1.7 for urine. These results are consistent with those previously reported (Walle, 1985, Lindner *et al.*, 1989).

Conclusion

This ELISA method using specific antibodies, validated against a reference HPLC method, allows the selective determination of the two propranolol enantiomers in biological fluids. It will thus be a suitable tool for pharmacokinetic and toxicological studies.

References

Barrett AM, Cullum VA (1968) The biological properties of the optical isomers of propranolol and their own cardiac arrhytmias. Br J Pharmacol 34: 43-55

Kawashima K, Levy A, Spector SJ (1976) Stereospecific radioimmunoassay for propranolol isomers. J Pharmacol Exp Ther 196: 517-523

Sahui-Gnassi A, Pham-Huy C, Galons H, Warnet JM, Claude JR, Duc HT (1993) Selective antibodies to propranolol enantiomers produced from a new conjugate. Chirality 15: 448-454

Pham-Huy C, Sahui-Gnassi A, Saada V, Gramond JP, Galons H., Ellouk-Achard S, Levresse V, Fompeydie D, Claude JR (1994) Microassay of propranolol enantiomers and conjugates in human plasma and urine by HPLC after chiral derivatization for pharmacokinetic study. J Pharm Biomed Anal. 12 : 1189-1198

Walle T (1985) Stereochemistry of the *in vivo* disposition and metabolism of propranolol in dog and man using deuterium-labeled pseudoracemates. Drug Metab Dispos 13: 279-282

Lindner W, Rath M, Stoschitzky K, Semmebrock HJ (1989) Pharmacokinetic data of propranolol enantiomers in a comparative study with (S)- and(R,S)-propranolol. Chirality 1: 10-13

Acknowledgements

The authors wish to thank Mrs M. RADIONOFF (Université René-Descartes Paris V) for preparation of the manuscript. Financial support was provided by A.D.R.T.E. (Association pour le Développement de la Recherche en Toxicologie Expérimentale), Paris.

References

Biological behavior of prestages of cancer cells

(Chair: Prof. Dr. R. Schulte-Hermann, Austria)

Biological Behavior of Prestages of Cancer Cells: An Introduction

R. Schulte-Hermann
Institut für Tumorbiologie-Krebsforschung, Borschkegasse 8a, A-1090 Vienna, Austria

Cancer formation occurs by different biological events including initiation, promotion and progression. This concept implies that malignant cells develop from normal ones via intermediate cell populations or prestages of cancer (Farber and Cameron 1980; Pitot and Sirica 1980; Schulte-Hermann 1985). Chemical compounds can interfere with these different biological events at different levels of the carcinogenic process. Such divergent interactions must be taken into account for risk assessment.

Cancer prestages are of considerable interest for toxicology. Some applications are shown in table 1.

Table 1. Relevance of cancer prestages in toxicology

* models to investigate mechanisms of tumor initiation and promotion
* discrimination between initiating and promoting properties of a carcinogenic chemical
* refinement of animal test for carcinogenicity:
  shorter duration
  less animals
  use of human tissues (exceptionally)

Initiation is triggered by genotoxic insults and, in most cases, occurs through mutation in the broad sense (i.e. at the DNA, chromosome or genome level). Initiation may also occur without known interaction of genotoxic noxes, namely „spontaneously". As a result compounds without initiating activity or potential may promote the development of spontaneous initiated lesions. This may be an important mechanism by which non-genotoxic carcinogens produce tumors (Schulte-Hermann 1985; Ward 1983; Kraupp-Grasl et al. 1991; Dietrich and Rasonyi 1995).

Initiation is a rare event and results in the formation of single initiated cells. These are the earliest prestages of cancer. In normal tissues they have little if any growth advantage over normal cells. Promotion stimulates their preferential or selective growth, giving rise to formation of clones of initiated cells. Growth stimulation or mitogenic activity seems to be a necessary but not sufficient condition for promoting activity of chemical agents. Thus, in rat liver different types of initiated cells appear to exist which are promotable specifically by different mitogenic compounds (non-genotoxic carcinogens), namely either by phenobarbital-type compounds or by peroxisome proliferators (Schulte-Hermann 1985; Kraupp-Grasl et al. 1990). Tumor promotion therefore seems to include induction of specific pleiotropic gene programs such as of enhanced drug metabolism and liver growth or of enhanced peroxisome function and liver growth.

With increasing size of clones of initiated cells the risk for progression to prestages more advanced towards malignancy increases. Finally malignant cells appear after an undefined number of intermediate stages.

Formation of prestages of cancer probably always precedes frank malignancy. In fact, prestages have been found in many organs, in animals and humans, during or before cancer appearance. However, the biological behavior, namely why they respond differently from normal cells to endogenous and exogenous tumor promoters and other signals, is not well understood. Molecular and cytogenetic studies have identified mutations associated with the development of prestages in some systems, e.g. colon cancer in humans, where 5 different steps have been found (Fearon and Vogelstein 1990). Such findings will help to understand and predict the biological behavior of prestages.

Rat liver offers one of the most intensely studied experimental models of stepwise cancer formation. Prestages of cancer have been discovered in this organ 40 years ago by Weiler. His pioneering work demonstrated for the first time the deficiency of certain antigens in „cell islands“ produced by administration of an aromatic amine (Weiler 1956). Today numerous markers have been found which are useful to identify putative prestages of cancer in rat liver (Huitfeld 1995). One of the most valuable markers is the expression of glutathione S-transferase P which may identify even single initiated cells (Satoh et al. 1989).

Foci markers reflect the altered phenotype of cancer prestages. They provide a key to understand the altered biological behavior of cancer prestages. Two different scenarios have been decribed connecting altered enzymic properties and preferential growth of liver cancer prestages. The resistent cell hypothesis predicts that cells in prestages are resistent to toxic damage due to enhanced expression of phase II drug metabolizing enzymes, mdr protein etc., which could provide a growth advantage during the action of toxic promoters (Solt et al. 1983). The over-response hypothesis suggests that prestages are over-sensitive to the pleiotropic effect of tumor promoting compounds. In this scenario prestages may preferentially respond to the mitogenic effects of (non-toxic) tumor promoters (Schulte-Hermann 1985; Schulte-Hermann et al. 1986). Similar prestages as in rat liver also occur in human liver where iron excluding cell foci were recently identified in patients with hemochromatosis (Deugnier et al. 1993). These foci may exemplify scenario 1 by their apparent resistance to cytotoxic effects of iron accumulation.

Another crucial characteristic of the behavior of cancer prestages relates to the regulation of cell replication and death. We and others have previously described that prestages in the liver show an enhanced basal mitotic activity and an excessive response to exogenous mitogenic signals (Schulte-Hermann et al. 1981). Later we found that prestages also exhibit enhanced apoptosis, a type of active cell death (Bursch et al. 1984). As a result of both enhanced cell birth and cell death rates liver cancer prestages show little net growth without promotion. Promoters were found to depress cell death thereby leading to accumulation of preneoplastic cells (Bursch et al. 1984; Schulte-Hermann et al. 1990). Apparently the tumor promoter acts as a survival factor for initiated cells.

In recent studies we have shown that cancer prestages in the liver are oversensitive to signals inducing apoptosis. Thus, food restriction to 60% for a period of 3 months eliminated 15% of normal hepatocytes but 85% of cancer prestages from the liver. Thereby a number of initiated (promotable) cell clones were extinguished (anti-initiating effect of apoptosis) (Grasl-Kraupp et al. 1994). Another situation where cancer prestages and even malignant tumors in the liver are more susceptible to induction of apoptosis as normal hepatocytes is after stop of treatment with a tumor promoting chemical. After prolonged treatment with the peroxisome proliferator Nafenopin rat liver showed considerable hyperplasia; in addition numerous adenoma and carcinoma were found. Cessation of treatment resulted in reduction of liver hyperplasia and in an apparently complete elimination of most of adenoma and carcinoma (Grasl-Kraupp et al., unpublished).

In conclusion cancer prestages at least in the liver are characterized by relatively high basal rates of cell replication and cell death both of which can be further increased by signals inducing cell multiplication or active cell death in normal liver. Prestages of cancer may therefore have an intrinsic defect in growth control rendering them more susceptible to endogenous and exogenous signals inducing birth or death of cells. Any disturbance of the balance between replication and death of cells in prestages, by tumor promoters or anti-promoters, will have much more dramatic effects as in normal liver and may result either in preferential accumulation or preferential elimination of preneoplastic cells. Eventually, formation of tumors or extinction of initiated clones may occur.

Conclusions

The multistage concept of carcinogenesis predicts that cancer formation occurs through a stepwise development of normal into malignant cells via several intermediate cell populations. Such prestages of cancer cells have been characterized in a number of animal tumor models and in some human organs. The elucidation of the deviations of preneoplastic cells from normal cells is of

Preneoplastic Lesions In Kidney And Carcinogenesis By Non-Genotoxic Compounds

D. R. Dietrich and T. Rasonyi
Institute of Toxicology, Swiss Federal Institute of Technology and University of Zurich, Schorenstr. 16, CH-8603 Schwerzenbach.

Carcinogenesis is known to be a multistep process which involves two or more genetic events and requires cell proliferation (Pitot 1986; Barrett and Wiseman 1987; Farber and Sarma 1987; Knudson 1987; Cerutti 1988; Lutz 1990). The genetic events can result from point mutations, chromosomal rearrangements (Croce and Klein 1990), insertions or deletions of genes, and gene amplification. Cell replication is required to convert DNA damage to mutations, irrespective of whether this damage is occurring spontaneously or induced by DNA damaging (genotoxic) compounds. Spontaneous mutations are assumed to stem from spontaneous genetic events such as depurination and deamination of DNA, formation of covalent DNA adducts by physiological reactive cellular constituents, DNA damage by oxygen free radicals produced in cellular metabolism, errors in DNA replication (Loeb 1989). Normally the cell has the possibility to either repair the damaged DNA, or, if the damage is irrepairable, to destroy itself via tumor suppressor gene product (e.g. p53) mediated apoptosis (Lane 1993). Abrogated DNA repair or apoptosis of damaged cells in conjunction with a cell proliferative stimulus can lead to the formation of clusters of mutated cells and clonal expansion. Continued enhanced proliferation and expansion of such clones increases the probability of additional genetic events within these populations (Knudson 1987) thus providing a basis for the formation of preneoplastic lesions and progression to veritable tumors (Swenberg et al. 1987; Dietrich and Swenberg 1991b). Although the sequence of events described above appears reasonable, little direct evidence has been put forward so far. However, some corroboration is provided by the observation that control animals from strains of laboratory rodents present with an appreciable incidence of spontaneous preneoplastic lesions and tumors (Dietrich and Swenberg 1991b). The incidence of spontaneous tumors varies greatly from tissue to tissue, strain to strain, sex to sex, and species to species (Swenberg and Short 1987).

Non-genotoxic compounds by definition do not induce DNA damage themselves. However, non-genotoxic carcinogens can induce enhanced cell proliferation and thus contribute to the fixation of spontaneous mutations as well as to the formation and progression of preneoplastic and neoplastic lesions. Similarly, non-genotoxic carcinogens could abrogate removal of damaged cells via inhibition of the apoptotic processes and, in conjunction with a simultaneously occurring proliferative stimulus, could promote formation of preneoplastic and neoplastic lesions. Indeed, a commonly observed feature of non-genotoxic compounds tested at high concentrations in carcinogenicity bioassays is the concurrent cytotoxicity and/or cell proliferation along with hyperplasia in the tissues that finally develop tumors.

It is the aim of this short overview to delineate the mechanisms by which non-genotoxic renal carcinogens induce the formation of preneoplastic and neoplastic lesions. In the first part, the possibly best studied mechanism involved in the chemically (e.g. d-limonene, unleaded gasoline) induced renal disease known as

Schulte-Hermann R, Ohde G, Schuppler J, Timmermann-Trosiener I (1981) Enhanced proliferation of putative preneoplastic cells following treatment with the tumor promoters phenobarbital, hexachlorocyclohexane, steroid compounds and nafenopin. Cancer Res 41:4140-4146

Schulte-Hermann R (1985) Tumor promotion in the liver. Arch Toxikol 57:147-215

Schulte-Hermann R, Timmermann-Trosiener I, Schuppler J (1986) Facilitated expression of adaptive response to phenobarbital in putative prestages of liver cancer. Carcinogenesis 7:1651-1655

Schulte-Hermann R, Timmermann-Trosiener I, Barthel G, Bursch W (1990) DNA synthesis, apoptosis and phenotypic expression as determinants of growth of altered foci in rat liver duribng phenobarbital promotion. Cancer Res 50:5127-5135

Solt DB, Cayama E, Tsuda H, Enomoto K, Lee G, Farber E (1983) Promotion of liver cancer development by brief exposure to dietary 2-acetylaminofluorene plus partial hepatectomy or carbon tetrachloride. Cancer Res 43:188-191

Ward JM (1983) Increased susceptibility of livers of aged F344/NCr rats to the effects of phenobarbital on the incidence, morphology, and histochemistry of hepatocellular foci and neoplasms. J Natl Cancer Inst 71:815-822

Weiler E (1956) Die Änderung der serologischen Spezifität von Leberzellen der Ratte während der Cancerogenese durch p-Dimethylaminoazobenzol. Z Naturforsch 11b:31-37

Preneoplastic Lesions In Kidney And Carcinogenesis By Non-Genotoxic Compounds

D. R. Dietrich and T. Rasonyi
Institute of Toxicology, Swiss Federal Institute of Technology and University of Zurich, Schorenstr. 16, CH-8603 Schwerzenbach.

Carcinogenesis is known to be a multistep process which involves two or more genetic events and requires cell proliferation (Pitot 1986; Barrett and Wiseman 1987; Farber and Sarma 1987; Knudson 1987; Cerutti 1988; Lutz 1990). The genetic events can result from point mutations, chromosomal rearrangements (Croce and Klein 1990), insertions or deletions of genes, and gene amplification. Cell replication is required to convert DNA damage to mutations, irrespective of whether this damage is occurring spontaneously or induced by DNA damaging (genotoxic) compounds. Spontaneous mutations are assumed to stem from spontaneous genetic events such as depurination and deamination of DNA, formation of covalent DNA adducts by physiological reactive cellular constituents, DNA damage by oxygen free radicals produced in cellular metabolism, errors in DNA replication (Loeb 1989). Normally the cell has the possibility to either repair the damaged DNA, or, if the damage is irrepairable, to destroy itself via tumor suppressor gene product (e.g. p53) mediated apoptosis (Lane 1993). Abrogated DNA repair or apoptosis of damaged cells in conjunction with a cell proliferative stimulus can lead to the formation of clusters of mutated cells and clonal expansion. Continued enhanced proliferation and expansion of such clones increases the probability of additional genetic events within these populations (Knudson 1987) thus providing a basis for the formation of preneoplastic lesions and progression to veritable tumors (Swenberg et al. 1987; Dietrich and Swenberg 1991b). Although the sequence of events described above appears reasonable, little direct evidence has been put forward so far. However, some corroboration is provided by the observation that control animals from strains of laboratory rodents present with an appreciable incidence of spontaneous preneoplastic lesions and tumors (Dietrich and Swenberg 1991b). The incidence of spontaneous tumors varies greatly from tissue to tissue, strain to strain, sex to sex, and species to species (Swenberg and Short 1987).

Non-genotoxic compounds by definition do not induce DNA damage themselves. However, non-genotoxic carcinogens can induce enhanced cell proliferation and thus contribute to the fixation of spontaneous mutations as well as to the formation and progression of preneoplastic and neoplastic lesions. Similarly, non-genotoxic carcinogens could abrogate removal of damaged cells via inhibition of the apoptotic processes and, in conjunction with a simultaneously occurring proliferative stimulus, could promote formation of preneoplastic and neoplastic lesions. Indeed, a commonly observed feature of non-genotoxic compounds tested at high concentrations in carcinogenicity bioassays is the concurrent cytotoxicity and/or cell proliferation along with hyperplasia in the tissues that finally develop tumors.

It is the aim of this short overview to delineate the mechanisms by which non-genotoxic renal carcinogens induce the formation of preneoplastic and neoplastic lesions. In the first part, the possibly best studied mechanism involved in the chemically (e.g. d-limonene, unleaded gasoline) induced renal disease known as

$\alpha 2\mu$-globulin nephropathy and subsequently increased cell proliferation and carcinogenesis will be reviewed. In the second part, the cytotoxicity and possible apoptosis inhibitive effect of nitrilotriacetic acid (NTA) will be discussed, while in the third some of the newest findings regarding the induction of renal tumors by the mycotoxin ochratoxin A will be presented. In the final part all of the above mechanisms will be discussed with regard to their importance in generating preneoplastic and neoplastic lesions in the kidney of rodents and the implications thereof for human cancer risk assessment. In addition, some emphasis will be laid on the comparison of rodent renal lesions with renal tumor types observed in humans.

Chemically induced $\alpha 2\mu$-globulin nephropathy and carcinogenesis

There is an ever growing list of chemicals that induce a male rat-specific renal disease known as $\alpha 2\mu$-globulin nephropathy. Neither female rats nor either sex of mice, guinea pigs, dogs, or monkeys developed this renal disease upon exposure to these chemicals (Dietrich 1994). Chronic exposure to a number of the $\alpha 2\mu$-globulin nephropathy-inducing agents resulted in a low but significant incidence of renal tumors in male rats, but not female rats, either sex of mice, or male NBR rats deficient for the male rat specific $\alpha 2\mu$-globulin (Dietrich and Swenberg 1991a; Dietrich and Swenberg 1991c).

$\alpha 2\mu$-Globulin is synthesized by the male rat liver, except in NBR rats, at the onset of puberty and excreted in great amounts via the urine. Approximately half of this protein is reabsorbed by the proximal tubules of the renal cortex and enzymatically hydrolized to amino acids and peptides via lysosomal digestion. Possibly due to the relatively high resistance of $\alpha 2\mu$-globulin toward catabolism, this protein has a natural tendency to accumulate in the P2 segment of the proximal tubules. Non-genotoxic chemicals e.g. d-limonene, unleaded gasoline, trimethylpentane, isophorone, or 1,4 dichlorobenzene and/or their metabolites have been shown to reversibly bind to $\alpha 2\mu$-globulin. This reversible binding of chemicals is thought to alter the proteins digestibility within the lysosomes (Rivera Torres et al. 1989; Borghoff et al. 1990), leading to an exacerbated accumulation of protein-chemical complex in the lysosomes and to increased cell necrosis within the P2 segment. The currently most plausible, however so far unproven and also most disputed hypothesis, explaining the observed increased cell necrosis, is that autolytic cytotoxicity may result from lysosomal dysfunction, leakage of lysosomal enzymes into the cytosol, and decreased membrane availability and turnover (Short et al. 1989). Male rats treated with d-limonene, PS-6 unleaded gasoline or trimethylpentane were shown to compensate for the enhanced loss of proximal cells by increasing the rate of cell proliferation in the P2 segment. This increase in cell proliferation exhibited a dose response (Short et al. 1989) and was found to remain sustained, in the case of chronic d-limonene exposure, at 4-5 fold the rate found in unexposed rats (Dietrich and Swenberg 1991c).

Given the observation that non-genotoxic chemicals increase cell turnover and that the assumption holds true that chronically exacerbated cell turnover can promote the formation of spontaneously mutated cells, one would expect to find tubules and hyperplastic nodules containing phenotypically altered cells, i.e. preneoplastic lesions (Dietrich and Swenberg 1991b), already after a relatively short time (4-12 months) of exposure to these non-genotoxic chemicals. The latter assumption was corroborated in an initation-promotion experiment (Dietrich and Swenberg 1991c). This study demonstrated a fairly high number of preneoplastic lesions in kidneys of rats treated with the non-genotoxic promoting agent, d-limonene, already following 30 weeks of promotion (Table 1).

Table 1. Preneoplastic lesions observed in kidneys of male F344 rats of an initation-promotion experiment (Dietrich and Swenberg 1991c), comprised of a 2 week initiation with the DNA-alkylating agent N-ethyl-N-hydroxyethylnitrosamine (EHEN) and a 30 week promotion with d-limonene. The number of preneoplastic lesions are expressed as the total number of lesions per treatment group and the number of lesions per animal. Nomenclature of preneoplastic lesions taken from (Dietrich and Swenberg 1991b); AT: atypical tubule, AH: atypical hyperplasia.

Compound	Total AT	AT/rat	Total AH	AH/rat	AT/AH
EHEN + diet	469	15.6 ± 1.5	37	1.2 ± 1.5	13
EHEN + d-limonene	2015	65.0± 3.4	482	15.5± 1.5	4
diet + d-limonene	330	10.7± 1.3	13	0.4 ± 0.1	25
Control diet	103	3.3 ± 0.7	0	0.0 ± 0.0	≥103

This study also demonstrated the importance of cell proliferation in fixing DNA damage to mutations and progression of mutated cells to preneoplastic lesions. Indeed, the number of atypical tubules was 4-5 fold higher in the rats initiated with N-ethyl-N-hydroxyethylnitrosamine (EHEN) and promoted with d-limonene than in rats treated with the initiating agent only. This factor 4-5 corresponds quite nicely with the observed 4-5 fold increase in proximal epithelial cell proliferation induced by d-limonene treatment. Although the total number of preneoplastic lesions found in rats treated with d-limonene only is distinctly lower than in rats treated with EHEN or EHEN and d-limonene, the three-fold and more than 13-fold higher number of atypical tubules and atypical hyperplasias in d-limonene promoted rats than in the corresponding control underlines the previous assumption that a higher number of spontaneous mutations can be fixed via enhanced cell turnover (Table 1). This observation, however, would also suggest that the incidence of these preneoplastic lesions is dependent on the rate of cell turnover. A dose dependent increase of atypical hyperplasias was indeed reported by Short et al. (1989) in male rats promoted with different concentrations of unleaded gasoline for 65 weeks.

As stated earlier, enhanced cell proliferation also appears to drive the progression from early preneoplastic lesions to veritable tumors possibly via additional mutational events. If these additional mutational events occur spontaneously i.e. via an endogenous mechanism, one would expect that preneoplastic lesions would progress to tumors only very slowly, despite an enhanced proliferative stimulus favoring fixation of these rare spontaneous mutations. This concept may be visualized by naively calculating how many atypical tubules it takes to produce one atypical hyperplasia. As demonstrated in Table 1, it takes more than 103 atypical tubules for the formation of even one atypical hyperplasia in control animals, while the proliferative stimulus provided by d-limonene exposure reduced this number already to approximately 25. Contrarily, genotoxic compounds characteristically induce DNA damage themselves and thus provide a strong basis for mutations, and in conjunction with a proliferative stimulus, favor the formation of fast growing preneoplastic lesions, as indicated by the fact that it takes only 4 atypical tubules for the formation of one atypical hyperplasia (Table 1). It is indeed observed that renal tumors induced by genotoxic compounds appear very early and in high incidence following treatment, in the case of EHEN initiation and d-limonene promotion already after 32 weeks (Dietrich and Swenberg 1991c), whereas renal tumors induced by non-genotoxic chemicals via the $\alpha 2\mu$-globulin mechanism appear very late in the life of the exposed rats and at relatively low incidence (Dietrich 1994).

NaNTA (nitrilotriacetate trisodium salt)

Chronic administration of high concentrations (2% in the diet) of NaNTA have been shown to induce sustained cytotoxicity, preneoplastic lesions and tumors (17% incidence) in the convoluted proximal tubule epithelial cells (PTEC) in either sex of mice and rats (Anderson et al. 1985). Acute and chronic dietary administration of NaNTA induces severe vacuolization and necrosis of the PTEC and proteinaceous casts at the cortico-medullary junction, followed by enhanced regenerative proliferation of PTEC. The mechanism by which cytotoxicity is induced is not entirely clear to date, however NTA, which is not metabolized in the mammalian system, is assumed to induce a redistribution of systemic zinc, leading to an increased resorption of zinc in the PTEC of the kidney (Anderson 1981). Both the acute (PTEC vacuolization) and the subchronic (PTEC vacuolization, necrosis and proliferation) nephrotoxicity were demonstrated to follow a dose response when the NTA-mediated intracellular zinc concentration was defined as the dose. Vacuolization appears to be the result of a change in cytoskeletal structure caused by micromolar concentrations of zinc which appear to alter the actin filaments. At higher zinc concentrations ($\geq 50 \mu M$) intercellular junctions as well as the intracellular actin filament and microtubular organization are affected to such an extent that the cells undergo necrosis (Mills et al. 1992). In view of the NTA/Zn mediated necrosis it is expected that the PTEC regenerate in order to maintain renal function. Based on the observation that hyperplastic nodules were present following subchronic and chronic oral administration of NTA (Alden and Kanerva 1982), it was assumed that the chronic insult of NTA would also exert an enhanced proliferative effect on the PTEC. However, as no cell proliferation studies have been carried out it is not a priori possible to assume that *i)* enhanced cell proliferation takes place, and *ii)* cell proliferation is increased to such an extent that a higher number of spontaneous mutations are fixed and propagated to preneoplastic lesions. Indeed, despite the consideration that vacuolization of PTEC is classically considered a degenerative change and was observed on average in more than 50% of the convoluted tubules of NTA treated animals, actual granular degeneration and necrosis of PTEC was observed only in a small number of treated rats i.e. the most severely affected animals (Alden et al. 1981). This observation suggests that chronically enhanced cell proliferation may not be at all involved in the generation of preneoplastic lesions. If, however, the latter assumption holds true, one must ask which "non-genotoxic" mechanism(s) could be responsible for the genetic changes and cellular transformations preceding overt preneoplastic lesions. One such pathway, although yet unproven and therefore purely hypothetical, may be that normally occurring apoptosis of renal epithelial cells could be abrogated and in conjunction with the observed slightly enhanced cell regeneration (wound healing) could lead to the survival and propagation of epithelial cells with "altered or damaged" DNA. Wound healing per se was observed earlier to provide a provisional matrix advantageous to the growth of transformed cells (Stevens and Jones 1990), thus corroborating the latter assumption. Abrogated apoptosis by micromolar concentrations of zinc has been reported in chronic lymphatic leukemia cells and in glucocorticoid hormone exposed thymocytes (Martin and Cotter 1994). Two mechanisms are proposed by which zinc can abrogate apoptosis, namely that zinc could block calcium channels (Büsselberg et al. 1990) and thus prevent calcium influx into the cells needed to activate calcium-dependent endonucleases and transglutaminases (Martin et al. 1994), or that zinc can inhibit the calcium-dependent endonucleases, therefore preventing the breakdown of DNA predominantly observed to be associated with apoptosis (Peitsch et al. 1994). If indeed apoptosis is inhibited, spontaneously occurring irrepairable DNA damage could persist and provide a source for an ever increasing number of transformed cells. It is important to emphasize here that from a genetic viewpoint the spontaneous DNA damage once fixed is considered irreversible. However, the survival of such genetically altered cells is highly dependent on the efficacy and presence of the mechanisms that regulate cell homeostasis such as apoptosis. This means that normal cells neighboring altered PTEC could recognize the altered state and induce apoptosis of the altered cell via cell-cell signalling. Consequently the organ should have the means to dispose of preneoplastic lesions via apoptosis, which is also indicated by the fact that not all preneoplastic lesions progress to veritable tumors (Dietrich and Swenberg 1991b). However, not all preneoplastic lesions may be susceptible to apoptosis, whether self-induced or induced by the surrounding normal tissue, susceptibility to apoptosis may largely depend on the nature and/or degree of cellular transformation of the respective preneoplastic lesion (Fig. 1).

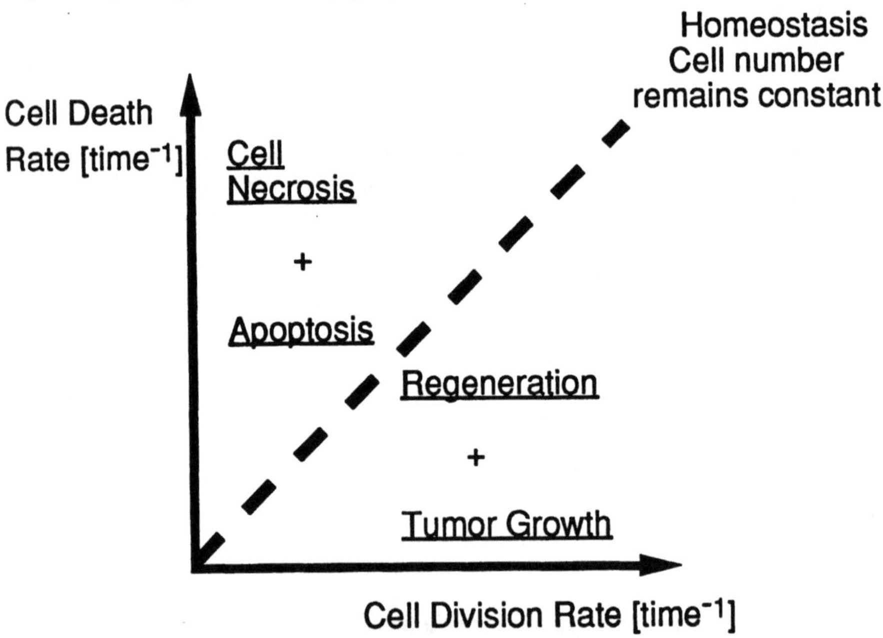

Fig. 1. Factors influencing the generation of altered cells and their propagation to preneoplastic lesions. The same factors affect the persistence of preneoplastic and neoplastic lesions and the probability of these lesions to become malignant and to metastasize

Furthermore, susceptibility to apoptosis of individual cells of preneoplastic lesions may also depend on the cell's location within the lesions, meaning that cells located in the innermost area of the lesion would not undergo apoptosis due to the lack of available stimuli.

Zinc-NTA was demonstrated to abrogate apoptosis thus immortalizing normally senescent Syrian hamster embryo cells in culture (LeBoeuf et al. 1990). In addition to longer viability thus treated Syrian hamster embryo cells appeared phenotypically transformed. Thus, in conjunction with a wound healing process observed following NTA treatment it appears reasonable to assume that transformed renal epithelial cells could be stimulated to divide and be propagated to preneoplastic lesions and tumors. Circumstantial evidence for the existence of such a mechanism in NTA induced renal tumors is provided by the observation that after 18 months of NTA treatment followed by a 6 month NTA-free diet a high number of the early preneoplastic lesions had regressed (Alden and Kanerva 1982). This observation suggests that cessation of NTA exposure reduced cellular zinc concentrations and thus allowed the kidney to gain back control over cell cycle homeostasis and possibly to delete preneoplastic lesions via apoptosis. The late appearance of relatively small tumors i.e. following a 24 month treatment with 2% NTA in the diet as well as the rather low tumor incidence of 17% indicates that these tumors develop very slowly, which may also support the idea of a non-highly proliferative environment and thus the presence of this yet hypothetical mechanism.

OTA

The mechanism by which ochratoxin A (OTA) can induce renal tumors in rodents is still an enigma to those involved in cancer research. Although OTA was negative in all routine test systems currently used for assessment of genotoxicity, reports have come forth stating that OTA can induce DNA damage in form of single strand breaks or OTA-DNA adducts (Dirheimer and Creppy 1991; Pfohl-Leszcowicz et al. 1991). From this it would appear that the renal tumors observed in mice and rats treated with OTA can be explained by the genotoxicity of OTA. However, it has to be emphasized that OTA induced renal tumors appear only in male mice but not in female mice and at a significantly higher incidence in male than in female rats. Furthermore, the daily dose needed to induce renal tumors in male mice in a 2-year bioassay is approximately 23-fold higher than the dose needed to induce a comparable tumor incidence in male rats. In order to characterize the tumor-induction potency of the DNA adducts found following OTA administration, these were compared to the potency of DNA adducts induced with aflatoxin B_1 in the liver of rats. 18 DNA adducts of aflatoxin/10^7 nucleotides (dose: 3 µg AFB_1/kg in the diet for 2 years) explain 50% of the tumors (Lutz 1986). Thus the 0.4 DNA adducts/10^7 nucleotides determined in renal DNA via P^{32} post-labelling following an i.p. dose of 2.5 mg OTA/kg (Pfohl-Leszcowicz et al. 1991) could explain at best only 1% of the tumors found given that a linear DNA-adduct to tumor formation relationship is applicable and only under the assumption that the OTA-adducts are as potent as aflatoxin B_1 in inducing tumors. The latter data demonstrate the discrepancy between the genotoxic potential of OTA and tumor induction, thus questioning the observation that the observed renal tumors can be explained merely by genotoxicity, especially when considering the vast sex and species differences in tumor susceptibility. In view of this it must be assumed that epigenetic mechanisms must play the predominant role in the aetiology of OTA induced renal tumors. Recent investigations regarding the site of OTA toxicity have shown that predominantly the pars recta proximal tubules (P3) are affected (Rasonyi et al. 1995). Increased cell exfoliation was observed in male and female rats treated with 210 and 1000 µg OTA/kg body weight for 4 weeks. One of the striking observation was that the predominant form of cell death was apoptosis and not necrosis. Whole epithelial cells still complete with brush border were seen in the tubular lumina presenting with varying stages of apoptosis at the light as well as on the electron microscopical level. Enhanced apoptosis may result from increased calcium influx into cells, as noted earlier. OTA was shown to deregulate cellular calcium homeostasis leading to an increased intracellular calcium concentration (Rahimtula and Chong 1991), thus providing an ideal environment for activating Ca-dependent endonucleases and transglutaminases. In conjunction with enhanced apoptosis, a high number of regenerating epithelial cells as well as enlarged nuclei (karyomegaly) were observed in the most affected areas, possibly indicating genetic changes in some of these cells (Rasonyi et al. 1995). Cells with karyomegaly could potentially represent the basis for preneoplastic lesions, provided a cell proliferative stimulus can propagate these cells to such lesions. Indeed, subsequent analysis of cell replication in male and female rats treated with 70, 210, and 1000 µg OTA/kg body weight for 4 weeks revealed a 6-fold increase in cell proliferation in the outer stripe of the outer zone of the medulla (P3-cells), compared to the corresponding control, in either sex of the 210 and 1000 (Table 2) but not the 70 µg OTA/kg groups (Rasonyi et al. 1995). These results demonstrate that a subchronic dose of 210 or 1000 µg OTA/kg can induce overt cell damage and regenerative cell proliferation in rats and thus provide a basis for the development and propagation of preneoplastic and neoplastic lesions. However, the latter study demonstrates no sex differences with regard to OTA toxicity, location of effects, and regenerative cell proliferation (Table 2). Thus the question remains to be answered why female rats have an approximately 10-fold lower carcinoma incidence than male rats given the same dose (210 µg OTA/kg) in the 2 year bioassay (Boorman 1989).

Table 2. Cell proliferation (S-phase labeling index (LI) in % of cells labeled) in F344 male and female rat renal pars recta proximal epithelial cells (P3) following 4 weeks continuous oral gavage with varying concentrations of OTA (from Rasonyi et al. 1995).

Sex	Control LI (%)	70 µg OTA/kg bw LI (%)	210 µg OTA/kg bw LI (%)	1000 µg OTA/kg bw LI (%)
Female	2.7 ± 0.8	2.0 ± 0.3	30.5 ± 9.0	38.8 ± 8.4
Male	5.9 ± 3.2	3.2 ± 2.3	25.9 ± 10.3	38.3 ± 4.6

Several possibilities can be imagined as to how this sex difference in tumor susceptibility can come about. One may be that female and male rats produce a different number of DNA adducts or adducts at a different genetic location leading to more serious genetic damage in the case of male rats. However, based on the fact that most DNA adduct studies have been conducted in male rats and mice and that only a negligible number of tumors can be explained with genotoxicity, this hypothesis seems a bit weak. Another may be that the cell proliferative response may ablate in female rats during chronic administration of OTA, whereas the cell proliferative stimulus might even increase in ageing male rats with the onset of chronic progressive nephropathy often observed to be more pronounced in male rats than in female rats (Dietrich and Swenberg 1991b). Studies looking at the chronic effect of OTA administration on cell proliferation, persistence of DNA adducts and formation of preneoplastic lesions in both sexes of rats are yet missing. With regard to tumor induction in mice and rats, the above data suggest that either the mechanism of OTA tumor induction is completely different in rats than in mice or that mice are much less susceptible to OTA than rats, thus questioning the extrapolatability of such data for human risk assessment without detailed knowledge of the mechanisms involved.

Types of preneoplastic lesions induced by non-genotoxic carcinogens

Sequential observations on the appearance of neoplastic lesions in rats and mice treated with alkylating agents, e.g. EHEN, Dimethylnitrosamine, Diethylnitrosamine, and N-Nitrosomorpholine led to the discovery of early lesions that shared morphological characteristics with renal cell tumors and to the identification of at least five different cell types, i.e. clear, acidophilic, basophilic, chromophobic, and oncocytic cells (Bannasch and Zerban 1990; Dietrich and Swenberg 1991b; Dietrich and Swenberg 1991c). Interestingly similar lesions and cell types were observed in rodents treated with non-genotoxic carcinogens such as d-limonene, NTA and OTA (Dietrich and Swenberg 1991b; Dietrich and Swenberg 1993). Furthermore, the presence of comparable cell types and tumors in humans (Stoerkel 1993) holds promise that the aetiology of renal tumors can be studied using the rodent as a surrogate. However, it is important to note that there is quite a discrepancy with regard to the predominance of tumor types in rodents and humans. While clear cell atypical tubules, atypical hyperplasias and tumors have been observed in rodent kidneys following treatment with alkylating agents, clear cell preneoplastic lesions but no clear cell tumors are reported in rodents treated with d-limonene, OTA or NTA. Indeed, clear cell lesions appear to change with increasing progression to either acidophilic and/or basophilic tumors as primarily basophilic tumors are reported in studies with OTA, NTA and d-limonene. Whether this morphological and biochemical change represents additional mutations that have been incurred during progression is a this point not clear. However, it is interesting to note that clear cell preneoplastic lesions induced with alkylating agents can persist and progress to veritable clear cell tumors (Dietrich and Swenberg 1991c), demonstrating that the morphological appearance does not give any indication of the genetic change that has occurred. Thus the morphological appearance of a preneoplastic lesion a priori does not allow deduction of the progressiveness and tendency for malignant conversion of the lesion. The rate of lesion growth and thus persistence and progression is not only determined by the genetic changes the clones have undergone, but is also a function of the proliferative potential of the individual lesions, i.e. the product of clone proliferation and susceptibility for clone necrosis/apoptosis and differentiation (Fig. 1), as well as of the environment of the lesion. The environment may allow the lesion to grow due to toxicant mediated chronic stimulation of cell proliferation (d-limonene and OTA) or abrogation of apoptosis (NTA). Thus, in order to characterize the aggressiveness of a preneoplastic lesion it would be necessary to determine its net growth. Although important, such studies are currently missing in rodent renal carcinogenesis. In contrast to tumors in the rodent where mostly basophilic and acidophilic adenomas, carcinomas and metastatic carcinomas are observed, e.g. OTA induced renal cell carcinomas, the predominant type of renal cell tumor in humans is the clear cell tumor with an incidence of 76% of renal tumors registered (Thoenes and Stoerkel 1991). This type of tumor also has the worst prognosis of the renal cell tumor types. This raises the question whether the aetiology of renal tumors in rodents is really comparable to the one in humans, especially considering the fact that the tumor cell type in rodents appears to be highly dependent on the carcinogenic agent, whereas the cause(s) for most human tumors is currently unknown. Furthermore, due to the late detection of renal tumors in humans it is virtually impossible to study the behaviour of preneoplastic lesions in humans and to compare these to those in rodents.

Conclusions

The mechanisms associated with the formation of preneoplastic and neoplastic lesions by non-genotoxic agents are known but for one group of compounds, namely those inducing $\alpha 2\mu$-globulin nephropathy. Due to the exclusive male rat specificity of $\alpha 2\mu$-globulin nephropathy, this mechanism is not extrapolatable for risk assessment to the human situation. The knowledge gained by studying the development of preneoplastic lesions induced by this mechanism, however, made it possible to understand how enhanced cell proliferation can provide a basis for renal tumor formation. Although viable hypotheses have been put forward as to the mechanisms involved in NTA and OTA induced renal tumors, the paucity of mechanistic data presently available make these hypotheses pure speculation. However, despite the above caveats it is possible to characterize the aetiology of renal tumors by studying the formation and behaviour of preneoplastic lesions, whether induced by genotoxic or non-genotoxic agents.

Acknowledgements

We would like to thank Regula Candrian for her magnificient help in the laboratory experiments as well as to extend our gratitude to Prof. Fritz Würgler, Prof. Werner Lutz, and PD Dr. Urs Bölsterli for critically reading this manuscript. Part of these studies were funded by BASF, Ludwigshafen FRG and by the Swiss Federal Office of Public Health (BAG) grant No. (FE 316 91-343).

References

Alden CL, Kanerva RL (1982) Reversibility of renal cortical lesions induced in rats by high doses of nitrilotriacetate in chronic feeding studies. Fd Chem Tox 20:935-937

Alden CL, Kanerva RL, Anderson RL, Adkins AG (1981) Short-term effects of dietary nitrilotriacetic acid in the male Charles River rat kidney. Vet Path 18: 549-559

Anderson RL (1981) The role of zinc in nitrilotriacetate (NTA) associated renal tubular cell toxicity. Fd Cosmet Toxicol 19:639-650

Anderson RL, Bishop WE, Campbell RL (1985) A review of the environmental and mammalian toxicology of nitrilotriacetic acid. CRC Crit Rev Toxicol 15(1):1-102

Bannasch P, Zerban H (1990) Animal models and renal carcinogenesis. In: Eble JN (ed) Contemporary issues of surgical pathology. New York, Churchill Livingstone. pp 1-34

Barrett JC, Wiseman RW (1987) Cellular and molecular mechanisms of multistep carcinogenesis: relevance to carcinogenic risk assessment. Environ Health Persp 76:65-70

Boorman G. (1989) Toxicology and carcinogenesis studies of ochratoxin A in F344/N rats. National Toxicology Program, Technical Report Series No. 358

Borghoff SJ, Short BG, Swenberg JA (1990) Biochemical mechanisms and pathobiology of alpha 2u-globulin nephropathy. Annu Rev Pharmacol Toxicol 30:349-67

Büsselberg D, Evans ML, Rahman H, Carpenter DO (1990) Zn2+ blocks the voltage activated calcium current of Aplysia neurons. Neuroscience Lett 117:117-122

Cerutti PA (1988) Commentary. Response modification creates promotability in multistage carcinogenesis. Carcinogenesis 9:519-526

Croce CM, Klein G (1990) Chromosomal-translocation and cancer. In: Schirrmacher V (ed) Cancer-Tumors, Cells, Genes. Heidelberg, Scientific American pp 102-109

Dietrich DR (1994) Alpha 2μ-Globulin: Species and sex specific protein synthesis and excretion and its association with chemically induced renal toxicity and neoplasia in the male rat, and its relevance in conjunction with human cancer risk assessment. Rev Biochem Toxicol 13:in press

Dietrich DR, Swenberg JA (1991a) NCI-Black-Reiter (NBR) male rats fail to develop renal disease following exposure to agents that induce alpha-2u-globulin (alpha 2u) nephropathy. Fundam Appl Toxicol 16(4):749-62

Dietrich DR, Swenberg JA (1991b) Preneoplastic lesions in rodent kidney induced spontaneously or by non-genotoxic agents: predictive nature and comparison to lesions induced by genotoxic carcinogens. Mut Res 248: 239-260

Dietrich DR, Swenberg JA (1991c) The presence of alpha 2u-globulin is necessary for d-limonene promotion of male rat kidney tumors. Cancer Res 51(13):3512-3521

Dietrich DR, Swenberg JA (1993) Renal carcinogenesis. In: Hook JB, Goldstein RS (eds) Toxicology of the Kidney. New York, Raven Press pp 495-537

Dirheimer G, Creppy EE (1991) Mechanism of action of ochratoxin A. In: Castergnaro M, Plestina R, Dirheimer G, Chernozemsky IM, Bartsch H (eds) Mycotoxins, Endemic Nephropathy and Urinary tract tumors. Lyon, International Agency for Research on Cancer, IARC. pp 171-186

Farber E, Sarma DRS (1987) Hepatocarcinogenesis: a dynamic cellular perspective. Lab Invest 56:4-22

Knudson AGJr (1987) A two-mutational model for human cancer. In: Klein G (ed) Advances in Viral Oncology. New York, Raven Press 1-17

Lane DP (1993) A death in the life of p53. Nature 362:786-787

LeBoeuf RA, Kerckaert GA, Aardema MJ, Poiley JA, Raineri R (1990) Enhanced morphological and neoplastic transformation of syrian hamster embryo cells cultured at pH 6.70. In: Mendelsohn ML, Albertini RJ (eds) Mutation and the Environment: Part D: Carcinogenesis. New York, Wiley-Lyss pp 219-228

Loeb LA (1989) Endogenous carcinogenesis: molecular oncology into the twenty-first century - Presidential Address. Cancer Res 49:5489-5496

Lutz WK (1986) Quantitative evaluation of DNA binding data for risk estimation and for classification of direct and indirect carcinogens. J Cancer Res Clin Oncol 112:85-91

Lutz WK (1990) Endogenous genotoxic agents and processes as a basis of spontaneous carcinogenesis. Mut Res 238:287-295

Martin SJ, Cotter TG (1994) Apoptosis of human leukemia: Induction, morphology, and molecular mechanisms. In: Tomei LD, Cope FO (eds) Apoptosis II: The molecular basis of apoptosis in disease. Cold Spring Harbor, Cold Spring Harbor Press pp 185-229

Martin SJ, Green DR, Cotter TG (1994) Dicing with death: dissecting the components of the apoptosis machinery. Trends in Biol Sci 19:26-30

Mills JW, Zhou J-H, Cardoza L, Ferm VH (1992) Zinc alters actin filaments in Madin-Darby canine kidney cells. Toxicol Appl Pharmacol 116:92-100

Peitsch MC, Mannherz HG, Tschopp J (1994) The apoptosis endonucleases: cleaning up after cell death. Trends in Cell Biol 4:37-41

Pfohl-Leszcowicz A, Chakor K, Creppy EE, Dirheimer G (1991). DNA adduct formation in mice treated with ochratoxin A. In: Castergnaro M, Plestina R, Dirheimer G, Chernozemsky IM, Bartsch H (eds) Mycotoxins, Endemic Nephropathy and Urinary tract tumors. Lyon, International Agency for Research on Cancer, IARC pp 245-253

Pitot HC (1986) The natural history of neoplastic development: Initiation and Promotion. In: Pitot HC (ed) Fundamentals of Oncology. New York, Marcel Dekker Inc

Rahimtula AD, Chong X (1991) Alterations in calcium homeostasis as a possible cause of ochratoxin A nephrotoxicity. In: Castergnaro M, Plestina R, Dirheimer G, Chernozemsky IM, Bartsch H (eds) Mycotoxins, Endemic Nephropathy and Urinary Tract Tumors. Lyon, International Agency for Research on Cancer, IARC pp 207-214

Rasonyi T, Dietrich DR, Schlatter J, Schlatter C (1995) Site specific toxicity and regenerative cell proliferation in male and female F344 rats treated for 4 weeks with different doses of ochratoxin A. Fundam Appl Toxicol; in preparation

Rivera Torres MI, Caudill D, Lehman-McKeeman LD (1989) Lysosomal degradation of alpha2u-globulin (A2u): Role of cysteine and aspartic acid proteinases and effects of d-limonene binding. Toxicologist 9(1):314

Short BG, Burnett VL, Swenberg JA (1989) Elevated proliferation of proximal tubule cells and localization of accumulated alpha 2u-globulin in F344 rats during chronic exposure to unleaded gasoline or 2,2,4-trimethylpentane. Toxicol Appl Pharmacol 101(3):414-31

Short BG, Steinhagen WH, Swenberg JA (1989) Promoting effects of unleaded gasoline and 2,2,4-trimethylpentane on the development of atypical cell foci and renal tubular cell tumors in rats exposed to N-ethyl-N-hydroxyethylnitrosamine. Cancer Res 49(22):6369-6378

Stevens JL, Jones TW (1990) The role of damage and proliferation in renal carcinogenesis. Toxicol Lett 53:121-126

Stoerkel S (1993) Carcinomas and oncocytomas of the kidney. In: Denk H, Dietel M, Fischer R, Hoefler H, Katenkamp D, Seifert G, Dhom G, Eder M, Heitz PU, Holzner H, Lennert K, Thoenes W (eds) Progress in Pathology. New York, Gustav Fischer Verlag pp 164

Swenberg JA, Richardson FC, Boucheron JA, Deal FH, Belinsky SA, Charbonneau M, Short BG (1987) High- to low-dose extrapolation: Critical determinants involved in the dose response of carcinogenic substances. Environ Health Persp 76:57-63

Swenberg JA, Short BG (1987) Influence of cytotoxicity on the induction of tumors. In: Butterworth BE, Slaga TJ (eds) Nongenotoxic mechanisms in carcinogenesis. Cold Spring Harbor, NY, Cold Spring Harbor Laboratory pp 151-161

Thoenes W, Stoerkel S (1991) Die Pathologie der benignen und malignen Nierenzelltumoren." Urologe-A 30:W41-W50

EARLY CELLULAR ALTERATIONS IN CHEMICAL LIVER CARCINOGENESIS AS ASSESSED BY IMMUNOHISTOCHEMICAL AND MOLECULAR BIOLOGICAL TECHNIQUES

Henrik S. Huitfeldt, Ellen Skarpen, Birgitte Lindeman, Hilde E. van Gijssel, Pål A. Nyhus, and Karen Leyshon-Sørland
Department of Environmental Medicine, National Institute of Public Health, N-0482 Oslo, Norway

Multistep liver carcinogenesis is characterized by the high metabolic activity of this organ. Decreased phase I activating and increased phase II detoxifying capasities of carcinogen-initiated cells render these cells resistant to the growth inhibitory effects of many xenobiotics. Furthermore, the liver is normally a quiescent organ, but proliferation can easily be induced. How altered metabolism is related to deregulated growth control during carcinogenesis is not clear. The metabolic alterations of initiated cells may be caused by a mutational event. But also growth induction by itself induce some of the same changes. A third possibility is that preneoplastic lesions are derived from oval cells or putative liver stem cells, which display a similar enzyme staining pattern. In the following, relations between alterations in metabolic and growt regulatory mechanisms in chemical rat liver carcinogenesis are discussed.

Multi-Step Liver Carcinogenesis

In the 1930s experimental rat liver carcinogenesis models were established, involving long-term continuous exposure to azo-dyes in the diet (Sasaki and Yoshida 1935). Cyclic exposure to liver carcinogens demonstrated that the induction of carcinomas was a multi-step process, with a sequential development from foci of altered liver cells, nodules compressing surrounding tissue, and finally true carcinomas with the capacity to metastasize and infiltrate. Later, the concept of two-stage carcinogenesis was introduced (Berenblum 1941). The first step, initiation, was shown to be a permanent, genetic alteration. It is today generally agreed that most chemical carcinogens cause initiation by binding to the genome, and subsequently induce mutations during DNA replication (Miller and Miller 1947; Pound 1968; Heidelberger and Davenport 1964). The following process, promotion, was considered an epigenetic event, involving a selection for initiated cells. Monoclonal expansion of these cells yields preneoplastic lesions such as altered liver foci and nodules. In later years it has been shown that the promotion phase consists of several steps, and also involves several genetic alterations, finally resulting in conversion to true malignancies (Yuspa and Poirier 1988). Compared to other experimental models, liver carcinogenesis is characterized by the high metabolic capacity of this organ, with the potential to activate a long range of carcinogens. Furthermore, adult liver is normally a quiescent organ almost without cell proliferation. Upon a proper growth stimulus, intense proliferation is induced. Liver carcinogenesis shares some fundamental, unresolved questions with most other experimental models: What are the initial and subsequent molecular alterations leading to benign and later malignant lesions; and what is the cellular origin of these lesions?

The Resistant Hepatocyte

About 20 years ago it was realized that the earliest preneoplastic liver lesions showed metabolic alterations compared to normal liver (Sherer et al. 1972; Farber et al. 1976). Generally, phase I activating enzymes were reduced, whereas phase II detoxifying enzymes showed increased activities in such lesions. Not only allowed this histochemical identification of foci, it also led to the "resistant hepatocyte" concept: An initiated liver cell was protected from the growth inhibitory effects of varying xenobiotics. Promotion could thus be viewed as growth inhibition of normal hepatocytes combined with a proliferation pressure, causing monoclonal expansion of initiated cells to enzyme altered foci and later nodules (Solat and Farber 1976; Roomi et al. 1985). Although the phenotypic variations were large, increased γ-glutamyl transpeptidase (GGT) and placental glutathione S-transferase (GSTp) expression combined with reduced canalicular ATPase, glucose 6-phosphatase and CYP isozyme activities correlated with increased growth capacities. These altered metabolic capacities can be demonstrated by the reduced content of aminofluorene-DNA adduct concentrations in enzyme altered foci during continuous 2-acetylaminofluorene feeding (Huitfeldt et al. 1986, Huitfeldt et al. 1988). It is plausible to conclude that reduced adduct accumulation contribute to a growth advantage during tumor promotion.

Basis of Metabolic Alterations

Several mechanisms are possible for the acquisition of a resistant phenotype. The first and obvious is a mutation in a gene or gene system regulating metabolic capacities. Since increasing deregulation of growth control is another hallmark of the carcinogenic development, it is conceivable that a regulatory system involved in the control of both growth and metabolism is affected. Although such an alteration has never been identified, this view fits well with initiation being a mutational event inducing a resistant phenotype and deregulated growth control.

Alternatively, altered metabolism is not necessarily a mutational event. It is well known that the metabolic capacities of the liver is reduced during regeneration. Expression of many liver enzymes depend upon a differentiated, growth-arrested cellular state. Since preneoplastic foci display increased growth, enzyme alterations in foci may at least partly reflect a proliferative cellular state. In the following some examples of such a mechanism are given:

By BrdU-pulse-labelling and two-color immunofluorescence of livers from AAF-fed rats, increased liver proliferation was demonstrated after 4 weeks of continuous AAF feeding (Huitfeldt et al.1990). Replicating cells were located either in foci or as individual cells. Foci contained undetectable amounts of DNA adducts, reflecting alterations in metabolic capasities of preneoplastic lesions. But also individually proliferating liver cells displayed much lower adduct concentrations than surrounding cells, and this phenomenon occurred prior to cellular initiation (Huitfeldt et al. 1991). Thus, both preneoplastic foci and unaltered liver cells are protected from adduct accumulation during DNA replication; either due to altered metabolism, or increased DNA repair.

Another example is the induction of the multidrug resistance (mdr) gene product P-glycoprotein during rat liver carcinogenesis. This has been considered closely related to malignant conversion (Thorgeirsson et al. 1987; Fairchild et al. 1987). P-glycoprotein expression is a rare example of a detoxifying factor present in normal liver, but not in bile ducts or oval cells, which are putative liver stem cells. Both AAF exposure and normal regeneration induced by partial hepatectomy increase the expression of P-glycoprotein in hepatocytes (Marino et al. 1990; Burt et al. 1988). Expression in early preneoplastic liver, although higher than in untreated liver, is lower than in surrounding liver tissue. This indicates that increased expression in early carcinogenesis is a response to proliferation and xenobiotic exposure, rather than a result of neoplastic conversion (Huitfeldt et al, submitted for publication). Another resistance factor, metallothionein, is normally not expressed in any of these cell populations, but is induced during cellular replication. These results demonstrate that increased expression of at least some resistance factors is caused by cellular adaptation or growth induction rather than by genetic alterations.

A third mechanism for metabolic alterations of preneoplastic foci is a stem cell derivation of these lesions (Sell 1990; Braun et al. 1989). In many liver carcinogenesis protocols, bile duct cell-like oval cells proliferate during inhibition of hepatocyte regeneration (Tatematsu et al. 1984). These cells share many of the metabolic characteristics of enzyme-altered foci; increased γ-glutamyl transpeptidase and placental glutathion S-transferase expression, and reduced carcinogen activation. There are indications that oval cells are stem cell progenies with the capacity to differentiate into hepatocytes. Possibly, they can also be carcinogen target cells and thus tumor progenitors.

Regulation of Growth and Differentiation

Xenobiotic metabolism may be regulated by signalling mechanisms involved in cellular growth or differentiation control. Receptor tyrosine kinase, and cAMP dependent protein kinase signalling pathways, are important regulatory mechanisms of the liver (Michalopoulos 1991). The distribution of such factors in normal and carcinogen-altered liver can possibly contribute to the understanding of altered growth regulation and metabolism. Transforming growth factor α and hepatocyte growth factor are major growth factors during rat liver regeneration and subsequent differentiation. Immunofluorescence studies have demonstrated that their receptors, Epidermal growth factor receptor and c-Met, were differently distributed in the liver: Epidermal growth factor receptor was strongly expressed along hepatocyte membranes, whereas c-Met was found in bile ducts and oval cells. Enzyme-altered foci display a hepatocyte-like phenotype, with expression of epidermal growth factor receptor but not c-Met (Huitfeldt et al. submitted for publication).

Norepinephrine, glucagon and vasopressin all appear to participate in the regulation of liver replication and metabolism through the cAMP-dependent protein kinase signalling pathway. There are two classes of the regulatory subunits of cAMP dependent protein kinase; RI and RII. These associates with C subunits to form two isozymes of cAMP dependent protein kinase, type I (RI_2C_2) and type II (RII_2C_2), respectively (Corbin et al. 1975). Type I cAMP dependent protein kinase has been suggested to be associated with proliferation, and type II with differentiation (Cho-Chung 1990). The RI/RII ratio appears to be high during cellular growth, and to fall during organ differentiation. Conversely, the ratio increases in malignancy, and this has been implicated to be a major event in experimental and clinical carcinogenesis. In rat liver we found type I expressed in bile ducts and oval cells, whereas hepatocytes and enzyme-altered foci contained both type I and type II (Skarpen et al. submitted for publication). Thus, preneoplastic liver lesions and hepatocytes contained a lower RI/RII ratio than bile ducts and oval cells, indicating a lower proliferative and more differentiated state.

Similarly, the liver-specific transcription factor C/EBP was immunolocalized to normal hepatocytes and preneoplastic lesions, but not bile ducts and oval cells

(Skarpen et al. submitted for publication). This transcription factor is important for cell maturation and growth arrest, and regulates functional genes like albumin, transthyretin, α_1-antitrypsin and fat metabolizing enzymes (Johnson 1990, Umek et al. 1991, Mischoulon et al 1992). Taken together, these data demonstrate that major growth and differentiation regulatory systems appear differently expressed in bile duct and oval cells compared to hepatocytes, whereas early preneoplastic lesions show similar characteristics as hepatocytes.

Table 1 Cellular markers, resistance factors and growth regulatory factors in rat liver

Marker	Hepatocytes	Bile ducts	Oval cells	Foci
Keratin K8	++	+++	+++	++/+++
Keratin K18	++	+++	+++	++/+++
Keratin K19	0	+++	+++	0
GSTp	0	+	+	0/++++
mdr	++/+++	0	0	++
Metallothionein	0 (inducible)	0 (inducible)	0 (inducible)	0 (inducible)
EGFr	+++	0	0	+++
c-Met	0	++	++	0
PKA type I	0	++	++	0
PKA type II	++	+	+	++
C/EBP	+++	0	0	+++

Distribution of some cell type markers, xenobiotic resistance factors, and signalling factors in normal and carcinogen-treated rat livers, as assessed by two-color immunofluorescence of frozen liver sections. Relative signal intensities of the different markers in liver cell populations were compared and scored on a scale from 0 (no staining) to +++ (maximum staining).

Conclusions

These and other results demonstrate a close relationship between hepatocytes and early preneoplastic lesions (Table 1). In fact, the only consistent alterations in foci identified this far are changed metabolism of xenobiotics. Even some of these alterations can be explained by cellular adaptation, or by recruition of liver cells from a resting G_0 state into the G_1 phase of the cell cycle. A concomitant phenotypic change in growth or differentiation regulatory mechanisms, or consistent mutation in cellular oncogenes, has not been identified. Thus, foci may represent hepatocytes with some disruption of differentiation or metabolic capacities. Although preneoplastic lesions thus appear to be of hepatocyte origin, a stem cell derivation can not be ruled out. Foci may represent initiated stem cells with retained capacity for a hepatocyte-like differentiation. Identification of the initial molecular events in liver carcinogenesis depend upon identification of a putative liver stem cell.

Acknowledgements

The expert technical help of Ms Edel Lilleaas, Ms Angelika B. Melhuus and Mr Hans Jørgen Dahlman is greatly appreciated. This work was supported by the Norwegian Cancer Society, the Norwegian Research Council, and Rakel and Otto Kr. Bruun's Legacy.

References

Berenblum I (1941) The co-carciogenic action of crotin resin. Cancer Res 1: 44-48

Braun L, Mikumo R, Fausto N (1989) Production of hepatocellular carcinoma by oval cells: Cell cycle expression of c-*myc* and p53 at different stages of oval cell transformation. Cancer Res 49: 1554-1561

Burt RK, Thorgeirsson SS (1988) Coinduction of MDR-1 multidrug-resistance and cytochrome P-450 genes in rat liver by xenobiotics. J Natl Cancer Inst 80: 1383-1386

Cho-Chung YS (1990) Role of cyclic AMP receptor proteins in growth, differentiation, and suppression of malignancy: New approaches to therapy. Cancer Res 50: 7093-7100

Corbin JD, Keely SL, Park CR (1975) The distribution and dissociation of cyclic adenosine 3':5'-monophosphate-dependent protein kinases in adipose, cardiac, and other tissues. J Biol Chem 250: 218-225

Fairchild CR, Ivy SP, Rushmore, T, Lee G, Koo P, Goldsmith ME, Myers CE, Farber E, Cowan KH (1987) Carcinogen-induced mdr overexpression is associated with xenobiotic resistance in rat preneoplastic liver nodules and hepatocellular carcinomas. Proc Natl Acad Sci USA 84: 7721-7725

Farber E, Parker S, Gruenstein M (1976) The resistance of putative premalignant liver cell populations, hyperplastic nodules, to the acute cytotoxic effects of some carcinogens. Cancer Res 36: 3878-3887

Heidelberger C, Davenport (1964) Studies on the molecular mechanisms of hydrocarbon carcinogenesis. J Cell Comp Physiol 64 (suppl. I): 129-148

Huitfeldt HS, Spangler EF, Hunt JM, Poirier MC (1986) Immunohistochemical localization of DNA adducts in rat liver tissue and phenotypically altered foci during oral administration of 2-acetylaminofluorene. Carcinogenesis 7: 123-129

Huitfeldt HS, Hunt JM, Pitot HC, Poirier MC (1988) Lack of acetylaminofluorene-DNA adduct formation in enzyme-altered foci of rat liver. Carcinogenesis 9: 647-652

Huitfeldt HS, Brandtzaeg P, Poirier MC (1990) Reduced DNA adduct formation in replicating cells during continuously feeding of a chemical carcinogen. Proc Natl Acad Sci USA 87: 5955-5958

Huitfeldt HS, Brandtzaeg P, Poirier MC (1991) Carcinogen-induced alterations in rat liver DNA adduct formation determined by computerized fluorescent image analysis. Lab Invest 64: 207-214

Johnson PF (1990) Transcriptional activators in hepatocytes. Cell Growth & Differ 1: 47-51

Marino PA, Gottesman MM, Pastan I (1990) Regulation of the multidrug resistance gene in regenerating rat liver. Cell Growth & Differ 1, 57-62

Michalopoulos G (1991) Control of hepatocyte proliferation in regeneration, augmentative hepatomegaly, and neoplasia. Prog Clin Biol Res 369: 227-236

Mischoulon D, Rana B, Bucher NLR, Farmer SR (1992) Growth-dependent inhibition of CCAAT enhancer-binding protein (C/EBPα) gene expression during hepatocyte proliferation in the regenerating liver and in culture. Mol Cell Biol 12: 2553-2560

Miller EC, Miller JA (1947) The presence and significance of bound aminoazo dyes in the livers of rats fed p-dimethylaminoazobenzene. Cancer Res 7:468-480

Pound AW (1968) Carcinogenesis and cell proliferation. NZ Med J 67(Suppl):88-89

Roomi MW, Ko RH, Sarma DSR, Farber E (1985) A common biochemical pattern in preneoplastic hepatocyte nodules generated in four different models in the rat. Cancer Res 45: 564-571

Sasaki T, Yoshida T (1935) Experimentelle erzeugung des lebercarcinoms durch futterung mit o-amidoazotoluol. Virchows Arch Pathol Anat Physiol Klin Med 295:175-200

Sell S (1990) Is there a liver stem cell? Cancer Res 50: 3811-3815

Solt D, Farber E (1976) New principle for the analysis of chemical carcinogenesis. Nature 263: 701-703

Scherer E, Hoffman M, Emmelot P, Friedrick-Freksa H (1972) Quantitative study of foci of altered liver cells induced in the rat by a single dose of diethylnitrosamine and partial hepatectomy. J Natl Cancer Inst 49: 93-106

Tatematsu M, Ho RH, Kaku T, Ekem JK, Farber E (1984) Studies on the proliferation and fate of oval cells in the liver of rats treated with 2-acetylaminofluorene and partial hepatectomy. Am J Pathol 114: 418-430

Thorgeirsson SS, Huber BE, Sorrell S, Fojo A, Pastan I, Gottesman MM (1987) Expression of the multidrug-resistant gene in hepatocarcinogenesis and regenerating rat liver. Science 236: 1120-1122

Umek RM, Friedman AD, McKnight SL (1991) CCAAT-enhancer binding protein: A component of a differentiation switch. Science 251: 288-292

Yuspa SH, Poirier MC (1988) Chemical carcinogenesis: from animal models to molecular models in one decade. Adv Cancer Res 50: 25-70

Late manuscript

The UK Approach to the Regulation and Evaluation of Novel Foods Produced by Biotechnology

David A Jonas

Novel Foods, Biotechnology and Food Technology Branch, Food Safety Directorate, Ministry of Agriculture Fisheries and Food, Room 204 Ergon House c/o Nobel House, 17 Smith Square, London SW1P 3JR, United Kingdom

Introduction

For the past ten years the UK has had in place procedures for the premarketing clearance of novel foods and novel food processes. Under these procedures the consumer safety of novel foods and foods produced by novel processes is evaluated by the UK Advisory Committee on Novel Foods and Processes (ACNFP). The terms of reference of the Committee are: "To advise Health and Agriculture Ministers of Great Britain and the Heads of the Departments of Health and Social Services and Agriculture for Northern Ireland on any matters relating to the irradiation of food or to the manufacture of novel foods or foods produced by novel processes, having regard where appropriate to the views of relevant expert bodies."

For the purposes of the UK clearance scheme novel foods are defined as: "foods are foods or food ingredients which have not hitherto been used for human consumption to a significant degree in the United Kingdom and/or which have been produced by extensively modified or entirely new food production processes." Examples of the types of products covered by the clearance scheme are given in Table 1.

Table 1. Examples of novel foods evaluated by the ACNFP

Product type	Example
Oilseeds/pulses	Lupins
Probiotics	*Lactobacillus casei*
Exotic oils	Passion fruit seed oil
Plant fibres	Sugar beet, lupins, onions
Synthetic oils	High behenic acid triglyceride
Fat replacers	Sucrose polyester
Products of gmos	Chymosin
Gmos	Brewing yeast

For the purposes of the UK clearance scheme all foods and food ingredients that are, or are obtained from, genetically modified organisms are regarded as novel foods and are evaluated by the ACNFP. This paper examines the main issues that need to be addressed in the food safety evaluation of these foods.

Discussion

Many foods and food ingredients produced using genetic modification will be variations of existing foods and food ingredients. Thus they might include new varieties of existing food crops, new strains of starter cultures and existing ingredients such as enzymes, produced in new source organisms. This relationship with existing, acceptable, foods and food ingredients facilitates the safety evaluation since the traditional counterpart provides a benchmark against which the biotechnologically produced food or food ingredient can be compared. For those biotechnologically produced products for which there is no traditional counterpart for comparison, an extensive safety evaluation may be necessary, possibly involving animal testing.

For those products of biotechnology for which there is a traditional counterpart the most practical approach to safety evaluation is to compare the biotechnologically produced product with the traditional counterpart and to focus the evaluation on any differences. Six of the aspects that will need to be considered as outlined below.

Consequential changes as a result of the modification

Consequential changes that might occur as a result of the modification include positional effects such as the interuption of coding or control sequences and feedback effects of targeted gene product(s) such as effects on organism's metabolism. These effects, particularly positional effects, are difficult if not impossible to predict. It is important therefore with all genetically modified organisms used as food or in the production of food or food ingredients to compare the modified organism with the host organism to establish that all differences are as might be predicted from the modification. The comparison should include a comparison of phenotypic properties, such as growth characteristics, and analytical data on major and minor components. Where possible, sequence data around the site of the genetic modification should be provided to show the absence of positional effects although, at the present time, meaningful data are only likely to be available for microorganisms.

If consequential effects are suspected to arise from the modification, an extensive evaluation is likely to be required for any foods or food ingredients obtained from the modified organism.

Targeted effects of the modification

A safety assessment of the targeted effects of the genetic modification is essential. This should include an assessment of all gene products the level of which is altered as a result of the modification. If the safety of a new gene product is infered from data on the isolated product it will be important to show identity with the product as expressed by the genetically modified organism and to show that there are no interactions with components of the modified organism.

It will be necessary to evaluate the safety of the targeted gene products within the context of the diet as a whole. Thus, for example, a viral coat protein expressed in a potato can only be assumed to be safe if it identical to that found in potatoes naturally infected with the virus and present at the same, or a lower, level.

A case by case evaluation of the targeted gene products is necessary to determine whether there are special factors that need investigation. Thus, in the case of the *npt*II gene which confers kanamycin resistance, reassurance of safety is provided by the knowledge that the enzyme requires cofactors to function and these are not present at high levels in the human gut. If the targeted genes code for known allergens or if they show homology with such genes or if they come from sources which are associated with food allergy then clearly the safety evaluation of foods and food ingredients obtained from the modified organisms will need to be screened for allergenicity. Special attention will need to be given to consequential changes if the targeted gene product is known or suspected to have metabolic effects on the modified organism.

Growth phase or site specificity of the targeted effects

For some targeted effects it will be necessary to demonstrate that they only occur at particular sites in the modified organism or at particular stages of growth. Thus if the modification is aimed at producing toxic products it may be necessary to show that these are only produced in parts of the organism that will not be used for food. Similarly, if toxin producing organisms are used to produce food ingredients it may be necessary to show that toxin production under production conditions is impossible. Analytical data are useful in this respect but an understanding of the biochemical pathways underlying toxin production will provide valuable additional reassurance.

Gene transfer from the gmo to gut cells or microflora

If the food or food ingredient contains intact DNA there is a possibility that gene transfer might occur to gut epithelial cells or to gut microflora. However, because of the high turn-over rate of epithelial cells, the safety implications of gene transfer to these cells are minimal. Gene transfer to gut microflora is not a possibility that is unique to genetically modified organisms but it assumes a new importance since genetically modified organisms might include DNA sequences not hitherto found in food organisms. In the case of foods containing intact DNA it will be necessary to show either that gene transfer does not occur or that if transfer does occur it will not be of greater safety significance than transfer from existing foods. The latter approach is usually adopted although this may change as *in vivo* and *in vitro* techniques are developed to measure gene transfer. Often it will be sufficient to demonstrate that although the food, as consumed, contains intact DNA, this is absent in the lower intestinal tract where the microorganisms are found.

Colonisation of the gut

Important in the case of genetically modified microorganisms, used perhaps as starter cultures or probiotics, is the possibility that they might colonise the gut. In all cases where potential colonisation is anticipated it will be necessary to determine the consequences including the effect on nutrient uptake and formation. Clearly if colonisation is a possibility then it is not acceptable for the microorganism to express products that might compromise the use of products such as antibiotics used in human therapy.

Intakes

The intake of a product produced using biotechnology will usually be similar to that of the food or ingredient produced from the traditional source, however, this may not always be the case. Where significant intake changes are anticipated it will be necessary to evaluate their impact on the diet, particularly in relation to natural toxins or nutrients present even if their level in the food is not directly affected by the genetic modification.

Conclusion

Over the past three years the ACNFP has evaluated, and is continuimg to evaluate, a number of foods and food ingredients produced using genetic modification including three chymosins from genetically modified microorganisms, a genetically modified baking yeast and beer produced using a genetically modified brewing yeast. The Committee's practical experience has led it to adopt, wherever possible, a comparative approach to the safety evaluation of foods produced using genetic modification. If there is an existing food with which to compare the biotechnologically produced food and if the safety implications of any differences can be determined, extensive toxicological testing of the new food is unlikely to be required.

Young Scientists Poster Award

The Young Scientists Poster Award is given to the first author of the best poster who must be younger than 35 years.

The winners are announced at the closing ceremony of the EUROTOX Congress. The abstracts are published in the congress proceedings.

The winners 1994 were U. Christen, Basel and N. Frey, Basel.

MOLECULAR MIMICRY OF CF_3CO-LYSINE BY LIPOIC ACID I: THE DIHYDROLIPOAMIDE ACETYLTRANSFERASE SUBUNIT OF THE HUMAN PYRUVATE DEHYDROGENASE AS AUTOANTIGEN IN HALOTHANE HEPATITIS

U. Christen, J. Gut
Department of Pharmacology, Biocenter of the University, CH-4056 Basel, Switzerland

Trifluoroacetylated proteins (CF_3CO-proteins), elicited in animals or humans exposed to halothane, were recognized by anti-CF_3CO antibody, monospecific N^6-trifluoroacetyl-L-lysine (CF_3CO-Lys). Anti-CF_3CO antibodies cross-reacted with the dihydrolipoamide acetyltransferase (E2 subunit) of pyruvate dehydrogenase (PDH), indicating that epitopes of the E2 subunit of PDH molecularly mimic those on CF_3CO-proteins. Lipoic acid, the prosthetic group of the E2 subunit was essential in that only the lipoylated (Llip) form the recombinantly expressed inner lipoyl domain of the human E2 subunit, but not the unlipoylated (Ulip) form, was recognized by anti-CF_3CO antibodies. Based on a high degree of structural relatedness, both CF_3CO-Lys and (6RS)-lipoic acid, as well as the lipoylated peptide ETDK (lipoyl)ATIG, inhibited the recognition by anti-CF_3CO-antibody of the E2 subunit of PDH and of human liver CF_3CO-proteins. In sera of patients with halothane hepatitis, autoantibodies with properties identical to those of anti-CF_3CO antibody were identified which could not discriminiate between CF_3CO-proteins and E2 subunit of PDH. These data suggest that the E2 subunit of PDH is an autoantigen in halothane hepatitis and that molecular mimicry of CF_3CO-proteins by the E2 subunit of PDH is due to the similar structures of CF_3CO-Lys and lipoic acid.

MOLECULAR MIMICRY OF CF_3CO-LYS BY LIPOIC ACID II: LIPOIC ACID BEARING PROTEINS AS AUTOANTIGENS IN HALOTHANE HEPATITIS

N. Frey, V. Koch, J. Gut
Department of Pharmacology, Biocenter of the University, CH-4056 Basel, Switzerland

The anesthetic agent halothane is metabolized by P450IIE1 via an acyl chloride intermediate, giving rise to trifluoroacetylated proteins (CF_3CO-proteins), which, in a small subset of susceptible individuals, are thought to precipitate halothane hepatitis. A monospecific antibody towards N^6-trifluoroacetyl-L-lysine (CF_3CO-Lys) recognizes such CF_3CO-proteins. In unexposed humans (and rats), anti-CF_3CO antibody crossreacts with constitutively expressed hepatic proteins of 52 kDa and 64 kDa. The latter was identified as the E2 subunit of the pyruvate dehydrogenase complex (PDC). Its prosthetic group, lipoic acid, was shown to be necessary and sufficient to molecularly mimic CF_3CO-proteins (i.e. CF_3CO-Lys). Additional constitutive proteins bearing lipoic acid are the E2 subunits of the 2-oxoglutarate (OGDC) and the branched chain ketoacid (BCOADC) dehydrogenase complexes, and protein X, a constituent of PDC, all of which were recognized by anti-CF_3CO antibody in a manner sensitive to competition by (6RS)-lipoic acid, Lipoyl-Lys and CF_3CO-Lys. Sera of halothane hepatitis patients recognize the E2 subunits of OGDC and BCOADC, and protein X, as autoantigens. A role of these lipoylated autoantigens in the etiology of halothane hepatitis is suggested by the fact that in liver biopsies of 5 out of 7 afflicted patients low levels of the E2 subunit of PDC were detectable by anti-CF_3CO antibody. In contrast, autoimmune sera towards the protein matrix of the E2 subunit detect normal levels, indicating an aberrancy in lipoylation of the E2 subunit of PDC in these patients.

Subject Index

Springer-Verlag and the Environment

We at Springer-Verlag firmly believe that an international science publisher has a special obligation to the environment, and our corporate policies consistently reflect this conviction.

We also expect our business partners – paper mills, printers, packaging manufacturers, etc. – to commit themselves to using environmentally friendly materials and production processes.

The paper in this book is made from low- or no-chlorine pulp and is acid free, in conformance with international standards for paper permanency.

MIX
Papier aus verantwortungsvollen Quellen
Paper from responsible sources
FSC® C105338

If you have any concerns about our products,
you can contact us on
ProductSafety@springernature.com

In case Publisher is established outside the EU,
the EU authorized representative is:
Springer Nature Customer Service Center GmbH
Europaplatz 3, 69115 Heidelberg, Germany

Printed by Libri Plureos GmbH
in Hamburg, Germany